The New Enclosures: Critical Perspectives on Corporate Land Deals

This collection explores the complex dynamics of corporate land deals from a broad agrarian political economy perspective, with a special focus on the implications for property and labour regimes, labour processes and structures of accumulation. This involves looking at ways in which existing patterns of rural social differentiation – in terms of class, gender, ethnicity and generation – are being shaped by changes in land use and property relations, as well as by the re-organization of production and exchange as rural communities and resources are incorporated into global commodity chains. It goes further than the descriptive 'what' and 'who' questions, in order to understand the 'how' and 'why' of these patterns. It is empirically solid and theoretically sophisticated, making it a robust and knowledge boundary-changing work. Contributors come from various scholarly disciplines. Covering nearly all regions of the world, the collection will be of interest to researchers from various disciplines, policy-makers and activists.

The book was originally published as a Special Issue of the *Journal of Peasant Studies.*

Ben White is Emeritus Professor of Rural Sociology at the International Institute of Social Studies (ISS), The Hague. His research focuses on agrarian change and the anthropology and history of childhood and youth, particularly in Indonesia. He is a founding member of the Land Deal Politics Initiative.

Saturnino M. Borras Jr. is Associate Professor at the International Institute of Social Studies (ISS), The Hague. He is also Adjunct Professor at China Agricultural University and Fellow of the Transnational Institute (TNI) and Food First. He is the Editor in Chief of the *Journal of Peasant Studies*, and is a founding member of the Land Deal Politics Initiative.

Ruth Hall is Associate Professor at the Institute for Poverty, Land and Agrarian Studies at the University of the Western Cape, South Africa. She coordinates the land theme of the Future Agricultures Consortium, is Reviews Editor of the *Journal of Peasant Studies*, and is a founding member of the Land Deal Politics Initiative.

Ian Scoones is a Professorial Fellow at the Institute of Development Studies at the University of Sussex. He is joint coordinator of the Future Agricultures Consortium and co-director of the STEPS Centre. He is a member of the editorial collective of the *Journal of Peasant Studies* and a founding member of the Land Deal Politics Initiative.

Wendy Wolford is the Robert A. and Ruth E. Polson Professor of Development Sociology at Cornell University. She is author of *This Land is Ours Now: Social Mobilization and the Meanings of Land in Brazil* (Duke Press, 2010) and is a member of both the editorial collective of the *Journal of Peasant Studies* and the Land Deal Politics Initiative.

Critical Agrarian Studies
Series Editor: Saturnino M. Borras Jr.

Critical Agrarian Studies is the new accompanying book series to the *Journal of Peasant Studies.* It publishes selected special issues of the journal and, occasionally, books that offer major contributions in the field of critical agrarian studies. The book series builds on the long and rich history of the journal and its former accompanying book series, the Library of Peasant Studies (1973–2008) which had published several important monographs and special-issues-as-books.

Critical Perspectives in Rural Development Studies
Edited by Saturnino M. Borras Jr.

The Politics of Biofuels, Land and Agrarian Change
Edited by Saturnino M. Borras Jr., Philip McMichael and Ian Scoones

New Frontiers of Land Control
Edited by Nancy Lee Peluso and Christian Lund

Outcomes of Post-2000 Fast Track Land Reform in Zimbabwe
Edited by Lionel Cliffe, Jocelyn Alexander, Ben Cousins and Rudo Gaidzanwa

Green Grabbing: A New Appropriation of Nature
Edited by James Fairhead, Melissa Leach and Ian Scoones

The New Enclosures: Critical Perspectives on Corporate Land Deals
Edited by Ben White, Saturnino M. Borras Jr., Ruth Hall, Ian Scoones and Wendy Wolford

The New Enclosures: Critical Perspectives on Corporate Land Deals

Edited by

Ben White, Saturnino M. Borras Jr., Ruth Hall, Ian Scoones and Wendy Wolford

Routledge
Taylor & Francis Group
LONDON AND NEW YORK

First published 2013
by Routledge
2 Park Square, Milton Park, Abingdon, Oxon, OX14 4RN

Simultaneously published in the USA and Canada
by Routledge
711 Third Avenue, New York, NY 10017

Routledge is an imprint of the Taylor & Francis Group, an informa business

This book is a reproduction of the *Journal of Peasant Studies*, vol. 39, issue 3-4. The Publisher requests to those authors who may be citing this book to state, also, the bibliographical details of the special issue on which the book was based.

British Library Cataloguing in Publication Data
A catalogue record for this book is available from the British Library

ISBN13: 978-0-415-82374-6

Typeset in Times New Roman
by Taylor & Francis Books

Publisher's Note
The publisher would like to make readers aware that the chapters in this book may be referred to as articles as they are identical to the articles published in the special issue. The publisher accepts responsibility for any inconsistencies that may have arisen in the course of preparing this volume for print.

Contents

Citation Information

The chapters in this book were originally published in the *Journal of Peasant Studies*, volume 39, issue 3-4 (July-October 2012). When citing this material, please use the original page numbering for each article, as follows:

Chapter 1
The new enclosures: critical perspectives on corporate land deals
Ben White, Saturnino M. Borras Jr., Ruth Hall, Ian Scoones and Wendy Wolford
Journal of Peasant Studies, volume 39, issue 3-4 (July-October 2012) pp. 619-648

Chapter 2
The international political economy of the global land rush: A critical appraisal of trends, scale, geography and drivers
Lorenzo Cotula
Journal of Peasant Studies, volume 39, issue 3-4 (July-October 2012) pp. 649-680

Chapter 3
The land grab and corporate food regime restructuring
Philip McMichael
Journal of Peasant Studies, volume 39, issue 3-4 (July-October 2012) pp. 681-702

Chapter 4
Situating private equity capital in the land grab debate
Shepard Daniel
Journal of Peasant Studies, volume 39, issue 3-4 (July-October 2012) pp. 703-730

Chapter 5
Global resource grabs, agribusiness concentration and the smallholder: two West African case studies
Kojo Sebastian Amanor
Journal of Peasant Studies, volume 39, issue 3-4 (July-October 2012) pp. 731-750

Chapter 6
Looking back to see forward: the legal niceties of land theft in land rushes
Liz Alden Wily
Journal of Peasant Studies, volume 39, issue 3-4 (July-October 2012) pp. 751-776

Chapter 7
New investment, old challenges. Land deals and the water constraint in African agriculture
Philip Woodhouse
Journal of Peasant Studies, volume 39, issue 3-4 (July-October 2012) pp. 777-794

Chapter 8
Patterns of agrarian transformation in Ethiopia: State-mediated commercialisation and the 'land grab'
Tom Lavers
Journal of Peasant Studies, volume 39, issue 3-4 (July-October 2012) pp. 795-822

Chapter 9
The next Great Trek? South African commercial farmers move north
Ruth Hall
Journal of Peasant Studies, volume 39, issue 3-4 (July-October 2012) pp. 823-844

Chapter 10
Land grabbing in Latin America and the Caribbean
Saturnino M. Borras Jr., Jennifer C. Franco, Sergio Gómez, Cristóbal Kay and Max Spoor
Journal of Peasant Studies, volume 39, issue 3-4 (July-October 2012) pp. 845-872

Chapter 11
Agrarian structure, foreign investment in land, and land prices in Brazil
Sérgio Sauer and Sergio Pereira Leite
Journal of Peasant Studies, volume 39, issue 3-4 (July-October 2012) pp. 873-898

Chapter 12
Oligarchs, megafarms and land reserves: understanding land grabbing in Russia
Oane Visser, Natalia Mamonova and Max Spoor
Journal of Peasant Studies, volume 39, issue 3-4 (July-October 2012) pp. 899-932

Chapter 13
The land question: special economic zones and the political economy of dispossession in India
Michael Levien
Journal of Peasant Studies, volume 39, issue 3-4 (July-October 2012) pp. 933-970

Chapter 14
Land expropriation and displacement in Bangladesh
Shelley Feldman and Charles Geisler
Journal of Peasant Studies, volume 39, issue 3-4 (July-October 2012) pp. 971-994

Chapter 15
Gendered experiences of dispossession: oil palm expansion in a Dayak Hibun community in West Kalimantan
Julia, with Ben White

Journal of Peasant Studies, volume 39, issue 3-4 (July-October 2012) pp. 995-1016

Chapter 16

Plantation rubber, land grabbing and social-property transformation in southern Laos
Miles Kenney-Lazar
Journal of Peasant Studies, volume 39, issue 3-4 (July-October 2012) pp. 1017-1038

Chapter 17

Large-scale land deals from the inside out: findings from Kenya's Tana Delta
Rebecca Smalley and Esteve Corbera
Journal of Peasant Studies, volume 39, issue 3-4 (July-October 2012) pp. 1039-1076

Chapter 18

Tree plantations, politics of possession and the absence of land grabs in Vietnam
Thomas Sikor
Journal of Peasant Studies, volume 39, issue 3-4 (July-October 2012) pp. 1077-1101

The new enclosures: critical perspectives on corporate land deals

Ben White, Saturnino M. Borras Jr., Ruth Hall, Ian Scoones and Wendy Wolford

The contributions to this collection use the tools of agrarian political economy to explore the rapid growth and complex dynamics of large-scale land deals in recent years, with a special focus on the implications of big land deals for property and labour regimes, labour processes and structures of accumulation. The first part of this introductory essay examines the implications of this agrarian political economy perspective. First we explore the continuities and contrasts between historical and contemporary land grabs, before examining the core underlying debate around large- versus small-scale farming futures. Next, we unpack the diverse contexts and causes of land grabbing today, highlighting six overlapping mechanisms. The following section turns to assessing the crisis narratives that frame the justifications for land deals, and the flaws in the argument around there being excess, empty or idle land available. Next the paper turns to an examination of the impacts of land deals, and the processes of inclusion and exclusion at play, before looking at patterns of resistance and constructions of alternatives. The final section introduces the papers in the collection.

Introduction: land deals in political economy perspective

The contributions to this collection use the tools of agrarian political economy to explore the rapid growth of large-scale land deals in recent years – the phenomenon popularly known as 'land grabbing', the large-scale acquisition of land or land-related rights and resources by corporate (business, non-profit or public)[1] entities.

This collection builds on two years of research collaboration through the Land Deal Politics Initiative (www.iss.nl/ldpi), a loose network initiated by the five authors of this introductory essay with the objective of promoting an agenda for more rigorous theoretical-empirical engaged research on the global land grab phenomenon. In 2009-2010 the 'LDPI five', in collaboration with the *Journal of Peasant Studies*, sponsored a small grants competition. The response was massive and positive, mostly coming from young researchers and PhD candidates. We provided a total of 40 grants, among them 20 on Africa supported by the Future Agricultures Consortium (www.future-agricultures.org) with each researcher receiving inputs and feedback from one or more of the LDPI five. The network has now expanded rapidly and widely. This initial phase of the LDPI initiative culminated with the International Conference on Global Land Grabbing held in April 2011 at the Institute of Development Studies at the University of Sussex, where 120 academic papers were discussed (selected from close to 400 abstracts submitted in response to our call for papers) (www.future-agricultures.org/land-grab.html). We have now initiated a second round of 40 new small research grants and are planning a second conference at Cornell University in October 2012. This introductory essay is the collective product of the LDPI five, with the first author taking main responsibility.

[1]'Non-profit' here includes such bodies as pension funds.

The focus is on the ways in which 'grabbing' creates specific kinds of property dynamics, namely dispossession of land, water, forests and other common property resources; their concentration, privatization and transaction as corporate (owned or leased) property; and in turn the transformation of agrarian labour regimes.

Since many contemporary large-scale land deals remain at the speculative stage, it is too early to see their longer-term impacts on local populations and agrarian structures. How extensive are these land deals? Are they occurring on the margins, or resulting in a significant restructuring of agrarian settings? Several groups in recent years have tried to document the location and extent of land deals (Anseeuw *et al.* 2012, von Braun and Meinzen-Dick 2009, Cotula et al. 2009, GRAIN 2008, Oakland 2011, Oxfam 2011,World Bank 2010). Even if there were consensus on the definition of land grabs, and on the methodology of counting them, large land deals are typically shrouded in secrecy and no one really knows how much land has been acquired. Lorenzo Cotula has critically reviewed the available body of evidence, pointing to sources of both over- and under-estimation, but confirming that 'the phenomenon is massive and growing' (Cotula, 2012, this volume)[2].

A large scale land deal is often no more than a framework, under which concrete deals between agribusiness corporations and local governments for the purchase or leasing of designated areas may or may not emerge, and these deals in turn may not result until many years later in the actual enclosure of land, dispossession of its previous users and establishment of new production and labour regimes. Potentially, however, these deals open the way to a truly wide-ranging global 'land reform' – in this case, a regressive land reform where governments take land from the poor and give (or sell or lease) it to the rich.

Faced with the rapid spate of big land deals in recent years and in all continents, there is plenty of reason for researchers, activists and policymakers to be concerned with the immediate problems of dispossession, exclusion and adverse incorporation that local communities face. Besides these immediate issues however, there is also a need to look for deeper understandings of the phenomenon and its longer-term implications for agricultural and rural futures, in other words to disentangle the immediate and the more fundamental dynamics at work.

Building on and complementing other recent and forthcoming paper collections on specific aspects of land deals,[3] this volume explores the complex dynamics of corporate land deals in a broad agrarian political economy perspective, with a special focus on the implications for property and labour regimes, labour processes and structures of accumulation. Among other things this can involve looking at ways in which existing patterns of rural social differentiation – in terms of class and gender, perhaps also ethnicity and generation – are being shaped by changes in land

[2]With some differences in definition and the time-span covered, recent estimates of the total area of large land deals world-wide range from 43 million ha (World Bank 2010) to 80 million ha (ILC 2011), and 227 million ha (Oxfam 2011). Among these and other reports there is reasonable consistency that at least 60 percent of total acquisitions are in sub-Saharan Africa.
[3]It builds on and complements other *Journal of Peasant Studies* special issues on *Biofuels, land and agrarian change* (McMichael and Scoones eds. 2010), *New frontiers of land control* (Peluso and Lund eds. 2011), and *Green Grabbing: A new appropriation of nature?* (Fairhead, Leach and Scoones eds. 2012), the *JPS* 'Forum on Global Land Grabbing' featuring contributions by Deininger, de Schutter and Li (Borras *et al.* 2011, Deininger 2011, De Schutter 2011, Li 2011), as well as the special issue under review at *Development and change* on *Governing global land deals: The role of the state in the rush for land* (Borras, Hall, Scoones, White and Wolford, eds. n.d.).

use and property relations, as well as by the re-organization of production and exchange as rural communities and resources are incorporated in global commodity chains. Also important (and sometimes forgotten by political economists) is the politics which steers all this: the role of mobilization, struggle or resistance in shaping, negotiating and challenging new land investments; the various competing policy and political narratives and discourses around the multiple crises of food, energy, climate and finance, and their impact on, or use in, land deal negotiations, the emerging or continuing dynamics of elite power and corruption, and the role of global land policies of international development agencies in facilitating and encouraging, or blocking and discouraging, land deals.

Although all the contributions to this collection are concerned with some aspect of large-scale land deals, the acquisition of land is not an end in itself. Some land deals (for example those made by hedge funds and pension funds) may be purely speculative, betting on rising global land values; investors put their money in land, as they might put it in gold, works of art, British Petroleum shares or pork-belly futures. Speculative investment aside however, the purpose of the great majority of corporate land grabs is to establish agricultural production (or other forms of extraction such as mining) on a large scale, and to guarantee access to its products. We are of course not the first to make this observation. As Marx noted in the remarks on the English enclosures in which he first coined the term 'land grabbing'

> Land grabbing on a great scale [...] is the first step in creating a field for the establishment of agriculture on a great scale. Hence this subversion of agriculture puts on, at first, more the appearance of a political revolution. (Marx 1906 [1867] *Capital*. Vol. I., p. 470)

What does this 'political revolution' (or social-economic-political revolution) involve? And what concepts do we need to analyze and describe it, and at what levels? What kinds of analytical tools do we need to understand relations between the level of the larger circuits of global (agribusiness) capital, and changes happening at the level of local communities, and at the various intermediate levels between? These questions are the agrarian manifestation of the more general issue that all social sciences have to confront: how do we frame and grasp the relations and dynamics between the local and the global, the individual, group and wider society, between 'actors' and 'structures'? And more specifically, how do global or national corporations and investors hope to organize the production and delivery of export crops? What alternatives are available to them in commanding land and labour? These issues can mostly be explored using very simple questions, and simple language, as in Bernstein's much-quoted 'Marxian *haiku*'[4] on key questions in agrarian political economy: 'who owns what? who does what? who gets what? what do they do with it?' (Bernstein 2010, 22). To these questions we would add two more: 'what do they do to each other? (How do social classes and groups in society and within the state interact with each other?)', to capture the relational and political side of property and labour regimes, labour processes and structures of accumulation, and 'How do changes in politics get shaped by dynamic ecologies and vice versa?', emphasising the political-ecological context of land deals.

Then of course we need to go further than these descriptive ('what?' and 'who?') questions, in order to understand the 'how' and 'why' of these patterns.

[4]The phrase is from Michael Watts (2012, 200).

These questions, we think, can be approached with the help of a nested hierarchy of 'umbrella' concepts: the (global) agro-commodity regime; (global) commodity chains or value chains, and (agrarian) labour regimes.

Food and other agro-commodity regimes[5] are political regimes of global value relations (Araghi 2003), which take on particular forms in different eras. The contemporary context of neo-liberal globalization means not just increasing world-wide connections, but a project to organize those connections in a certain way:

> The globalization project is an attempt to fashion the world around a central principle through powerful political and financial institutions. ...
> an emerging vision of the world and its resources as a globally organized and managed free trade/free enterprise economy pursued by a largely unaccountable political and economic elite. (McMichael 2000, 241, 354)

Of course we (and McMichael) are not the first to make this observation.

> Chased around the world by its burning desire for ever-expanding markets, the bourgeoisie has no choice but to settle everywhere; cultivate everywhere; establish connections everywhere ... in a nutshell, it creates the world in its own image. (Marx and Engels, *The Communist Manifesto*, 1848)

> Literally before our eyes the world's agriculture [...] is being more and more drawn into the circulation of the world economy, and the centres of capitalism are more and more subordinating it to their leadership. (Chayanov 1966 [1925], 257)

The food regime concept, developed initially by Friedmann and McMichael (1989), allows us to refocus 'from the commodity as object to the commodity as relation, with definite geo-political, social, ecological and nutritional relations at significant historical moments' (McMichael 2009, 63). Analysis of evolving agro-commodity regimes thus has to encompass many levels, from local to global, if like Chayanov we hope to understand 'the social threads that bind Sidor Karpov's farm, lost in the Perm' forests, to the London banks and oblige him to feel the effects of changes in the pulse rate of the London stock exchange' (Chayanov 1966 [1925], 257). These commodity-specific 'social threads' are nowadays analysed as global commodity chains: '... all the activities that connect the production of commodities with their final consumption [...] and the actors and institutions, relations and practices, that structure those journeys' (Bernstein 2010, 125). The local producers, distant consumers and many other agents linked in these chains all claim a share of the 'added value' at various points in the chain, and their inclusion or exclusion and the strength of their claims are highly influenced by the exercise of power at various levels.

At the local end of these value chains, labour regime analysis is a useful tool for analyzing agrarian structures and comparing them across space and time, and identifying points and processes of contestation, conflict and negotiation. Henry Bernstein's definition – 'specific methods of mobilizing labour and organizing it in production, and their particular social, economic and political conditions' (Bernstein 1988, 31-2) – emphasizes that the emergence of specific labour regimes is not

[5]Since not only food crops but also feed, fibre and fuel crops are involved, this awkward formulation is more accurate than 'food regime'.

inevitable but the product of politics.[6] A modern and flexible agrarian political economy also incorporates, in its exploration of these dynamic relationships and processes, both material and ideological aspects, as well as dimensions that were relatively neglected in classical agrarian studies such as the dynamics of gender, ethnicity, livelihoods diversity, mobility and rural-urban links.

In the following sections we examine the implications of this agrarian political economy perspective on large-scale large deals. First we explore the continuities and contrasts between contemporary processes and historical experiences, before examining the core underlying debate around small- versus large-scale farming futures. Next, we unpack the diverse, overlapping causes of land grabbing today in different contexts, highlighting six overlapping mechanisms. The following section turns to assessing the crisis narratives that frame the justifications for land deals, and the flaws in the argument around there being 'excess', 'empty' or 'idle' land available. Next the paper turns to an examination of the impacts of land deals, and the processes of inclusion and exclusion at play, followed by brief observations on regulation, resistance and constructions of alternatives. The final section introduces the papers in the collection.

Historical dimensions: continuities and contrasts

> ... what is so new [...] about 'land grabbing'? Dutch, French, Spanish, Portuguese, and English colonizers were heavily engaged in both land grabbing and the creation of private property [...] The ways officials implemented land controls differed from one colony to another, or differed by crop, by region of a colony, or by legal logic, and often changed over colonial times. What is new in the land grabs today are the new mechanisms of land control, their justifications and alliances for 'taking back' the land, as well as the political economic context of neoliberalism. (Peluso and Lund 2011, 672)

Most regions of the global South, as well as the global North, have a long history of land 'grabbing' on a large scale. In the North, the best-known episodes have been the European and particularly the British enclosures. But we should also recall the dispossession of the native peoples of North America and Australasia, and, arguably, some historical episodes of forced socialist collectivization. In many regions of the global South land was first grabbed by pre-colonial rulers in chronic territorial wars with each other, then by colonial governments and increasingly by foreign or domestic corporations. Through the nineteenth century, outright sale by colonial powers of large tracts of land held under customary tenure was quite common, but this was often prohibited or limited by later colonial legislation so that long, renewable leases became the norm. Liz Alden Wily (2012, this volume) argues that the current spate of land deals is 'less a new phenomenon than a significant surge in the continuing capture of ordinary people's rights and assets by capital-led and class-creating social transformation'. Using the examples of the Irish and English enclosures of the seventeenth to nineteenth centuries, the North American dispossession of native populations, and the three great African land rushes of 1885-

[6]We prefer this older, broader definition to those in Bernstein's more recent publications – most recently 'different methods of recruiting labour and their connections within how labour is organized in production (labour process) and how it secures its subsistence' (Bernstein 2010, 53) – which are heavier on the structure side and appear to omit the essential aspects of agency and power.

1915, 1919-1939, and 1945-1955, she underlines one key common aspect of all these land-grabbing episodes, the legal manipulations that 'render untitled (but traditionally occupied and used) lands as unowned, and the state, by default, their legal owner' (Wily 2012, this volume).

In the late-colonial and post-colonial decades both governments and civil society groups in many countries attempted to correct some of these historical distortions by land reforms or other means of breaking up large private or corporate estates and their redistribution to smallholders. Some of these initiatives were modest reformist measures intended to stem the radicalisation of the rural poor as a political force, as in Kenya's Swynnerton Plan of 1954, while others were adopted by newly independent post-colonial states engaged in nationalist projects of indigenisation (El-Ghonemy 2001, Heyer *et al.* 1981, 102–103, Sorenson 1967). The World Bank also was a major proponent of the breakup of large corporate plantations in favour of small farmer based agricultural development strategies through the second half of the twentieth century. Radical land reforms, in the context of revolutionary regime change, saw the division of large privately-owned land-holdings and their redistribution to tenants and landless workers, sometimes with the goal of creating a stable and productive mass of relatively homogeneous 'family farms' (Ghose 1983), and in other cases as a first step towards eventual socialist collectivisation.

Paradoxically, all these policies are now being reversed as governments and international development organisations support the acquisition of great expanses of land by large corporations, both foreign and domestic – usually in the form of long-term concessions or leases rather than outright purchase – in the name of development. Thus large-scale capitalist farming, which Jan Douwe van der Ploeg argues had formerly 'nearly disappeared, among other things through the many land reform processes that swept the world', is now 're-emerging everywhere under the aegis of the agro-export model' (van der Ploeg 2008, 2). This reversal predates the current land rush of the past decade; rather, its conditions were laid partly in structural adjustment and neo-liberal reforms that indebted developing countries were subjected to during the 1980s, and the former Soviet Union countries during the 1990s, which promoted market-driven enclosure in some regions and the privatisation of state farms in others. As Kojo Amanor argues, 'many agribusiness companies have been at the forefront of large-scale land acquisitions for many years, and [...] the dynamic towards land grabbing arises internally from the logic of agribusiness accumulation rather than from exogenous developments' (2012, this volume).

While large-scale land acquisitions, linked to advocacy of large farm models for production, have their precursors in the colonial plantations and estates, the dualistic economies of the former settler colonies and the latifundia of Latin America there are important contrasts with the contemporary rush for land. The (potentially) enormous scale and speed of expansion of the current deals, which may be more rapid than previous land 'rushes' in colonial or post-colonial history, may mean that they will have correspondingly greater impacts in radically restructuring agrarian economies, transforming livelihoods and rural social relations and, with this, changing the power dynamics in the countryside across the global South, with major implications for national, and indeed regional and international, politics. Contemporary forms of agrarian transition involve investments and dispossession that expel people from agriculture without absorbing their labour in manufactures or elsewhere in the economy, and create an 'agrarian question of labour' (Bernstein 2004) involving large 'surplus populations' of the dispossessed (Li 2009, 2010,

Araghi 2010). This probably represents a majority of historical capitalist agrarian transitions worldwide, the main exceptions being the 'classic' agrarian transitions of (parts of) industrializing Europe and East Asia. Thus, as Tanner concludes for Mozambique

> There are clear signs that [...] the Mozambican enclosures could produce the same result as their predecessors in Europe – a dispossessed rural majority, and migration to towns. Yet, unlike Europe, this will be in a country that is not about to embark on a labour-intensive industrial revolution generating thousands of new jobs for the dispossessed peasant farmers and their families. (Tanner 2010, 125)

'Surplus population' here has nothing to do with Malthusian or neo-Malthusian theories of absolute overpopulation (the notion that human populations have some natural tendency to grow until they outstrip the limits of subsistence), but a historically-specific relative surplus population, itself the result of capital accumulation and technical progress, which is 'surplus' not to society's capacity to provide subsistence but to capital's requirements for labour, resulting in low wages of the employed and pauperism of the un- and underemployed even in contexts of rapid economic development, producing 'the wonderful phenomenon ... that a nation must starve to death from sheer wealth and abundance' (Marx 1977, 591f.). Nowadays it is not only agriculture, but also many other sectors whose 'development', through capital investment and technical change, involves the shedding, rather than the absorption, of labour.

The underlying debate: large- vs. small-scale agricultural futures

Underlying historical moves towards enclosure and industrial farming on the one hand and campaigns for the retention of smallholder farming on the other, is the broader issue of the relative superiority of large-scale versus small-scale farming. This has its origins in the debates between Marxists and agrarian populists (Lenin 1960 [1899], Chayanov 1966 [1925], Kautsky 1988 [1899]) which have framed much of the critical agrarian studies and research in the 1970s-80s, including the urban versus rural bias debate between Byres and Lipton (Kay 2009, 109-115) and which intersect with more specific arguments on the so-called 'inverse relationship' between farm size and productivity (see also Byres 2004, Bernstein 2010, Griffin *et al* 2002, Lipton 2009, Hazell *et al.* 2010, Wiggins *et al.* 2010). In reaction to the contemporary phenomenon of land grabbing and the negative assumptions that are made about the consequences of large-scale agriculture for food security and employment, Peter Gibbon has reviewed the historical evidence on the productivity and employment generation record of 'plantation farming and large-scale farming' in twentieth century sub-Saharan Africa. He concludes that large-scale farms have not always been less productive, or less labour absorbing, than small, but that they are generally marked by relatively 'low quality of employment' (Gibbon 2011).

Advocacy of a large farm model for agricultural growth, linked to assumed comparative advantages in a globalised economy, is often at the centre of justifications for large-scale corporate land deals (Collier 2008). A 'dualistic' agrarian economy is envisaged, with large scale farms engaged in capitalist production often for export, while smallholder farms gradually disappear or are incorporated as part of contract farming arrangements, with the former peasantry proletarianised, providing low paid labour to the new estates and plantations. This is the trajectory envisaged in the World Bank's much-cited, and much criticized, report on *Agriculture for Development* (World Bank 2008)[7]. In the increasingly integrated value chains of global agricultural production, it is argued, only large farms or smallholder outgrowers hooked into large agribusiness nuclei can compete, and meet the kinds of standards required for successful export. An agribusiness-oriented vision for agriculture, with large-scale farms at the core, even if linked through outgrower schemes to smallholders, is one that some see as the logical and perhaps inevitable extension of global capital into rural economies; a culmination of processes of agrarian change and capitalist transformation (Sender and Johnston 2004). This argument has been taken on by national governments, investors and donor agencies alike. This has made many policy actors, eager to promote private sector driven investments but on the other hand committed to a discourse of support for smallholder farming, decidedly ambivalent about the process of global land grabbing.

The World Bank for example, while arguing forcefully for private sector investment in land and agriculture, has also made a cogent argument for smallholder based land reform (Binswanger and Deininger 1999), even though advocating a flawed 'willing buyer-willing seller' model (cf. Borras, Kay and Lahiff 2007 for a critique). Indeed the Bank, which for some years has juggled 'politically correct messages about the need to support small farmers' with policy and financial support for agribusiness (Oya 2009, 595, cf. Akram-Lodhi 2008) can hardly oppose large-scale corporate investment in farmland when it is itself an important promoter and facilitator of these investments. A parallel track is envisaged, with large-scale export-oriented farms in some areas, while smaller-scale production persists elsewhere. Recent experiences in Zimbabwe show how such a dualistic model of agrarian structure is unsustainable on both economic and justice grounds in the long term and that, following land reform, many smallholders have been successful at 'accumulating from below' and engaging in markets, including for export, despite numerous constraints (Scoones *et al.* 2010). In some ways, the 'win-win' proposition in the current land grab policy discourse builds on the assumption that small farms can, and should, be linked to large farms through a variety of contract farming schemes, as long as these are transparent and follow a broad set of principles of responsible investments (Deininger 2011).

'Large versus small' in fact may not be the most crucial point in envisaging farming futures. It is at least possible to imagine a large-scale farm production unit which is based on low-input and mixed cropping rather than high-input monocrop cultivation, using artisanal and employment-enhancing rather than capital-intensive and labour-shedding techniques, which is water-saving, earth-cooling and sustainable and maintains the dignity and food security of those who work in it, and based on principles of agrarian and environmental justice rather than capitalist profit maximization. Only it is rather hard to imagine that socially responsible large farming of this type would be based on corporate capital rather than collective or cooperative ownership.

Contexts, causes and mechanisms

Large-scale land deals, while sensational, are only the most recent and visible manifestation of a growing agrarian crisis around the world, reflected in heightened

[7]Among many critical views on the World Bank Report, see particularly Akram-Lodhi 2008 and Li 2009.

levels of food insecurity, poverty, landlessness and environmental degradation (Patnaik 2010, Weis 2010, Holt-Gimenez and Shattuck 2011, McMichael 2012, this volume). High world food and fuel prices in 2007-08 led to a wave of protests and anti-government riots in more than 60 countries (Holt-Gimenez and Patel with Shattuck 2009, Bello 2009), precipitating protectionist measures by those with food production capacities and expansionist strategies by those without. The combined effects of global climate change, agro-industrial development, natural resource extraction, neo-liberal austerity policies and rapid urbanization have increased insecurity and vulnerability in rural areas across the globe. Several decades of neoliberal policy prescriptions have resulted in a diminution of state capacity, the privatisation of previously public assets and the financialisation of economic networks, linked to a new form of global capitalism.

This historical juncture has profound impacts on rural economies and agrarian livelihoods. The opening of markets, combined with the indebtedness of developing country nations, has opened up opportunities for land grabbing. Alliances between state officials, local political elites and domestic and foreign investors enable them to seize these opportunities for the appropriation of resources. For the state, often in dire fiscal straits, such a moment also provides an opportunity for extending its reach, exerting power over marginal areas and peoples, as well as extracting rent from such 'unruly' places (cf. Scott 2009). Corporate land grabbing is thus linked to a new expansion of state authority and control (Borras *et al.* n.d.).

David Harvey (2003, 2005) argues that contemporary political-economic relations facilitate a process of 'accumulation by dispossession', whereby public assets are enclosed, with some able to make a profit from these resources, while others are excluded in an important extension of the classic process of primitive accumulation described by Marx. We might therefore see land grabbing as an outcome of the inter-relating processes of privatization and financialization which Harvey identifies as central to neo-liberal capitalism, as well as being fuelled by 'the management and manipulation of crises' – and particularly the discursive emphases on multiple 'scarcities' in policy debates – and what he terms 'state redistributions', whereby the state allies with capitalist business interests, offering concessionary terms, tax breaks and subsidies to support investment.

In this contemporary context, we argue (drawing on Wolford 2010 and Safransky and Wolford 2011) that the rapid increase in large-scale land and resource deals can be linked to six trends which promote various mechanisms of accumulation through land investment, some of which repeat historical manoeuvres from the colonial handbook while others involve new configurations of old relations. Different mechanisms of accumulation result in different forms of dispossession, each with different consequences for rural livelihoods and social and political relations.

First, the *global anticipation of food insecurity* has prompted widespread corporate investment in food crops (some destined as human foods, and some as feed for the growing livestock industry) (Foresight 2010, Godfray *et al.* 2010). Although not new, investment levels have increased in the past ten years, particularly in areas where industrial inputs are readily available, markets relatively accessible, and land costs low. The expanding volume and changing diet of fast growing large economies such as China and India has triggered major shifts in how much and what kind of food crops are produced, and how. The 'meatification' of changing diets (Weis 2010) has profound impact on the feed complex (soya, corn, and so on) with has direct consequences for land use. A significant part of land deals in Latin

America has been for soya expansion. Food importing countries such as the Gulf States and South Korea felt that they could not rely any longer on market sourcing of food through the surplus-based food regime that existed in the recent past (McMichael 2009), and looked for avenues of more direct control of food supply, resulting in their efforts to gain direct control of off-shore land and in turn off-shore food production (as in the case of Libya in Mali, South Korea in Madagascar, Saudi Arabia in Sudan).

Second, fears of rising and volatile fuel prices and peak oil or loss of national sovereignty through 'foreignization' of energy resources have promoted increasing reliance on, and demand for, *new forms of resource extraction for fuel security* and national development (Zoomers 2010, White and Dasgupta 2010). In this context Borras, McMichael and Scoones identified a 'biofuel complex', arguing that the biofuels revolution is a response to 'an assumed "energy crisis", as the cost of capital inputs (production, processing, transport) rises in an age of peaking oil supplies. In addition, a desire to reduce dependence on Middle-Eastern oil drives governments to develop an industrial biofuels complex that delivers "energy security". At the same time, biofuels represent a new profitability frontier for agribusiness and energy sectors beset with declining productivity and/or rising costs' (2010, 576). The alliances being formed blur the boundaries between food, feed and fuel in vertically integrated agribusinesses, often in 'triangular' relationships between North and South (Dauvergne and Neville 2010). For example, major multi-national players such as Cargill and Monsanto are involved in a 'feed-fuel alliance', using genetically modified maize, soy and rapeseed (TNI 2007). In the 'palm-oil alliance' the Indonesian palm oil trade is dominated by Cargill (the world's largest private company), ADM-Kuck-Wilmar (the world's largest biofuels manufacturer), and Synergy Drive, a large Malaysian government company. In addition, an 'ethanol alliance', involving the US, Brazil, Argentina and large multinational players, links with India, China, Mozambique and South Africa in new sugar-soy production enterprises, promoted by EU and US subsidies and trade preferences (Borras *et al.* 2010). Thus the standard narrative that those involved in such deals are simply foreign companies from the North appropriating land in the South is inaccurate. There are South-South alliances, as well as relationships between Northern and Southern actors, both public and private. National governments in the South, as well as national and local elites, are intimately involved (Dauvergne and Neville 2010, Cotula 2012, this volume).

This reconfiguration of the global food-feed-fuel industry creates new political-economic configurations, but also new types of farming. Crops, for example, are envisaged in new ways. In Latin America and the Caribbean 'flex crops' – crops that have multiple and flexible uses within the global food, feed and fuel complex – have led to a relatively safer, and so more attractive investment, because they somehow allow for 'diversification within a single crop': sell as sugar when ethanol price is low, sell as ethanol when sugar price is low; sell palm oil as cooking oil when biodiesel market is not yet ready, sell as biodiesel when supply for cooking oil is high, and so on (Borras *et al.* 2012). Yet the biofuel revolution is not all large-scale plantations, run by conglomerates of multinational companies in alliance with national governments. There are also many small-scale operations, involving NGOs, responding to climate change targets (Hunsberger 2010, Ariza *et al.* 2010) and smaller private companies, investing in crops such as *Jatropha* (Sulle and Nelson 2009). These have different consequences and impacts, yet remain caught up in the wider political economy of the biofuel complex.

The third contextual trend promoting accumulation through land acquisitions is *new environmental imperatives and tools*, both of which have helped generate what John Vidal calls the 'Great Green Land Grab' – the appropriation of land and resources for environmental ends (see Fairhead *et al.* 2012). The application of the market paradigm to the environment ('market environmentalism') has become an important trend which reinforces a growing 'metabolic rift', and so the separation of people and nature (Schneider and McMichael 2010, Moore 2011). The 'neoliberalization' of nature and the environment creates new alliances of actors, new patterns of commoditisation and new forms of governance (Castree 2008a, 2008b, McCarthy and Prudham 2004, Heynen *et al.* 2007). New non-governmental organizations have formed to negotiate the complicated but potentially lucrative rules of carbon cap and trade programs or REDD initiatives – Reducing Emissions from Deforestation and Degradation. Motivated by these opportunities to match conservation with income generation opportunities, NGOs such as the World Land Trust, Cool Earth and the WildLands have purchased hundreds of thousands of hectares of so-called 'empty' land globally. Such processes, as Fairhead, Leach and Scoones (2012) explain, build on long histories of colonial and neo-colonial resource alienation in the name of the environment, which have resulted in the creation of forest reserves, national parks and often draconian interventions to reduce assumed degradation by local people. The new market environmentalism, however, also involves new dynamics of commodification, with the environment valued for 'products' and 'services' in new markets. This process results in an important mechanism of accumulation, driven by new values for carbon, wildlife, ecosystem services and landscapes more broadly (Corbera *et al.* 2007).

A fourth emerging mechanism of accumulation involves the establishment of *extensive infrastructure corridors* and *Special Economic Zones*. These are focused on infrastructure investments and have been constructed to link extractive frontiers to metropolitan areas and foreign markets. Funded by the Inter-American Development Bank, the International Monetary Fund, the World Bank and Chinese and other transnational investors, infrastructure corridors throughout Africa, Asia and Latin America are designed to serve as the many super highways that will help open as-yet untapped areas of the developing world to private foreign investment, agribusiness expansion, and natural resource extraction. Special Economic Zones have mushroomed in many fast growing economies, e.g. India and China. This has led to expropriation of lands, especially in peri-urban areas, for the industrial and commercial complexes and the corresponding urban sprawl. This has led to massive land conflicts in these countries (see e.g. Levien 2012, this collection).

The fifth contextual development involves the *creation of new financial instruments* intended to reduce market risk while allowing third party investors to profit from widespread concerns that food is running out. Private equity groups such as TIAA-CREF (a US pension fund), Berkshire-Hathaway (led by Warren Buffet), the Pharos Group (led by George Soros) and Black River Asset Management (led by Cargill) manage a variety of investments that increasingly include pieces of foreign territory. There are many new players involved in land investments globally (Pearce 2012). This involves big money, with (at least claimed) high returns. There is much speculative activity, involving a range of sovereign, hedge and asset funds. As highlighted by the Oakland Institute (2011), pension funds from US universities are players in land deals, as well as many others. These investments often involve complex operations. One instructive example is Emergent Asset Management, a

UK-South African financial management firm which previously focused on hedge funds but now specialises in farmland investments in Africa, including the African Agricultural Land Fund. A partnership with the top South African company, Grainvest, resulted in the Emvest Agricultural Corporation, which is now a vehicle for South African, UK and other investors to diversify their investments into African agriculture. Currently fourteen countries are being targeted, and the fund has a value of over US$500m. The fund's long-term objective according to its website is 'to secure food production across a diverse range of soft commodities managed across sub-Saharan Africa and throughout the value chain. Africa is one of the final frontiers for alpha generation, an un-crowded market reflecting the Emergent ethos of acting "pro alia" [sic] or "before others"'. It goes on to explain how 'EmVest is establishing farming hubs throughout the region, working with national and local authorities to develop large scale agriculture on a commercial basis, within a socially responsible framework, at a consistently high level of execution' and offers investors 'target risk-adjusted returns [of] +25% per annum from combined soft commodity production yields and land price appreciation. Such yield enhancement is based on the introduction of modern farming techniques and technologies to increase yields, while agglomerating farms to increase efficiency and generate economies of scale'. The website argues that 'Agriculture is an excellent defensive investment in the long term with expected real price appreciation of soft commodities and land in Africa'.[8] This is one among many examples of investment in land and agriculture, based on a large-farm and high-tech model, supported by investment finance through a range of vehicles.[9]

Finally, another development promoting land deals is the emerging set of *rules, regulations and incentives provided by the international community*, enshrined in international legal frameworks and facilitated by aid and lending programmes. Multilateral organizations from the United Nations to the international development banks are generating both supply and demand in the global rush for resources. Panicked estimates of increasing demand for food cited at high-level agro-investment conferences fuel speculation. At the same time, international agencies such as the World Bank and USAID have worked to create a hospitable environment for large-scale investment, funnelling money into Africa and Latin America for the purposes of 'rural development' and 'improving rural markets'. Contradictory signals and accompanying discourses often exist within and across organisations. Within the World Bank Group, support for smallholder farming exists alongside investment support for large-scale land deals from the International Finance Corporation and insurance to cover land investments from the Multilateral Investment Guarantee Agency (MIGA). For example, MIGA recently supplied Chayton Atlas Investments with 'political risk insurance' for a series of planned agriculture investments in southern Africa, including a USS$50m investment in Zambia involving the acquisition of 20,000 hectares of irrigated farmland[10]. The MIGA website explains that its mission is 'to promote foreign direct investment into underserved markets'.

[8]www.emergentasset.com (accessed October 19 2011); http://blogs.reuters.com/africanews/2010/08/24/african-agricultural-finance-under-the-spotlight/ (accessed 7 November 2011).
[9]Agriculture and land investment conferences are now regularly held, offering investors access to information and advice on land investments globally (see (http://www.aginvestconference.com/london/index.shtml and http://www.terrapinn.com/2012/agri/index.stm, accessed 5 February 2012).
[10]http://www.miga.org/news/index.cfm?aid=2637 (accessed 12 February 2012).

The Chayton Atlas support 'is a clear illustration of how we can help investors put in place the risk-mitigation instruments needed to meet sub-Saharan Africa's vast investment needs in a time of global economic turmoil'. This contract, it explains is 'ongoing support to private equity funds', whereby the contract can be used by fund managers 'to raise capital from institutional investors'[11]. While such instruments help unleash much needed investment financing in agriculture, the focus and impacts do not seem to be subject to rigorous scrutiny, and in important respects the emphasis runs counter to policy advice and analysis from elsewhere in the World Bank Group.[12]

Imbalances between investors and recipients are exacerbated by the legal frameworks for land investments which often result in highly inequitable deals, with benefits accruing to the investor and the risks being carried by the host government or community. Bilateral investment treaties often provide the framework for transnational land deals. As Malik (2011) explains, stabilisation clauses in these treaties are highly restrictive, and mean that states are locked in to particular agreements over long periods, and cannot impose new laws as they may be subject to disputes, and potentially massive compensation payments. Globally there are over 3000 bilateral investment treaties across 183 counties, with many emerging from European countries, with Germany, Netherlands and the UK being particularly significant (Malik 2011). These treaties allow private individuals or companies to bring claims against governments in international tribunals, where the scales are balanced very much in favour of the investor. In the case of Pakistan, Malik shows how large scale investments have occurred in what the state defined as 'barren lands'. Allocating large areas for new agricultural investments meant that investors required water in large quantities, although water access had not been covered in the deal, and investors had to increase extraction, affecting neighbouring land users. There was no prior legislation on water use, so through the new land contract, new private investors had essentially uncontrolled access to highly valuable public water resources, with potentially serious long-term consequences for the wider ecosystem and people's livelihoods. The nature of the contract, involving long-term leasing of land (over 99 years), meant that investor rights overrode all other rights and any subsequent legislation over this period. Thus any new legislation on, for example, environmental management, labour rights, export restrictions and so on could legitimately be challenged by investors under the terms of the contract, thus limiting opportunities for changing the terms of deal over a long period.

Crisis narratives and empty lands

The justification for land investments on a large scale is often presented around a series of 'crisis narratives', linked to growing scarcity and impending catastrophe. The underlying assumption is that the solution to such food, energy and climate 'crises' lies in capturing the potentials of so-called' marginal, empty, and available' lands across the globe.

---

[11]http://www.miga.org/news/index.cfm?aid=2635 (accessed 12 February 2012)

[12]The World Bank Group of course is not the only agency where 'joined up' thinking, advice and investment is absent. This is replicated across most of the aid agencies, where confusion abounds as to how to match a focus on private sector led investment and 'pro poor', 'inclusive' growth.

The World Bank, for example, undertook an influential assessment of the world's 'available land' and concluded that 'The currently non-cultivated area suitable for cropping that is non-forested, non-protected, and populated with less than 25 persons/km2 (or 20 ha/household) amounts to 446 million ha. This is equivalent to almost a third of globally cropped land which is 1.5 billion ha' (World Bank 2010, xv-xvi). Potential land availability was then plotted against the potential for increasing yields (the 'yield gap') for all countries to reveal four 'types' of countries: ones with little land for expansion and a low yield gap, those with suitable land available and a low yield gap (such as Brazil, Argentina and Ukraine), those with little land available and a high yield gap (such as Malawi, Rwanda, India or Egypt) and those where there is deemed to be suitable land available and a high yield gap (such as Sudan, Tanzania, Zambia and Mozambique). The report describes these countries as including 'countries with large tracts of suitable land, but also a large proportion of smallholders with very low productivity. If labour supply constrains smallholder expansion and in-migration is limited, larger farm sizes enabled through mechanization could be a viable strategy. This situation could create opportunities for outside investors' (World Bank 2010, xviii). The report goes on to argue that commodity-level analysis shows how 'in many African countries with large amounts of suitable but currently uncultivated land, transfers of technology could provide large benefits to local populations' (World Bank 2010, xviii).

This now familiar diagnosis misses many critical points. First, measures of 'potential production' under 'best practices' give a misleading picture of what is possible in such areas, with poor infrastructure, limited input supplies and inadequate market connections. The yield gaps resulting are therefore wildly inflated, giving unrealistic expectations of potential. Second, what may appear 'available' may in fact be being used for a range of livelihood activities, from small-scale farm production to pastoralism to shifting cultivation. 'Availability' suggests that there are no claims over the land and, although the World Bank report discusses issues of land governance and tenure, this is not part of the assessment of potential land supply. Lorenzo Cotula shows the inaccuracy of the rhetoric on targeting of 'marginal' lands for investment: 'investor interest often focuses on the best land in terms of water availability and irrigation potential, soil fertility, proximity to markets or availability of infrastructure' (Cotula 2012, this volume). Third, in those countries deemed to have plentiful land and a high yield gap, a scenario involving large-scale farms, mechanisation and technological investments in crop production is outlined, in the absence of any critical assessment of the experience of past initiatives of this kind in similar areas. Memories of past grand experiments to mechanise farming through large-scale production – whether the irrigation schemes of the Gezira in Sudan or the groundnut programmes of Senegal or Tanzania – seem to have been lost.

While clearly deeply flawed, the justificatory discourse for urgent action and focused solution is nevertheless a powerful one, and frames a particular political economy of intervention which garners support among national governments, private sector investors and international banks and donor agencies.

Terms of inclusion: impacts on agrarian societies

Many of the countries being targeted for land deals have been starved of investment in their agricultural sector for decades, as both state and donor attention waned

(World Bank 2010). Boosting capital investment, improving infrastructure, facilitating access to new technologies and providing much-needed rural employment are all facets of land deals that are potentially positive. Then why, despite the theoretical potentials, do large-scale commercial land deals so often result in negative consequences, not only for those immediately affected, but also for the wider economy? Why are the much-touted win-win solutions so difficult to realise?

These questions underline the need for a contextual understanding of the political economy of the new enclosures and the labour regimes that emerge from them. Outcomes critically depend on the 'terms of inclusion' of local people in land deals (McCarthy 2010). Describing the expansion of oil palm plantations in Indonesia, John McCarthy argues that factors such as the functioning of smallholder development schemes, the form of democratic control over village institutions, the spatial pattern of villages and investments and the way land tenure systems and informal land markets work are all critical factors. Thus, depending on the context, local people may be included or excluded from land deals, and those included may be incorporated with either beneficial or adverse consequences, often with quite divergent outcomes occurring between one project and the next.

This relates to the way local political processes influence access to and control over land and production, and what Derek Hall *et al.* (2011) refer to as the 'powers of exclusion' – the interacting processes of regulation, force, the market, and legitimation that include or exclude people from land. Property and access are not, or not only, matters of land title and contract, but fundamentally 'matters of power and authority' (Sikor and Lund 2009, 1, cf. Peluso and Lund 2011).

The micro-politics of negotiations of land control, access and exclusion are played out at the local level, and this is where the livelihood consequences of land deals must be assessed. There has been less work at this scale, although the papers in this collection offer important insights. However, in its global survey, the World Bank (2010, xiv) reports that 'in most cases the expected job creation and net investment were very low', a finding shared by many other studies (e.g. Anseeuw *et al.* 2012, Oxfam 2012, Oakland 2011, IIED 2011). Claims about employment generation are often inflated, as in the case of oil palm plantations in Indonesia (Li 2011). Most of the emerging plantation enclaves are large-scale, industrial monocrop ventures that are generally labour-expelling/labour-saving, in contrast to earlier episodes of plantation agriculture which depended on the availability of large quantities of cheap manual labour.

As Lorenzo Cotula and his colleagues have described in a series of studies, land deal contracts are enormously variable, although with the common feature that local people and communities are rarely partners to the deal, and rarely consulted or even aware of them before they happen.[13] The distribution of benefits, costs and risks of any deal is also highly dependent on the nature of the 'business model',[14] and the associated implications for land ownership, access, control, employment and labour rights (Cotula 2011, Cotula and Leonard 2010, Vermeulen and Cotula 2010).

[13]Cotula and his team analyzed the contents of contracts for twelve land deals in seven African countries (Cameroon, Ethiopia, Liberia, Madagascar, Mali, Senegal and Sudan); the contracts were generally made between host governments and foreign investors, with local landholders or communities, or their representatives, having no formal role at all (Cotula 2011).

[14]'Business model' is more or less the business-studies and business consultants' equivalent of 'labour regime'.

Borras and Franco categorize the direction and consequences of land deals across four broad 'types' of transformation in land use – food to food, food to biofuels, non-food to food, non-food to biofuels – each with different consequences for agrarian settings and different flows of land-based wealth and power. Some deals involve forms of redistribution, others distribution, others non-(re)distribution and others (re)concentration (Borras and Franco 2012). Using this schema, Ruth Hall (2011) shows for southern Africa that most deals involved a transition in land use from non-food to biofuels and a concentration of land, wealth and power in private companies.

Ruth Hall (2011) identifies a five-fold typology of institutional arrangements through which land deals are structured. All have important historical precedents, as well as contemporary incarnations; and all have different impacts and consequences in different contexts. First is an *extraction model* which involves the simple stripping of resources, a model that, like mining, is unsustainable in the long-term (Mackenzie 2006, 2009). Second is an *enclave model* involving outright takeover of land and related resources (perhaps displacing others) and the construction of related infrastructure, partly to provide inputs to and process output of a commercial enterprise, but also to provide the social and physical infrastructure required for commercial operations (Kydd and Christiansen 1982, Loewenson 1992). Such 'enclave economies' (Ferguson 2006) are poorly integrated into their surrounding society and economy,[15] but such isolation may act to focus investment and support in particular 'special' areas or zones with the idea being that in such areas growth will take off. Third is a *colonist model* involving the introduction of commercial operators who take over a block or area, as in the case of Zimbabwean farmers in Nigeria or Mozambique (Ariyo and Mortimore 2011, Hammar 2010). Such a model is familiar to the former settler colonies, where European farmers were allocated land as part of colonial occupation. Fourth is an *outgrower model* involving the development of processing facilities (usually with a commercially operated 'nucleus' estate), through which small producers under contract are incorporated into value chains (Porter and Phillips Howard 1997, Amanor 2008). Fifth, and finally, is a model of *commercialisation in situ*, in which small producers and other land users are incorporated into commercial value chains in the absence of any core estate or sometimes even any processing facility (Chalfin 2004, Amanor-Wilks 2009).

Each of these institutional arrangements differs dramatically, and there are many variations between these ideal types. They each have different implications for processes of incorporation or displacement of local land users; each involves different requirements for investment in infrastructure like roads and irrigation systems, as well as social investments such as schools and clinics; each has different requirements for labour, and so social and economic relationships between local land holders and each suggests different patterns of accumulation, appropriation and reinvestment in production. The last two models do not necessarily result in dispossession of local cultivators or changes in the forms of their access to land (although they may do, as in Julia and White 2012, this volume); they may rather amount to 'control grabs' without enclosure, in which investors source products from local producers under contract, rather than grabbing their land and establishing large-scale, industrial, wage-labour farms.

[15]The classic statement on agrarian (plantation) enclave models is Beckford (1972).

While for some purposes it is useful to extend the concept of 'land deals' to embrace deals governing (only) land use and not land access, in our view there are also important reasons not to downplay the importance and implications of enclosure and the actual corporate acquisition of land. Deals involving the large-scale purchase or long-term leasing of land are essentially permanent – when leasehold contracts of 35 or even 99 years expire, the land rights normally revert not to local people but to the state – and as already noted, local people are normally not even formal parties in these deals; contract farming deals, in contrast, do involve local cultivators as parties to the deal, and are not permanent. Such land grabs, therefore, close off irrevocably the option of a future return to smallholder tenure and smallholder farming, in the absence of mass 'grab land back' movements and redistributive land reforms. This is why, in a world where the penetration of corporate capital into agri-commodity chains may be inevitable and in some cases the only available source of needed investments, it is in our view very important to explore how this can be promoted and negotiated without the actual acquisition of land. As Chandrasekhar reminds us 'there are many different ways to release and extract resources and commodities [...] and appropriate surpluses from the less developed regions needed for accumulation of capital in the metropolitan centres' (2012, 612).

In this context, UN Rapporteur on the Right to Food Olivier de Schutter has argued consistently that 'the most pressing issue regarding investment in agriculture is not how much, but how: what we need is not to regulate land grabbing as if this were inevitable, but to put forward an alternative programme for agricultural investment' (De Schutter 2011, 250). These alternatives, he argues, must not be based on speculative large-scale acquisition of farmland, nor on the creation of a market for land rights based on individual titling, but on security of tenure, agrarian reform where land concentration has become excessive, and reorientation of agricultural systems towards agroecological modes of production that are both productive, sustainable and contribute to the progressive realization of the human right to adequate food. De Schutter therefore arrives at quite different principles for responsible agro-investment from those recommended by the World Bank or FAO (which we consider below).

> Land investments implying an important shift in land rights should represent *the last and least desirable option*, acceptable only if no other investment model can achieve a similar contribution to local development and improve the livelihoods within the local communities concerned (UN General Assembly 2010, 20. Emphasis added).

For these reasons it is important to explore the possibilities, advantages and disadvantages of various actually-existing 'business models' (labour regimes) linking local populations to agribusiness but not involving large-scale land acquisition, as Lorenzo Cotula and his colleagues have recently done. They have studied cases of contract farming, management and lease contracts, tenant farming and share-cropping, joint ventures, farmer-owned businesses, and upstream/downstream business links, looking at how value and influence are shared between the business partners, explored in four closely interlinked aspects: 'ownership' (of the business and of key assets), 'voice' (who takes/influences business decisions and how), 'risk' (how supply, production, market and other risks are shared between the parties) and 'reward' (sharing of economic costs and benefits, including market access, price setting, finance arrangements). They conclude that the impact on smallholders depends not so much on the form of the business-smallholder relationship as on how it functions in specific contexts: 'The willingness of the company to engage with more inclusive business models, as a genuine economic component of their business, rather than as a part of corporate responsibility programmes, is a key ingredient for more inclusive business models to work', and also – once again underlining the crucial role of politics and power - that 'the negotiating power of smallholders in their relations with government and agribusiness is also key' (Vermeulen and Cotula 2010, 7; see also Cotula and Leonard 2010).

In this context relations of labour are central: 'the truncated trajectory of agrarian transition in much of the global South, one in which there is no pathway from country to city, agriculture to industry, or even a clear pathway into stable plantation work that pays a living wage, is the crucial scale at which to review the land grab debate' (Li 2010, 66). The availability and quality of employment following land investments has major implications for the impacts of land deals on social differentiation, gender and age relations. This applies to all types of institutional arrangements following land deals, but perhaps especially to the 'outgrower' contract model. Past studies identify multiple impacts following the establishment of such agrarian contracts. These include widespread changes in rural social and class relations; the creation of new 'middle farmer' classes at the centre of new schemes; processes of differential accumulation and dispossession, resulting in the eviction of some and land size increase of others; highly gendered patterns of labour participation and wage levels, with women systematically disadvantaged; patterns of income increase but simultaneously declines in human development, including higher morbidity and lower nutrition due to lack of services; and broader processes of social dislocation from the places where people have historically come from (see contributions in Little and Watts 1994, White 1997, Julia and White 2012, this volume).

Asking searching questions of the sort identified above, and developing useful, operational typologies of both types and impacts of land deals will help move the debate about contemporary 'land grabbing' forward. This will hopefully promote a more firmly grounded analytical basis for the discussion, to date too often framed in terms of highly contrasting and polarized views, as we discuss in the next section.

Regulation and resistance

There is currently much policy and political discussion about how to tackle the challenges brought about by global land grabs. There are two opposing positions (and a great diversity of positions in between). At one pole are those who think that corporate land acquisitions can be harnessed as opportunity for development (see, e.g. World Bank 2010, Deininger 2011). They are aware of the multiple problems that follow land deals: corruption, shady deals, people getting dispossessed and displaced, unfulfilled promises for jobs, and so on. However, it is assumed that these problems can be subject to regulation. Hence, the World Bank, FAO, UNCTAD and IFAD (2010) have proposed to adopt a set of principles for 'responsible' land investments. The World Bank group's seven 'Principles for Responsible Agro-Investment', for example, clustered along social, political and environmental issues are the following: respecting land and resource rights; ensuring food security; ensuring transparency, good governance, and a proper enabling environment; consultation and participation; responsible agro-enterprise investing; social sustainability; and environmental sustainability (World Bank 2010, x, 68-91).

But in the particular contexts where land deals are unfolding, will these make any difference at all? Why should we expect corporate agribusiness to act on a basis of voluntary corporate 'social responsibility'? Capitalist firms are not Boy Scouts, and they are unlikely to place moral codes and 'good governance' above the interests and demands of their owners or shareholders (White n.d.). Saturnino Borras Jr. and Jennifer Franco (2010) argue that the 'codes of conduct' position does not question the root causes of land grabbing, accepts that land grabbing is inevitable, assumes that large-scale investments benefit the poor and that large farms benefit small farms, that corporations will voluntarily self-regulate to advance social and environmental justice, and that transparent and participatory land grabs are better than non-transparent and non-participatory ones. Most fundamentally, as the UN Special Rapporteur on the Right to Food has argued, to see the solution in terms of guidelines for the 'responsible' expansion of large-scale, capital-intensive farming is to narrow the terms of the debate and close the door to other alternatives (De Schutter 2011).

Large land deals most frequently occur where people do not have formal land rights, but they also occur across the whole range of property rights regimes: in state, public, community, common property and private lands, including in many recent land reformed lands as in the example of 11,000 ha of land in Isabela, Philippines acquired from land reform beneficiaries for the country's largest sugarcane ethanol plantation (Franco, Carranza and Fernandez 2011). This implies that securing land rights through formal private land title is no guarantee against land grabbing as is sometimes casually assumed.

The opposite pole is exemplified by *La Via Campesina*, today's largest and most important transnational agrarian movement representing small-scale farmers and poor peasants in the global South and North. In November 2011 *Via Campesina* and other allied civil society groups (NGOs, research think tanks, and so on) convened the international conference: 'Stop Land Grabbing Now!'. The resulting Mali Declaration against land grabbing underscores the critique of the fundamental assumptions and orientation of current land grabbing and large-scale industrial model of agriculture and the flawed assumption about the efficacy of corporate voluntary self-regulation (La Via Campesina 2011). Key to this position is the accompanying attempt to reframe the discourse around land investment which in the mainstream discourse has been interpreted only as 'corporate land investment'. Via Campesina and its allies are trying to bring two other types of land investments back on the agenda – state/public and small farmer land investments – to address what they see as a flawed 'there is no alternative' assumption. Like all agrarian struggles, this movement has multiple levels, and the local-national processes will prove to be key. Unfortunately, in all regions of the world it is at the local level that organized social movements are relatively thin and weak. While there is a need for more comprehensive and systematic empirical research on resistance, initial evidence from the field suggests a significant disconnect between the base of organized social movements and the key sites of land grabs. This often leads to competing and conflicting positions between organized social movements and their allied civil society groups on the one hand, and ordinary villagers actually affected by land deals on the other. Painstaking local community organization and mobilization work is perhaps the most urgent and difficult challenge.

There are numerous initiatives and groups navigating somewhere in between these two positions. One of these is the initiative around the now-approved

FAO-anchored Voluntary Guidelines for the Responsible Governance of Tenure of Land, Fisheries and Forests in the Context of National Food Security.[16] Negotiations for these Voluntary Guidelines have been politically complex, involving key state actors for or against stronger guidelines, as well as a range of civil society organizations and social movements with differentiated positions on land grabbing. Without complementary action from outside, it is likely that the Guidelines will be much weaker than hoped for by civil society groups. Their implementation will become a key arena of political contestation between key actors wanting to influence the institutional rules that govern land deals.

In trying to understand the dynamics between various positions among state and non-state actors, additional areas of contention are also important. Basic reform of investment frameworks which shift the balance between investors, national authorities and local land users is a priority. This will require a review of, for example, bilateral investment treaty legislation, as well as a more stringent application of equity, justice and development principles in investment financing and support by public aid agencies and banks.

Such policies also have a bearing on human rights. A rights-based framework, it is argued, must pay attention to rights of land holders subject to forced evictions, the foreclosure of land, the denial of information about deals, compensation arrangements and the prevention of local participation in political decisions all violate land users' human rights (De Schutter 2011, HLPE 2011). One strategy is to reorient agricultural investment away from land deals for large plantations or estates and towards small-scale family agriculture and markets, supported by stronger farmer voices (De Schutter 2011).

Thus besides analysing land deals and their consequences, parallel debates are needed on ways to increase the options and potentials of smallholder farming. This requires going beyond simplistic and flawed 'yield gap' calculations (which compare the productivity of long-neglected, resource- and infrastructure-starved peasantries with that of well-supported high-input agriculture) to the really relevant comparison: comparing the productivity, employment potential and long-term sustainability of industrial capitalist agriculture with that of a well-supported smallholder farming. A small farmer focus of course is not without its own problems; agrarian structures based on small-scale ('peasant') farming are inherently unstable under conditions of commodity economy, due to the in-built mechanisms of land concentration and agrarian differentiation, which many authors, from Lenin onwards, have described.[17] But these problems are not impossible to overcome, once we move away from fixations on private ownership titling to other forms of secure individual tenure, subject to maximum holdings and periodic redistribution.

As we have shown, the current policy debate has become polarised between adherents of an often simplistic 'win-win' scenario, where investors and local farmers benefit through new contracting arrangements, with land investments being seen as more an opportunity than a threat (von Braun and Meinzen-Dick 2009, 3–4, World Bank 2010) and those who see only danger and potential disaster, across economic, ecological and social dimensions (FIAN 2011, La Via Campesina 2011). The reality, as we have discussed, often fits neither of these poles, and a more nuanced analysis is

[16]For details see http://www.csm4cfs.org/policy_working_groups-6/land_tenure-6/ (accessed 30 April 2012).
[17]Summarised in Bernstein 2010, Ch. 7.

required, informed by the kind of agrarian political economy perspectives, and associated typologies, which we have outlined above. The contributions to this collection, which are introduced in the next section, offer examples of such an approach

The contribution of this collection

The contributions to this collection cover a great diversity of the changes in broad agrarian structures that are fostering land deals; the 'newness' of land deals versus their continuity with earlier historical forms of appropriation; the ways in which existing patterns and processes of rural social differentiation – in terms of class, gender, ethnicity – are being shaped by changes in land use and land property relations, as well as by the re-organization of production and exchange; the impact of land deals on local level and national food security; the role of struggle or resistance and mobilization in shaping, negotiating and challenging the new land investment; the issues uniting or dividing the rural poor, organized movements, and rural communities around the issue of land deals; the various competing policy and political narratives and discourses around the multiple crises of food, energy, climate and finance, and their impact on or use in land deal negotiations; the ways in which competing narratives on land and property been deployed by various camps around the contested meanings of 'marginal lands' (or, idle', 'waste', 'unoccupied' lands); the emerging or continuing dynamics of power, elites and corruption, where land serves as a source of patronage; and the role of global land policies from different overseas development agencies – including the World Bank, Food and Agricultural Organisation, European Union, International Fund for Agricultural Development, and so on) in facilitating and encouraging, or blocking and discouraging, land deals.

Two framing papers explore global trends and how to understand them. Lorenzo Cotula (2012, this volume) addresses the 'geo-economics' of land deals, critically analyzing evidence of trends in scale and geography. While this analysis broadly corroborates some widespread assumptions, it also points to shifts in global-to-local economic and geopolitical relations linking sovereign states, global finance, agribusiness through to local groups. Understanding these, he argues, can help identify levers and pressure points for policy responses. Philip McMichael (2012, this volume) reflects on his own and his earlier co-authored work on food regimes, arguing that the 'new enclosures' are an expression of food regime restructuring, symptomatic of a crisis of accumulation in the neoliberal globalization project. Drawing on 'global ecology' and political economy, McMichael situates the land grab in the conjunctural crisis of capitalist ecology, expressed in climate, energy and food crises, which in turn are transforming the food and fuel regime.

Shepard Daniel (2012, this volume) addresses a largely unresearched issue, the international financing of land deals, providing a detailed analysis of the involvement of private equity funds in land deals. As well as documenting investment flows, she draws attention to the ways in which development finance institutions, especially the World Bank's International Finance Corporation, are promoting such deals, and reflects on the implications of speculative capital moving into land. She explores particularly the power relations embedded in this trend, exploring the information asymmetry and 'creative destruction' inherent in private equity finance, and their implications for governance and labour.

Four papers – all with an African focus – locate contemporary land deals within broader agricultural development strategies. Kojo Amanor (2012, this volume)

analyses critically the prevalent tendency of neoliberal understandings of large-scale estate agriculture in Africa, as an 'aberration' brought about by Asian and Middle Eastern investors. Using the cases of cocoa in Côte d'Ivoire and pineapples in Ghana from the 1970s onward, he argues that land grabbing arises from the internal logic of agribusiness accumulation rather than from exogenous developments. Liz Alden Wily (2012, this volume) argues that both past and present land rushes have been based on, added to, and legitimized by legal manipulations which fail to recognize customary interests as real property rights, thereby legalizing the theft of lands of the poor or subject peoples. Phil Woodhouse (2012, this volume) focuses on the supply side of land deals in Africa specifically, placing the current enclosures in a longer historical context. He points out the ways in which the two dominant institutional models – large-scale mechanised farms and small-scale outgrower contract farming – respond to real productivity constraints, including water constraints facing rainfed agriculture and market-related risks. His contribution draws attention to the centrality of water considerations in land deals, and to the need for solutions to these constraints. Tom Lavers (2012, this volume), addressing the case of Ethiopia, contests the understanding of the 'land grab' as a neo-colonial process, and instead locates the expansion of foreign land-based investment within the Ethiopian government's export-led development strategy. His paper shows how macro benefits in terms of increased foreign exchange earnings come at the cost of increased risks at the micro level to people living in the vicinity of new investments, particularly politically marginal pastoral populations in remote regions.

Four papers highlight the intra-regional character of land deals, suggesting again that the widespread characterization of the new enclosures as a repeat of the colonial experience may be missing the mark. Rather, new actors, particularly emerging economies, are seeking out control of territory and natural resources in neighbouring states, often within the legitimating framework and discourse of regional integration. Oane Visser, Max Spoor and Natalia Mamonova point out that the existing literature has focused on the 'developing world' and has largely ignored the scale and significance of land deals in the former Soviet Union (2012, this volume). They contribute to rectifying this gap by documenting land deals in Russia, the legal and illegal acquisition strategies pursued by a range of investors, and they place these processes within the context of post-Soviet land reform and the forms of privatization which were key to enabling this later concentration of corporate property. Saturnino Borras Jr., Jennifer Franco, Sergio Gómez, Cristóbal Kay and Max Spoor (2012, this volume), summarize and reinterpret FAO studies from 17 countries in the Latin American and Caribbean region to show that land grabbing is occurring in this region to a greater extent than previously assumed. Key contexts for land deals include energy/biofuel ventures and climate mitigation strategies, and the resource demands of various newer hubs of global capital. Hallmarks of the process in this region are its intra-regional character, with (Trans) Lartin American Corporations playing a key role alongside traditional TNCs, and the rise of 'flex crops' such as sugar, soya and oil palm. Ruth Hall analyses the shifting role of South African farmers, agribusiness and capital on the rest of the continent, exploring the ideologies and discourses of legitimation employed in favour of them (2012, this volume). She documents the AgriSA farmers' move into Congo Brazzaville, and acquisitions by the two South African sugar giants, Illovo and Tongaat-Hulett, for outgrower and estate expansion elsewhere in the region. She argues that South Africa is no longer merely exporting its farmers, but also its value chains, to the rest of the continent, and that these deals that appear to be 'intra-regional' are increasingly reliant on external financing through transnational partnerships. Miles Kenney-Lazar (2012, this volume) documents the role of a Vietnamese corporation in what he argues is 'primitive accumulation' in Laos, as enclosure of 'freed up' land for timber extraction and plantation development prompts the semi-proletarianization of dispossessed communities.

Several papers explore the socially differentiated impacts of the new enclosures. Julia and Ben White (2012, this volume) document the gendered politics of monocrop oil-palm expansion in West Kalimantan, Indonesia. They show how dispossession from customary landholdings and the new contract farming system, based on registration of smallholder plots to 'household heads', has eroded women's independent tenure and rendered them a class of plantation labour – on the one hand offering the attractions of cash incomes and on the other hand locating them in new exploitative household and commodity relations. Sérgio Sauer and Sergio Pereira Leite show how the Brazilian government has responded to the global rush for farmland by re-establishing a legal mechanism for 'controlling' foreign investment in land deals (2012, this volume). They address the drivers of these investments and their consequences for both social relations and land values. Michael Levien argues that India's Special Economic Zones (SEZs) have become the epicentres of the country's 'land wars', with farmers resisting the state's attempts to transfer their land to private companies for developing these hyper-liberalized enclaves (2012, this volume). His analysis of this 'accumulation by dispossession' challenges David Harvey's (2003) understanding of the process as a means of absorbing over-accumulated capital in the global economy; rather, he argues that it is an extra-economic process through which states act as land brokers for capital, using eminent domain to overcome the barriers to accumulation posed by insufficiently capitalist rural land markets. While such development only minimally and precariously absorbs the labour of dispossessed farmers, it creates a peculiar agrarian transformation through land speculation that absorbs fractions of the rural elite, drastically amplifies existing inequalities, and fuels non-productive and pre-capitalist economic activity. Given the minimal benefits for rural India in this model of development, farmer resistance to land dispossession is likely to continue. Shelley Feldman and Charles Geisler's analysis of *in-situ* and *ex-situ* displacement in the *char* riverine and coastal regions and peri-urban regions in Bangladesh draws attention first to contestation over access to the rich alluvial soils of newly formed islands or *chars*, and susceptibility of small producers to dispossession from these shifting landscapes; and second, to new patterns of land capture in peri-urban areas, by elites who engage gangs, corrupt public servants and the military to coerce smallholders to relinquish title to their valuable lands. The multiple articulations of domestic land grabbing force us to examine displacement as an 'everyday', but widespread, long-term process rather than a cataclysmic event (Feldman and Geisler 2012, this volume).

Finally, as an antidote to all these analyses of land deals, dispossession and displacement, two contributions provide instructive 'dog that didn't bark'[18] cases of land grabs that have not happened. Using the lens of access and discourse theory Rebecca Smalley and Esteve Corbera (2012, this volume) argue that domestic actors have a greater influence on large-scale land deals than is assumed in neo-colonialist

[18]"Is there any other point to which you would wish to draw my attention?". "To the curious incident of the dog in the night-time." "The dog did nothing in the night-time".

arguments. In the case of Kenya's Tana Delta, they show that top-down processes of land acquisition by powerful state-linked actors and supported by policy discourse can be stalled by local resistance, if rural people have strong customary claims and forge links with broader opposition. Thomas Sikor asks why private plantation companies and transnational corporations have not been able to find a foothold in Vietnam's growing tree plantation sector, which continues to be dominated by smallholders, supported by state policies which promote their access to land, finance and wood markets (Sikor 2012, this volume). It is both refreshing and strategic, as Olivier De Schutter (above) has underlined, not to close off the options and to remind ourselves that there are workable alternatives to corporate land accumulation and large-scale industrial farming.

References

Akram-Lodhi, H. 2008. (Re)imagining agrarian relations? The World Development Report 2008: Agriculture for Development. *Development and Change*, 39(6), 1145–1161.

Alden-Wily, L. 2012. Looking back to see forward: the legal niceties of land theft in land rushes. *Journal of Peasant Studies*, 39(3&4).

Amanor, K. 2008. Sustainable development, corporate accumulation and community expropriation: Land and natural resources in West Africa. In: K. Amanor and S. Moyo, eds. *Land and sustainable development in Africa*. London and New York: Zed Books.

Amanor, K. 2012. Global resource grabs, agribusiness concentration, and the smallholder: two West African case studies. *Journal of Peasant Studies*, 39(3&4).

Amanor-Wilks, D-E. 2009. Land, labour and gendered livelihoods in a 'peasant' and a 'settler' economy. *Feminist Africa*, 12, 31–50.

Anseeuw, W., L. Alden Wily, L. Cotula, and M. Taylor. 2012. *Land rights and the rush for land: Findings of the global commercial pressures on land research project*. Rome: International Land Coalition.

Araghi, F. 2003. Food regimes and the production of value: Some methodological issues. *Journal of Peasant Studies*, 30(2), 41–70.

Araghi, F. 2010. The end of 'cheap ecology' and the crisis of 'Long Keynesianism'. *Economic and Political Weekly*, xlv(4), 39–40.

Ariyo, J. and M. Mortimore. 2011. *Land deals for commercial agriculture in Shonga, Nigeria: The policy challenge*. Policy brief. Nairobi/Brighton: Future Agricultures Consortium.

Ariza-Montobbio, P., S. Lele, G. Kallis and J. Martinez-Alier. 2010. The political ecology of *Jatropha* plantations for biodiesel in Tamil Nadu, India. *Journal of Peasant Studies* 37(4), 875–897.

Beckford, G. 1972. *Persistent poverty: underdevelopment in the plantation economies of the Third World*. New York: Oxford University Press.

Bello, W. 2009. *The Food Wars*. New York: Verso.

Bernstein, H. 1998. Labour regimes and social change under colonialism. In: Ben Crow, Mary Thorpe *et al*., eds. *Survival and change in the third world*. Cambdridge: Polity Press, pp. 30–49.

Bernstein, H. 2010. *Class dynamics of agrarian change*. Halifax: Fernwood Publishing; Sterling, VA: Kumarian Press.

Binswanger, H. and K. Deininger 1999. The evolution of the world bank's land policy: principles, experience, and future challenges. *The World Bank Research Observer*, 14(2), 247–276.

Borras, S. Jr. and J.C. Franco. 2010. From threat to opportunity? Problems with the idea of a 'code of conduct' for land-grabbing. *Yale Human Rights and Development Law Journal*, 13(2), 507–523.

Borras, S. Jr. and J.C. Franco. 2012. Global land grabbing and trajectories of agrarian change: a preliminary analysis. *Journal of Agrarian Change*, 11(2), 34–59.

Borras, S., J. Franco, S. Gomez, C. Kay and M. Spoor. 2012. Land grabbing in Latin America and the Caribbean. *Journal of Peasant Studies* 39(3&4).

Borras, S. Jr., R. Hall, I. Scoones, B. White and W. Wolford. 2011. Towards a better understanding of global land grabbing: an editorial introduction. *Journal of Peasant Studies*, 38(2), 209–216.
Borras, S., C. Kay and E. Lahiff, eds. 2007. *Market-led agrarian reform: a critique*. London: Routledge.
Borras, S., P. McMichael and I. Scoones. 2010. The politics of biofuels, land and agrarian change: editors' introduction. *Journal of Peasant Studies*, 37(4), 575–592.
Borras, S. Jr., R. Hall, I. Scoones, B. White and W. Wolford, eds. n.d. Governing global land deals: the role of the state in the rush for land. Special Issue, *Development and Change* (under review).
Byres, T. 2004. Neo-classical neo-Populism 25 years on: déjà vu and deja passe. towards a critique. *Journal of Agrarian Change*, 4(1–2), 17–44.
Castree, N. 2008a. Neoliberalising nature I: the logics of de- and re-regulation. *Environment and Planning*, 40(1), 131–152.
Castree, N. 2008b. Neo-liberalising nature II: Processes, outcomes and effects. *Environment and Planning A*, 40(1), 153–173.
Chalfin, B. 2004. Old commodities in new niches: The shea economy as frontier. *African Economic History*, 32, 51–63.
Chandrasekhar, C.P. 2012. Critical perspectives on the great recession. *Development and Change*, 43(2), 603–613.
Chayanov, A.V. 1966 [1925]. *The Theory of Peasant Economy*. Homewood: American Economic Association.
Collier, P. 2008. Politics of hunger: How illusion and greed fan the food crisis. *Foreign Affairs*, November/December.
Conan Doyle, A. 1950. Silver blaze. In: *The Memoirs of Sherlock Holmes*. London, Penguin Books, pp. 7–34.
Corbera, E., K. Brown and W.N. Adger. 2007. The equity and legitimacy of markets for ecosystem services. *Development and Change*, 38(4), 587–613.
Cotula, L. 2011. *Land deals in Africa: What is in the contracts?* London: International Institute of Environment and Development.
Cotula, L. 2012. The international political economy of the global land rush: A critical appraisal of trends, scale, geography and drivers. *Journal of Peasant Studies*, 39(3&4).
Cotula, L., S. Vermeulen, R. Leonard and J. Keeley. 2009. *Land grab or development opportunity? Agricultural investment and international land deals in Africa*. London/Rome: International Institute for Environment and Development, Food and Agricultural Organization of the United Nations, and International Fund for Agricultural Development.
Cotula, L. and R. Leonard. 2010. *Alternatives to land acquisitions: Agricultural investment and collaborative business models*. London: International Institute for Environment and Development.
Daniel, S. 2012. Situating private equity capital in the land grab debate. *Journal of Peasant Studies*, 39(3&4).
Dauvergne, P. and K. Neville 2010. The changing north–south and south–south political economy of biofuels. *Third World Quarterly*, 30(6), 1087–1102.
De Schutter, O. 2011. How not to think of land-grabbing: three critiques of large-scale investments in farmland. *Journal of Peasant Studies* 38 (2), 249–279.
Deininger, K. 2011. Challenges posed by the new wave of farmland investment. *Journal of Peasant Studies*, 38(2), 217–247.
El-Ghonemy, M.R. 2001. The political economy of market-based land reform. In: K.B. Ghimire, ed. *Land reform and peasant livelihoods: the social dynamics of rural poverty and agrarian reforms in developing countries*. London, ITDG Publishing, pp. 105–133.
Fairhead, J., M. Leach and I. Scoones, eds. 2012. *Green grabbing: A new appropriation of nature?* Special Issue, *The Journal of Peasant Studies*, 39(2).
Feldman, S. and C. Geisler. 2012. Land expropriation and displacement in Bangladesh. *Journal of Peasant Sudies*, 39(3&4).
Ferguson, J. 2006. *Global shadows: Africa in the neoliberal world order*. Durham: Duke University Press.

FIAN. 2011. Dakar appeal against the land grab. World Social Forum. Available from: http://www.fian.org/news/press-releases/dakar-appeal-against-the-land-grab/pdf (accessed 2 March 2011).
Foresight. 2011. *The future of food and farming*. Final Project Report. The Government Office for Science, UK.
Franco, J., D. Carranza and J. Fernandez. 2011. New biofuel project in Isabela, Philippines: Boon or Bane for Local People? TNI Discussion Paper October 2011. Available from: http://www.tni.org/article/new-biofuel-project-isabela (downloaded 4 December 2011).
Friedmann, H. and P. McMichael. 1989. Agriculture and the state system: the rise and fall of national agricultures, 1870 to the present. *Sociologia Ruralis*, 29(2), 93–117.
Ghose, A.K. 1983. *Agrarian reform in contemporary developing countries*. London: Palgrave MacMillan.
Gibbon, P. 2011. Experiences of plantation and large-scale farming in 20th century Africa. *Copenhagen: Danish Institute for International Studies Working Paper* 2011:20.
Godfray, H., Beddington, J.R., Crute, I.R., Haddad, L., Lawrence, D., Muir, J.F., Pretty, J., Robinson, S., Thomas, S.M., Toulmin, C. 2010. Food security: the challenge of feeding 9 billion people. *Science*, 327 No 5967, 812–818.
GRAIN. 2008. Seized: The 2008 land grab for food and financial security. Barcelona: GRAIN.
Griffin, K., A. Rahman Khan and A. Ickowitz. 2002. Poverty and distribution of land. *Journal of Agrarian Change*, 2(3), 279–330.
Hall, R. 2011. Land grabbing in southern Africa: The many faces of the investor rush. *Review of African Political Economy*, 128, 193–214.
Hall, R. 2012. The next Great Trek? South African commercial farmers move north. *Journal of Peasant Studies*, 39(3&4).
Hall, D., P. Hirsch and T.M. Li. 2010. *Powers of exclusion: Land dilemmas in southeast Asia.* Singapore: National University of Singapore Press; Honolulu: University of Hawaii Press.
Hammar, A. 2010. Ambivalent mobilities: Zimbabwean commercial farmers in Mozambique. *Journal of Southern African Studies*, 36(2), 395–416.
Harvey, D. 2003. *The new imperialism*. Oxford: Oxford University Press.
Harvey, D. 2005. *A brief history of neoliberalism*. Oxford: Oxford University Press.
Hazell, P., C. Poulton, S. Wiggins and A.R. Dorward. 2007. *The future of small farms for poverty reduction and growth.* 2020 Vision Discussion Paper 42. Washington DC: International Food Policy Research Institute.
Heyer, J., P. Roberts and G. Williams, eds. 1981. *Rural development in tropical africa*. London: Macmillan.
Heynen, N., J. McCarthy, S. Prudham and P. Robbins, eds. 2007. *Neoliberal environments*. London: Routledge.
HLPE 2011. Land tenure and international investments in land: a report by the High Level Panel of Experts on Food Security and Nutrition, authored by Camilla Toulmin, Saturnino M. Borras Jr., Prem Bindraban, Esther Mwangi and Sergio Sauer. Rome: FAO.
Holt-Gimenez, E. and R. Patel with A. Shattuck. 2009. *Food rebellions: Crisis and the hunger for justice*. Oakland: Food First.
Holt-Gimenez, E. and A. Shattuck. 2011. Food crises, food regimes and food movements: rumblings of reform or tides of transformation? *Journal of Peasant Studies*, 38(1), 109–144.
Hunsberger, C. 2010. The politics of Jatropha-based biofuels in Kenya: convergence and divergence among NGOs, donors, government officials and farmers. *Journal of Peasant Studies*, 37(4), 939–962.
ILC. 2011. Land rights and the rush for land: a report. Rome: ILC.
Julia and B. White. 2012. Gendered experiences of dispossession: oil palm expansion in a Dayak Hibun community in West Kalimantan. *Journal of Peasant Studies*, 39(3&4).
Kay, C. 2009. Development strategies and rural development: Exploring synergies, eradicating poverty. *Journal of Peasant Studies*, 36(1). 103–138.
Kautsky, K. 1988 [1899]. *The agrarian question*. Winchester: Zwan Publications.
Kenney-Lazar, M. 2012 Plantation rubber, land grabbing, and social property transformation in southern Laos. *Journal of Peasant Studies*, 39(3&4).

Kydd, J. and R. Christiansen. 1982. Structural change in Malawi since independence: Consequences of a development strategy based on large-scale agriculture. *World Development*, 10(5), 355–375.
Lavers, T. 2012. Patterns of agrarian transformation in Ethiopia: state-mediated commercialisation and the 'land grab'. *Journal of Peasant Studies*, 39(3&4).
La Via Campesina. 2011. G20 Agriculture: No to 'Food grabbing' by the richest nations of the world! Press release. Jakarta: La Via Campesina. 16 June 2011.
Lenin, V.I. 1960 [1899]. *The development of capitalism in russia. Vol. 3 of Collected Works.* London: Lawrence and Wishart.
Levien, M. 2012. The land question: special economic zones and the political economy of dispossession in India. *Journal of Peasant Studies*, 39(3&4).
Li, T.M., ed. 2009. Forum: Reading the *World Development Report: Agriculture for Development*. *Journal of Peasant Studies*, 36(3), 591–661.
Li, T.M. 2010. To make live or let die? Rural dispossession and the protection of surplus population. *Antipode*, 41(supplement 1), 66–93.
Li, T.M. 2011. Centering labour in the land grab debate. *Journal of Peasant Studies*, 38 (2), 281–298.
Lipton, M. 2009. *Land reform in developing countries: Property rights and wrongs*. London: Routledge.
Little, P. and M. Watts. 1994. *Living under contract: Contract farming and agrarian transformation in sub-saharan Africa.* Madison: University of Wisconsin Press.
Loewenson, R. 1992. *Modern plantation agriculture: Corporate wealth and labour squalor.* London: Zed Books.
Malik, M. 2011. Foreign investment into agriculture: Investment Treaties and the ability of governments to balance rights and obligations between foreign investors and local communities. Paper presented at the International Conference on Global Land Grabbing 6–8 April 2011, organised by the Land Deals Politics Initiative (LDPI) in collaboration with the Journal of Peasant Studies and hosted by the Future Agricultures Consortium at the Institute of Development Studies, University of Sussex.
Marx, K. 1977 [1867]. *Capital*. Vol. I. London: Lawrence and Wishart.
McCarthy, J. 2010. Processes of inclusion and adverse incorporation: oil palm and agrarian change in Sumatra, Indonesia. *Journal of Peasant Studies*, 37(4), 821–850.
McKenzie, C. 2006. *Forest governance in Zambezia, Mozambique: Chinese takeaway!*. Final Report for FONGZA.
McKenzie, C. 2009. *Tristezas tropicais: Further observations of forest governance in Zambezia.* Final Report for FONGZA.
McCarthy, J. and S. Prudham. 2004. Neoliberal nature and the nature of neoliberalism. *Geoforum* 35, 275–283.
McMichael, P. 2000. *Development and social change: A global perspective*. London: Sage Publications.
McMichael, P. 2009. A food regime genealogy. *Journal of Peasant Studies*, 36(1), 139–169.
McMichael, P. 2012. The land grab and corporate food regime restructuring. *Journal of Peasant Studies*, 39(3&4).
McMichael, P. and I. Scoones, eds. 2010. *Biofuels, land and agrarian change*. Special Issue, *Journal of Peasant Studies*, 37(4).
Moore, J.W. 2011. Transcending the metabolic rift: a theory of crises in the capitalist world-ecology. *Journal of Peasant Studies*, 38(1), 1–46.
Oakland Institute. 2011. Understanding land investment deals in Africa. Available from: http://www.oaklandinstitute.org/land-deals-africa. Accessed 12 February 2012
Oxfam. 2011. *Land and power: The growing scandal surrounding the new wave of investments in land.* Oxford: Oxfam International Briefing Paper 51.
Oya, C. 2009. The *World Development Report 2008*: inconsistencies, silences, and the myth of 'win-win' scenarios. *Journal of Peasant Studies*, 36 (3), 593–601.
Patnaik, U. 2010. Origins of the food crisis in India and developing countries. In: F. Magdoff and B. Tokar, eds. *Agriculture and Food in Crisis: Conflict, resistance and Renewal.* New York: Monthly Review Press, pp. 85–101.
Pearce, F. 2012. *The land grabbers: The new fight over who owns the earth.* London: Transworld and Boston: Beacon Press.

Peluso, N.L. and C. Lund. 2011. New frontiers of land control: Introduction. *Journal of Peasant Studies*, 38(4), 667–681.

Peluso, N.L. and C. Lund, eds. 2011. *New frontiers of land control.* Special Issue, *Journal of Peasant Studies*, 38(4), 667–681.

Ploeg, J.-D. van der. 2008. *The new peasantries: Struggles for autonomy and sustainability in an era of empire and globalization.* London: Earthscan.

Porter, G. and K. Phillips Howard. 1997. Comparing contracts: An evaluation of contract farming schemes in Africa. *World Development*, 25(2), 227–238.

Safransky, S. and W. Wolford. 2011. Contemporary land grabs and their alternatives in the Americas. Paper presented at the International Conference on Global Land Grabbing 6–8 April 2011, organised by the Land Deals Politics Initiative (LDPI) in collaboration with the Journal of Peasant Studies and hosted by the Future Agricultures Consortium at the Institute of Development Studies, University of Sussex.

Sassen, S. 2010. A savage sorting of winners and losers: Contemporary versions of primitive accumulation. *Globalizations*, 7(1–2), 23–50.

Sauer, S. and S. Leite Pereira. 2012. Agrarian structure, foreign investments on land, and land price in Brazil. *Journal of Peasant Studies*, 39(3&4).

Schneider, M. and P. McMichael. 2010. Deepening, and repairing, the metabolic rift. *Journal of Peasant Studies*, 37(3), 461–484.

Scott, J.C. 2009. *The art of not being governed. An anarchist history of upland southeast Asia.* New Haven: Yale University Press.

Scoones, I., N. Marongwe, B. Mavedzenge, F. Murimbarimba, J. Mahenehene, and C. Sukume. 2010. *Zimbabwe's land reform: Myths and realities.* Oxford: James Currey; Harare: Weaver Press; Johannesburg: Jacana.

Sender, J. and D. Johnston. 2004. Searching for a weapon of mass production in rural Africa: Unconvincing arguments for land reform. *Journal of Agrarian Change*, 4(1–2), 142–164.

Sikor, T. 2012. Tree plantations, politics of possession and the absence of land grabs in Vietnam. *Journal of Peasant Studies*, 39(3&4).

Sikor, T. and C. Lund. 2009. *The politics of possession: Property, authority, and access to natural resources.* London: Wiley-Blackwell.

Smalley, R. and E. Corbera. 2012. Large-scale land deals from the inside out: insights from Kenya's Tana Delta. *Journal of Peasant Studies*, 39(3&4).

Sorenson, MPK. 1967. *Land reform in the kikuyu country: A study in government policy.* Nairobi: Oxford University Press.

Sulle, E. and F. Nelson. 2009. *Biofuels, land access and rural livelihoods in Tanzania.* London: IIED.

Tanner, C. 2010. Land rights and enclosures: Implementing the Mozambican land law in practice. In: W. Anseeuw and C. Alden, eds. *Struggles over Land in Africa: Conflicts, Politics and Change.* Cape Town: Human Sciences Research Council, pp. 105–130.

UN General Assembly. 2010. *Report of the special rapporteur on the right to food.* New York: United Nations General Assembly, 65th Session, August 2010.

Vermeulen, S. and L. Cotula. 2010. *Making the most of agricultural investment: A survey of business models that provide opportunities for smallholders.* London: International Institute of Environment and Development.

Visser, O. and M. Spoor. 2011. Land grabbing in post-Soviet Eurasia: the world's largest agricultural land reserves at stake. *Journal of Peasant Studies*, 38(2), 299–323.

Visser, O., M. Spoor and N. Mamonova. 2012. Land grabbing in Russia: The post-Soviet political economy of insecurity and dispossession. *Journal of Peasant Studies*, 39(3&4).

Von Braun, J. and R. Meinzen-Dick. 2009. *'Land grabbing' by foreign investors in developing countries: Risks and opportunities.* Policy Brief 13. April 2009. Washington DC: International Food Policy Research Institute.

Watts, M. 2012. Book review of: *Class dynamics of agrarian change* by Henry Bernstein. *Journal of Peasant Studies*, 39(1), 199–204.

Weis, T. 2010. The accelerating biophysical contradictions of industrial capitalist agriculture. *Journal of Agrarian Change*, 10(3), 315–341.

White, B. 1997. Agroindustry and contract farmers in upland West Java. *Journal of Peasant Studies*, 24(3), 100–136.

White, Ben. n.d. Corporate land deals, dispossession and the future of farming. In: S.J.T.M. Evers, C. Seagle and F. Krijtenburg, eds. *Africa for sale: Analysing and theorizing foreign land claims and acquisition*. Leiden: Brill.

White, B. and A. Dasgupta. 2010. Agrofuels capitalism: a view from political economy. *Journal of Peasant Studies*, 37 (4), 593–607.

Wiggins, S., J. Kirsten and L. Llambi. 2010. The future of small farms. *World Development*. 38(10), 1341–1526.

Wolford, W. 2010. Contemporary land grabs in Latin America. Presentation at the FAO Committee on Food Security 36th Session, October 2010, Rome. www.tni.org. Accessed 15 January 2012.

Woodhouse, P. 2012. New investment, old challenges: land deals and the water constraint in African agriculture. *Journal of Peasant Studies*, 39(3&4).

World Bank. 2008. *World Development Report 2008: Agriculture for development*. Washington DC: World Bank.

World Bank. 2010. *Rising global interest in farmland: Can it yield sustainable and equitable results?*. Washington DC: World Bank.

World Bank, UNCTAD, FAO and IFAD. 2010. Principles of responsible agricultural investments. Washington DC: World Bank.

Zoomers, A. 2010. Globalisation and the foreignization of space: Seven processes driving the current global land grab. *Journal of Peasant Studies*, 37(2), 429–44.

Saturnino M. Borras Jr. is Associate Professor at the International Institute of Social Studies (ISS) in The Hague, Adjunct Professor at China Agricultural University in Beijing, and Fellow of the Transnational Institute (TNI) and Food First. He has carried out research on land grabbing in Southeast Asia, China, Southern Africa and Southern America. He is the Editor-in-Chief of the Journal of Peasant Studies.

Ruth Hall is Associate Professor at the Institute for Poverty, Land and Agrarian Studies (PLAAS) at the University of the Western Cape in South Africa. Her research interests include land and agrarian reforms, rural labour markets and farm worker rights, agricultural commodity chains, and the politics of rural development. She is Book Reviews Section co-editor of the Journal of Peasant Studies.

Ian Scoones is a Professorial Fellow at the Institute of Development Studies. He is co-director of the STEPS Centre (www.steps-centre.org) and joint coordinator of the Future Agricultures Consortium (www.future-agricultures.org). He works on land, agriculture and agrarian change issues in Africa, and his most recent book is *Zimbabwe's Land Reform: Myths and Realities* (James Currey, 2010). He is a member of the editorial collective of the Journal of Peasant Studies.

Ben White is Emeritus Professor of Rural Sociology at the International Institute of Social Studies, The Hague. He works on agrarian change and the anthropology and history of childhood and youth, mainly in Indonesia.

Wendy Wolford is Polson Professor of Development Sociology at Cornell University, USA. Her research interests include the political economy of development, social movements, land distribution and agrarian societies. She is a member of the editorial collective of the Journal of Peasant Studies.

The international political economy of the global land rush: A critical appraisal of trends, scale, geography and drivers

Lorenzo Cotula

Over the past few years, agribusiness, investment funds and government agencies have been acquiring long-term rights over large areas of farmland in lower income countries. It is widely thought that private sector expectations of higher agricultural commodity prices and government concerns about longer-term food and energy security underpin much recent land acquisition for agricultural investments. These processes are expected to have lasting and far-reaching implications for world agriculture and for livelihoods and food security in recipient countries. This paper critically examines evidence of trends, scale, geography and drivers in the global land rush. While this analysis broadly corroborates some widespread assumptions, it also points to a more complex set of drivers that reflect fundamental shifts in economic and geopolitical relations linking sovereign states, global finance, and agribusiness through to local groups. Only a solid understanding of these fundamental drivers can help identify levers and pressure points for policy responses to address the challenges raised by large-scale land acquisitions.

1. Introduction

Recent years have witnessed growing interest in farmland as an economic asset. Large-scale acquisitions of farmland in Africa, Asia and Latin America have made headlines in media reports across the world. These acquisitions involve outright land purchases or, more commonly, long-term leases mainly on government-owned land. It is widely thought that private sector expectations of higher agricultural commodity prices and government concerns about longer-term food and energy security underpin much recent land acquisition for agricultural investments.

Dubbed ‘land grabs’ in the media, land acquisitions have kindled much international debate, in which strong positions are taken on the impacts of such investments on environment, rights, sovereignty, livelihoods, development and conflict at local, national and international levels. Some commentators have welcomed these trends as bearers of new livelihood opportunities in lower income countries, and as an important step towards ensuring food security for a growing world population. Others have raised concerns about the possible social and environmental impacts of large land deals, including loss of land for rural people, and, more generally, about the risk that large-scale investments may marginalise family farming.

This paper critically examines evidence of trends, scale, geography and drivers in the global land rush. While this analysis broadly corroborates some widespread assumptions, it also points to a more complex set of drivers that reflect fundamental shifts in economic and geopolitical relations linking sovereign states, global finance, and agribusiness through to local groups. Only a solid understanding of these fundamental drivers can help identify levers and pressure points for policy responses to address the challenges raised by large-scale land acquisitions.

Ongoing research by the International Land Coalition (ILC) shows that 'commercial pressures on land' are increasing in many parts of the world as a result of multiple forces beyond agriculture – including extractive industries, tourism and natural parks.[1] Such a holistic approach is crucial to understanding the land pressures faced by the rural poor worldwide. However, this essay focuses on agriculture, broadly defined to include agrifood, bioenergy, agro-industrial crops (e.g. rubber) and tree plantations. In agriculture, land is a crucial means of production. On the other hand, extractive industry projects often involve the taking of land, but the primary interest of the company is in subsoil resources. And while extractive industry developments can exacerbate pressures on land, they raise different issues to those at stake in the global rush to farmland. Water is a major driver in recent trends in agricultural FDI – acquiring land in arid and semi-arid areas would be of no use without corresponding water rights. Some of the discussion of the drivers developed in this paper with regard to land would also be relevant to water. However, water is not explicitly covered as it is discussed in another paper, by Philip Woodhouse (Woodhouse 2012).

Finally, this contribution does not discuss the social, environmental and economic implications of land deals for recipient countries and communities. There is a growing body of case studies on these aspects, carried out by, among others, the World Bank (Deininger *et al.* 2011), the International Institute for Environment and Development (IIED) (Sulle and Nelson 2009, Nhantumbo and Salomão 2010), the Deutsche Gesellschaft für Internationale Zusammenarbeit (GIZ) (Görgen *et al.* 2009), the International Land Coalition (ILC) (Anseeuw *et al.* forthcoming) the Center for International Forestry Research (CIFOR) (e.g. German *et al.* 2011), the Centre de Coopération Internationale en Recherche Agronomique pour le Développement (CIRAD) (e.g. Burnod *et al.* 2011), civil society groups (e.g. FIAN 2010), the Oakland Institute (Oakland Institute 2011), and in the run up to an international conference hosted by the Institute for Development Studies and this journal.[2] A legal analysis of a number of contracts for land acquisitions is also available (Cotula 2011a).

The next section discusses scale and geography; Section 3 analyses the main drivers underpinning the global land rush, and Section 4 discusses the implications of this paper's analysis.

2. Scale and geography

2.1. How much land has been acquired, and where?

Three years after the media spotlight turned on large-scale land acquisitions, a growing body of evidence is emerging on the scale, geography, players and key characteristics of the phenomenon. Sustained media reporting has played a key role in raising public awareness about large-scale land acquisitions, and has generated an impressive amount of data. Online databases of media reports run by GRAIN (farmlandgrab.org 2011)

[1]Commercial Pressures on Land. International Land Coalition. Available from: http://www.landcoalition.org/cpl [accessed on 25 August 2011].

[2]Scores of papers available from the International Conference on Land Grabbing: http://www.future-agricultures.org/index.php?option=com_docman&task=cat_view&gid=1552&Itemid=971 [accessed on 25 August 2011].

and the ILC (2011) contain extensive information on publicly known deals.[3] This information provides useful insights on trends in the global land rush.

Quantitative analyses drawing on media-based datasets reach somewhat different findings, but present a broadly consistent picture in terms of the significant scale of the phenomenon. Friis and Reenberg (2010) reviewed the media reports featured on the blog of the ILC for the period 2008–2010. They found that land deals in Africa alone affected between some 51 and 63 million hectares (ha). An inventory of media reports on the GRAIN blog, carried out by the World Bank (Deininger *et al.* 2011), documented land acquisitions for 56.6 million ha worldwide – roughly the size of a country like Ukraine – over a period of one year between 2008 and 2009. According to this inventory, two-thirds of the land area transacted globally was in Africa, with Southeast Asia also being an important recipient area (about 40 million ha in Africa, and more than eight million in Southeast Asia). Finally, a 'Land Matrix' featuring land deals reported in the media or discussed in published research is being established by an international consortium of organisations led by the International Land Coalition and Oxfam. Differently from earlier media-based inventories, the Matrix also involves triangulation of reported deals. Drawing on preliminary findings from the Matrix, Oxfam (2011) refers to reported deals for 227 million ha worldwide over the period 2001–2010. Of these, deals for about 67 million ha have been cross-checked through triangulation (Oxfam 2011). Similar figures appear in Anseeuw *et al.* (forthcoming), drawing on the same Matrix dataset. While at first sight the Matrix figure for reported deals may appear higher than earlier estimates, this is not so once two circumstances are considered. First, the period covered is significantly longer than that of the other two inventories, though deals farther back in time may be expected to be significantly under-reported (not least because media attention really took off from 2007 onwards). Second, in addition to agricultural investments, Matrix figures include mining and timber concessions as well as land acquisitions for tourism; these sectors are excluded from the other two inventories. It is also likely, however, that the Matrix dataset greatly underestimates mining, forestry and tourism concessions due to the lower priority that seems to have been attached to these sectors and perhaps the lower levels of media interest. Table 1 presents key features and findings of these three inventories.

Media reports suggest that Sudan, Ethiopia, Madagascar and Mozambique are among the key recipients of land-based investments in Africa. Outside Africa, Southeast Asia (Cambodia, Laos, Philippines, Indonesia) and parts of Eurasia (e.g. Ukraine and Russia) appear to be significant recipient countries. For example, based on media reports (or research drawing on media reports), Visser and Spoor (2011) documented the significant level of FDI interest in farmland in Ukraine, Russia and – to a lesser extent – Kazakhstan. Argentina and Brazil are relevant countries in Latin America, though acquisitions here may more commonly involve buying shares in companies that hold land, rather than buying land directly.

However, media reports are not always reliable, and data from these sources must be treated with caution. Also, media reports tend to emphasise global over national processes. For instance, attention usually focuses on international players, to the detriment of reporting acquisitions by national elites. Similarly, the reported regional distribution of deals may reflect the strong media interest in African deals as

[3]See http://www.commercialpressuresonland.org/ and http://farmlandgrab.org/ [accessed on 25 August 2011].

Table 1. Aggregate land areas acquired, based on media reports.

Land area (ha)	Coverage	Time period	Source	Method
51–63 million	27 countries in Africa	2008–2010	Friis & Reenberg (2010)	Systematic inventory of media reports included in the ILC blog
56.6 million	Global (81 countries covered)	1 October 2008–31 August 2009	Deininger *et al.* (2011)	Systematic inventory of media reports included in the GRAIN blog
About 67 million	Global	2001–2010	Oxfam (2011); Anseeuw *et al.* (forthcoming)	Systematic inventory of media and research reports, triangulated through a cross-checking process; includes mining, forestry and tourism

Source: Toulmin *et al.* (2011), with changes and integrations.

much as genuine, real-world differences in volumes of transactions. For example, some African countries that are or were recipients of food aid have attracted extensive media reporting (e.g. Ethiopia and Sudan). Anecdotal evidence from private sector operators suggests that there has been strong investor interest in Eastern Europe, Australia or North America, and that investor circles view Africa as a 'frontier' rather than mainstream market. In other words, it is possible that media reports have overemphasised the role of Africa as a recipient region.

A number of systematic national inventories of approved land deals based on in-country research have also been carried out by IIED together with FAO and IFAD (Cotula *et al.* 2009), by GIZ (Görgen *et al.* 2009) and by the World Bank (Deininger *et al.* 2011). Typically, these inventories draw on data from government agencies responsible for investment, land or agriculture, and on interviews with third-party sources to cross-check data. Figures from these inventories tend to be more conservative than media-based figures. National inventories must be treated with caution, however, as they may underestimate scale due to constrained data access and exclusion of deals still under negotiation. Defining what constitutes an 'approved' deal is also not straightforward – for example, where a Convention of Establishment has been signed that commits the host government to make land available, but no land lease has actually been granted, or the lease only concerns a smaller land area. Conversely, some contracts allow the investor to acquire additional land in future if certain conditions are in place. These difficulties may partly explain discrepancies among datasets – together with conflicting data sources. For example, while land ministries tend to have figures for lands actually transferred, investment promotion agencies under pressure to show success in attracting investment may refer to the usually larger land areas featured in the MoU or business plan. More generally, most existing systematic inventories cover the period ending in 2009, and are therefore rapidly becoming outdated.

Aggregate figures for approved land deals in selected countries based on systematic national inventories through in-country research are presented in Table 2. Methods used in the inventory exercises differed somewhat. For example, Cotula

Table 2. Aggregate land areas acquired in selected countries (ha).

Source	Deininger *et al.* (2011), 2004–2009	Görgen *et al.* (2009)	Cotula *et al.* (2009), 2004–2009	Media sources as compiled by Friis and Reenberg (2010), 2008–2010
Country				
Cambodia	958,000	943,000		
Ethiopia	1,190,000		602,760	2,892,000–3,524,000
Ghana			452,000	89,000
Lao PDR (two provinces)		417,075		
Liberia	1,602,000			421,000
Madagascar		1,720,300	803,414	2,745,000
Mali		159,505	162,850	2,417,000
Mozambique	2,670,000			10,305,000
Nigeria	793,000			
Sudan	3,965,000			3,171,000–4,899,000

Source: compiled by the author based on datasets cited in the table.
Notes: World Bank data for Liberia includes renegotiations of pre-existing concessions. In Ghana, Cotula *et al.* (2009) data refers to land-based investments registered with the Free Zones Board only; leases may be concluded directly with customary chiefs and are therefore difficult to track in a systematic way.

et al. (2009) only include data for projects above 1,000 ha, and Görgen *et al.* (2009) figures refer to land 'demanded' – not necessarily approved deals. Deininger *et al.* (2011) data for Ethiopia includes land allocations by regional government agencies, while Cotula *et al.* (2009) only include allocations by federal government agencies and by one regional government (Oromia). Where available and for comparative purposes, Table 2 also includes country-specific figures based on media reports, as compiled by Friis and Reenberg (2010).

Additional quantitative evidence not featured in Table 2 includes an official inventory of land acquisitions in Mali's Office du Niger area (Office du Niger 2009), which is where land acquisitions in Mali are concentrated according to country-wide inventories (Cotula *et al.* 2009, Görgen *et al.* 2009). This official compilation provides higher figures than those emerging from research-based inventories, though it does not indicate timeframes. Its total figure of 242,577 ha includes 49,304 ha leased by the Office du Niger authority and 193,273 ha promised by the central government through Conventions of Establishment (for which the Office du Niger will then have to allocate land leases). According to Office du Niger (2009), letters of intent were also issued for 402,682 ha, but these expire if the investor does not undertake feasibility studies within a year, and unfruitful expiry is reportedly common (a circumstance that seems to be confirmed by the roughly ten-to-one ratio between areas under letters of intent and actual leases). In Senegal, Faye *et al.* (2011) documented land acquisitions for a total of 409,363 ha, mainly drawing on a literature review. Fieldwork in Ghana by Schoneveld *et al.* (2010) documented land acquisitions for biofuels only totalling 1,075,000 ha. However, none of the nine projects covered by Schoneveld *et al.* (2010), out of a total of 17 identified by that research, had actually obtained leasehold titles and would therefore be considered as 'approved' under the Cotula *et al.* (2009) inventory.

With some important exceptions (Ghana, Liberia, Sudan), research-based figures for 2004–2009 are much lower than those suggested by media reports published in the period 2008–early 2010, even using top-end inventory figures. For example, about 250,000 ha have been acquired in Mali (about 650,000 ha if one-year letters of intent are included), compared to media reports of 2,417,000 ha; 1,190,000 ha in Ethiopia, compared to a media-based figure ranging between 2,892,000 and 3,524,000; and 2,670,000 ha in Mozambique, compared to media reports of 10,305,000 ha. The figures of Deininger *et al.* (2011) for Liberia include renegotiations of pre-existing concessions, which may explain the discrepancy with media reports.

Also, although the size of single land acquisitions can be very large (e.g. a 220,000-ha project in Liberia and two 100,000-ha projects in Ethiopia and Mali[4]), the average sizes of projects above 1,000 ha are much smaller than what is suggested by media reports: in Ethiopia a mean of 7,500 ha and in Mali a mean of 22,000 ha (figures based on data collected for Cotula *et al.* 2009). These figures can be contrasted with the average sizes suggested by media reports: a mean ranging between 111,000 and 135,000 ha for Ethiopia and of about 186,000 ha for Mali.[5] Deininger *et al.* (2011) also found that large areas for individual deals are significantly below the scale suggested by media reports, though that report also found that land area size distribution was heavily skewed. In Ethiopia, for instance, five large projects accounted for half the land area leased out by the government (Deininger *et al.* 2011).

Even in the larger deals, implementation usually begins on a much smaller scale, and is phased up to full capacity over relatively long periods of time. For example, a 100,000-ha investment contract in Mali has resulted in an actual lease for 25,000 ha (Office du Niger 2009), with upscaling being subject to successful implementation. In addition, many approved deals have had only limited implementation, often due to greater-than-expected difficulties on the ground or to difficulties in financing (Anseeuw *et al.* forthcoming; and, for examples of biofuels projects in Mozambique and Tanzania, see Nhantumbo and Salomão 2010, and Sulle and Nelson 2009, respectively). Deininger *et al.* (2011) found that, in Mozambique, more than 50% of approved projects had not started any activity or were significantly behind schedule. Even so, deals approved but not implemented may have significant opportunity costs and may exacerbate local land pressures, as alternative land uses are prevented or delayed.

The figures presented in Table 2 are not of great help in analysing the international distribution of land acquisitions. Even in empirically grounded systematic inventories, country selection was much informed by media-based perceptions of geographic distribution – key recipient countries were usually deliberately targeted. It is possible that other countries are experiencing substantial land acquisitions but are falling below the media radar. The above-mentioned finding by Faye *et al.* (2011) that 409,363 ha have been acquired in Senegal is a case in point.

Media- and research-based datasets do not fully capture how much land is being acquired in ways other than direct land acquisition for greenfield investments. Conversations with private sector operators suggest that many investors prefer to take over the management of existing farms, which were often established or run by

[4]Contracts available from LEITI (2011) and farmlandgrab.org (2011).
[5]Mean calculated by the author on the basis of aggregate land areas and number of reported deals compiled by Friis and Reenberg (2010).

parastatals, and to rehabilitate existing irrigation and other infrastructure. For example, a South African sugar company has been taking over government estates in several Southern African countries since the 1990s (Richardson 2010). Some deals reported in the media may refer to takeovers of existing plantations, rather than new land acquisitions, while some such takeovers may not be reflected in media or country-based inventories. A related corporate strategy to access land involves acquiring equity participations in companies that already have plantations. Conversations with private sector operators indicate that this strategy has been used by some agribusiness companies that are expanding from a particular segment of the value chain (e.g. traders, processors) to agricultural production. In principle, the takeover of existing farms or of companies controlling existing farms does not involve new land acquisitions, though some takeovers of existing plantations are for larger areas than the original farm: for example, the above-mentioned deal in Liberia concerns a pre-existing 120,000-ha farm and a new land allocation of 100,000 ha.

While media reports appear to overestimate scale compared to figures based on in-country research, national inventories confirm that the phenomenon is massive and growing. According to World Bank data (Deininger *et al.* 2011), about ten million ha were acquired in five African countries alone (Ethiopia, Liberia, Mozambique, Nigeria and Sudan) in a five-year period ending in 2009. Although quantitative inventories suggest that acquired land areas usually account for relatively small proportions of land suitable for rain-fed agriculture in any given country (Cotula *et al.* 2009), the effects of land acquisition processes on competition for land are increased by two factors.

First, despite much rhetoric on targeting 'marginal' lands, investor interest often focuses on the best land in terms of water availability and irrigation potential, soil fertility, proximity to markets or availability of infrastructure. For example, land acquisitions in Mali are heavily concentrated in the irrigable areas of the Ségou Region (Cotula *et al.* 2009, Görgen *et al.* 2009). Similarly, investor interest in Senegal has focused on the high-irrigation potential areas of the Senegal River valley (Faye *et al.* 2011). Initiatives to support agricultural development corridors in Mozambique (Beira Agricultural Growth Corridor initiative) and Tanzania (Southern Agricultural Development Corridor of Tanzania) also suggest that investor interest may concentrate along strategic transport and communication axes (Kaarhus 2011). In Mozambique's Manica Province, which is located on the Beira Corridor and has long experienced interest from outside commercial agriculture players due to its proximity to Zimbabwe, data provided to the author by the provincial cadastral office in early 2010 suggests that approved land leases increased from 562 ha in 2007 to 21,334 ha in 2008 and 58,880 ha in 2009, while applications for 367,165 ha were pending as of January 2010. Some of the reports published by the ILC also suggest that much investor interest targets higher-value lands in peri-urban areas, for example with regard to more fertile land in river-basin marshlands near the Rwandan capital Kigali (Veldman and Lankhorst 2011, Anseeuw *et al.* forthcoming).

Second, while media attention has focused on agriculture for food or fuel, pressures on land are growing as a result of a wider set of factors – both endogenous, such as strong demographic growth in many recipient countries, and exogenous, like investments in tree plantations or agro-industrial crops like rubber, and, outside the agricultural sector, in petroleum, mining and tourism. Globally, tree plantations are expanding fast (at a rate of 2.5 million ha per year in 1990–2005), and this expansion is likely to continue in future (Deininger *et al.* 2011). As will be discussed, this trend is partly driven by the expansion of biomass energy capacity in some major energy consuming countries, for example through the actual or planned construction of new energy plants fired with woodchips and wood pellets in some European countries (Cotula *et al.* 2011). Carbon markets are a growing driver for land acquisitions involving forest conservation and/or tree or biofuel plantations.[6] In the latter cases, carbon credits provide a complementary revenue stream to cash flow from sales of wood or feedstock.

The relevance of natural resource investments outside the agricultural sector is illustrated by both quantitative and qualitative data. In Liberia, for example, in addition to the 1,602,000 ha acquired through farmland concessions, mining exploration or development concessions have been granted for 1,195,894 ha since 2004.[7] And the case of a failed 30,000-ha biofuel project in an area of Mozambique where villagers were being resettled from a newly established natural park illustrates how, from the perspective of local people, farmland acquisitions are only one of the multiple sources of pressure (FIAN 2010, Nhantumbo and Salomão 2010).[8]

2.2. Where do investments come from?

Media attention has focused on government-backed entities from the Gulf and East Asia and on Western investment funds as the main land acquirers. However, empirical research highlights the central role of national elites in national acquisitions (Cotula *et al.* 2009, Görgen *et al.* 2009, Deininger *et al.* 2011). In Ethiopia, for example, domestic investors account for over 60% of the land area acquired in the period 2004–2009.[9] The World Bank study found that nationals accounted for 97% of the land area acquired in Nigeria, and for about half or more in Sudan (78%), Cambodia (70%), Mozambique (53%) and Ethiopia (49%) – though only 7% in Liberia (Deininger *et al.* 2011). Similarly, Faye *et al.* (2011) found that in Senegal acquisitions by nationals accounted for 61% of acquired land areas.[10]

Nationals have also been acquiring land in countries that have received less interest from international investors. A study from Benin, Burkina Faso and Niger found that over 95% of the investors involved in land deals were nationals, including professional farmers and urban groups like civil servants, traders and politicians (Hilhorst *et al.* 2011). While the average size of these plots is very small relative to some international deals (the average size was 85 ha), these farms are still considerable relative to local average plot sizes (Hilhorst *et al.* 2011), and the aggregate land area acquired by many small deals can be larger than that involved in fewer, larger deals (as suggested by the findings of the inventories).

The diaspora – local nationals living overseas – is also a player in land acquisitions (Anseeuw *et al.* forthcoming, Zoomers 2010). This is not a new

---

[6]On carbon markets as a driver for land acquisitions in Mozambique, see Nhantumbo (forthcoming). One of the contracts analysed by Cotula (2011a), from South Sudan, is for a tree plantation and carbon credit project.
[7]Author's calculation based on contracts database available at LEITI (2011). Data include renegotiation of existing concessions.
[8]For a discussion of natural parks as a new frontier of enclosure and privatisation, see Peluso and Lund (2011).
[9]Calculated by the author based on data reported in Cotula *et al.* (2009).
[10]Calculated by the author based on data reported in Faye *et al.* (2011).

phenomenon; use of international remittances to acquire land in the home countries has been documented for a while (see for example Cotula and Toulmin 2004). It is difficult to quantify the relative scale of this phenomenon – it has received very little media attention and it was not fully covered by the in-country inventories. Land contracts published by the government of Ethiopia in May 2011 (for a total of 350,099 ha, which is substantially lower than the figures from research-based inventories) include several leases involving the diaspora (6 out of 23), though the aggregate land area acquired by international migrants was small (less than 5% of the total).[11]

Where foreign investment is at stake, much of it is intra-regional – a circumstance highlighted by research carried out by the ILC (Anseeuw *et al.* forthcoming). In the Mekong region, intra-regional investment from ASEAN countries (Vietnam, Thailand) and from China dominates land acquisitions in Lao PDR and in Cambodia (Görgen *et al.* 2009, Ravanera 2011). In Africa, South Africa plays a key role in investment flows relating to land. For example, AgriSA, a body representing commercial South African farmers, is reported to have acquired 200,000 ha of land in the Republic of Congo, and to be negotiating with several other African governments (Hall 2011). South African sugar companies have acquired land in several Southern African countries, and more recently elsewhere (Richardson 2010, Hall 2011). Libya has also been involved in some large land acquisitions in sub-Saharan Africa – including Liberia and Mali.[12]

Where cross-regional foreign investment is involved, evidence points to Europe and North America as key regions of origin, in addition to the more publicised role of Gulf states (Saudi Arabia, UAE, Qatar), East Asian countries (China, South Korea) and India. Investment from different regions is distributed unevenly, with some investor countries preferring some recipient countries over others in light of considerations ranging from geographical proximity, to cultural and political links, through to perceived land availability or market potential – as illustrated by Middle Eastern investments in a band of recipient countries around the Gulf (Sudan, Pakistan, Central Asia).

Western companies have been a key player in the global land rush. They are dominant players in biofuels. For example, all of the biofuels projects reviewed by Nhantumbo and Salomão (2010) and by Sulle and Nelson (2009) in Mozambique and Tanzania, respectively, were run by European companies, though in one case capital appeared to be mainly held by South African interests. Similarly, Visser and Spoor (2011) documented the key role played by European (UK, Swedish, Danish, German) companies in farmland acquisitions in Russia, Ukraine and Kazakhstan. Western companies also seem to lead renewed interest in tree plantations for biomass energy (Cotula *et al.* 2011). And a recent survey found that most investment funds involved with farmland investments are based in Europe and North America (OECD 2010), a circumstance that is confirmed by Buxton *et al.* (forthcoming).

On the other hand, the role of some Middle Eastern or East Asian countries seems to have been overstated or misunderstood. Middle Eastern operators have certainly been active participants in the global land rush. Saudi Arabia is the largest investor country in Sudan, accounting for about half the land area acquired by

[11]The contracts were accessed on 12 May 2011 from EAPortal (2011) the link was no longer working in late August 2011.
[12]Contracts available from LEITI (2011) and farmlandgrab.org (2011).

foreign investors in Sudan (Deininger *et al.* 2011). However, Sudan is the only country in the World Bank's inventories where Middle Eastern countries accounted for a majority of foreign projects (Deininger *et al.* 2011). And while there have been reports in the media and in some studies that Kuwait has been a leading land acquirer, there is no real evidence of any actual land deal involving the Kuwait Investment Authority.

South Korea has also featured prominently in public discourse, but this seems mainly due to a reported very large deal by a South Korean company in Madagascar; this deal failed. Evidence suggests that South Korea has been active in land deals in Southeast Asia, for example in the Mekong Region (Görgen *et al.* 2009) and in Indonesia (Cotula *et al.* 2011); but media reports seem to vastly overstate South Korea's role in the global land rush.

China is an active player in the global land rush, but again it is possible that public perceptions overstate its role, especially with regard to Africa. China does have a 'Going Global' policy that promotes investment overseas, which is further discussed below. China is the main country of origin for land-based agricultural FDI in Cambodia and Lao PDR, particularly for rubber and rice, with investor countries like South Korea, Thailand, the US and Vietnam following suit (Görgen *et al.* 2009). China is reportedly active in the Asian part of Russia (while European investors dominate farmland investments in the Western Black Earth area of Russia and Ukraine; Visser and Spoor 2011). Media reports of land acquisitions in Africa by Chinese government-controlled or private operators have multiplied since a high-level China-Africa conference in 2006. The important role of China is therefore likely to be reflected in datasets that significantly rely on media reports, like the Land Matrix, discussed above.[13] However, beyond media reports, hard evidence of a key role played by China in land deals in Africa has so far been more difficult to come by. In Ethiopia, officially disclosed contracts include one deal with a Chinese operator (for 25,000 ha, i.e. about 7% of the total land acquired through the contracts disclosed).[14] A Chinese deal has also been documented in Mali, involving a 20,000-ha extension of an existing sugar cane plantation (Cotula 2011a). Conversations with informed stakeholders suggest that China is also active in Sudan, for example in connection with sugar cane. But overall, Chinese companies operating in Africa seem more interested in subsoil resources – and, with regard to agriculture, in a range of upstream and downstream business opportunities (Bräutigam 2011), including, for example, using joint training centres as an entry for the distribution of seeds and other inputs (Buckley 2011).

On the other hand, Southeast Asian companies have received far less media attention for their investments in Africa, but have been very active – as illustrated by two 220,000-ha palm oil plantations by companies from Singapore and Malaysia (Reuters 2011a).[15] Indian companies have also been active. Among the contracts disclosed by the Ethiopian government in 2011, eight (out of 23) are with Indian companies, constituting a staggering 71% of the aggregate land areas acquired through the contracts disclosed.[16] It must be borne in mind, however, that the

[13]Given China's substantial involvement in extractive industry investments in Africa, the relative importance of China's role in the Land Matrix is also likely to be affected by the inclusion of mining in the Matrix dataset.
[14]Author's calculations based on data from EAPortal (2011).
[15]One of the two deals is available from LEITI (2011).
[16]Author's calculations based on data from EAPortal (2011).

disclosed contracts only account for about 30% of the land area acquired according to the World Bank inventory, which covers a longer timeframe (Deininger *et al.* 2011). It is possible that Indian contracts featured particularly prominently in the official release. India's role in land acquisitions in Ethiopia is borne out by evidence from other countries. In Madagascar, a 230,000-ha deal involved an Indian company (Ullenberg 2009), though it is not clear whether this deal is going ahead. Rowden (2011) suggests that more than 80 Indian companies have invested about US$ 2.4 billion in buying or leasing plantations in Africa, for instance in Ethiopia, Kenya, Madagascar, Senegal and Mozambique, and specifically lists 20 examples.

Discussion of investor origin has usually focused on the land acquirer. However, the implementation of large land deals typically involves a range of players – possibly including lenders, insurers, contractors and suppliers. Therefore, the nationality of the land acquirer does not fully represent the geography of the interests at stake. A large Libyan deal in Mali reportedly involved contracting out construction work to a Chinese company, for example. Similarly, South African consulting engineers have been involved with contracts to build sugar mills and ethanol plants in different parts of Africa (Hall 2011). And some European or North American farmland investments in Africa involve leveraging agricultural know-how from Brazilian expertise (OECD 2010). Lenders and insurers may be located elsewhere. In addition, acquiring companies may be headquartered in a country, but capital mainly sourced from other countries. For example, several biofuels companies active in Africa are listed on London's AIM – which is is 'the London Stock Exchange's international market for smaller growing companies' (AIM 2011); but capital invested in these companies may originate from all over the world. So different geographies of interests may be involved in a single investment project.

In addition, borderlines between international, regional and national investments are fluid. Some African countries act as strategic transit countries through which investments from outside the region are channelled into third African countries. In particular, some investments are channelled through South Africa, due to geographical proximity and established expertise in African agriculture. For example, a UK-based asset management firm acquires and runs its land-based investments in Africa through a joint-venture company incorporated in South Africa. This company has operations in Mozambique, Swaziland, South Africa, Zimbabwe and Zambia (EmVest 2011). In addition, in 2006 a UK food and retail conglomerate acting through a subsidiary completed the purchase of a 51% stake in a South African sugar company that has ongoing operations in South Africa, Malawi, Zambia, Swaziland, Tanzania and Mozambique (Associated British Foods plc 2010, AB Sugar 2012a, 2012b, Illovo Sugar 2011). There have also been reports of an emerging partnership between China and AgriSA for collaboration in land-based investment in Africa (Hall 2011).

Mauritius also seems to be a strategic transit country, most likely due to its tax regime[17] and its sizable number of bilateral investment treaties concluded with other African countries (so that investments in these third countries channelled via Mauritius would be protected by applicable investment treaties).[18] The contract for a land deal in Mozambique, which is publicly available on the Oakland Institute

[17]CAS (2011) lists Mauritius among offshore financial centres with regard to taxation.
[18]Eleven of the 23 publicly available BITs signed by Mauritius are with African countries; see UNCTAD (2011).

website (Oakland Institute 2011), involves companies registered in Mauritius. The role of Mauritius is also linked to that government's efforts to attract foreign capital to finance the acquisition of land in Africa (mainly Mozambique) so as to ensure the country's long-term food security. These efforts have reportedly attracted interest from Australian investors acting through a Singapore-based vehicle (GRAIN 2009, Reuters 2010).[19] Another recent deal in Mozambique by a Mauritius-based firm in partnership with a UK agribusiness was recently reported in the media (Reuters 2011b).

Similar to the relationship between international and intra-regional investment, the borderline between national and foreign investment may also be blurred (Anseeuw *et al.* forthcoming). As many foreign investors operate through a locally incorporated subsidiary, it is difficult to assess corporate control of national companies. Writing about Madagascar, Burnod *et al.* (2011) show how national operators may provide intermediation services to help foreign investors acquire land, for example by assisting with navigating formal procedures or negotiating deals with local communities. It is also possible that national elites may be acquiring land with a view to then entering into an agreement with a foreign investor.

2.3. Governments or companies?

While data on the nature of the investors remains limited, available evidence suggests that private companies, rather than government entities, account for much land acquisition, though government policy plays a crucial role in supporting agribusiness-led acquisitions. A systematic inventory of media reports carried out by the World Bank (Deininger *et al.* 2011) found that agribusiness accounted for the largest share of investors, with investment funds also being key players. This picture is borne out by national inventories based on in-country research. For example, private agribusiness deals account for about 90% of the aggregate land area acquired in Ethiopia, Ghana, Madagascar and Mali, with government-owned investments making up the remainder (calculated on the basis of data generated for Cotula *et al.* 2009).

However, the divide between private and government-backed land deals should not be overestimated. The home country governments of investors can play a major supportive role for private sector-led initiatives, providing diplomatic, financial and other support to private deals; the role of public policy in supporting land deals is further discussed below. Also, the very borderline between public and private investors may be fluid, as the implementation of deals signed between governments may be driven by private operators. For example, a deal between the governments of Syria and Sudan concerning an agricultural investment in Sudan enabled the government of Syria to delegate implementation to the private sector, subject to this being cleared with the government of Sudan (this contract is discussed in Cotula 2011a).

The Chinese case illustrates how the boundaries between 'state' and 'non-state' enterprises may be fuzzy. There are two aspects to this discussion: state ownership and state influence (Cotula *et al.* 2009). In China, corporations emerging from the centrally planned economy such as COFCO (China National Cereals, Oils and Foodstuffs Import and Export Company) are clear state-owned enterprises: senior

[19]Mauritius has a bilateral investment treaty with Singapore.

staff are appointed by the state, and chief executive officers have ministerial level rank. In other cases, however, it is less easy to distinguish whether a Chinese firm is 'public' or 'private'. Many companies do not disclose clear information on equity structure, which makes it difficult for outsiders to be precise about ownership. An apparently private company may be controlled by a state-owned, unlisted parent company. In addition, there is likely to be significant state influence over strategic private firms, as private companies may flourish because of their formal and informal links to key state agencies. Such companies benefit from access to special credit lines, tax breaks, and possibly favourable interpretation of regulations and priority in allocation of key contracts (Cotula *et al.* 2009).

Where home governments are directly involved in land acquisitions, they usually rely on investment vehicles that are not under direct civil service control, such as state-owned enterprises and sovereign wealth funds (SWFs). More rarely, governments have acquired land abroad directly. For example, two large land deals in Mali and Sudan were signed by the ministries for agriculture of the host and home country governments (both contracts are discussed in Cotula 2011a).

Contrary to some public perceptions, borderlines between the motivations driving private companies and state-owned enterprises or SWFs are not clear-cut. Some commentators see SWFs as mainly motivated by public policy concerns, particularly with regard to national food security. But SWF managers are also likely to be interested in returns through capital appreciation and agricultural production. Also, in recent debates about concerns that the growing role of SWFs in strategic sectors of some Western economies might be manipulated for political ends, SWFs have pointed to their track record of market-based operations, and two Middle Eastern SWFs issued a statement that formally disavowed 'geopolitical goals' (Cohen 2009).

3. Drivers

3.1. A brief historical contextualisation

Large-scale acquisitions of land are not new. In the nineteenth and twentieth centuries, many agricultural investments in developing countries, led by companies based in Europe, the United States and Japan, involved the establishment of large-scale plantations. Yet, from the 1960s onwards, with decolonisation and the ensuing nationalisations in Africa, and with land redistribution programmes in some Latin American countries, agribusiness companies shifted away from the plantation model, and moved towards developing long-term contractual relationships with local suppliers (UNCTAD 2009). Increasing unionisation of estate labour forces and stricter labour legislation also encouraged a move away from plantations (Tiffen and Mortimore 1990).

In addition to political factors, economic forces prompted the shift away from direct involvement by agribusiness in production. The distribution of risks and returns plays a crucial role in business decisions about the degree of vertical integration. For much of the past few decades, agricultural value chains have tended to concentrate returns in processing and distribution, while the risks fell mainly on primary production (Selby 2009). This situation created incentives for agribusiness companies to concentrate on activities upstream (provision of inputs, seeds and machinery) and downstream (processing and distribution), and to source agricultural production from local suppliers. Sourcing produce through long-term contracts rather than plantations also offered greater flexibility in responding to fluctuating commodity prices – as renegotiating or even terminating contractual relations is easier than divesting land ownership (Tiffen and Mortimore 1990). These factors led to a shift away from plantations in contexts as diverse as banana farming in Central America or tea in East Africa (UNCTAD 2009).

The evidence discussed in the previous sections suggests that a new shift is taking place – toward greater agribusiness involvement in agricultural production, through direct land acquisitions. The drivers underpinning this shift involve a combination of both policy and market forces, and are discussed in this section.

3.2. Market forces – Changing global supply and demand for agricultural commodities

There are widespread perceptions that population growth, changing diets, growing energy demand and increasing rates of urbanisation (which expand the share of the world's population that depends on food purchases) are pushing up global demand for food and fuel from agriculture. Given supply constraints in parts of the world, including declining production and productivity (in the Gulf, for example), this is likely to put upward pressure on agricultural commodity prices in the longer term, and create new incentives to acquire land as an investment option.

From the early 1980s to the early 2000s, food prices were on a long-term trajectory of decline, reflecting the expansion of agricultural frontiers and agricultural trade, increasing concentration in the retail sector that generated economies of scale and drove down farm gate prices, as well as innovations in production. However, the food price hike of 2008 shook the assumption that the world will continue to experience low food prices. Maize and wheat prices doubled between 2003 and 2008 (von Braun 2008). Grain and other food prices started dropping after the summer of 2008, in conjunction with the onset of the global economic downturn. But in the spring of 2011, global prices had again reached the levels of 2008 (FAO *et al.* 2011), though short-term price volatility has also increased (OECD/FAO 2010) and the distribution of price increases among commodities and products differed from that of 2008 (FAO *et al.* 2011).

There has been much debate about the nature and causes of the price hike of 2008. More informed analyses indicate that peak price levels in 2008 were, in real terms, at similar levels as prices in the early 1980s, and that sharp increases in food prices intervened after a period when prices were at an all-time low (Headey and Fan 2008). Informed analyses also question some of the widespread assumptions about the underlying causes of the 2008 price hike. While many commentators have suggested an important role of rising meat consumption in China and India, there is no evidence to back up this perception – and both countries have long been largely self-sufficient in food (Headey and Fan 2008). Contingent explanations of the crisis, like reported bad harvests due to climatic conditions in Australia, have also been questioned, as global wheat production declined more substantially in 2000/2001 (11%) than in the crisis period (5%, 2006/2007) (Headey and Fan 2008). On the other hand, there is strong evidence to suggest that the biofuel boom played an important role in food price hikes, mainly via the diversion of the US corn crop, and that export restrictions on crops like rice once prices started to rise exacerbated price increases by affecting global supplies (evidence reviewed by Headey and Fan 2008).

Overall, the evidence points to a prospect of higher food prices in the medium to longer term. By 2050, the world is expected to host about nine billion people.

Demand for food is expected to increase more than proportionally, due to the additional effect caused by growing incomes on changing diets. Expert estimates suggest that, for a 40% increase in world population, food production would need to increase by 70% (Deininger *et al.* 2011). Biofuel production will also put upward pressure on food prices. Oil prices will affect food prices both directly (through the cost of transport and fertilisers, for example) and indirectly (through affecting incentives for biofuel production) (FAO *et al.* 2011). The OECD-FAO Agricultural Outlook predicts prices for the decade 2010–2019 to be higher in both nominal and real terms than the decade prior to the 2008 peak (OECD/FAO 2010).

These forces make agriculture an increasingly attractive investment option. This includes the acquisition of shares in companies holding land, producing fertilisers, providing management services or otherwise involved in upstream or downstream agricultural activities (The Economist 2009). It also affects the attractiveness of land as an investment option. Changing agricultural commodity prices are shifting the distribution of risks and returns along the agricultural value chain, by increasing the downstream risks to processors and distributors, concerned about the security of their supplies, and boosting returns from production (Selby 2009). As resource constraints are increasing and advances in technology slowing down, it is likely that incentives to expand cultivated land areas, including through land-based investments, will continue to grow (Deininger *et al.* 2011).

Biofuels have also been a main driver in the global land rush – mainly linked to public policies, discussed below, but also in connection with price effects, particularly with regard to changing oil prices (Comité Technique Foncier et Développement 2010). As a result of profitability prospects underpinned by public policies and market forces, agribusiness, energy and biotech companies have developed strategic partnerships to invest in biofuels projects, including by acquiring land in the global South (Borras *et al.* 2010). Drawing on Land Matrix data, Anseeuw *et al.* (forthcoming) found that biofuels accounted for 37.2% of cross-referenced land areas acquired worldwide in 2001–2010, food accounted for 11.3%, and the rest was split between agro-industrial crops, forestry, mining and other land uses. The importance of biofuels in the global land rush is confirmed by country, which documented high levels of interest in biofuel projects: for instance, 798,578 ha of land were found to have been requested for biofuel projects in Tanzania (author's calculations based on data presented in Sulle and Nelson 2009), and 2,327,296 ha were found to have been requested in Mozambique (Nhantumbo and Salomão 2010); in both cases, land areas actually allocated to biofuels projects were much smaller.

The borderline between food and fuel is blurred, as the same crop may be used for both or the same plantation may involve multiple crops, and as investment plans may evolve over project duration to respond to changing international prices and other incentives. Also, it is possible that the relative importance of food and fuel in the rush to land has changed over time. International interest in biofuel production started before the food price hikes of 2008, which sparked the renewed momentum for agrifood projects. For example, several of the biofuel projects documented by Nhantumbo and Salomão (2010) and Sulle and Nelson (2009) were initiated in 2006–2007. Yet, more recently, some biofuels projects have run into financial difficulties as a result of the ongoing economic crisis (which has made project financing more difficult) and of changing world oil prices. Two of the four biofuels projects studied by Sulle and Nelson (2009) in Tanzania have run into such difficulties; one of the three projects examined by Nhantumbo and Salomão (2010) in Mozambique has since had its land allocation withdrawn due to failure to invest (the government is reportedly planning to tender out the project to a new investor), while another one has shifted from a biofuel to a forestry project.

While much discussion has focused on global processes and while media reports have emphasised export markets as the main outlet for agricultural investments in poorer countries, informal exchanges with private sector operators point to expectations linked to the growing potential of domestic food and energy markets as a key consideration in land-based investments. These expectations are exemplified by an influential report published by the McKinsey Global Institute in 2010, Lions on the Move (Roxburgh *et al.* 2010). The report identified the African agrifood market as a high-growth market due to rapid economic, demographic and household income growth and to rapid urbanisation. Although reliable statistics are hard to come by, anecdotal evidence points to a large number of land-based investments that mainly target the domestic market for food and fuel. This is particularly the case for some contexts and crops. For example, it is estimated that 90% of global rice production is concentrated in Asia; while global trade is estimated to account for only 8% of global rice production, over 85% of that trade involves imports by Africa and Asia (Olam 2011). West Africa is a significant rice importer due to the role that this staple plays in diets. Informal conversations with private sector operators suggest that some see land-based investments for rice production aimed at domestic consumption in West African markets as an attractive business opportunity. Other crops for which local markets may be a central target include sugar and, in oil-importing countries, feedstock for biofuels.

Beyond food and fuel, agricultural commodities for industrial use, such as rubber, are also key drivers. Rubber accounts for a significant share of the land areas acquired by Chinese investments in Lao PDR, for example (Görgen *et al.* 2009). Cotton, alone or in combination with other crops, accounts for 9 of the 23 land deals released by the Ethiopian government in 2011, and for about 17% of the land area acquired through those deals.[20] Tree plantations are also thought to account for a large share of land acquisitions, though reliable statistics are scarce. An ongoing expansion of biomass energy capacity (especially plants to produce electricity from wood chips and pellets) in some OECD countries (e.g. in the UK), driven by renewable energy policies, is likely to substantially expand global demand for wood chips and pellets and may increase pressures on land in the global South (Cotula *et al.* 2011). Finally, carbon markets are also fostering the global land rush: cheap land values in Africa allow the acquisition of large areas of land; with minimal investment commitments, revenues can be generated through carbon credits from reduced land use change (on Mozambique, see Nhantumbo, forthcoming). Carbon credits may also increase the profitability of investments like tree and biofuels plantations, thereby compounding incentives for land acquisitions in these sectors. A land deal involving a tree plantation and carbon credit scheme is discussed in Cotula (2011a).

Within the context of these global processes, from the perspective of individual agribusiness companies the decision to acquire land for direct production activities (as opposed to increasing coordination with suppliers in the value chain, for example) is a function of a number of factors. These include the evolving

[20]Author's calculations based on data from EAPortal (2011).

comparative advantages of ownership of productive assets versus coordination along the value chain. While some companies (supermarkets, for instance) may be particularly efficient at coordinating the value chain, others may acquire a distinct comparative advantage in forms of more direct involvement in agricultural production. Volume and security of supplies are also key concerns (De Schutter 2011). An international agricultural commodity trader that has long worked with smallholder farmers in Africa is now revising its business model by acquiring land and getting into direct agricultural production; the rationale is linked to the sheer volume of supplies needed and to side-selling problems in contract farming relations, whereby reliability of supplies is jeopardised where farmers sell to competing traders/ processors. The need to guarantee a minimum level of throughput for processing plants may push the company to establish more direct control over the production stage. These factors are weighed against the greater political risks involved in land acquisitions, and the longer-term investment required by it. In addition, the need to comply with traceability requirements and with food quality and safety standards increases incentives for greater vertical coordination along the value chain – and even for outright vertical integration. Context-specific factors may also play a role in the economic considerations, for instance where new agricultural investments bring into production land areas with low population densities and weak local capacity to undertake agricultural production activities. In these cases, long-term contracts with local suppliers may be perceived as not economically viable.[21]

Strategic considerations may influence company choices, with regard to both vertical relations along the value chain and horizontal relations among competing firms. For companies seeking to increase control over the value chain, land is a strategic asset – and control over land can significantly influence negotiating power along the value chain. In outgrower schemes, for instance, the negotiating power of outgrowers would be expected to vary depending on whether they cultivate their own land or blocks subleased by the company. Also, evidence of investors acquiring larger land areas than they can cultivate has been explained as an attempt to 'lock in very favourable terms of land access and eliminate future competition' (Deininger *et al.* 2011, 63). Finally, land is cheap in many lower-income countries, and many investors see capital appreciation (increasing land values) as an important component of their business model. This issue is further discussed in the next section.

3.3. Market forces – The 'financialisation' of agriculture

The changes in global supply and demand for agricultural commodities, discussed in the previous section, point to increasing land values in the longer term, particularly in places where land is now cheap. Combined with other factors, this circumstance has increased the attractiveness of land as an investment option not only for agribusiness and energy companies interested in direct production, but also for financial operators interested in increasing returns and lowering risks for their portfolios. This process is part of a wider trend involving greater interest by financial investors in the agriculture sector as a whole (FAO 2010), linked to the factors discussed in the previous section. Some have referred to this trend as the 'financialisation' of agriculture (Anseeuw *et al.* 2011).

[21]This paragraph draws heavily on conversations with industry operators.

A review of the role of financial investors in agriculture (undertaken for Buxton *et al.* forthcoming), documented 66 investments funds specialised in agriculture and investing in farmland, including both private equity and listed funds, with the sector being projected to grow fast. While some funds specialise in farmland, others pursue a 'farm-to-fork' strategy involving investment along the agricultural value chain – from input supply to production and processing, through to storage and distribution (Hawkins 2010, FAO 2010). Money can also be allocated to land and agriculture in other, less visible ways – for example, under a property portfolio, or through a fund of funds that invests in other funds involved with land-based investments.

Interest in land from financial investors is linked to a range of different factors. First, the economic fundamentals discussed in the previous section create expectations of growing returns from agriculture. Some return forecasts are as high as 25% (Hawkins 2010). These returns are expected to be generated by a combination of capital appreciation (increased land values) and income flowing from increased productivity in agricultural production on acquired land. Historical trends indicate that there is potential for significant returns from capital appreciation. Historically, farmland values have tended to increase, and farmland prices in the US and the UK have considerably outperformed stock markets on a 10-year timescale (Hawkins 2010). Wilkes and Bailey (2011) present similar findings for the past 15-year period in both the US and the UK, though residential property was found to outperform farmland. But unlike residential property, farmland in certain regions is seen as being 'undervalued' (Hawkins 2010). Indeed, farmland prices are very low in many parts of Africa (some of the contracts reviewed by Cotula 2011a involved no land fees at all), and potential for capital appreciation is significant. And while returns from agricultural production tend to be highly volatile due to production risks (e.g. weather, pests), land is seen as providing stable returns (Wilkes and Bailey 2011).

A second set of factors is linked to risk management, and has acquired greater relevance following the global financial crisis that started in 2008. In this context, land is seen a hedge against inflation. Also, returns on land have low correlation with (volatile) equity markets. Therefore, land is seen as an increasingly important 'asset class' for portfolio diversification purposes. In this respect, considerations about land as an asset class are not too dissimilar to those underpinning growing interest in commodities (gold, agricultural commodities). Also, the collapse in equity and bond markets in 2008 reduced the appeal of these asset classes and precipitated a resurgence of interest in land and commodities (UNCTAD 2009).

But while the crisis of 2008 created new momentum for financial investor interest in land, it also constrained the expansion of the sector, at least in the short term. Asset management firms would launch 'farmland funds' with ambitious financing targets and much media publicity, but institutional investors (e.g. pension funds, private equity and endowments) would have limited liquidity to inject (FAO 2010). In addition, potential investors may need to divest elsewhere, so the downturn in stock values esperienced in 2008 increased challenges in capital mobilisation. Farmland funds also typically involve locking in capital for a number of years: while legal structures can be designed in ways that create an exit strategy, these funds tend to be relatively illiquid. In conditions of prevailing market uncertainty, some investors have preferred to stick to the more familiar asset classes, including equities in agriculture companies. As a result, some farmland funds have struggled to attract the levels of investment sought.[22] In addition, the deployment of capital raised takes time, partly because of the complex due diligence involved with farmland investments. A 'sizable amount' of capital managed by the investment funds interviewed by OECD (2010) was yet to be deployed.

These circumstances mean that ambitious and publicised announcements concerning the establishment of farmland funds do not necessarily translate into proportional levels of land investment on the ground. For example, an investment management firm dumped plans for a $387 million USD farmland fund it first announced in 2008 (Pensions & Investments 2010). OECD (2010) found that 20 out of 25 investment funds interviewed were in the process of raising capital. And to put things in perspective, industry experts estimate that farmland investments account for less than 0.1% of the portfolio of European pension funds involved in the business (Pensions & Investments 2010).

While much public attention has focused on the involvement of investment funds in Africa, evidence suggests that location choices are a function of investor goals concerning the balance between risk and reward. Investors mainly concerned about wealth preservation (hedge against inflation, portfolio diversification) tend to invest in low-risk, low-return locations in Europe, North America and Australia, where existing land prices are already high; while higher returns (but also higher risks) are seen as possible in the 'emerging' and 'frontier' markets of South America and Africa (Hawkins 2010, Wilkes and Bailey 2011). A recent survey of investment funds suggests that interest in Latin America and Africa is growing, and that although Africa still accounts for a very small share of farmland acquired institutionally, both regions are attracting increasing amounts of capital. On the other hand, there appears to be little institutional investor activity in Asia (OECD 2010).

Investor strategies with regard to land management vary. In many cases, interest is in pure landholding; the land acquired by the fund is leased out for agricultural production to a farm operator (Wilkes and Bailey 2011). In some cases, however, investors pursue more 'hands-on' management strategies. For example, some firms operate through a corporate structure that includes an asset management company for fund management and an agricultural operator that undertakes agricultural activities to drive productivity increases, though industry experts estimate that 'at the heart of the strategy [of these firms] is the relative cheapness of this land' (Hawkins 2010, 6).

3.4. The crucial role of public policy

It is not just market forces that underpin the global land rush. Public policy plays a central role in a number of ways, in investor and recipient countries as well as globally. While these policies respond to specific concerns (about food or energy security, for instance), they also reflect a more general shift in perceptions about the roles of markets and states. Indeed, the financial crisis of 2008 has shaken the confidence in markets as a way to structure economies, and has created renewed momentum for state interventions, not just in agriculture. This section briefly discusses some of the key policies that influence large-scale land acquisitions.

As for investor country policy, the more direct policy instruments involve explicit support to land acquisitions overseas. For example, Saudi Arabia's 'King Abdullah

[22]This paragraph draws heavily on personal communications from industry insiders.

Initiative for Saudi Agricultural Investment Abroad' supports agricultural investments by Saudi companies in countries with high agricultural potential. The scheme includes credit facilities to private operators, and strategic crops include rice, wheat, barley, corn, sugar and green fodders, in addition to animal and fish resources (Saudi Arabia Ministry of Foreign Affairs 2009). This scheme seems to mainly respond to concerns about national food security, in line with broader trends in the Gulf region. Indeed, while cereal agriculture in the Gulf countries is in decline, the population of the region is expected to double from 30 million in 2000 to nearly 60 million by 2030. Dependence on food imports, now at 60% of total demand, will grow as a result (Woertz 2009). This situation, coupled with widespread expectations of rising food and agricultural commodity prices linked to fundamentals of global supply and demand (see Section 3.2.1), have raised major concerns about national food security in some Gulf states. Food price rises are particularly problematic in relation to the large migrant blue-collar workforce that is present in some of these states. In other words, skewed wealth and income distribution coupled with a concern to avoid social unrest and ensure political stability are arguably among the root drivers of the policies that some Gulf countries have established to support agricultural investments overseas as a means to ensure continued access to cheap food for their population. Some commentators (Johnstone and Mazo 2011) have argued that increases in food prices played an important role in the 'Arab Spring' that swept some Middle Eastern and Northern African countries in 2010–2011.

Public policies supporting agricultural investments overseas may also be driven by considerations about business opportunities, macroeconomics or geopolitics, rather than national food security. For example, China has been pursuing a 'Going Global' (or 'Going Out') strategy since 1999. The strategy encourages Chinese firms to invest abroad, with a view to: creating business opportunities for Chinese companies that operate overseas (particularly in light of a perceived disadvantage of Chinese firms vis-à-vis the operational capability and branding of Western TNCs); stimulating the export of goods and services; securing access to natural resources where Chinese demand outstrips domestic supply; and diversifying the country's investment portfolio.[23] A range of incentives such as tax breaks, low-interest loans and customs preferences, allied to high-level diplomatic support, support the implementation of the Going Global strategy (Anderlini 2008, Xinhua News Agency 2008). Geopolitical considerations appear to play a role in China's promotion of investments in Southeast Asia, or in Libya's investments in the Northern part of Africa South of the Sahara, for example. In these contexts, agricultural investments may be part of larger, cross-sectoral deals also involving infrastructure or extractive industries, for instance, or export of investor country labour. Alternatively, they may be part of wider strategies to increase political influence in regions that the investor country sees as strategic.

Europe has lacked recent direct policies on foreign land acquisition for agriculture. The predominant policy driver for large-scale land investments has been the European Union (EU) policy to promote renewable energy. The 2003 Biofuels Directive, now repealed, set a biofuels consumption target of 5.75% of all petrol and diesel for transport by 31 December 2010 (EU 2003, article 3b(ii)). The more recent 2009 EU Renewable Energy Directive (RED) sets the target of

[23]http://www.gov.cn/node_11140/2006-03/15/content_227686.htm, accessed in Chinese on 23 August 2011, read through translation from Google Translate.

increasing the share of energy from renewable sources to at least 20% of gross final consumption and at least 10% of the final consumption of energy in transport, all by 2020; EU member states are required to adopt national Renewable Energy Action Plans (nREAP) to promote compliance with this Directive (EU 2009, articles 3 and 4). With the expectation that biofuels are likely to be central to meeting the RED targets, European firms have responded to the promise of a guaranteed market with widespread investment in production of biofuel feedstocks, not only in Europe, but also in Asia, Africa and South America. The United States (US) Renewable Fuel Standard provides an equivalent mandate and set of financial incentives for US firms, which are sourcing feedstock predominantly from the US and Brazil. Such renewable fuel targets provide a commercial incentive for investment in biofuel feedstock production and associated land acquisition that would not be driven by market forces alone (Dufey *et al.* 2007). Climate change mitigation is often presented as a key policy goal underpinning biofuels promotion policies, as shifting energy sources in high-consumption countries is seen as politically more palatable than reducing consumption levels. Other compelling reasons for governments to pursue a switch from oil to biofuels include energy security (diversifying energy sources from Russia and the Middle East, addressing concerns linked to fluctuating global oil prices – Borras *et al.* 2010) and rural development (as biofuels can provide income-generating opportunities for European farmers).

While biofuel promotion policies have attracted much public attention, policies in OECD countries have also promoted other forms of renewable energy that in the longer term may foster the global land rush. In Europe, for example, the 2009 Renewable Energy Directive is increasing momentum for biomass energy, particularly energy plants fired (or co-fired) with wood chips or pellets. Some nREAPs set ambitious targets for biomass energy. In the United Kingdom, a new strategy for biomass energy is under discussion as part of the nREAP process, with planning permission having been granted to more than 7 GW of biomass power plants. Although operators will initially look to tap the temperate woodlands of developed countries, there are significant timber growth rate advantages that may lead them to turn to the tropics and sub-tropics to fill their biomass gap in the near future. Already there is evidence of foreign investors acquiring land in Africa, South America and Southeast Asia to establish tree plantations for biomass energy (Cotula *et al.* 2011).

Public policies in the area of carbon trading are also fostering the global land rush, as land is acquired for tree plantations or biofuels projects that involve carbon credit as a main or complementary revenue stream (Nhantumbo forthcoming). While existing markets in carbon offsets are largely voluntary, global climate change negotiations concerning REDD (Reduced Emissions from Deforestation and Forest Degradation) are strengthening expectations that the role of carbon markets is likely to increase.

While investor-country and global policies play a central role in the global land rush, host-country policies also matter – particularly measures to increase the attractiveness of the policy environment for agricultural investment. These policy efforts include revising investment legislation to increase incentives for foreign investment (with measures including tax breaks through to streamlining investment promotion agencies); reforming land legislation to facilitate foreign investors' access to land; and, more generally, macro-economic measures to remove policy distortions penalising agriculture (including overvalued exchange rates that lowered real agricultural prices, export taxes, or high taxation of agriculture; Deininger *et al.* 2011). Some governments (e.g. Mozambique, Tanzania and Ethiopia) have also made specific efforts to identify 'idle' lands within their territory, with a view to allocating them to agribusiness operators.

Some commentators have emphasised the role of the 'privatisation' of land in the global land rush. However, land remains state-owned in many key recipient countries, particularly in Africa and Southeast Asia. In other words, wider economic liberalisation has entailed a shift towards recognising private enterprise as a driver of economic development, but the state retains a central role in making land and natural resources available to private operators. Although policy features linked to centralised land control are not 'drivers' of the global land rush as such (they have been in existence for a very long time, and well before the latest wave of land acquisitions), using terminology developed by Anseeuw *et al.* (forthcoming) they have been crucial 'enablers' – in other words, factors that are facilitating the acquisition of very large areas of land at rapid pace. It is worth discussing this issue in greater depth.

In most African countries, much land is owned by the state. For example, land is nationalised in Ethiopia and Mozambique, where outright private land ownership is outlawed and only long-term land leases may be acquired. Other countries do allow private ownership, which may be acquired through land registration procedures (for instance, in Cameroon and Mali). But even in these cases, costly and cumbersome procedures mean that very few rural people hold ownership rights (Egbe 2001, on Cameroon, Djiré 2007, on Mali). In addition, where customary tenure systems are functioning and perceived as legitimate, local resource users may feel they have sufficient tenure security under these systems without needing to seek formal title. In Cameroon, for example, only about 3% of the land has been formally registered and is held under private ownership (Egbe 2001), mainly by urban elites such as politicians, civil servants and businessmen (Firmin-Sellers and Sellers 1999). As in many jurisdictions all untitled land is owned or otherwise held by the state, governments end up controlling much rural land even where the statute books devote numerous provisions to regulating private ownership.

On lands owned or administered by the state, rural people usually access lands and resources through 'customary' rights. The extent to which these rights enjoy legal recognition and protection under national law varies across countries and depending on resource use (e.g. farming versus pastoralism). A recurrent challenge, however, is that the protection provided by national law is weakened by productive use requirements (which undermine local claims to rangelands, hunting-gathering grounds, sacred sites and land reserves set aside for future generations); by wide state powers of eminent domain (whereby investment projects are considered to be for a 'public purpose' that enables the compulsory taking of local rights); by weak compensation requirements that limit compensation to improvements like trees and crops, to the exclusion of land values; and by absent or inadequate local consultation requirements.[24]

Where land is owned or controlled by the state, legal devices enable the government to make land available to outside investors. Quantitative inventories of land acquisitions in a range of different jurisdictions suggest that most deals involve

[24]For a more detailed legal analysis of these aspects, see Cotula (2011a, b) and Alden Wily (2011).

long-term land leases or concessions on state-owned land (Cotula *et al.* 2009, Görgen *et al.* 2009, Deininger *et al.* 2011). The World Bank study (Deininger *et al.* 2011) found a statistically significant correlation between weak protection of local land rights and levels of agricultural investment – a finding that compounds the notion that land tenure regimes are 'enablers' of the global land rush.[25]

The central role of the state in land relations, legal devices for the state to allocate land rights to investors, and varying, but overall limited, protection for local resource rights respond to the perceived need for poorer countries to attract investment as a way in which to promote economic development, create employment and generate public revenues. But benevolent strategic choices to promote economic development are only part of the story. The legal features of national legal systems are rooted in the colonial system, when colonisers treated conquered lands without visible developments as being empty (*terres vacantes et sans maître*, in French) and brought them under state ownership, and in decades of post-independence law-making shaped by single-party regimes or military dictatorships. They also reflect Africa's integration into the world economy mainly as a supplier of natural resource-based commodities, and respond to the political economy of the state in Africa, which is discussed by authors such as Bayart (1993) and Chabal and Daloz (1999). In this context, attracting international capital provides national elites with opportunities for business activities, political patronage and personal gain. The central role of the state in natural resource relations enables national elites to control resources through their control over state institutions; conversely, it allows them to maintain their grip on state institutions by using resource allocation as a tool for political patronage. High-level government officials may benefit personally from large land deals (Comité Technique Foncier et Développement 2010). Keeping local resource rights in check facilitates the unhindered deployment of these strategies. This is particularly so in rural areas, while politically more vocal, urban-based groups may be better placed to use the costly and cumbersome procedures provided by the law to secure property rights – as illustrated by the predominance of urban groups in ownership of registered land in Cameroon and Mali (according to Firmin-Sellers and Sellers 1999, on Cameroon, and Djiré 2007, on Mali).

4. Discussion

The global land rush reflects profound economic and social transformations in agriculture. The projected mismatch between global demand and supply in agricultural commodities has created expectations of growing commercial returns from agriculture. The global restructuring of the food industry has created incentives towards greater vertical integration in agriculture, while economic considerations have increased the attractiveness of land as an asset class for financial players. These global processes have also triggered policy change in countries concerned about their long-term food or energy security, about business opportunities overseas or about geopolitical considerations. High-consumption countries have adopted policies that aim to mitigate climate change without, however, adversely affecting consumption

[25]The finding must be treated with some caution, however, as regression analysis used a database of land acquisitions originating from media sources. As discussed above, some African countries have attracted particularly strong media attention, and it is possible that this circumstance may have affected the finding of the Bank's report.

levels – for example, through promoting shifts from fossil fuel to biofuels or biomass energy. Recipient countries have reformed their policies on land and on investment to increase the attractiveness of their investment climate. This combination of policy and market forces has made land – particularly Africa's 'abundant' and cheap land – a more attractive investment option.

The growing body of evidence on the scale and geography of the global land rush still presents important inconsistencies and lacunae – for instance, as datasets rely on different sources and methods and present different figures. New reports have been and will be published with yet different figures. But an excessive focus on pinning down the exact quantity of land that is being transacted is unlikely to pay off: some inconsistencies in datasets are rooted in conceptual and methodological problems that are very difficult to overcome. The most important point, though, is that all evidence consistently indicates that the land rush is happening on a very large scale, even more so when the growing interest in farmland is related to non-agricultural sources of pressures on the land (from extractive industries to natural parks). Therefore, the global land rush must be taken very seriously in both policy and practice.

Seen from a historical perspective (Huggins 2011), these trends are only the latest stage of a long-term process of commercial penetration in lower-income countries, including Africa, which dates back to the colonial project and involves ensuring continued supply of agricultural commodities to Northern (and, increasingly, Southern) markets. In other words, the global land rush confirms a long-standing international division of labour that casts Africa as a provider of commodities. However, three features of the global land rush importantly quality this picture. The first feature is the sheer scale of the phenomenon, which is unprecedented in many recipient countries (Anseeuw *et al.* forthcoming). Large land deals are not new – many such deals have been concluded since colonialism – and indeed the global land rush has been framed by some as a form of 'neo-colonialism'. But with the important exception of settler colonies, in most key recipient countries even colonialism did not involve the direct acquisition of land on a comparable scale. And after independence, for reasons discussed in this essay, many agribusinesses focused on upstream and downstream operations, leaving agricultural production to local farmers. In this respect, the global land rush is a new turning point in relations between North and South.

The second feature is the growing role of emerging economies like China and India in agriculture FDI (if not necessarily in land acquisitions). But while this aspect has received much media attention, the patterns of actual involvement may have been misunderstood in public perceptions. Also, the growing involvement of China and India is merely a reflection of the ongoing shift in the global geo-economic equilibrium, and it should not distract policy-makers in the West from the important role that Western companies are playing in the global land rush. More generally, geographical origin of investments and North-South divides are becoming less straightforward – as different geographies of interests may be at stake in a single investment project, and investments may be channelled through transit countries.

The third feature is the importance, largely underestimated in international debates, of national and local actors and dynamics. This applies to the role of domestic markets in investment strategies, as evidence increasingly points to market- as well as resource-oriented land acquisitions. In other words, it is becoming increasingly clear that a still unspecified share of land acquisitions are for investments that target domestic rather than export markets. It also applies to the central role played by national elites (politicians, senior civil servants, business people) in land acquisition processes. The role of these players, both as land acquirers and as intermediaries and strategic allies working with international capital, is the continuation of a longer-term process whereby national elites have become increasingly interested in rural lands – a process that was documented well before international debates on the global land rush started (for example, on the role of 'new land actors' in Burkina Faso, see M. Ouédraogo 2003, and S. Ouédraogo 2006). The role of nationals must be understood in light of different and often more localised factors, including the importance of land in local investment options, strategic positioning and political patronage; but also, increasingly, in light of its links to international capital and global processes.

Continuities and ruptures with long-term historical trajectories are also reflected in the 'actors, contexts and dynamics' that are involved in the global land rush (Peluso and Lund 2011, 668). There is nothing new in the acquisition of land in the global South to produce crops for export to the global North, and some crops have a long history in tropical plantations (e.g. rubber). However, the biofuels boom has brought new actors (e.g. constellations of energy, agribusiness and biotech companies) and new crops (e.g. jatropha), or has reinvigorated interest in longstanding crops (e.g. sugar cane, palm oil). Similarly, tree plantations are not new, but the ongoing push for biomass energy in the global North has created new market outlets for wood products. Carbon markets are changing the nature of the financial returns at stake and, more fundamentally, the very relationship between humankind and nature. The growing participation of financial players in the global land rush brings to the land arena a new set of players, motivations and investment models. These considerations also point to the great diversity of actors, contexts and dynamics in the global land rush. In the words of Peluso and Lund (2011, 669), 'there is no one grand land grab, but a series of changing contexts, emergent processes and forces, and contestations that are producing new conditions and facilitating shifts in both *de jure* and *de facto* land control'.

However one looks at it, the land rush, if sustained over the next few years, will have profound implications for the future of world agriculture, including the roles of states and markets, of agribusiness and family farming, and of the global trading system. For example, as companies increase their degree of vertical integration and as governments acquire land overseas to import agricultural produce, a growing share of world agricultural trade will occur within the 'close circuit' of corporate or country systems – deepening a trend that has emerged over the past few decades (Anseeuw *et al.* 2011). In turn, this trend may have repercussions in multiple directions: to name a few, greater vertical integration of value chains can squeeze local operators; intra-firm transactions may increase opportunities for tax avoidance through transfer pricing; and mercantilist approaches to outsourcing agricultural production for national food security can ultimately undermine the multilateral trading system. The land rush also signals a shift away from family farming, which has long constituted the backbone of agricultural systems in many recipient countries, including in much of Africa, and towards large-scale, mechanised agriculture.

The global land rush has triggered a lively, if polarised, public debate. A number of international policy responses have emerged, including 'soft' principles to make agricultural investments more 'responsible', and work to promote models of agricultural investment that do not involve land acquisitions. Some commentators, NGOs and farmer organisations have expressed scepticism about the ability of voluntary principles to result in tangible change (Borras and Franco 2010). Drawing on historical research, others have pointed to the important inequalities that exist within family farming and in collaborative models that bring together smallholders and international capital (Amanor 2011). In addition to these considerations, any discussion of policy responses to the global land rush must take account of two key issues.

The first involves going beyond individual investment projects, and considering processes and impacts in aggregate terms. There is no shortage of international guidance for investors who want to 'do the right thing'. The OECD Guidelines for Multinational Enterprises were revised in May 2011, and so were IFC's Performance Standards, including those on land acquisitions. In June 2011, the UN Human Rights Council unanimously endorsed the Guiding Principles on Business and Human Rights developed by the Special Representative to the Secretary-General on Business and Human Rights. Specifically in relation to agriculture, the World Bank, FAO, IFAD and UNCTAD have proposed Principles on Responsible Agricultural Investment, while various sets of principles and standards have been developed by commodity-based bodies like the Roundtable on Sustainable Palm Oil or the Roundtable on Sustainable Biofuels. But even if all individual investment projects were to comply with international guidance (a scenario that is unlikely, not least given the non-binding nature of these instruments), the global land rush would still exacerbate pressures on resources. As a result, there would still be a real risk that local people will lose out – especially those with weaker rights and negotiating power. This situation calls for a more strategic approach to regulating the increased competition for natural resources, and well thought-out host government regulation plays a central role within that – including through sustained investment in securing land access for local groups. International guidance on the governance of natural resources, such as the Voluntary Guidelines for the Responsible Governance of Tenure of Land, Fisheries, and Forests in the Context of National Food Security, can help governments sharpen national regulatory frameworks. Regional organisations can also play an important role, not only by developing and supporting the implementation of policy guidance (e.g. the Africa Union's Framework and Guidelines on Land Policy), but also by providing arrangements for the joint management of shared resources. The latter consideration is particularly relevant in relation to the water implications of land deals, and is illustrated by the impacts that cumulative large-scale land acquisitions for irrigated agriculture in Mali's Office du Niger area may have on downstream countries in the River Niger basin.[26]

The second issue to be factored in relates to the need for any policy response to build on a solid understanding of the fundamental drivers that underpin the global land rush. Promotion of investment models alternative to land acquisitions is unlikely to go far in the absence of well thought-out policies that reverse existing incentives that favour land-based investments. Much experimentation with business models that involve collaboration between companies and local farmers or communities is taking place in countries where regulation has created strong incentives for companies to work with local groups. In South Africa, for example, the land restitution process has started to bite in rural areas. As land changes hands

[26]This issue and case is discussed in Cotula (2011a).

from companies to local people, companies are forced to work with communities to keep their businesses going (Makhathini 2010). In other words, policy interventions can only be effective if they properly address the policy- and market-rooted structural factors that create incentives for land acquisitions.

In addition, policies are of little use if they are not backed by determined political will to make them work. The belief that large-scale plantations are needed to 'modernise' agriculture is dominant in many host government circles, and governments are vying to attract land-based investment. According to recent media reports, for example, governments in Mozambique and Ethiopia have made public offers for millions of ha of land in the hope of attracting foreign investors. These trends contrast with the mounting evidence of failed land deals – projects that take land and then fail to deliver on investment promises due to financing difficulties, changed economic circumstances or over-ambitious business plans (Anseeuw *et al.* forthcoming). They also contrast with evidence showing that, when put in a condition to work, family farmers can be highly dynamic and competitive on global markets, and that small farm development is feasible and desirable for its impacts on poverty reduction (e.g. Wiggins *et al.* 2010). For those working to generate evidence and channel findings into policy, questioning this predominant paradigm and exploring workable alternatives to it must be key priorities for the years to come.

References

AB Sugar. 2012a. Our group: Africa. Available from: http://www.britishsugargroup.com/Our-Group/Africa [Accessed 21 July 2011].

AB Sugar. 2012b. About us: timeline. Available from: http://www.britishsugargroup.com/About-Us/Timeline [Accessed 21 July 2011].

AIM, 2011. www.londonstockexchange.com/companies-and-advisors/aim/aim/aim.htm [Accessed 22 August 2011].

Alden Wily, L. 2011. *The tragedy of public lands: the fate of the commons under global commercial pressure.* Rome: International Land Coalition. Available from: http://www.landcoalition.org/publications/accelerate-legal-recognition-commons-group-owned-private-property-limit-involuntary-lan [Accessed 25 August 2011].

Amanor, K.S. 2011. Global landgrabs, agribusiness and the commercial smallholder: a West African perspective. Paper presented at the International Conference on Global Land Grabbing, Brighton, 6–8 April 2011. Available from: http://www.future-agricultures.org/index.php?option=com_docman&task=cat_view&gid=1552&limit=10&limitstart=30&order=hits&dir=DESC&Itemid=971 [Accessed 25 August 2011].

Anderlini, J. 2008. China eyes overseas land in food push. *Financial Times*, 8 May.

Anseeuw, W., L. Alden Wily, L. Cotula and M. Taylor, forthcoming. *Land rights and the rush for land: findings of the Global Commercial Pressures on Land Research Project.* Rome: International Land Coalition.

Anseeuw, W., A. Ducastel and J.-J. Gabas. 2011. The end of the African peasant? From investment funds and finance value chains to peasant related questions. Paper presented at the International Conference on Global Land Grabbing, Brighton, 6–8 April 2011. Available from: http://www.future-agricultures.org/index.php?option=com_docman&task=cat_view&gid=1552&limit=10&limitstart=30&order=hits&dir=DESC&Itemid=971 [Accessed 25 August 2011].

Associated British Foods plc. 2010. Our Companies. Available from: http://www.abf.co.uk/our-companies.aspx [Accessed 21 July 2011].

Bayart, J.-F. 1993. *The state in Africa: the politics of the belly.* London: Longman.

Borras, S. and J.S. Franco. 2010. From threat to opportunity? Problems with the idea of a 'code of conduct' for land grabbing. *Yale Human Rights and Development Law Journal*, 13(1).

Borras, S., P. McMichael and I. Scoones. 2010. The politics of biofuels, land and agrarian change: editors' introduction. *Journal of Peasant Studies*, 37(4), 575–592.

Bräutigam, D. 2011. *China in Africa: what can Western donors learn?*. Report prepared for Norfund. Available from: http://www.norfund.no/images/stories/publikasjoner/andre_publikasjoner/Norfund_China_in_Africa.pdf [Accessed 25 August 2011].

Buckley, L. 2011, Eating bitter to taste sweet: an ethnographic sketch of a Chinese agriculture project in Senegal. Paper presented at the International Conference on Global Land Grabbing, Brighton, 6–8 April 2011. Available from: http://www.future-agricultures.org/index.php?option=com_docman&task=cat_view&gid=1552&limit=10&limitstart=30&order=hits&dir=DESC&Itemid=971 [Accessed 25 August 2011].

Burnod, P., M. Gingembre, R. Andrianirina Ratsialonana and R. Ratovoarinony. 2011. From international land deals to local informal agreements: regulations of and local reactions to agricultural investments in Madagascar. Paper presented at the International Conference on Global Land Grabbing, Brighton, 6–8 April 2011. Available from: http://www.future-agricultures.org/index.php?option=com_docman&task=cat_view&gid=1552&limit=10&limitstart=30&order=hits&dir=DESC&Itemid=971 [Accessed 25 August 2011].

Buxton, A., M. Campanale and L. Cotula, forthcoming. *Farms and funds: investment funds in the global land rush*. London: International Institute for Environment and Development.

CAS. 2011. Centres Financiers Offshore et Système Bancaire «Fantôme». Paris: Centre d'Analyse Stratégique, Note d'analyse No. 222, May. Available from: http://www.strategie.gouv.fr/content/note-d%E2%80%99analyse-222-centres-financiers-offshore-et-systeme-bancaire-fantome [Accessed 25 August 2011].

Chabal, P. and J.-P. Daloz. 1999. *Africa works: disorder as political instrument*. Oxford: James Currey.

Cohen, B. 2009. Sovereign wealth funds and national security: the great tradeoff. *International Affairs*, 85(4), 713–731.

Comité Technique Foncier et Développement. 2010. Les appropriations de terres à grande échelle: Analyse du phénomène et propositions d'orientations. Paris: MAEE/AFD. Available from: http://www.foncier-developpement.fr/wp-content/uploads/appropriation_finale1.pdf [Accessed 25 August 2011].

Cotula, L. 2011a. *Land deals in Africa – what is in the contracts?*. London: International Institute for Environment and Development (IIED). Available from: http://pubs.iied.org/12568IIED.html [Accessed 25 August 2011].

Cotula, L. 2011b. *Human rights, natural resource and investment law in a globalised world: shades of grey in the shadow of the law*. London and New York: Routledge.

Cotula, L., L. Finnegan and D. Macqueen. 2011. *Biomass energy: another driver of land acquisitions?* London: IIED.

Cotula, L. and C. Toulmin, eds. 2004. *Till to tiller: international migration, remittances and land rights in West Africa*. London: IIED. Drylands Issue Paper No. 132. Available from: http://pubs.iied.org/9508IIED.html [Accessed 25 August 2011].

Cotula, L., S. Vermeulen, R. Leonard and J. Keeley. 2009. Land grab or development opportunity? Agricultural investment and international land deals in Africa. Rome/London: Food and Agriculture Organization of the UN (FAO)/International Fund for Agricultural Development (IFAD)/International Institute for Environment and Development (IIED). Available from: http://www.iied.org/pubs/display.php?o=12561IIED [Accessed 25 August 2011].

Deininger, K. and D. Byerlee, with J. Lindsay, A. Norton, H. Selod and M. Stickler. 2011. *Rising global interest in farmland: can it yield sustainable and equitable benefits?*. Washington DC: World Bank. Available from: http://econ.worldbank.org/external/default/main?pagePK=64165259&theSitePK=469382&piPK=64165421&menuPK=64166322&entityID=000334955_20110208033706 [Accessed 25 August 2011].

De Schutter, O. 2011. How not to think of land-grabbing: three critiques of large-scale investments in farmland. *Journal of Peasant Studies*, 38(2), 249–279.

Djiré, M. 2007. *Land registration in Mali – no land ownership for farmers? Observations from peri-urban Bamako*. London: IIED. Drylands Issue Paper No. 144. Available from: http://www.iied.org/pubs/display.php?o=12538IIED [Accessed 25 August 2011].

Dufey, A., S. Vermeulen and W. Vorley. 2007. *Biofuels: strategic choices for commodity dependent developing countries*. Amsterdam: Common Fund for Commodities.

EAPortal, 2011. Ethiopia Agriculture Portal, www.ethiopian-gateway.com/eaportal/nore/386 [Accessed 30 May 2011].

EmVest, 2011. Our Produce Map, www.emvest.com/Our_produce_map.aspx [Accessed 21 July 2011].

EU, 2003. Directive 2003/30/EC of the European Parliament and of the Council of 8 May 2003 on the Promotion of the Use of Biofuels or Other Renewable Fuels for Transport.

EU, 2009. Directive 2009/28/EC of the European Parliament and of the Council of 23 April 2009 on the Promotion of the Use of Energy from Renewable Sources.

Egbe, S.E. 2001. The concept of community forestry under Cameroonian law. *Journal of African Law*, 45, 25–50.

farmlandgrab.org, 2011. Food Crisis and the Global Land Grab. http://www.farmlandgrab.org/ [Accessed 25 August 2011].

FAO. 2010. *Agricultural investment funds for developing countries*. Rome: FAO. Hardman & Co.

FAO, IFAD, IMF, OECD, UNCTAD, WFP, the World Bank, the WTO, IFPRI and the UN HLTF. 2011. Price volatility in food and agricultural markets: policy responses, 2 June. http://www.oecd.org/document/20/0,3746,en_2649_37401_48152724_1_1_1_37401,00.html [Accessed 25 August 2011].

Faye, I.M., A. Benkahla, O. Touré, S.M. Seck and C.O. Ba. 2011. *Les acquisitions de terres à grande echelle au Sénegal: description d'un nouveau phénomène*. Dakar: Initiative Prospective Agricole et Rurale.

FIAN. 2010. *Land grabbing in Kenya and Mozambique*. Heidelberg: FIAN. Available from: http://www.fian.org/resources/documents/others/land-grabbing-in-kenyaand-mozambique.

Firmin-Sellers, K. and P. Sellers. 1999. Expected failures and unexpected successes of land titling in Africa. *World Development*, 27, 1115–1128.

Friis, C. and A. Reenberg. 2010, *Land grab in Africa: emerging land system drivers in a teleconnected world*. University of Copenhagen, GLP-IPO, GLP Report No. 1. Available from: http://www.globallandproject.org/Documents/GLP_report_01.pdf [Accessed 25 August 2011].

German, L., G. Schoneveld and E. Mwangi. 2011. Processes of large-scale land acquisition by investors: case studies from sub-Saharan Africa. Paper presented at the International Conference on Global Land Grabbing, Brighton, 6–8 April 2011. Available from: http://www.future-agricultures.org/index.php?option=com_docman&task=cat_view&gid=1552&limit=10&limitstart=30&order=hits&dir=DESC&Itemid=971 [Accessed 25 August 2011].

Görgen, M., B. Rudloff, J. Simons, A. Üllenberg, S. Väth and L. Wimmer. 2009. *Foreign direct investment (FDI) in land in developing countries*. Eschborn: Deutsche Gesellschaft für Technische Zusammenarbeit (GTZ). Available from: http://www2.gtz.de/urbanet/library/detail1.asp?number=7529 [Accessed 25 August 2011].

GRAIN. 2009. Mauritius leads land grabs for rice in Mozambique, 1 September. Available from: http://www.grain.org/hybridrice/?lid=221 and http://farmlandgrab.org/7231 [Accessed 25 August 2011].

Hall, R. 2011. The next great trek? South African commercial farmers move north. Paper presented at the International Conference on Global Land Grabbing, Brighton, 6–8 April 2011. Available from: http://www.future-agricultures.org/index.php?option=com_docman&task=cat_view&gid=1552&limit=10&limitstart=30&order=hits&dir=DESC&Itemid=971 [Accessed 25 August 2011].

Hawkins, D. 2010. The world agriculture industry: a study in falling supply and rising demand. Hardman & Co, 26 May. Available from: www.hardmanandco.com/Research/Agriculture%20Funds.pdf [Accessed 25 August 2011].

Headey, D. and S. Fan. 2008. Anatomy of a crisis: the causes and consequences of surging food prices. *Agricultural Economics*, 39, 375–391.

Hilhorst, T., J. Nelen and N. Traoré. 2011. Agrarian change below the radar screen: rising farmland acquisitions by domestic investors in West Africa – Results from a survey in Benin, Burkina Faso and Niger. Paper presented at the International Conference on Global Land Grabbing, Brighton, 6–8 April 2011. Available from: http://www.future-agricultures.org/index.php?option=com_docman&task=cat_view&gid=1552&limit=10&limitstart=30&order=hits&dir=DESC&Itemid=971 [Accessed 25 August 2011].

Huggins, C. 2011. A historical perspective on the 'global land rush'. Rome: International Land Coalition. Available from: http://www.landcoalition.org/publications/historical-perspective-global-land-rush [Accessed 25 August 2011].

ILC, 2011. Commercial Pressures on Land. http://www.commercialpressuresonland.org/ [Accessed 25 August 2011].
Illovo Sugar. 2011. Fast Facts. Available from: http://www.illovosugar.com/About_Us/Fast_Facts.aspx [Accessed 21 July 2011].
Johnstone, S. and Mazo, J. 2011. Global warming and the Arab spring. *Survival: Global Politics and Strategy*, 53(2), 11–17. Available from: http://www.iiss.org/publications/survival/survival-2011/year-2011-issue-2/global-warming-and-the-arab-spring/ [Accessed 25 August 2011].
Kaarhus, R., 2011, Agricultural growth corridors equals land-grabbing? Models, roles and Accountabilities in a Mozambican case. Paper presented at the International Conference on Global Land Grabbing, Brighton, 6–8 April 2011. Available from: http://www.future-agricultures.org/index.php?option=com_docman&task=cat_view&gid=1552&limit=10&limitstart=30&order=hits&dir=DESC&Itemid=971 [Accessed 25 August 2011].
LEITI, 2011. Liberia Extractive Industries Transparency Initiative. www.leiti.org.lr [Accessed on 25 August 2011].
Makhathini, M. 2010. Mondi's 'sale and leaseback' model in South Africa. In: L. Cotula and R. Leonard, eds. *Alternatives to land acquisitions: agricultural investment and collaborative business models*. London/Bern/Rome/Maputo: International Institute for Environment and Development/Swiss Agency for Development and Cooperation/ International Fund for Agricultural Development/Centro Terra Viva, pp. 19–32.
Nhantumbo, I. forthcoming, *REDD+: ready to engage private investors?*. London: International Institute for Environment and Development.
Nhantumbo, I. and A. Salomão. 2010. *Biofuels, land access and rural livelihoods in Mozambique*. London: IIED. Available from: http://pubs.iied.org/12563IIED.html .
Oakland Institute, 2011. Understanding Land Investment Deals in Africa, set of reports and briefs. Available from: http://media.oaklandinstitute.org/special-investigation-understanding-land-investment-deals-africa [Accessed 25 August 2011].
OECD. 2010. Private financial sector investment in farmland and agricultural infrastructure. Paris: OECD, working party on agricultural policies and markets. Available from: http://www.oecd.org/officialdocuments/publicdisplaydocumentpdf/?cote=TAD/CA/APM/WP(2010)11/FINAL&docLanguage=En [Accessed 25 August 2011].
OECD/FAO. 2010. *Agricultural outlook 2010–2019*. Paris/Rome. OECD/FAO.
Olam, 2011. www.olamonline.com/globalbusiness/rice.asp [Accessed 22 August 2011].
Office du Niger. 2009. *Note technique sur la situation des aménagements et attributions de terres à l'Office du Niger*. Ségou, Direction Générale de l'Office du Niger, unpublished document on file with the author.
Ouédraogo, M. 2003. *New stakeholders and the promotion of agro-silvo-pastoral activities in Southern Burkina Faso: false start or inexperience?*. London: IIED. Drylands Issue Paper No. 118. Available from: http://pubs.iied.org/9236IIED.html [Accessed 25 August 2011].
Ouédraogo, S. 2006 *New actors and land acquisition around Lake Bazèga. Burkina Faso*. London: IIED. Drylands Issue Paper No. 138. Available from: http://pubs.iied.org/12525IIED.html [Accessed 25 August 2011].
Oxfam. 2011. *Land and power: the growing scandal surrounding the new wave of investments in land*. Oxford: Oxfam. Available from: http://www.oxfam.org/sites/www.oxfam.org/files/bp151-land-power-rights-acquisitions-220911-en.pdf [Accessed 25 August 2011].
Peluso, N.L. and C. Lund. 2011. New frontiers of land control: introduction. *Journal of Peasant Studies*, 38(4), 667–681.
Pensions & Investments. 2010. Fertile ground for investment – European pension funds target farmland in emerging markets. *Pensions & Investments*, 19 April.
Ravanera, R. 2011. *Commercial pressures on land in Asia: an overview*. Rome: International Land Coalition. Available from: http://www.landcoalition.org/publications/commercial-pressures-land-asia-overview [Accessed 25 August 2011].
Reuters. 2010. Mauritius hunts investors for Mozambique land deal, *Reuters*, 27 January. Available from: http://af.reuters.com/article/investingNews/idAFJOE60Q0L820100127?pageNumber=2&virtualBrandChannel=0&sp=true [Accessed 25 August 2011].
Reuters. 2011a. UPDATE 1 – Sime Darby to invest $3.1 bln in Liberia project, *Reuters*, 19 May. Available from: http://af.reuters.com/article/liberiaNews/idAFN1925797320110519.

Reuters. 2011b. Mauritius firm to develop land in Mozambique, 3 June. Available from: http://af.reuters.com/article/mozambiqueNews/idAFLDE75219M20110603 [Accessed 25 August 2011].

Richardson, B. 2010. Big sugar in Southern Africa: rural development and the perverted potential of sugar/ethanol exports. *Journal of Peasant Studies*, 37(4), 917–938.

Rowden, R. 2011. India's role in the new global farmland grab: an examination of the role of the Indian government and Indian companies engaged in overseas agricultural land acquisitions in developing countries. Paper produced in collaboration with GRAIN and the Economics Research Foundation. Available from: http://www.networkideas.org/featart/aug2011/Rick_Rowden.pdf [Accessed 25 August 2011].

Roxburgh, C., N. Dörr, A. Leke, A. TAzi-Riffi, A. van Wamelen, S. Lund, M. Chironga, T. Alatovik, C. Atkins, N. Terfus and T. Zeino-Mahmalat. 2010. *Lions on the move: the progress and potential of African economies*. McKinsey Global Institute. Available from: http://www.mckinsey.com/mgi/publications/progress_and_potential_of_african_economies/index.asp [Accessed 25 August 2011].

Saudi Arabia Ministry of Foreign Affairs. 2009. Custodian of the Two Holy Mosques receives Minister of Commerce and Industry, 27 Jan. Available from: http://www.mofa.gov.sa/sites/mofaen/ServicesAndInformation/news/statements/Pages/NewsArticleID88796.aspx [Accessed 23 August 2011].

Schoneveld, G.C., L.A. German and E. Nutakor. 2010. Towards sustainable biofuel development: assessing the local impacts of large-scale foreign land acquisitions in Ghana. Paper presented at the World Bank Land Governance Conference, 26–27 April.

Selby, A. 2009. Institutional investment into agricultural activities: potential benefits and pitfalls. Paper presented at the conference 'Land governance in support of the MDGs: responding to New Challenges'. Washington, DC, World Bank, 9–10 March, PowerPoint presentation on file with the author.

Sulle, E., and Nelson, F. 2009. *Biofuels, land access and rural livelihoods in Tanzania*. London: IIED. Available from: www.iied.org/pubs/pdfs/12560IIED.pdf [Accessed 25 August 2011].

The Economist. 2009. Green shoots - no matter how bad things get, people still need to eat, 19 May. Available from: http://www.economist.com/business/displayStory.cfm?story_id=13331189 [Accessed 25 August 2011].

Tiffen, M. and M. Mortimore. 1990. *Theory and practice in plantation agriculture: an economic review*. London: ODI.

Toulmin, C., P. Bindraban, J. Borras, E. Mwangi and S. Sauer. 2011. *Land tenure and international investments in agriculture*. Report prepared for the High Level Panel of Experts on Food Security and Nutrition, Rome, Committee on World Food Security.

Ullenberg, A. 2009. *Foreign direct investment (FDI) in land in Madagascar*. Eschborn: Deutsche Gesellschaft für Technische Zusammenarbeit (GTZ). Available from: http://www2.gtz.de/urbanet/library/detail1.asp?number=7529 [Accessed 25 August 2011].

UNCTAD. 2009. *World investment report 2009: transnational corporations, agricultural production and development*. Geneva: United Nations Conference on Trade and Development.

UNCTAD. 2011. UNCTAD Investment Treaties Online. www.unctadxi.org/templates/docsearch_779.aspx [Accessed 21 July 2011].

Veldman, M. and M. Lankhorst. 2011. Socio-economic impact of commercial exploitation of Rwandan marshes: a case study of sugar cane production in rural Kigali. Rome: International Land Coalition. Available from: http://www.landcoalition.org/sites/default/files/publication/908/RCN_Rwanda_web_11.03.11.pdf [Accessed 25 August 2011].

Visser, O. and M. Spoor. 2011. Land grabbing in post-Soviet Eurasia: the world's largest agricultural land reserves at stake. *Journal of Peasant Studies*, 38(2), 299–323.

von Braun, J. 2008. *Food and financial crises: implications for agriculture and the poor*. Washington, DC, International Food Policy Research Institute (IFPRI), Food Policy Report No. 20.

Wiggins, S., J. Kirsten and L. Llambí. 2010. The future of small farms. *World Development*, 38(10), 1341–1348.

Wilkes, H. and I. Bailey. 2011. International farmland markets. Savills Plc. Available from: http://www.aginvestconference.com/usa/savills.pdf [Accessed 25 August 2011].

Woertz, E. 2009. Gulf food security needs delicate diplomacy. *Financial Times*, March 4.

Woodhouse, P. 2012. "New investment, old challenges. Land deals and the water constraint in African agriculture." *Journal of Peasant Studies*. iFirst. Available from http://www.tandfonline.com/action/showAxaArticles?journalCode=fjps20
Xinhua News Agency. 2008. Fagaiwei: Wu Haiwai Duntian Jihua (NDRC: China has no plan to acquire land overseas, in Chinese). Available from: http://news.xinhuanet.com/fortune/2008-11/13/content_10351772.htm [Accessed 25 August 2011].
Zoomers, A. 2010. Globalisation and the foreignisation of space: seven processes driving the current global land grab. *Journal of Peasant Studies*, 37(2), 429–447.

Lorenzo Cotula is a senior researcher at the International Institute for Environment and Development (IIED), based in the UK. At IIED, Lorenzo leads the Land Rights Team and the Investment Team. He undertakes research, capacity building, advocacy and advisory work at field, national and international levels.

The land grab and corporate food regime restructuring

Philip McMichael

Land grab appears to be a phenomenal expression of deepening contradictions in the corporate food regime. In particular, the end of cheap food (signaled in the 2008 'food crisis') has generated renewed interest in agriculture for development on the part of the development industry, matched by a rising interest in offshore land investments, driven by governments securing food and fuel exports and financiers speculating on commodity futures and land price inflation. This paper interprets these developments as illusory solutions to a fundamental accumulation crisis of the neoliberal project. While this new (and final?) enclosure registers a restructuring of the food regime, as its geopolitical relations and productive content re-centers on Southern land and an emergent bioeconomic imperative, it is likely to only buy time (and space) in the short run for political and economic elites and a global consuming class. In the longer run, the attempt to resolve food regime contradictions by a spatial fix may well be catastrophic.

Introduction

Land grabbing is nothing new, and yet the recent 'land rush' has its own distinctive features.[1] If land grabbing under colonialism was tragedy, it repeats now as farce. I argue here that this rush to acquire land – however varied (in origin, destination and impact) and inconclusive – is symptomatic of a crisis of accumulation in the neoliberal globalization project. The term 'grab' invokes a long history of violent enclosure of common lands to accommodate world capitalist expansion, but it sits uneasily with the 'free market' rhetoric of neoliberal ideology. Even so, as Marx and Polanyi have both reminded us, markets are instituted politically as expressions of property relations. In that sense, one might argue that host governments accommodating local and foreign investors' land acquisitions, facilitated by development agency assessments and emerging investor codes, sustain the deepening of commodity relations. Neoliberalism has never been free of contradiction.

This paper interprets these developments as illusory solutions to a fundamental accumulation crisis precipitated by the neoliberal globalization project (as a political response to US hegemonic decline). That is, land grabbing is understood here as a reflex of changing conditions of accumulation: first, as capital's costs of production

The author is grateful to three anonymous reviewers for invaluable feedback on an earlier version of this essay.

[1]While the so-called 'land grab' is quite heterogeneous, including urban expansion, tourism complexes, conservation initiatives, and smallholder grabbing (Zoomers 2010, Corson 2011, Hall 2011), this essay focuses on land grabs for agro-industrial purposes.

(energy) and reproduction (wage-foods) rise in tandem; and second, as finance capital capitalizes offshore agro-food zones as (speculative) substitutes for ecologically exhausted Northern crop lands and as energy crop sites. As such, the land grab provides a lens on the contradictory dynamics of the food regime, which, at one and the same time, *situates* the land grab as something other than simply a contemporary enclosure of land for capitalist expansion.

The food regime context

The 'food regime' concept situates the global ordering of international food production, circulation and consumption relations within specific institutionalized world-historical conjunctures (McMichael 2009a). The 'corporate food regime' (1980s–present) specifies a neoliberal project of agricultural liberalization via structural adjustment mechanisms and WTO rules encouraging universal agro-exporting and requiring states in the global South to open their economies to the Northern-dominated international food trade, dismantle farm sector protections and adopt intellectual property protections. All of these rules have institutionalized market and property relations privileging agribusiness in the name of production 'efficiencies', 'free trade' and global 'food security'. At the same time, the corporate food regime embodies mercantilist practices from the previous US-centered food regime in the form of WTO-institutionalized subsidies for Northern energy-intensive agribusiness production and export of artificially cheapened foodstuffs – at the competitive expense of Southern farmers in particular and global food security in general (McMichael 2005, Pritchard 2009). This subsidy-driven cheap food regime has destabilized small and medium-sized farms across the world, resulting in cycles of depeasantization and casualization of a global labor force for capital (McMichael 2005).[2] The key effect has been to service an unproductive class relationship of global capital accumulation via the redistribution of value from an under-reproduced social periphery to an over-consuming social core (Araghi 2009).

Southern state opposition to the hypocrisy of Northern 'food dumping' contributed to the breakdown of the Doha Round and WTO paralysis in the first decade of the twenty-first century, accompanied by a mushrooming 'food sovereignty' countermovement with an alternative vision of democratic food security arrangements embedded in socio-ecological relations on local and regional scales. These institutional and political-economic contradictions combined with the reversal of food price trends in 2007–2008 as cheap food came to an end. Compounded by agribusiness monopoly pricing, rising food prices – related to rising energy costs and fuel crop substitutes – were transmitted globally under the liberalized terms of finance and trade associated with neoliberal policies (McMichael 2009c). The end of cheap food, in context of rising energy prices, is at the same time a signal crisis of capital accumulation, insofar as capital is ill-disposed to translate its financial power into new productive forms of investment other than speculative acquisitions, including cheap land in the global South (Arrighi 1994, Moore 2010).

[2]For example, the FAO estimated, conservatively, that 20–30 million peasants were displaced in the 1990s following the institution of the WTO, and in Mexico upward of two million *campesinos* lost land through the destabilizing impacts of NAFTA (Madeley 2000, 75, Carlsen 2003).

In this context, the land grab is both a response to food price reversals generating export bans and government initiatives to secure offshore food and biofuel supplies and reflects a speculative interest in food and biofuel futures and associated land price inflation on the part of finance capital. Such offshore designs override the WTO free trade architecture and signal a relocation of agro-industry from North to South as the coordinates of the corporate food regime shift to take advantage of widening land costs between North and South.[3]

The changing geography of the corporate food regime coincides with, and consolidates, a new targeting of Southern smallholders for incorporation into value-chain agribusiness in a potential new frontier of capitalization of farming as conduit of commercial inputs and outputs. Under the guise of rhetoric about 'agriculture for development' (World Bank 2007), institutional and philanthropic (notably the Gates Foundation's Alliance for a Green Revolution in Africa/AGRA) developers combine a variant of (smallholder) land grabbing with a counter-mobilization to the movements for food sovereignty expressed in terms of 'Responsible Agricultural Investment' principles designed to justify and enable the enclosure of smallholder and common lands alike (Borras and Franco 2010). The unifying ideology is that lands occupied (farms) or accessed (commons) by smallholders and pastoralists are low-yield and underutilized lands that, with capitalization, can improve rural incomes and address the global food security problem underscored by the current 'food crisis'.

Arguably, the land grab is one expression of food regime restructuring. The associated politics of investment codes at the FAO's Committee on Food Security represents a new arena of struggle over governance between quite distinctive coalitions such as La Vía Campesina, the International Planning Committee for Food Sovereignty (IPC), FoodFirst Information and Action Network (FIAN), the International Land Coalition (ILC) and the International Federation of Agricultural Producers (IFAP) (Borras 2010). Such struggle is part of a broader agrarian mobilization around opposing (self) organizing principles of food sovereignty, agro-ecology, 're-peasantization', slow food and the like, symbolized by La Vía Campesina's global campaign for a UN Peasants' Charter (Edelman and Carwill 2011). In this sense, the contradictions of the corporate food regime have sharpened, deepening the terms of struggle as the land grab – which includes initiatives such as AGRA, the agrofuels project, Reduced Emissions from Degradation and Deforestation (REDD+) and US and transnational corporation promotion of GMO crops in the name of security – occupies the agrarian world.

The land grab, in fact, deepens the dynamic underlying the climax of the metropolitan agrarian crisis identified by Kautsky in the late nineteenth century. This crisis involved cheap grains from the New World flooding European markets at the expense of European capitalist farmers (Kautsky 1988, 243), and underwriting European industrialization as an integral part of the British-centered food regime (Friedmann 1978, Friedmann and McMichael 1989). The successor, US-centered food regime resolved the metropolitan crisis temporarily via agricultural mercantilism, with managed overproduction of farm products sold at concessional prices to strategic post-colonial states – a relationship adopted by Europe and generalized as

[3]For example, arable land prices in the US rose 13 percent in 2007, and over 10.5 percent in 2008, while in the UK prices rose 28 percent in late 2007, and by more than 10 percent in the first quarter of 2008 (Berthelot 2009, 16).

Northern food dumping in Southern markets (McMichael 2009d). As Kautsky forewarned, the agrarian crisis today is deepening, globally, as land give-aways, subsidies and financialization associated with the land grab express declining Northern productivity and extend the (cheap) land frontier to its ecological limits. As he put it,

> Those tropical countries which are not suited to wheat cultivation – Central America, Northern Brazil, large parts of Africa, India, South Eastern Asia – would then also join the ranks of the European grain farmers' competitors. Eventually, this competition will have to lose its ruinous character. The surface of the earth is finite... And if the capitalist burdens which once depressed agriculture in Western Europe now begin to do the same to its competitors in the USA, Russia, and so on, this is not proof that the crisis in Western European agriculture is coming to an end. It simply proves that the crisis is extending its grip. (Kautsky 1988, 252)

In this sense, the long-term agrarian crisis associated with offshoring agriculture deepens, as the land grab signals a restructuring of the food regime through acquisition of cheap land, water and labor in the global South. This 'spatial fix' represents a short-term attempt to resolve the contradictions of rising agro-industrial costs on the one hand,[4] and rising (food) costs of reproduction of labor on the other, but under conditions of agribusiness as usual that will only accelerate ecological and social contradictions (cf Araghi 2009), notably in African regions already vulnerable to the depredations of climate change (Toulmin 2009). As an expression of food regime restructuring, the land grab reveals a new threshold in the conversion of farming and farm land into a source of food, feed, agrofuels and general biomass to serve the needs of a (minority) global class of consumers distributed across an increasingly multi-centric global food system (as the G-20 displaces the G-7 and its privileges in the WTO-centered regime). Fundamental to this transition is the role of finance, as financial speculation renders land and crops increasingly fungible as governed by the price form – at the expense of a rational farming of the land for social and ecological sustainability.

Political economy and global ecology

Less than a decade ago, the international peasant coalition, *La Vía Campesina* claimed that agribusiness power no longer resided in control over land, rather in the relations that surround agricultural production – those that 'control loans, materials supply, the dissemination of new technologies, such as transgenic products, on the one hand, and those that control national and international product warehousing systems, transportation, distribution and retail sales to the consumer, on the other hand, have real power' (2004, 5). In the meantime, the so-called triple crisis (of finance, energy and food) has altered the landscape, so to speak, such that with the

[4]Weis refers to the 'accelerating biophysical contradictions' of agro-industry (2010) under conditions of rising costs of 'biophysical override' (2007). Here it is worth noting that a recent Global Citizen's Report on the State of GMOs expresses serious concern with the 'increased use of synthetic chemicals to control pests despite biotech companies' justification that GM-engineered crops would reduce insecticide use... Soya growers in Argentina and Brazil have been found to use twice as much herbicide on their GM as on conventional crops, and a survey by Navdanya International, in India, showed that pesticide use rose 13-fold after Bt cotton was introduced.... Ten common weeds have now developed resistance in at least 22 US states, with about 6m hectares of soya, cotton and corn affected' (Vidal 2011, 7).

prospect of rising energy and food prices, land is back on the investment agenda, but this time as a speculative venture and hedge against food and fuel supply shortfalls.

This realignment of interest in land (and water)[5] requires examining the land grab through the lens of 'global ecology' and political economy. Wolfgang Sachs (1993, 20) defined *global ecology* as the 'rational planning of the planet for Northern security', following the Earth Summit in 1992, where Southern forests, for example, were to be managed as carbon sinks and for biodiversity preservation – bioregions of intrinsic (ecological) value to a Northern-led accumulation drive. Sachs (1993, 13) noted that, 'Far from "protecting the earth", environmental diplomacy which works within a developmentalist frame cannot but concentrate its efforts on rationing what is left of nature'. At the same time, by classifying the atmosphere and biodiversity as a 'global commons', the Bank's Global Environmental Facility 'was able to override the local claims of those who rely on local commons and effectively assert that everyone has a right of access to them, that local people have no more claim to them than a corporation based on the other side of the globe' (Hildyard 1993, 34). The terms of reference of the contemporary 'land rush' are similar, namely, that global food security and ecological security (via 'green' biofuels) depend on 'global' access to land for offshore food, fuels and general biomass production. 'Northern security' includes the North in the South now, and is associated with the global crisis of neoliberal capitalism, as peak oil, peak soil and food riots stalk the landscape, invoking green solutions. Thus the World Bank (2010, vii) avers

> ...although deforestation associated with the expansion of the agricultural frontier has been a serious problem (and one of the world's larges contributors to greenhouse gas emissions), our analysis shows that the projected increase in the demand for agricultural commodities over the next decade could be met, *without cutting down forests, by increasing productivity and farmland expansion in non-forested areas.* [emphasis added]

From the *political economy* angle, the crisis of accumulation under neoliberalism expresses itself in 'rationing what is left of nature' through historically specific mechanisms. Food regime restructuring is anticipated by a prospective large-scale relocation of agro-industry to the global South, accelerated by world-market 'override' (export bans, offshore investment) following the 2007–2008 'food crisis'. This migration of agricultural production has several drivers, including soil depletion in the 'breadbasket' regions and rising costs of compensatory inputs, cheap land in the South increasingly accessible through new forms of 'environmental diplomacy' (offset protocols/Clean Development Mechanism, World Bank governance interventions), climate and food crises spurring biofuel and 'agriculture for development' solutions implicating Southern land, and financialization. Facilitating this transition are accommodating policies of host governments, public–private biofuel complexes and public authority governance mandates regarding land titling – legitimating new initiatives for the development industry (*cf.* Da Costa and McMichael 2005).

As above, the crisis of neoliberal accumulation is deeply rooted in rising ecological contradictions (anthropocentric climate change, fossil fuel depletion), which, combined with rising food prices, increase the reproductive costs of capital

[5]Woodhouse and Ganho (2011), for example, offer a nuanced evaluation of the extent to which land grabbing includes accessing water flows, most likely through monopolistic infrastructural developments that alter and deprive access to water by smallholders.

and infrastructural[6] costs of states. Not only does this crisis propel, and justify, investment in land offshore in the name of addressing food shortages and alternative energy, but also agro-industrial capitalization opens an investment frontier for capital in an era of financialization. To the extent that agriculture and its products are absorbed into financial chains, the mix of physical crops becomes increasingly irrelevant to the financial profit calculus. That is, production decisions reflect a boardroom financial calculus with little concern for allocations between crops for food or fuels, and/or environmental integrity.[7] In this sense, the so-called rational planning of planetary resources such as land (and water)[8] is driven as much by financial goals as by material considerations.

Here there is an apparent Jekyll and Hyde operation, ultimately substitutable, between food and fuel claims. Land grabbing for food justified by the world food crisis is undertaken by combinations of development agencies (World Bank, FAO), investment banks (Goldman Sachs), funds (Carlyle Group) and philanthropists (Soros, Gates Foundation), and was sanctioned by the World Food Summit of 2008. Land grabbing for biofuels and/or biomass, on the other hand, is undertaken largely by private investors and sometimes by governments through State Firms and Sovereign Wealth Funds (McMichael 2009b, ETC 2010). The controversy over biofuel claims (energy return on investment, emissions reduction?) and impacts (land clearance, displacement, class disparities) has stigmatized biofuels,[9] symbolized in UN Human Rights *Rapporteur* Jean Ziegler's 2007 charge that they are a 'crime against humanity' (quoted in Ferrett 2007).

---

[6]Here, the externalized costs of capitalism, absorbed by states and under-reproduced communities alike, are not only rising, but becoming more visible with climatic transformations and ecosystem degradation.

[7]As Merian Research (2010, 7) reports, greenwashing claims by investors about their associations with environmental NGOs are often bogus. For this reason, and reasons of legitimacy under pressure from civil society, the development agencies are engaged in formulating (voluntary) codes of conduct regarding land acquisition and use.

[8]'The obvious motives for the deals are the spike in food prices and the subsequent decision of governments in several key producer countries to restrict their exports, threatening the food security of food importing countries such as the Gulf states, China and South Korea (the main participants in the deals). However, water shortages are another, hidden driver. Peter Brabeck-Letmathe, the chairman of Nestlé, claims: "The purchases weren't about land, but water. For with the land comes the right to withdraw the water linked to it, in most countries essentially a freebie that increasingly could be the most valuable part of the deal." He calls it "the great water grab"' Green (2011).

[9]An International Energy Agency (IEA) Report, *From 1st- to 2nd-Generation Biofuel Technologies* (2008), acknowledges shortcomings of first-generation biofuels, noting growing 'interest in developing biofuels produced from non-food biomass. Feedstocks from lingo-cellulosic materials include cereal straw, bargasse, forest residues, and purpose-grown energy crops such as vegetative grasses and short rotation forests'. Such second-generation biofuels are expected to avoid many of the concerns with first-generation biofuels, including cost, but are not expected to be widely deployed until 2020. Even so, the Report's observation that the lack of risk assessment means 'poor policy decisions could result in negative unexpected consequences for GHG emissions, the environment, biodiversity, land ownership, and producer and consumer welfare' (IEA 2008, 34). As Wetter (2009) notes: 'Impacts of increased residue removal will include impoverished soils (requiring more industrial fertilizers) and dangerous increases in soil erosion. We will see vast increases in pesticide- and herbicide-use. Removal of dead and dying trees from forests will increase biodiversity losses and decrease forest carbon-sequestration capacity. Additionally, many plants identified as good candidates for second-generation agrofuels are harmful to the environment as invasive species (e.g., miscanthus, switch grass, reed canary grass)'.

Whether for food or fuel, land grabbing forces institutional oversight, including rules and certification schemes (such as the Roundtable on Sustainable Biofuels), and Voluntary Guidelines on Responsible Governance of Tenure of Land, Fisheries and Forests (FAO). Attempts by the World Bank (and IFPRI) to elaborate 'Principles for Responsible Agricultural Investment' (RAI) have been met with current UN Human Rights *Rapporteur* Olivier de Schutter's charge of 'Responsibly destroying the world's peasantry' (2010), and a civil society-led attempt to construct more democratic Voluntary Guidelines through the FAO and its Committee on Food Security (see CFS 2011). Food regime dynamics are by definition framed by tensions between distinctive organizing principles, and the current contestation over how to value land is no exception.

The new bioeconomy

The land grab includes the prospective development of the bioeconomy (Levidow 2011). In particular, it is perhaps the clearest manifestation of the pressing value question surrounding land, driven by a 'neoliberalization of nature' (Birch *et al.* 2010). The 'new bioeconomy' marks 'just the beginning of converting the liquid fuel market to biomass' (ETC 2010, 3). As ETC (2010, 6) notes:

> The new bioeconomy as currently envisioned by foresters, agribusiness, biotech, energy and chemical firms furthers the ongoing enclosure and degradation of the natural world by appropriating plant matter for transformation into industrial commodities, engineering cells so they perform as industrial factories, and redefining and refitting ecosystems to provide industrial support 'services'.

And proponents of the new bioeconomy target the global South, as Stephen Chu, US Secretary of Energy, observed in 2006: 'Land best suited for biomass generation (Latin America, Sub-Saharan Africa) is the least utilized' (quoted in ETC 2010, 15). A European report claimed in 2004: 'A prerequisite for the bioenergy potential in all regions is . . .that the present inefficient and low-intensive agricultural management systems are replaced in 2050 by the best practice agricultural management systems and technologies' (Smeets *et al.* 2004). This observation echoes World Bank rhetoric about 'yield gaps' as justification for the introduction of value-chain agriculture. Whether the commons or peasant farms, land and its living carbon bounty is the new target for the biomasters, as the limits of dead carbon (fossil fuel) become apparent. Rachel Smolker (2008, 519) notes that, beyond biofuels:

> Biomass is also increasingly in demand for heat and electricity production, chemicals, manufacturing and industrial processing, as well as an ever-expanding range of materials and products. *Agriculture is thus poised uniquely at both 'ends' of the debates on food and energy policies, as both a source of, and a solution to, the problems at hand.* [emphasis added]

The land grab, then, anticipates the rising value of living biomass as the source of inputs into the new bioeconomy, where 'innovation in synthetic biology is allowing companies to retrofit the hydrocarbon economy to accommodate carbohydrate feedstocks' (ETC 2010, 11). The US Department of Energy claims: 'there are very few products that are made today from a petroleum base, including paints, inks, adhesives, plastics and other value-added products, that cannot be produced from biomass' (quoted in Smolker 2008, 520).

On the face of it, this discourse suggests that the land grab is a vehicle of food/fuel regime restructuring, whereby the (profitability) projections and technologies of the new bioeconomy depend on increasing access to offshore production of biomass to power affluent economies. At the same time, previous geopolitical relations of the corporate food regime, anchored in the EU/US agro-exporting of bulk commodities, are shifting as Northern farm sectors lose their competitive advantage in a world market governed by new forms of neo-mercantilism, with subsidies favoring offshore agriculture where land, water and labor are substantially cheaper. Departing from, or complementing, previous patterns of investment in high-value export crops, the new investment patterns in the global South favor bulk commodities – thus, for Southeast Asia, '83% of the farmland being acquired or leased on a long-term basis is dedicated to the production of major row crops (soft oilseeds, corn, wheat and feed grains)' (Borras and Franco 2010, 31).[10] In an echo of Kautsky's predictions, the European food sovereignty movement has warned: if 'Europeans want to maintain an agricultural production in Europe, they need a European agricultural policy. Otherwise Brazil or other countries will produce the base of our food' (CPE 2006).

In short, the land grab registers an ongoing transformation of industrial agriculture and its postwar political coordinates, framed in a security discourse of feeding the world and saving the planet.[11] The emerging bioenergy economy, fusing global ecology and political economy depends on the enabling role of financialization in managing a spatio-sectoral shift in capital accumulation toward a new extractive food/fuel/biomass regime enclosing the world's remaining land and water. Whether and to what extent such a shift can underwrite a new revolution in capital accumulation is a matter of speculation (cf Moore 2010), particularly as climatic changes threaten the durability of any such developments.

Financialization and the land grab

The land grab coincides with the era of financialization – a conjuncture in which investors prefer to hold capital in liquid (rather than illiquid/asset) forms. Arrighi explains financialization as a consequence of last-ditch efforts by the US government during the 1980s, instituting rules promoting liberal capital markets and deregulating banking to attract capital flows to the US with rising interest rates in order to overcome the relative decline in its industrial productive capacity (2007, 145). Here, 'financialization' is symptomatic of a declining hegemon losing its geo-economic competitive edge, as its industrial capitalists switch investment from fixed capital into financial ventures. Preference for liquidity, intensified institutionally by neoliberal-led financial deregulation, has encouraged securitization (consolidating and selling debt), mergers (including firm acquisition by private equity companies that unbundle unprofitable units for financial gain) and general financial speculation.[12]

[10]Cotula (2011, 37) notes that some contracts (Sudan, Mali) 'appear to create no safeguards to ensure that local food security needs are met' – at odds with claims by host governments to improve domestic food security.

[11]Thus, 'The spectre of a hungry world is being used to push the agenda for industrial agriculture, but in reality, the majority of the land is used for producing animal feed and agrofuels, as well as land speculation, rather than food crops. A World Bank report on land acquisitions shows that only 37% of this land is used to grow food' (Henriques 2011).

[12]Parallel deregulation in the financial services industry thus enabled cross-over investments by banks, in addition to a process of concentration and centralization, such that between 1980

Financialization parallels a global decline in productivity outside of the information and communications technology sectors. Over the past four decades, manufacturing has steadily relocated to Southern regions of cheap labor (and land) through export processing, assembly and subcontracting systems. Northern consumption of such offshore products was sustained for a while by the banking revolution, involving profligate mortgage lending and rising consumer debt. By the twenty-first century, declining industrial productivity combined with a collapse of the financial derivatives market to generate an accumulation crisis. Investment capital shifted significantly into speculative ventures in land, food and biofuels – venture capital investment in biofuels increasing by 800 percent between 2004–2007 (Holt-Giménez 2007,10). Meanwhile, trade in agricultural futures and other derivatives increased in 2007 by 32 percent, and 'the value of commodities derivatives that are traded *over the Counter*, i.e. off-market, rose by nearly 160 percent between June 2005 and June 2007. The number of futures between October 2007 and the end of March 2008 increased by 65 percent on the Chicago Mercantile Exchange, without a corresponding increase in real production' (Bank for International Settlements, cited in Ernst and Wahl 2010, 13).[13] The resulting food price inflation, generating the 'food crisis' of 2007–2008, deepened investor attention to offshore crop-land.

Food speculation intensified via commodity index funds, whereby investors targeted 'agrofutures' (alongside energy and industrial metals)[14] as agricultural contracts were converted into derivatives, under pressure by financiers on legislators to deregulate the commodity contract business in the 1990s. Henceforth, speculators joined handlers of agricultural products in an agro-futures market. Thus what was once a market in food converted to a self-driven market in food contracts, counting on rising derivative prices, as futures traded multiple times. Formerly a mechanism of hedging risks on food prices for producers and consumers, which tended to reduce volatility, agro-futures went virtual, as financiers constructed commodity index funds allowing no-risk profit from price volatility – by shifting most clients' index fund investments to safer ventures and then profiting from rising, or declining, food prices (Kaufman 2010, 30–31). Buying and selling food futures, then, developed into a derivative market, which in turn inflated food prices. Speculation, enabled by computer automation, intensified in the mid-2000s as the real estate market crisis unfolded. At that point, investors shifted funds into commodity futures: between 2003 and 2008 commodity index holdings increased from \$13 billion to \$317 billion (Kaufman 2010, 32). This speculative spike resulted from the process whereby 'the mechanism created to stabilize grain prices had been reassembled into a mechanism to inflate grain prices' (Kaufman 2010, 34).[15]

---

and 1998 some 8,000 bank mergers occurred, accounting for assets of over \$2.4 trillion (Shattuck 2008).

[13]Thus, 'the excessive speculation in the financial commodity markets has seen a parallel increase in food prices. The increase between March 2003 and March 2008 of the agricultural commodities futures has been in parallel with the price increases during the same period for coffee with 167%, for soybean oil with 199% and for wheat with 314%' (Kerckhoffs *et al.* 2010, 7).

[14]As of July 2008, the Standard & Poor's-Goldman Sachs Commodity Index accounted for about 63 percent of the index fund market share, and a 32 percent share was held by the Dow Jones-AIG index – with agricultural commodities accounting for about 30 percent of these indices, and the rest in energy, base metals and precious metals (IATP 2008).

[15]Financial speculation compounded food price inflation, which spiked in 2008: rice prices

The general accumulation crisis, expressed through the conjunction of food, energy and financial crises, has resulted in international capital markets gravitating towards agriculture as a relatively safe investment haven for the relatively long-term. The trumping of 'hard commodities' (non-renewables, such as oil and metals) by 'soft commodities' (renewable crops) in the commodities investment market, in 2007, was one indicator of this conjuncture, driven as it was by a rising interest in bio-economic products (Daniel 2009, 5). In addition to food futures and crops, land and agriculture have constituted a new investment frontier in recent years:

> Financial investors have unleashed a wave of funds in the past three years, raising capital to invest across the entire agricultural value chain, from greenfield land sites to farmland to agribusiness and agro-processing all over Africa. This has taken the investment thesis way beyond portfolio investment in listed debt and equity markets, and from primary capital markets issuance. Making money from land and agricultural production demands a longer-dated approach – all of the new land funds have lock-ups, in some cases out to 10 years – but the returns to be had are potentially mouth-watering, from the mid-teens to upwards of 25 percent per annum.
>
> As well as financial investors aggressively focused on generating outsize returns, inflows into African land have attracted a significant volume of funds from a large number of multilateral development organizations such as the International Finance Corp (IFC), African Development Bank (AfDB) and OPEC Fund for International Development (OFID); single-country (bilateral) development finance agencies; food multinationals; foundations from around the world, as well as SRI funds. (Mullin 2011)

Under the banner, 'Food is Gold, So Billions Invested in Farming', *The New York Times* (Henriques 2008) detailed plans to capitalize farmland in the US, England, Argentina and Brazil. Sub-Saharan Africa was the target of Emergent Asset Management (UK), investing $450–750 million for food and biofuels. CEO Susan Payne explained Africa was targeted because 'land values are very, very inexpensive, compared to other agriculture-based economies'. In June, 2009, the Executive Director of JP Morgan, stated: 'Physical agriculture's assets are the new focus in longer term investments as institutional investors explore opportunities in everything from raw land to grain elevators to food processing plants' (quoted in Gillam 2009). The point about financialization is that it is not simply wealthy investors like Bill Gates, Warren Buffett, and George Soros and other financial interests (e.g. Louis Dreyfus, Merrill Lynch, and sovereign and pension funds) investing in agrofuels, but conglomerates in traditional sectors like oil, auto, chemicals and agribusiness[16] that deploy their financial resources to capitalize (on) the new fuel frontier.

---

surging by 31 percent on 27 March 27, and wheat prices by 29 percent on 25 February 2008. Diana Henriques wrote in *The New York Times* (22 April 2008), 'This price boom has attracted a torrent of new investment from Wall Street, estimated to be as much as $300 billion;' with the Commodity Futures Trading Commission noting that 'Wall Street funds control a fifth to a half of the futures contracts for commodities like corn, wheat and live cattle on Chicago, Kansas City and New York exchanges. On the Chicago exchanges… the funds make up 47 percent of long-term contracts for live hog futures, 40 percent in wheat, 36 percent in live cattle and 21 percent in corn' (quoted in Berthelot 2008).

[16]For example, Total, Shell, BP, Exxon-Mobil, Petrobras, ADM, Cargill, Bunge, Monsanto, Syngenta, Dow Chemicals, Bayer, DuPont, BASF, etc. (Houtart 2010, 131-2).

These reports signal a significant deepening of agro-industrialization in a crisis conjuncture.[17] While agro-industrialization has concentrated in the global North, powering a cheap (corporate) food regime, declining sustainability and rising costs have propelled movement offshore to exploit cheap inputs.[18] Just as industrial agriculture has experienced a declining biophysical productivity, with soil depletion and a drop in efficiency of nitrogen use from 60 to 20 percent from the 1950s to the 1990s (van der Ploeg 2010, 100), the expense of 'biophysical override' rises with the price of commodity inputs (Weis 2007), exacerbated by increased energy and irrigation requirements due to declining efficiency (van der Ploeg 2010, 100, Holt-Giménez 2007b, 10).[19] As agri-capital centralizes in response, augmented by financialization, it prefigures a move offshore to take advantage of cost reducing investments in Southern land, water and labor. This contributes to the land grab, spurred by financial speculation and anticipation of risk evident in climate change[20] – a risk which will only deepen given *extant* climate change projections for Africa (Toulmin 2009).

The consolidation of a 'world agriculture' (McMichael 2005) or the 'interchangeability of large agricultural systems' (van der Ploeg 2010, 101) based on industrial crops, or high-input contract farming, is the result. The familiar contours of the food regime, anchored as it has been in subsidized Northern agro-industry, at the expense of Southern farming, are reconfiguring as agro-industrial restructuring re-spatializes the food regime through expanding offshore food/fuel supply zones.

Capital's frontier

Rising food prices, peaking oil, emission mandates and stalled investment funds find material resolution in the land grab, accompanied by an ideology of enclosure ('global ecology') in the name of humanity (food) and the environment (green fuel). Whether agricultural investments can resolve the general crisis of capital accumulation is in question, but the short answer may be that the logic of financialization is to privilege futures over productivity gains. Certainly there is a development agency rhetoric regarding the 'yield gap' between attainable and potential yields in agriculture on Southern lands. For example, the World Bank claims 'none of the African countries of most interest to investors is now achieving more than 30 percent of the potential yield on currently cultivated areas' (2010, vii). And the European Commission advocates land reforms to address this gap:

---

[17]Weis (2010, 327) notes: 'Just under half of the world's total grain production (48 percent) is directly consumed by humans, while 35 percent is fed to livestock and 17 percent to biofuel production. The surge in the latter two comes at a time when the yield gains associated with the Green Revolution have effectively maxed out, and the volume of per capita grain production on a global scale has been level since peaking in 1986'.

[18]Steven Blank (1998) was one of the first to draw attention to this.

[19]By 2030, it is predicted that the global supply of phosphorus, a key plant nutrient, will peak. In combination with rising oil prices, the upward trending of inorganic fertilizer will consolidate (Cordell 2009).

[20]Thus a GRAIN researcher notes: 'Rich countries are eyeing Africa not just for a healthy return on capital, but also as an insurance policy. Food shortages and riots in 28 countries, declining water supplies, climate change and huge population growth have together made land attractive' (Vidal 2010).

> Secure access to land and secure land tenure and use rights are prerequisites for higher productivity of small holder farmers. Effective national land policies and laws are essential, requiring governments to take priority action on land. Where countries develop policies on agriculture, land, and biofuels, the EU and its Member States should advocate that these policies address concerns over availability and access to food and stimulate the integration of smallholder farmers in production chains. (Quoted in Borras and Franco 2011, 40)

The question of a 'yield gap' is a euphemism for an extractive form of agriculture that caters to (inefficient, climate-threatening) overconsumption by a global minority and would further jeopardize the 'under-reproduction' of smallholding populations (Araghi 2009). Such 'biomass' is produced as a world/corporate product, not for local or domestic food/fuel sovereignty. Whether or not this results in a significant yield increase (even using biotechnology), insofar as it promotes food exports for consumers elsewhere, it is not 'development' for producing regions so much as for investors. And the assumption that high-input value-chain agriculture will resolve the 'yield gap' is grossly misleading. There is a growing international scientific consensus that small-scale farming is as, if not more, productive than industrial agriculture (*cf.* Pretty *et al.* 2006, Badgley and Perfecto 2007, Hamer and Anslow 2010), and, further, smallholder farming along agro-ecological lines is more environmentally sustainable than industrial agriculture (Altieri 2010, Altieri and Toledo 2011). Nevertheless, land and agriculture appear to be an immediate answer to the accumulation crisis, but tellingly it depends fundamentally on Northern subsidies to agribusiness, energy and transport companies, and Southern concessions to investors.[21]

Arguably, it does not appear that agrofuels themselves, and indeed biomass production in general, will resolve the accumulation crisis (Moore 2010, 497). To begin with, capitalizing grass- and forest-land with agro-inputs degrades the natural foundations of production. Global fertilizer production has increased over 31 percent since 1996 – a trend now intensified by agrofuels and the removal of cellulose fiber from fields (ETC 2009). In addition, it is questionable whether there is sufficient biomass to convert into renewable chemicals, plastics and fuels to realize the open-ended claims of the bio-economic vision of such entities as governments, the US military and the chemicals and power industries (ETC 2009). By 2030, the International Energy Agency estimates agrofuels will 'barely offset the yearly increase in global oil demand' (Holt-Giménez 2007a), and all renewables, including agrofuels, will amount to only nine percent of global energy consumption (GRAIN 2007, 6). When agrofuels displacement of food crops is paired with financial speculation on agricultural commodities in general, the possibility for capital using the land grab frontier as an open-ended source of cheap energy and food resources to

[21]Houtart's suggestion that agrofuels 'have come just in time to revive the prices of agricultural products and their role as a financial refuge in times of crisis' (2010, 128) is a partial explanation of this phenomenon. Agribusiness profited substantially from the 'cheap food regime' – providing low priced commodities for subsidized trading and processing, contributing to a process of concentration and centralization of agri-capital. Agrofuels, heavily subsidized with public monies, offer alternative financial outlets, simultaneously raising the price of foodstuffs. Arguably, agrofuels amplify the centralization of agri-food capital, via financial agglomeration and recombinant capital, as energy, chemical, auto and biotechnology capitals join the rush for cheap land. In 2007, biofuels were the fastest growing segment of the world agricultural market (ETC 2007, 2).

reduce the costs of production and reproduction, respectively, is likely to be short-lived (but nonetheless devastating).

Land may be cheap to investors, but it has its own value for its local inhabitants – in particular, occupied and/or common lands provide the possibility of subsistence and socio-ecological resilience for millions across the world, in addition to having ancestral and spiritual value. Once the concept of a 'global commons' becomes the *modus operandi*, agency, government and investor acquisition of land devalues its local cultural and ecological functions. Eviction of 'unproductive' populations becomes the basis of 'rational planning' – driven by claims for increased productivity, debt-reduction, export enhancement and rural development.[22] In one example in September 2011, Oxfam reported the eviction (with government support) of 20,000 inhabitants in Uganda to enable the New Forests Company (UK) to plant pine and eucalyptus forests on their land – as a source of carbon credits to sell to polluters elsewhere. Converting habitats into profits for the false economy of climate mitigation exemplifies the imposition of a 'global ecology' rationale for land colonization, financed in this case by investors such as the World Bank's IFC, and the Hongkong and Shanghai Banking Corporation, HSBC (Kron 2011).

The Ugandan eviction is emblematic of *state-managed* enclosure, extending subsidies of cheap/free land to investors at the expense of the social reproduction rights of smallholders. Biomass production, or carbon offsets, may represent capital's new frontier, but as with most frontier expansions it is subsidized by home and host governments.[23] Public subsidies for land grabbing contribute to a composite set of 'externalized' environmental, social, cultural and human rights costs. Displacing the social and intrinsic value of such habitat eventually recycles as monetary costs of resettlement, food shortages and ecosystem depletion for governments and development agencies. In other words, the 'external' costs of doing agribusiness multiply, with global warming and ecosystem degradation – following the capitalization of nature (via land grabs, oil palm plantations, GM seeds, etc) – combining to undermine the conditions of capital accumulation in the long run.[24]

In these senses, the land grab is not merely a reflex to resolve an accumulation crisis via investment fund management within a subsidy regime. It sacrifices land and its inhabitants to a financial calculus represented as a necessary global good (food yields, green fuels, and even carbon offsets). As expressed in innumerable reports in the media, journals, and NGO outlets, the land grab effectively authorizes large-scale removal of rural populations from ancestral lands to install 'agriculture without farmers' – as the international peasant coalition *La Vía Campesina* calls agro-industrialization. In this process, biomass-driven land grabbing substitutes management of an accumulation crisis for the sustainability of human and natural ecology.

[22]For example, in Colombia between 2001–2005, 263,000 peasant families were expropriated from 2.6 million hectares by agrobusiness and/or paramilitaries interested primarily in oil palm development (Houtart 2010, 107; see also Grajales 2011). Houtart (2010, 119) claims that 60 million people risk expulsion by biofuels.

[23]According to Friends of the Earth and EarthTrack, the combination of the Renewable Fuels Standard Mandate (which provides a market for biofuels) with tax credits would subsidize the US biofuels industry to the tune of $400 billion through 2022 (www.foe.org/biofuelsubsidies). Analyst Bloomberg New Energy Finance reported that 'in 2009 governments provided subsidies worth between $43bn (£27bn) and $46bn to renewable energy and biofuel industries, including support provided through feed-in tariffs, renewable energy credits, tax credits, cash grants and other direct subsidies' (Business Green 2010).

[24]For O'Connor (1998) this process represents the 'second contradiction of capitalism'.

At the same time, it foreshadows a 'biomass regime' in the making, based in a general offshoring of agriculture, and the emergence of new South–North, East–North and East–South,[25] and South–South transfers of agri-products.

Land grab ideology and governance as food regime infrastructuring

Assisted by World Bank policy and proliferating forms of governance, the land grab is represented as a form of development, insofar as land 'development' is associated with productivity gains and employment, and indebted governments in the global South stand to receive foreign investment and hard currency from conversion of their land and forests into agro-export platforms. Cotula's research suggests that host states expect investment in infrastructure to develop landed property, noting however that leasing land for free or at less than market rents encourages speculative investment given the long contracts and lease transferability (2011, 22–24). In other words, integrated rural development is neither the intention nor is it likely to be the result.

Inherent in the development narrative is the notion that subsistence or near-subsistence producers are necessarily poor and would benefit from jobs. Cash is viewed as the currency of modernity, identifying wealth with money, rather than intact habitat and common lands and the security of landholding. The World Bank's report, *Rising Global Interest in Farmland,* views large-scale land acquisition as a vehicle for poverty reduction via rural employment, contract farming, and selling or renting (World Bank 2010). Li's critique of these claims, with the Bank's own data on employment, shows that while the report claims oil palm employs 1.7–3 million people on 6 million hectares, 'field data indicates that an established plantation uses only one worker per four to ten hectares of land, depending on the efficiency and stage of production' (2011, 284). For fuel crops, estimates are that in tropical regions, '100 hectares dedicated to family farming generates 35 jobs. Oil-palm and sugarcane provide 10 jobs, eucalyptus two, and soybeans a scant half-job per 100 hectares, all poorly paid... Hundreds of thousands [of smallholders] have already been displaced by the soybean plantations in the "Republic of Soy", a 50m hectare area in southern Brazil, northern Argentina, Paraguay, and eastern Bolivia' (Holt-Giménez 2007,10). And Cotula's research documents the vagueness about local employment contracts (2011, 25–26).

Land enclosure in the global South revitalizes a long-standing (but institutionally dormant) modernization trope, namely that industrialization of agriculture is necessary to development. In the development era, the industrial bias was in part shaped by the intervening food regimes, which subsidized Third World manufacturing with cheap food imports from the US and European 'breadbaskets', undermining peasant farming in general, alongside targeting strategic states (e.g. India, Pakistan, the Philippines, Indonesia, Vietnam, Egypt, Turkey, Brazil, Mexico, Argentina) with a green revolution that selected for farmers with the resources to adopt the technological package. Elsewhere (particularly Africa), food dependence expanded as food corporations obtained privileged access to domestic markets via

[25]Visser and Spoor (2011) rightly underscore the strategic significance of Eastern European lands (notably in Kazakhstan, the Ukraine and Siberia) where Northern investors and Southern states grab large agricultural land reserves – allowing land grabbing food regime circuits their managed universality.

WTO rules (McMichael 2005). Provisioning through the global market, in a neoliberal form of 'food security', is now in question, given the protectionism arising from the 2007–2008 food crisis. This interruption, along with the shift of financial capital from manufacturing into agro-food futures and land and agriculture, not only provides the development industry with a new crusade, but also portends a reconfiguration of the familiar patterns of the so-called 'cheap food regime'.

The World Bank's 2008 *World Development Report*,[26] advocating 'agriculture for development', was the first time in 25 years that agriculture commanded the attention of this key development institution. Here, the urgency of the food and energy crises has refocused the global political-economic elite's vision around mobilizing agricultural resources to offset food, water and fuel shortages. Southern agricultural land is a particular target for 'productivity increase' via technification. Thus Susan Payne, CEO of Emergent Asset Management (a UK investment fund planning to spend $50 million on African land), declared, 'Farmland in sub-Saharan Africa is giving 25% returns a year and new technology can treble crop yields in short time frames... Agricultural development is not only sustainable, it is our future. If we do not pay great care and attention now to increase food production by over 50% before 2050, we will face serious food shortages globally' (quoted in Vidal 2010).

In Africa and Asia, most land is state land but communally held, and as such is subject to government designation as 'idle' land,[27] given potential rewards of commercialization. Unsurprisingly, international development and financial institutions are working behind the scenes on privatizing land relations to attract and enable foreign investment in African land. The US government's Millennium Challenge Corporation (MCC) encourages investment, disbursing money in the form of grants to particular countries on condition that they meet certain neo-liberal economic criteria. Most MCC Compacts signed with African countries focus on agriculture, with a central land privatization component, supporting 'market-based solutions to food security'. Such provisions include certifying outgrowers for food exports, constructing infrastructure to gain access to world markets, and partnering with The Alliance for a Green Revolution in Africa (AGRA) to provide inputs to farmers in their first year (GRAIN 2010). The Gates Foundation (financing AGRA) suggests that enabling the commercial development of African agriculture 'will require some degree of *land mobility* and a lower percentage of total employment involved in direct agricultural production' – foretelling an eviction trajectory (quoted in Xcroc 2009). The modernization trope of agricultural productivity, enforcing a narrative of depeasantization, deepens via the radical decoupling of urbanization from industrialization, intensifying the 'planet of slums' phenomenon (Davis 2006) at a time when global development (framed as 'global ecology') via the land grab overrides any remaining pretense of national development.

Displacement of rural peoples is anticipated as a byproduct of land acquisition and subject to management according to IFC Performance Standards on Social and Environmental Sustainability. For example, Performance Standard #5: Land Acquisition and Involuntary Resettlement refers 'both to physical displacement (relocation of loss of shelter) and to economic displacement (loss of assets or access to assets that leads to loss of income sources or means of livelihood) as a result of project-related land acquisition' (quoted in Daniel 2010, 49). However, Performance

[26]For critical reviews, see the *Journal of Agrarian Change*, 39, 6 (2008).
[27]This is also the case elsewhere, such as in Southeast Asia – see e.g. Cotula *et al.* (2008).

Standard #7 suggests that 'Private sector projects may create opportunities for Indigenous Peoples to participate in, and benefit from, project-related activities that may help them fulfill their aspiration for economic and social development' (Daniel 2010, 50). Here, eviction is a prelude to opportunity in the cash economy, affirming the violence of development. In sanctioning the violence of dispossession via development, such guidelines also consolidate the privatization of states, which in cases such as Colombia involve direct deployment of paramilitary violence to accomplish this process (see Grajales 2011).

Promoters of the global land grab include the World Bank, its International Finance Corporation (IFC),[28] the International Rice Research Institute (IRRI) of the Consultative Group on International Agricultural Research (CGIAR), the European Bank for Reconstruction and Development (IBRD) and others, with particular focus on Sub-Saharan Africa. In addition to direct investment by the Bank's IFC in agribusiness operations, the IFC's partner, The Foreign Investment Advisory Service (FIAS), targets 'investment climates' in foreign markets, creating land registries and easing the process of land titling, leasing and foreign investment. The greater possibility of accomplishing this, according to the International Institute for Environment and Development (IIED), stems from countries lacking 'sufficient mechanisms to protect local rights and take account of local interests, livelihoods, and welfare' (quoted in Houtart 2010, 17).

The rush to acquire land is matched by a rush to institute para-statal and private systems of governance – all focused on an extractive model of development justified by managed principles of comparative advantage. Thus the FIAS initiative, Investing Across Borders (IAB), conducted project surveys in 87 countries in 2009, targeting information regarding technical regulatory and licensing information only – disregarding potential human impact, and focusing on mapping of comparative investment climates with respect to land holding patterns, power structures and state capacities (Daniel 2010, 15, 17–18).

Such development 'services' constitute a broad infrastructural complex supporting land grabbing – both material and ideological. Insofar as a food regime has an institutional framework, governed by implicit rules appealing to normative understandings of a developmentalist ordering of the world (Friedmann 2005, 234), these 'services', with emerging 'guidelines', register an institutional updating of the corporate food regime.

WTO rules institutionalized a 'cheap food regime' that sanctioned corporate subsidies (hidden in 'boxes' protocols), legitimizing continuation of Northern food dumping from the previous food-aid regime. However, current institutional trends suggest a restructured framework, with two key dimensions. First, a multi-centric complex of rules and codes of conduct emerging via the development community at large (including influential NGOs), but centered in the UN organizations (notably the FAO) and the International Financial Institutions (notably the World Bank) concerning management of farmland acquisition and technical assistance. These are complemented by public–private partnerships to finance agribusiness; bilateral agreements on land access; emerging climate protocols such as the EU's Emission Trading System, the Clean Development Mechanism and REDD+, which sanction appropriation of land and forests

[28]IFC expenditures in Sub-Saharan Africa rose from $167 million in 2003 to $1.8 billion in 2009 (Daniel 2010, 12).

in the global South in particular as carbon sinks (distinct forms of land alienation[29]); and platforms for green fuels (including Round Tables for certification). A second dimension of this emerging framework involves promoting a reversal of patterns of circulation centered on Southern agro-exporting of food, fuel and general biomass. Access to cheap land, water and labor is the foundation of such a regime and its normative vision of agricultural modernization, enhanced food production, smallholder incorporation into value chains, rural employment and smart agro-technologies (McMichael and Schneider 2011). The food regime and its 'global ecology' ideology thus sanctions a final enclosure of the commons in the name of food security and saving the planet from emissions and land degradation by under-resourced peasants.

Conclusion

This paper situates the land grab in a broader restructuring of the corporate food regime – from food-surplus to food-deficit relations. While the former relations impoverished peasant cultures via an ethos of cheap food for the world ('food security'), the latter promises to accelerate dispossession in the name of managing endemic food insecurity ('food crisis') resulting from the cheap food regime's neglect of domestic food security mechanisms and destabilization of populations, environments and the climate. The land grab is the medium through which the development agencies attempt to renew their legitimacy (constructing voluntary codes of conduct) in the face of a rising food sovereignty movement, and through which finance capital can profit even as capitalism enters a profound crisis of political legitimacy, and energy and environmental limits.

Stemming rising costs of inputs (especially energy and food) depends on a landed frontier of accumulation. In the new bio-economy, the basic ingredients are indiscriminate crop production via a process of enhanced enclosure of Southern land, accompanied by a normative appeal to securing world food and green fuel supplies at a time of crisis. The Bank's 'agriculture for development' is an afterthought that presents as a new development strategy, even as it sanctions land grabbing for the security of 'capital' in the name of a reformulated global ecology. 'Rational planning of the planet' for security purposes remains the basic rationale, although as argued earlier, there is no such security to be had, and the land grab – to the extent that it is incapable of recognizing the salience of low-carbon bio-diverse agriculture – is the ultimate death wish as industrial biofuels and value-added agriculture will not resolve the combined problems of climate change and food insecurity. They will only buy time (and space!) in the short run for political and economic elites and consumers with purchasing power. In this scenario the longer run is destined to be catastrophic.

References

Akram-Lodhi, H. 2008. (Re)imagining agrarian relations? The World Development Report 2008: Agriculture for Development. *Development and Change*, 39(6), 1145–1611.

[29]Analogously, biodiversity conservation projects alienate land and resource rights from forest-dwellers (see e.g. Corson 2011 and Kelly 2011).

Altieri, M. 2010. Scaling up agroecological approaches for food sovereignty in Latin America. *In:* H. Wittman, A.A. Desmarais and N. Wiebe, eds. *Food sovereignty. Reconnecting food, nature and community*. Halifax: Fernwood.

Araghi, F. 2000. The great global enclosure of our times: peasants and the agrarian question at the end of the twentieth century. *In:* F. Magdoff, J.B. Foster and F.H. Buttel, eds. *Hungry for profit: the agribusiness threat to farmers, food, and the environment*. New York: Monthly Review Press.

Araghi, F. 2009. Accumulation by displacement: global enclosures, the food crisis, and the ecological contradictions of capitalism. *Review*, XXXII(1), 113–146.

Arrighi, G. 1994. *The long twentieth century. Money, power and the origins of our times*. London: Verso.

Arrighi, G. 2007. *Adam Smith in Beijing. Lineages of the twenty-first century*. London: Verso.

Badgley, C. and I. Perfecto. 2007. Can organic agriculture feed the world? *Renewable Agriculture and Food Systems*, 22(2), 80–85.

Berthelot, J. 2008. Sorting the truth out from the lies about the explosion of world agricultural prices, *Solidarité*, 18 May. Available from: http://solidarite.asso.fr. [Accessed 12 January 2011]

Berthelot, J. 2009. Agribusiness' headlong flight to agrofuels and their impact on food security. *Solidarité*. Available from: http://solidarite.asso.fr. [Accessed 12 January 2011]

Birch, K., L. Levidow and T. Papaioannou. 2010. Sustainable capital? The neoliberalization of nature and knowledge in the European knowledge-based bio-economy. *Sustainability*, 2, 2898–2918.

Blank, S.C. 1998. *The end of agriculture in the American portfolio*. Westport, CT: Quorum Books.

Borras, S. Jr. 2010. The politics of transnational agrarian movements. *Development and Change*, 41(5), 771–803.

Borras, S. Jr. and J. Franco. 2010. From threat to opportunity? Problems with the idea of a 'code of conduct' for land-grabbing. *Yale Human Rights & Development L.J.*, 13, 507–523.

Business Green. 2010. Fossil fuel subsidies are 10 times those of renewables, figures show. *The Guardian*, Aug 3.

Carlsen, L. 2003. The Mexican farmers' movement: Exposing the myths of free trade. *In: Americas Program Policy Report*. Silver city, NM: Interhemispheric Resource Center. Available from: www.americaspolicy.org. [Accessed 15 December 2010]

Committee on World Food Security (CFS), 2011. Land tenure and international investments in agriculture. A Report by the High Level Panel of Experts on Food Security and Nutrition. July. Rome: Committee on World Food Security (FAO).

Cordell, D. 2009. A new global challenge for food security: peak phosphorus. *The Organic Way Magazine*, 2, 10–12.

Corson, C. 2011. Territorialization, enclosure and neoliberalism: non-state influence in struggles over Madagascar's forests. *The Journal of Peasant Studies*, 38(4), 703–726.

Cotula, L. 2011. *Land deals. What's in the contracts?* London: IIED.

Cotula, L., N. Dyer and S. Vermeulen. 2008. *Fuelling exclusion? The biofuels boom and poor people's access to land*. International Institute for Environment and Development (IIED) and Food and Agricultural Organization (FAO).

Daniel, S. 2009. *The great land grab. rush for world's farmland threatens food security for the poor*. The Oakland Institute.

Daniel, S. 2010. *(Mis)Investment in agriculture. The role of the international finance corporation in global land grabs*. Oakland, CA: The Oakland Institute.

Davis, M. 2006. *Planet of slums*. London & New York: Verso.

De Schutter, O. 2010. Responsibly destroying the world's peasantry, 4 June. Available from: http://farmlandgrab.org/13528. [Accessed 12 January 2011]

Edelman, M. and J. Carwill. 2011. Peasants' rights and the UN system: Quixotic struggle? Or emancipator idea whose time has come? *The Journal of Peasant Studies*, 38(1), 81–108.

Ernst, C., and P. Wahl. 2010. *Simply collateral damage? The financial crisis and the developing countries*. Berlin: World Economics, Environment & Development.

ETC, 2007. 'Peak soil + peak oil = peak spoils,' etc group Communique, #96. November/December, 1–14.

ETC, 2009. Biomass power grab highlighted as biotech industry meets in Montreal. *News Release*. Available from: http://www.etcgroup.org/en/node/4696 [Accessed 12 January 2011]

ETC, 2010. The new biomasters. Synthetic biology and the next assault on biodiversity and livelihoods. Available from: www.etcgroup.org [Accessed 12 January 2011]

Fargione, J., J. Hill, D. Tilman, S. Polasky and P. Hawthorne. 2008. Land clearing and the biofuel carbon debt, *Science*, 319(5867), 1235–1238.

Ferrett, G. 2007. Biofuels 'crime against humanity'. *BBC News*, October 27. Available from: http://news.bbc.co.uk/2/hi/7065061.stm. [Accessed 3 June 2009]

Fraser, A. 2009. Harnessing agriculture for development. Oxfam International Research Report, September. Available from: http://www.oxfam.org/en/policy/harnessing-agriculture-development. [Accessed 15 December 2010]

Friedmann, H. 1978. World market, state, and family farm: social bases of household production in an era of wage labor. *Comparative Studies in Society and History*, 20(4), 545–586.

Friedmann, H. 2005. From colonialism to green capitalism: social movements and emergence of food regimes. *In:* F.H. Buttel and P. McMichael, eds. *New directions in the sociology of global development*. Amsterdam: Elsevier, pp. 229–267.

Friedmann, H. and P. McMichael. 1989. Agriculture and the state system: The rise and fall of national agricultures, 1870 to the present. *Sociologica Ruralis*, 29(2), 93–117.

Gallagher, E. 2008. The Gallagher review of the indirect effects of biofuels production. UK Government: Renewable Fuels Agency.

Gillam, C. 2009. Investors eye global ag sector for boost, *Reuters*. 21 June. Available from: http://globlalandgrab.org/5695. [Accessed 12 January 2011]

Grajales, J. 2011. The rifle and the title: paramilitary violence, land grab and land control in Colombia. *The Journal of Peasant Studies*, 38(4), 771–791.

Gouverneur, C. 2009. The palm oil land grab, *Le Monde diplomatique*, 5 December.

GRAIN, 2007. Agrofuels (special issue). *Seedling*, July. Available from: http://www.grain.org/seedling/?type=68&l=0 [Accessed 19 February 2009].

GRAIN, 2008. Seized: The 2008 land grab for food and financial security. *GRAIN Briefings*, October. Available from: http://www.grain.org/briefings/?id = 212. [Accessed 3 June 2009]

GRAIN, 2010a. Turning African farmland over to big business: The US's Millennium Challenge Corporation, *Seedling*, 3–5 April.

GRAIN, 2010b. The World Bank in the hot seat, *Against the Grain*, 4 May. Available from: www.grain.org/articles/?id=64 [Accessed 9 August 2010].

Green, D. 2011. The latest (big) numbers on land grabs, and some powerful case studies. Weblog entry. From poverty to power. 22 Sept. Available from: http://www.oxfamblogs.org/fp2p/?cat=42 [Accessed 16 January 2012].

Greenpeace, 2007. How the palm oil industry is cooking the climate. Available from: www.greenpeace.org [Accessed 8 January 2010].

Hall, D. 2011. Land grabs, land control, and Southeast Asian crop booms. *The Journal of Peasant Studies*, 38(4), 837–858.

Hamer, E. and M. Anslow. 2010. Ten reasons why organic can feed the world. *The Ecologist*, 1 March. Available from: www.theecologist.org. [Accessed 12 January 2011]

Henriques, D. 2008. Price volatility adds to worry on U.S. farms. *The New York Times*, April 22.

Henriques, D. 2008. Food is gold, so billions invested in farming. *The New York Times*, June 5.

Hildyard, N. 1993. Foxes in charge of the chickens. *In:* W. Sachs, ed. *Global ecology*. London: Zed Press, pp. 22–35.

Holt-Giménez, E. 2007a. Biofuels: myths of the agro-fuels transition. *Food First Backgrounder*, 13(2). Available from: http://www.foodfirst.org/node/1711 [Accessed 19 February 2009].

Holt-Giménez, E. 2007b. Exploding the biofuel myths. *Le Monde diplomatique*, July 2007, 10–11.

Holt-Giménez, E. and I. Kenfield. 2008. When renewable isn't sustainable: agrofuels and the inconvenient truth behind the 2007 US Energy Independence and Security Act. Policy Brief, No 13, Oakland, CA: Institute for Food and Development Policy.

Houtart, F. 2010. *Agrofuels. Big profits, ruined lives and ecological destruction*. London & New York: Pluto Press.

IATP, 2008. Commodities market speculation: The risk to food security and agriculture. Institute for Agriculture and Trade Policy (IATP) Trade and Governance Program: 7.
International Energy Agency, 2008. From 1st- to 2nd-generation biofuel technologies. An overview of current industry and RD&D activities. Paris: OECD/IEA.
Kaufman, F. 2010. The food bubble. How Wall Street starved millions and got away with it. *Harper's Magazine*, July, 27–34.
Kautsky, K. 1988. *The agrarian question*, Vol. 2. (1899), London: Zwan Publications.
Kelly, A.B. 2011. Conservation practice as primitive accumulation. *The Journal of Peasant Studies*, 38(4), 683–701.
Kerckhoffs, T., R. van Os and M. Vander Stichele. 2010. *Financing food. Financialization and financial actors in agricultural commodity markets.* Amsterdam: Stichting Onderzoek Multinationale Ondernemingen.
Kron, J. 2011. In scramble for land, group says, company pushed Ugandans out. *The New York Times*, September 21. Available from: http://www.nytimes.com/2011/09/22/world/africa/in-scramble-for-land-oxfam-says-ugandans-were-pushed-out.html?_r=1&emc=eta1. [Accessed 21 September 2011]
Levidow, L. 2011. *Agricultural innovation: sustaining what agriculture? For what European bio-economy?* Co-operative Research on Environmental Problems in Europe (CREPE). Open University.Available from: http://dpp.open.ac.uk. [Accessed 11 August 2011]
Li, T.M. 2011. Centering labor in the land grab debate. *The Journal of Peasant Studies*, 38(2), 281–298.
MacFarquhar, N. 2010. African farmers displaced as investors move in. *The New York Times*, 22 December, A1, A4.
Madeley. J. 2000. *Hungry for trade*. London: Zed Press.
Mathiason, N. 2010. Shell hooks up in Brazil on biofuels. *The Guardian Weekly*, 12, February, 43.
McMichael, P. 2005. Global development and the corporate food regime. *In:* F.H. Buttel and P. McMichael, eds. *New directions in the sociology of global development*. Amsterdam: Elsevier, pp. 265–300.
McMichael, P. 2009a. A food regime genealogy. *The Journal of Peasant Studies*, 36(1), 139–170.
McMichael, P. 2009b. The agrofuels project at large. *Critical Sociology*, 35(6), 825–839.
McMichael, P. 2009c. A food regime analysis of the 'world food crisis'. *Agriculture and Human Values*, 26(4), 281–295
McMichael, P. 2009d. Food sovereignty, social reproduction, and the agrarian question. *In:* A. Haroon Akram-Lodhi and Cristobal Kay, eds. *Peasants and globalization, political economy, rural transformation and the agrarian question*. Routledge, pp. 288–312.
McMichael, P. 2010. Agrofuels in the food regime. *The Journal of Peasant Studies*, 37(4), 609–630.
McMichael, P. and M. Schneider. 2011. Food security politics and the Millennium Development Goals. *Third World Quarterly*, 32(1), 119–139.
Merian Research. 2010. The vultures of land grabbing. The involvement of European financial companies in large-scale acquisition abroad. Available from: www.regulatefinanceforde-velopment.org. [Accessed 12 January 2011]
Moore, J. 2010. The end of the road? *Journal of Agrarian Change*, 10(3), 389–413.
Mullin, K. 2011. Land grab or opportunity of a lifetime? *Reuters*, January 30. Available from: http://farmlandgrab.org/post/view/18078. [Accessed 17 August 2011]
Murphy, S. 2008. Will free trade solve the food crisis? *Food Ethics Council*, 3(2). Available from: www.foodethicscouncil.org. [Accessed 22 March 2011]
O'Connor, J. 1998. *Natural causes*. New York: Guilford Press.
Oxfam. 2007. Bio-fuelling poverty. Why the ET renewable-fuel target may be disastrous for poor people. *Oxfam Briefing Note*, November 1.
Patel, R. 2007. *Stuffed and starved: markets, power, and the hidden battle for the world food system*. London: Portobello Books
Philpott, T. 2007. Bad wrap. *Grist*, February 22.
Pritchard, B. 2009. The long hangover from the second food regime: a world-historical interpretation of the collapse of the WTO Doha Round. *Agriculture and Human Values*, 26(4), 297–307.

Rosset, P. 2008. Food sovereignty and the contemporary food crisis. *Development*, 51(4), 460–463.
Sachs, W., ed. 1993. *Global ecology*. London: Zed Press.
Shattuck, A. 2008. The financial crisis and the food crisis: two sides of the same coin. *Food First*, 24 September.
Smeets, E. *et al.* 2004. A quickscan of global bio-energy potentials to 2050. *Bio-EnergyTrade*, March. Available from: www.bioenergytrade.org/downloads/smeetsglobalquickscan2050.pdf. [Accessed 12 January 2011]
Smolker, R. 2008. The new bioeconomy and the future of agriculture. *Development*, 51(4), 519–526.
TNI [Transnational Institute]. 2007. Agrofuels: Towards a reality check in nine key areas. June. Available from: http://archive.corporateeurope.org/docs/AgrofuelsRealityCheck.pdf [Accessed 19 February 2009].
Toulmin, C. 2009. *Climate change in Africa*. London: Zed Press.
van der Ploeg, J.D. 2009. *The new peasantries: struggles for autonomy and sustainability in an era of empire and globalization*. London: Earthscan.
van der Ploeg, J.D. 2010. The food crisis, industrialized farming and the imperial regime. *Journal of Agrarian Change*, 10(1), 98–106.
Vidal, J. 2007. Climate change and shortages of fuel signal global food crisis. *The Guardian Weekly*, 11 September, p. 3.
Vidal, J. 2010. How food and water are driving a 21st-century African land grab. *The Observer*, 7 March.
Vidal, J. 2011. Global effects of GM crops questioned. *The Guardian Weekly*, 28 October, p. 7.
Visser, O. and M. Spoor. 2011. Land grabbing in post-Soviet Eurasia: the world's largest agricultural land reserves at stake. *The Journal of Peasant Studies*, 38(2), 299–324.
Weis, T. 2007. *The global food economy. The battle for the future of farming*. London & New York: Zed Press.
Weis, T. 2010. The accelerating biophysical contradictions of industrial capitalist agriculture. *Journal of Agrarian Change*, 10(3), 315–41.
Wetter, K.J. 2009. Synthetic biology may usher in a post-petroleum era, but it won't be green. *GeneWatch*. Available from: http://www.councilforresponsiblegenetics.org/GeneWatch/GeneWatchPage.aspx?pageId=45&archive=yes [Accessed 12 January 2011]
Woodhouse, P. and A.S. Ganho. 2011. Is water the hidden agenda of agricultural land acquisition in sub-Saharan Africa? Paper presented at the International Conference on Global Land Grabbing, University of Sussex, 6–8 April 2011.
World Bank, 2007. World Development Report 2008: agriculture for development. Washington, DC: World Bank.
World Bank. 2010. Rising global interest in farmland. Can it yield sustainable and equitable benefits? Washington, DC: World Bank.
Xcroc. 2009. AGRA & Monsanto & Gates, green washing & poor washing. *Crossed Crocodiles*, April 6. Available from: http://crossedcrocodiles.wordpress.com/2009/04/06/agra-monsanto-gates-green-washing-poor-washing/.
Zoomers, A. 2010. Globalisation and the foreignisation of space: seven processes driving the current global land grab. *The Journal of Peasant Studies*, 37(2), 429–448.

Philip McMichael is a Professor of Development Sociology at Cornell University. His research is on the contemporary agrarian question, including land grabbing and its legitimating discourses, and agrarian resistances. He has authored *Development and social change: a global perspective* (Sage, 2012, 5th edn.), and edited *Contesting development: critical struggles for social change* (Routledge, 2010). He has also worked with the FAO, UNRISD, the IPC for Food Sovereignty and La Vía Campesina.

Situating private equity capital in the land grab debate

Shepard Daniel

This paper examines the private equity investment landscape in African land and agriculture markets by exploring two underlying questions: (1) Within the global land grab trend, how is power manifested for private equity investors? and (2) What are the development implications of private equity-backed investments? The essay begins with an overview of the recent rise of private equity in African land markets. It then analyzes the power relations embedded in this trend, particularly exploring the issues of information asymmetry and 'creative destruction' inherent in private equity finance, and their implications for governance and labour. Finally, the paper examines the power dynamics that characterize the relatively new relationship between private equity groups and development finance. Drawing on several illustrations of the World Bank Group's involvement in private equity-backed land investments, this paper demonstrates how the overlapping interests between development finance and private equity groups, in part, explain private equity's presence in emerging farmland markets.

1. Introduction

The phrase 'global land grab' has come to describe the current explosion of large scale (trans)national commercial land transactions, which are largely concentrated in the developing world. Given the speed and magnitude of this trend, the implications of the massive amount of capital being directed to the agricultural sectors of emerging markets cannot be understated. Estimates reveal that more than USD 100 billion has been invested in buying farmland since 2008.[1] To characterize these capital flows, four main investment types have been observed among reported international land deals (Cotula *et al.* 2009). First, direct investments in foreign lands have been carried out by sovereign wealth funds (SWFs). Though the number of land investments through SWFs is relatively low, SWFs are also indirectly involved in land deals through equity participation in more directly engaged companies. Second, state-owned enterprises with sectoral expertise in agribusiness are investing in primary agricultural production in foreign countries. For example, the Zad

[1]A Spanish NGO, GRAIN, has tracked the recent demand for large-scale tracts of farmland through its online blog. Elements of these data have been used by a number of research institutions (Braun and Meinzen-Dick 2009) and interested parties (Deininger 2011, Uellenberg 2009) to make inferences on the size of the 'land rush'.

Holding Company, a state-owned firm from Qatar, is reported to be involved in the formation of a joint holding company to produce food in Sudan for export to Arab markets (Cotula *et al.* 2009, 36). Third, a number of land investments involve government-to-government deals, where negotiations are carried out directly between two state governments, rather than through subsidiary bodies like SWFs.

Most reported international land deals, however, involve a fourth and final investment type – private sector investments (FAO 2010, Cotula *et al.* 2009, 37). These include agribusiness and agrifood companies, biofuels developers, and increasingly, private institutional investors. Defined as financial organizations that invest large sums of money in securities, real estate, and other assets on behalf of third parties, 'private institutional investors' include mutual funds, banks, pension funds, hedge funds and private equity funds (McNellis 2009). This paper examines the latter group, as private equity investors have come to represent a significant, yet relatively unexamined, actor within the land grab trend.

To characterize this actor, it is first necessary to define the private equity investment process. Private equity is an investment vehicle whereby limited partners (university endowments, pension funds, high net worth individuals, etc.) can invest equity in a managed fund. Limited partners have no management authority and share only in their fund's profits and liabilities. A private equity firm sets a fundraising target for a given fund and goes 'on the road' to attract limited partner capital. This process can take a few months or years and is primarily dependent on the firm's return on investment record and the fund's perceived opportunity (Makhene 2009). Private equity firm managers, known as general partners, are also expected to make substantial investment in the fund – ordinarily one to five percent – to demonstrate commitment to the fund's performance. General partners have unlimited liability and receive an annual management fee (two percent of the fund) for making strategic decisions and overseeing fund operations (Makhene 2009).

Once the fundraising target is accomplished, the fund is 'closed' to further investment from limited partners, and general partners are responsible for scouting 'target companies' that match the fund's investment philosophy. General partners can either purchase targets outright or acquire an equity stake, together with other investors, in the target company. Before completing the transaction process, general partners perform due diligence on the target company, which involves assessing the target firm's financial statements, weighing strengths and weaknesses of the company's business model, evaluating market opportunities, market trends and projecting strategic fit with the fund's portfolio of acquired target companies. If due diligence provides enough evidence of the target firm's intrinsic and synergistic value, then the next step is acquisition or 'buyout'.[2] In a typical leveraged buyout transaction, a private equity firm buys majority control of an existing or mature firm – a 'portfolio company' – thereby aligning investor and management goals (Kaplan and Strömberg 2008). Private equity funds typically have a five- to seven-year

[2]There are two types of buyouts in private equity – management buyouts (MBO) and leveraged buyouts (LBO). MBOs are initiated by the management team of the target company and typically involve management buying out shareholders' equity and taking the public company private. The primary goal of MBOs is to align management strategy with ownership interests. A leveraged buyout (LBO) is the acquisition of a target company through a high debt-to-equity ratio transaction. Successful LBOs allow the private equity fund to assume control of the target company – known as a portfolio company upon acquisition – aligning investor and management goals. For more information, see Makhene (2009).

horizon, after which time the fund is exited – whether through an Initial Public Offering (IPO)[3] or a sale – for a profit or carried interest. General partners receive 20 percent of the carried interest or fund profit after the fund is exited (Makhene 2009).

This paper examines the private equity investment landscape in African land and agriculture markets by exploring two underlying questions: (1) Within the global land grab trend, how is power manifested for private equity investors? and (2) What are the development implications of private equity-backed investments? The paper first characterizes the increasing trend of private-equity backed investments in African land markets, and then analyzes the power relations embedded in this trend. It then goes on to examine the power dynamics that characterize the relatively new relationship between private equity groups and development finance, largely drawing on evidence of World Bank Group's promotion of private equity in emerging markets.

2. Placing private equity-backed investments in the land grab context

2.1. A growing trend

Private equity is not a new phenomenon in Africa, but recently, it has drawn increasing attention as 'an innovative means of private-sector development on the continent' (OECD 2008). Until a few years ago, frontier investing in sub-Saharan Africa was the focus area of only a few specialized funds, and while the African private equity market has yet to achieve the status of markets in other emerging regions (such as Asia and Latin America), the benefits of investing in Africa-focused funds cannot be ignored. Indeed, with six of the ten fastest-growing economies in the world in mid-2011,[4] Africa has attracted the interest of fund managers, especially those eager to gain 'first mover advantage' by arriving at an early stage in the development of economies (de Sa'Pinto 2011). For a number of investors, the sub-Saharan region, along with the Middle East and North Africa nations, represent the next leg in the successful expansion of market economies across the globe (McNellis 2009).

To date, the aggregate capital sought by currently active Africa-focused funds eclipses targets set by funds closing in 2009–2010, suggesting that the private equity market in the region is quickly expanding (Prequin 2010). An Emerging Markets Private Equity survey reported that the weighted 'attractiveness' rating for sub-Saharan Africa (including South Africa), in terms of its relative desirability as an investment destination, nearly doubled from 2010 to 2011 (Coller Capital and EMPEA 2011). As of 2010, there were 71 Africa-focused private equity funds involved in fund-raising, seeking a combined aggregate of USD 24.9 billion in capital commitments. On a global scale, 172 fund managers include Africa in their regional focus (with just over 20 percent of these firms headquartered in the region) (Preqin 2010).

[3]An Initial Public Offering (IPO) is a procedure by which a private company raises capital by selling company shares to the public. A private company can only have one IPO, unless the company subsequently goes private. Once a private company goes public, it is listed on a public stock exchange.
[4]Measured in terms of GDP growth, these countries include Ghana, Liberia, Angola, Ethiopia, Mozambique, and Timor Leste.

Within this trend, a significant number of private equity funds are looking to farmland as a key asset class within emerging markets (FAO 2010). Following years of underinvestment due to low financial margins and inherent risks, farmland has emerged in recent years as a strong alternative asset class owing to its strong fundamentals (e.g. growing world population, dietary changes and energy tendencies), strong inflation hedge qualities, and potential for high returns. As of September 2010, over 190 private equity firms were globally investing in agriculture. Another 63 firms were raising capital for private equity investments in the sector, with an aggregate target of USD 13.3 billion (Prequin 2010).

Africa possesses myriad advantages for private equity funds seeking farmland. First, productive arable land is considered to be widely available. While figures vary, it is frequently claimed that Africa comprises 18 percent of the world's arable land, and only 10 percent of Africa's land is currently cultivated (Deininger 2011), a yield gap which promises huge opportunity for investment. World Bank economist Klaus Deininger argues that, given its sparsely populated regions, the suitability of its land for cultivation, and its relatively high yield gap, sub-Saharan African countries present perhaps the most appropriate 'country type' for outside investment to foster local development (Deininger 2011). In addition, the cost of arable land in Africa is a fraction of that of comparable land in Europe, South America, and North America, and Africa's range of climates and micro-climates allow for agricultural diversity (Mullen 2010). Fund managers also recognize that African governments are seeking out agricultural investment, as many countries have put an increasing emphasis on enhancing agricultural productivity and expanding the area of land under production (Davies 2011). African countries, therefore, offer numerous incentives to attract investment in farmland and other key sectors, including duty exemptions, full or partial tax holidays, or tax rate reductions for specific types of activities (Daniel 2011a).

The amount of private institutional capital being invested in African farmland is impressive. In 2010, *The Wall Street Journal* identified 45 private equity groups looking to deploy over USD 2 billion into African agriculture over the next two to four years, and industry representatives have affirmed that this represents only the tip of the iceberg (Davies 2011). A number of Africa-focused funds have reported ambitious earnings targets. For example, EmVest Asset Management, a joint venture between UK-based Emergent Asset Management and Grainvest, is an agricultural investment company operating in sub-Saharan Africa, whose African Agricultural Land Fund is targeting USD 2.7 billion with target risk-adjusted returns at 25 percent per annum from combined soft commodity production yields and land price appreciation (EmVest 2011). Similarly, UK-based SilverStreet Capital is raising capital for the Silverlands Fund, a private equity fund that is investing in African agricultural businesses across the value chain in Southern and Central Africa. SilverStreet's targeted fund size is USD 350 million, and target return is 20 to 25 percent per annum (Mullen 2010). Table 1 lists a number of the large private equity funds focusing on land and agriculture in Africa. It illustrates the increased focus of private equity groups on African farmland and their efforts to capitalize on increasing world demand for agricultural commodities. It also demonstrates the impressive size of capital flows into Africa, given the relatively low amount of private equity investments in Africa prior to 2008.

Fund managers of private equity groups adopt varying strategies to farmland investment, depending on investors' expectations and investment philosophies.

Table 1. Sample of private equity funds focused on African farmland in agriculture.

Firm/Fund	Legal Base	Target	Investment Countries	Investment Strategy
Phatisa/African Agriculture Fund (AAF)	South Africa	USD300m	pan-Africa	Phatisa is targeting investments in farms, agribusinesses and infrastructure. The manager seeks to bring modern management skills to commercial farming operations to boost expansion. To achieve optimal diversification within the sector, the Fund will invest across the value chain (from primary production to processing and tertiary services) (Phatisa 2010).
SilverStreet Capital/ Silverland Fund	UK	USD450m	Malawi, Mozambique, South Africa, Tanzania, Uganda, and Zambia	Launched in 2010, Silverland will invest across the agricultural value chain, deploying a minimum of USD 2 million to each deal. The fund is targeting annual returns of between 15 and 20 percent, with a term of up to nine years. The fund focuses on primary production, backing businesses that farm grain, soya, fruits, vegetables, sugar, tea, and coffee (Private Equity Africa 2011a).
Citadel Capital/ Wafra	Egypt	-	Sudan	Wafra is Citadel Capital's Platform Company for agricultural production in Sudan and includes the rights to more than 200,000 hectares (ha) of land through investments held under Portfolio Companies Sabina (106,680 ha in northern Sudan) and Concord, previously known as Sudanese Egyptian Agricultural Crops Company – SEAC (105,000 ha in

(*continued*)

Table 1. (*Continued*).

Firm/Fund	Legal Base	Target	Investment Countries	Investment Strategy
				southern Sudan). These projects will engage in large-scale cultivation of cash crops including grain sorghum, maize, sunflower, rice and various grain legumes and together comprise one of the largest agricultural projects in Sudan. In April 2011, the company successfully completed its first harvest, with local prices ranging from 25 to 30 percent higher than international prices of wheat. SEAC will be ready to seed 4,000 acres of land by the onset of the rainy season in mid-2011; the entire area will be planted with maize for sale in the local market (Citadel Capital 2011).
Vital Capital Investments/ Vital Capital Fund I	Switzerland	USD500m	Angola, Congo Brazzaville, Gabon, Ghana, Uganda and Mozambique	An Africa-focused agriculture and real estate fund, Vital Capital Fund I seeks to invest in companies at all stages of development, investing a minimum of USD 10 million per transaction to attain a controlling stake in the targeted companies (Private Equity Africa 2011b).
Emergent Asset Management/ African AgriLand Fund	UK	USD2.7b	initial focus on South Africa (where the firm has a second office) and Mozambique. Later in the fund life it expects to expand	Its long-term objective is to secure food production across a diverse range of soft commodities managed across sub-Saharan Africa and throughout the agriculture value chain. Target risk-adjusted returns are +25 percent per

(*continued*)

Table 1. (*Continued*).

Firm/Fund	Legal Base	Target	Investment Countries	Investment Strategy
			further through the African continent.	annum from combined soft commodity production yields and land price appreciation. Such yield enhancement is based on the introduction of modern farming techniques and technologies to increase yields, while agglomerating farms to increase efficiency and generate economies of scale (Emergent 2011).
Chayton Capital/ Chayton Atlas Agricultural Company	UK	USD150m	Zambia, Botswana, Malawi, Mozambique, and Tanzania	The Fund's strategy is to assemble and cultivate primary production hubs and to establish service businesses that connect with primary production sites to achieve economies of scale and associated operational efficiencies. Investments span the agriculture value chain (Trade Invest Africa 2012).
Carlyle Group	US	USD750m	South Africa, Nigeria, Zimbabwe	Carlyle believes that Sub-Saharan Africa (SSA) is one of the fastest growing regions in the world, driven by favorable demographics, expanding domestic industries and an improving political environment. The fund will initially target the consumer goods, financial services, agriculture, infrastructure and energy sectors. Carlyle is one of the largest private equity investors, globally, with USD 97.7 billion of assets under management committed to 76 funds as of September 2010 (Private Equity Africa 2011c).

(*continued*)

Table 1. (*Continued*).

Firm/Fund	Legal Base	Target	Investment Countries	Investment Strategy
Agri-Vie / Agribusiness Private Equity Fund (backed by South African private equity investors, SP-aktif and Sanlam Private Equity)	South Africa	USD100m	East and South Africa	The fund targets food and agribusiness companies in Sub-Sahara Africa, and is particularly interested in businesses producing agricultural inputs, food, beverage, fiber, timber and aquaculture products. Agri-Vie also backs food processors and companies offering marketing and distribution services. As of November 2010, about 30 percent of the fund had already been invested in a number of companies across East Africa and South Africa, and the rest of the capital is expected to be invested through mid-2012 (Private Equity Africa 2010a).
Catalyst Principal Partners/ Catalyst Fund I	Mauritius	USD100m	Dem. Rep. of the Congo, Ethiopia, Zambia	The fund will make equity investments of between $5m and $15m. Focus will be on companies producing consumer goods such as agribusiness companies, packaging companies, and telecoms service providers (Private Equity Africa 2010b).

(*continued*)

Table 1. (*Continued*).

Firm/Fund	Legal Base	Target	Investment Countries	Investment Strategy
Emerging Capital Partners/Africa Fund III	US	USD1b	pan-Africa	ECP Africa Fund III will focus on companies pursuing a regional growth strategy, particularly in the telecommunications, natural resources, financial services, agriculture, transportation and utility sectors. The fund's mission is to generate above-market returns by taking controlling stakes or influential minority positions in high-growth companies through equity and quasi-equity investments such as convertible debt (Emerging Capital Partners 2011).
African Agricultural Capital/AAC Fund	Uganda	USD25m	East Africa	The fund is focused on providing capital to small growing businesses (SGBs) operating in the agriculture value chain in East Africa. The fund will invest between USD 200,000 and USD 2 million in each business, using a range of equity and quasi-equity instruments. AAC says its success criteria are to earn a minimum gross return of 12 percent per annum on funds invested, and to mobilize increased investment capital of at least an additional USD 5 million into the East African agricultural sector through partnerships with other investors (AAC 2011).

(*continued*)

Table 1. (*Continued*).

Firm/Fund	Legal Base	Target	Investment Countries	Investment Strategy
Advanced Finance and Investment Group/Atlantic Coast Regional Fund (ACRF)	Senegal	USD150m	Core investment countries: Nigeria, Senegal, Côte d'Ivoire, Ghana, Cameroon, Gabon, DRC and Angola	ACRF focuses on mid-size, strong growth companies with a regional scope. The fund will make investments ranging between USD 3 million and USD 15 million. The sector focus will be agribusiness, transportation and logistics, financia services, telecoms, mining and natura resources and manufacturing companies (Mullen 2010).
Symbiotics/Regional Micro, Small and Medium Enterprises Investment Fund for Sub-Saharan Africa (REGMIFA)	Switzerland	USD150m	pan-Africa	This micro-finance fund, targeting African small and medium sized enterprises, will provide long-term debt and equity financing, as well as subordinated local currency loans (Private Equity Africa 2010c).
Blackrock Global Funds/BGF World Agriculture Fund	Luxembourg	USD700m	pan-Africa and international	The companies they will target will be those involved with agriculture-related chemical products, equipment and infrastructure, as well as soft commodities and food, biofuels, forestry, agricultural sciences and arable land. Its largest holdings as of July 2011 include Monsanto, Potash Corp, John Deere, and Syngenta (Citywire 2010).

(*continued*)

Table 1. (*Continued*).

Firm/Fund	Legal Base	Target	Investment Countries	Investment Strategy
Cru Investment Management/ Africa Agriculture Fund	UK	USD 800m	Malawi	Cru Investment's agriculture fund, which was launched in 2008 benefitted from significant exposure in commercial agriculture in Malawi with over 2,500 ha of land under its own control and another 4,000 ha in outgrower schemes. Its returns on investment were in the range of 30 to 40 percent per annum. The fund was suspended in September 2009 (Gulf Daily News 2008).
Pharos Financial Group/Pharos Global Agricultural Fund	Moscow, Dubai	USD 350m	Tanzania	The Pharos Global Agriculture Fund is focused on acquiring and managing agricultural land holdings in the Eastern Europe, Eurasia and Africa. The Fund has recently undertaken a deal along with AgriSol Energy LLC to invest in over 325,000 ha in Tanzania. The Fund targets annualized returns of 13 to 20 percent for investors through the acquisition of agricultural land, agricultural infrastructure assets and distressed or underperforming operational farm holdings (Pharos 2011).

Source: Author's compilation.

Davies (2011) distinguishes between two main types of farmland investments. The first is an investment directly in the land, whereby land is purchased and rented to an established operator. Profits are based on appreciation of land value as well as rental income. These investments are relatively low-risk, present low volatility, generally provide consistent long-term asset appreciation, and allow a store of value in times of inflation and recession. They further provide an annual cash return, with targeted returns at approximately 12 to 14 percent. The second way to gain exposure in the sector is by purchasing a controlling stake in an agricultural company and seeking to increase its value (although, agribusiness investments can, and do, involve the acquisition of land resources as well). These investments are relatively higher-risk and present higher volatility, but they allow investors greater control in the management of the company. Targeted returns are higher, at about 20 percent, and profits are based on increases in food commodity prices. Many investment managers are engaged in both areas.

2.2. Distinguishing private equity-backed deals from other land deals

Proponents of land deals often cite the potential developmental value of large-scale land acquisitions, pointing out that such investments will bring an influx of capital, create jobs, support infrastructural development, and promote transfers of knowledge, skills, and technology (see for example IFAD 2009). Yet, land deal critics point to the profit motive of investors, arguing that development benefits cannot be equitably distributed if investors are merely seeking assets with the highest returns potential (Oakland Institute 2011b). Indeed, critics have sought to draw a distinction between 'productive' land deals and mere 'speculative' land deals, arguing the latter provide limited potential for achieving land deals' purported benefits, as they seek short-term financial gains rather than promote long-term development for host countries (De Schutter 2011).

Importantly, private equity-backed investments are speculative by nature, as private equity groups bet on the ability to invest and quickly increase the value of their target investments. EmVest, for example, buys agricultural land in Africa and then develops industrial agricultural projects that produce grain crops, biofuels, fruits, vegetables, livestock, aquaculture, tea, timber, and nuts, primarily for export (Oakland Institute, 2011a). The fund sees a return on their investment when EmVest's operations begin to make a profit and the value of the purchased land increases (Walsh 2008). Others funds do not engage in any production whatsoever, but rather sit on their land, waiting for its value to appreciate. Charles Allison, managing director of Prudential Agricultural Investments, has stated, 'It is about safety. Farmland is a great place to store our wealth' (Reuters 2010b).

Private equity-backed investments are also distinguishable from other types of land deals due to the particularly low level of disclosure and transparency characteristic of the private equity industry. The scant analysis to date regarding the role of private equity in land grabs is reflective of this. Private equity fund managers are not required to make full public disclosure of their investments, and they operate with little regulatory oversight (McNellis 2009). Indeed, until recently,[5]

[5]In June 2010, the US Securities and Exchange Commission (SEC) approved new reporting rules for investment managers, which will require firms that manage more than $150 million to disclose the size of their funds and the type of clients who invest in them. The funds must now

this investor group has been able to operate in an almost confidential manner, as they often provide even their fund investors with only an opaque overview of their activities (McNellis 2009). There are several general reasons for this. First, a fund manager's dissemination of information may compromise the fund's competitive advantage. For example, the release of information on patents, products, and profitability to potential competitors may hurt the portfolio companies' competitive position. Furthermore, information about a company's value may disturb the fund manager's bargaining position during transactions. Second, investors avoid information disclosure for fear of increased litigation, particularly with respect to claims of a fund's alleged mismanagement. Finally, firms simply want to avoid the cost and time required to prepare and disseminate information, as well as the costs associated with potential misinterpretation of that information. Thus, as a result of private equity's high level of general nondisclosure, tracking private equity investments is problematic not only for investors in the fund, but also for third parties, including regulatory authorities. This information asymmetry between fund managers vis-à-vis limited partners or host governments creates cause for concern on various grounds.

Finally, private equity-backed deals are unique due to the inherent goals of this investment vehicle, namely, private equity's systematic aims to grow, innovate, and increase the efficiency of its target companies – a process otherwise known as 'creative destruction.' This is the idea that private equity identifies and eliminates poor performance and inefficiencies, ultimately with the goal of increasing the competitiveness of and adding value to its target companies. Schumpeter (1975) has described creative destruction as 'the fundamental impulse that sets and keeps the capitalist engine in motion... the new consumers, goods, the new methods of production or transportation, the new markets, the new forms of industrial organization that the capitalist enterprise creates'. Indeed, this process requires the incessant revolutionizing of economic structures from within – constantly destroying the old ways of doing things and constantly creating new ones (Schumpeter 1975). Porter (1990, 73) argues that this process is vital to the innovative capacity, and ultimately, the competitiveness of nations, stating, 'The basis of competition has shifted more and more to the creation and assimilation of knowledge'.

As Mathis (2008) points out, private equity groups continually engage in the process Schumpeter and Porter describe. Creative destruction, as applied by the private equity industry, represents the productive and competitive use of knowledge in strategic ways to gain a competitive edge. Private equity is essentially concerned with efficiency and competition. Applying creative destruction to private equity-backed land and agribusiness acquisitions, therefore, will essentially seek to introduce new methods and new technologies, and consequently eliminate inefficiencies – including the obsolete skills of the rural poor. This leads to important considerations for agrarian labour regimes, as will be explored in the following section.

also name their 'gatekeepers' – the auditors, prime brokers and marketers that service the funds. However, this rule will not go into effect until March 2012, and as expressed by one industry lawyer, 'the rules, at the end of the day, are not enormously onerous' (Protess and Rusli 2011).

3. Implications of private equity-backed land deals

3.1. Power dynamics at play

To be sure, North-South power relations between private equity groups and local actors come into play during land deal transactions. A classic example of power asymmetry in an international political economy context is the relationship between small and ill-informed African governments and large, sophisticated international firms in business negotiations (Collier 2006). The dominant discourse surrounding land deals speaks to a win-win scenario for investors and host governments, stressing host countries' desperate need for foreign capital and the many development benefits of land deals (increased employment, transfer of skills and technology, and infrastructural development). Thus, host country governments tend to embrace land investments, ignoring potential risks to local communities or to long-term economic development. This process is illustrative of Lukes' (1974) conception of power, namely that power can shape the ways the powerless perceive their wants, such that 'A exercises power over B when A affects B in a manner contrary to B's interests'. In Lukes' view, this conception includes the complex idea of the hegemony of the powerful and the production of a 'false consciousness' among subordinate groups who appear to be in consensus with systems that oppress them (Mosse 2004).

Such power imbalances in land deal negotiations are perhaps encouraged by the extremely low cost of African farmland as compared to other emerging market regions, as well as the extent of investor incentives offered by host governments in order to attract capital. I have argued elsewhere (Daniel 2011a) that by providing generous fiscal incentives, host governments act in a manner contrary to their own interests by potentially crowding out opportunities for direct fiscal support to local taxpayers or for domestically financed investment. Indeed, the public resources forgone due to investor incentives can severely undermine a country's tax base.

Yet, low land prices and investors incentives are reasons why private equity groups are now flocking to African markets in increasing numbers. Susan Payne, CEO of Emergent Asset Management, has stated,

> In South Africa and sub-Saharan Africa the cost of agriland, arable, good agriland that we're buying is one-seventh of the price of similar land in Argentina, Brazil and America. That alone is an arbitrage opportunity. We could be moronic and not grow anything and we think we will make money over the next decade.

3.2. Information and governance

Information asymmetry is another contributing factor to the abuse of power in land deal negotiations. Bachrach and Baratz (1962) note that power includes situations where individuals or groups are able to limit the scope of public debate and decisions to those issues that are relatively harmless, thereby maintaining their advantageous position. Due to limited debate, state governments as well as individual land owners are often willing to lease their land for negligible amounts due to their lack of a clear understanding of the benefits the leaseholder will retain or the long-term effects of a third party's use of the land. Information asymmetry can thus lead to inequitable contract negotiations and irresponsible land investments. In this context, transparency and standardized disclosure have been cited as necessary to reverse this power imbalance and help stakeholders more effectively evaluate risks and potential development benefits of land deals (see Deininger 2011, Cotula *et al.* 2009).

There remains a considerable gap, however, between the existing institutional and governance conditions in host states and the framework that should be established in order for large-scale investments to truly render development benefits (De Schutter 2011). Host states face a number of significant obstacles in the regulation and monitoring of land investment deals, including states' weak institutional capacity to manage investments and regulate wide-ranging impacts, difficulty in mapping available land, difficulty in strengthening security of tenure, and weak legal frameworks to ensure social and environmental protections. As De Schutter has observed, these obstacles appear almost insuperable, at least for the foreseeable future (2011, 267).

Given these difficulties, alternative mechanisms of monitoring and governance have been proposed, including international and regional agreements, and voluntary agreements on the part of the investor. Such agreements include the International Food Policy Research Institute's (IFPRI) 'Code of Conduct' and the Principles on Responsible Agricultural Investment (RAI), a joint undertaking and position by the World Bank, the UN Food and Agriculture Organization (FAO), the International Fund for Agricultural Development (IFAD), and UNCTAD. These are efforts to encourage investors to invest in a manner which ensures transparency and food security, respects land and resource rights, and promotes social and environmental sustainability (RAI 2011). Both rely on the assumption that corporate entities can and will self-regulate, and that these types of institutional mechanisms will encourage forms of internal regulation, allowing investors themselves to engage in more responsible investment practices.

Many have pointed to problems with these governance approaches, including their voluntary nature, their apparent failure to connect perceived risks with realistic outcomes, and the fact that they fail to incorporate a pro-poor, social-justice driven analysis (e.g. Borras and Franco 2010). As we specifically consider private equity capital, the voluntary nature of such codes of conduct poses substantial concern. Indeed, voluntary governance mechanisms are rendered futile if we consider the nontransparent nature of private equity-backed investments. Internal regulation of private equity-backed investment is implausible due to the fact that limited partners themselves have little information by which to hold fund managers accountable, for a number of reasons previously established. To be sure, general partners will always have more information than limited partners, thereby preventing limited partners from holding fund managers accountable to an efficient and/or responsible allocation of funds (Diem 2002). While leading players in the industry have, in fact, established voluntary mechanisms through which basic information enters the public domain[6], it is clearly to the benefit of fund managers to keep private any and all information that may affect their bargaining power.

[6]In November 2007, a working group formed by The British Private Equity and Venture Capital Association ('BVCA') and led by Sir David Walker issued the Guidelines for Disclosure and Transparency in Private Equity. That publication, which is also known as the 'Walker Report,' makes specific recommendations for improving the level of public disclosure by private equity firms operating in the United Kingdom.

Thus, the highly disproportionate level of power of private equity fund managers – both in terms of capital and information, vis-à-vis host country actors, regulators, and even their own limited partners – greatly complicates both the evaluation of risks and benefits of private equity-backed land investments as well as the design and implementation of appropriate governance mechanisms. Where decision-makers advocate for voluntary instruments to promote responsible land investment, private equity firms and other institutional investors will likely continue to pass under the radar, as host countries continue to provide them with incentives and regulatory oversight remains minimal.

3.3. Labour considerations

Another key set of implications of the growth of private-equity backed investments is their potential impact on agrarian labour. Agriculture is the primary activity of more than 60 percent of the African population, accounting for more than 30 percent of the Gross Domestic Product (GDP) of many sub-Saharan African countries (GRET 2011). African wage earners, on the other hand, represent only 11 percent of the total population, and they face a number of unique challenges, among them low skills and insufficient opportunities to enhance them, a narrow range of formal sector jobs, and a generally declining role for unions in protecting workers' interests (Taylor 2009).

A general argument put forward in defense of land deals is that farmland investment will help to overcome some of these challenges through employment generation and other development benefits, such as new technologies and infrastructure. In particular, private-equity proponents, such as the World Bank Group, claim that private equity-backed investments have significant development potential due to private equity's combined capital and expertise (World Bank Group 2010). However, the fact that a goal of private equity is to sustain its own competitive advantage, achieved through the constant upgrading and improvement of acquired firms, calls this allegation into serious question, as most wage earners in Africa possess insufficient skills to add value to existing agribusiness companies. Furthermore, private equity capital is targeted at larger-scale projects, while the majority of African farmers are concentrated in the small and medium enterprise sectors. The large amount of untapped capital need, or the 'missing middle', is not addressed by an increased influx of private equity capital, which primarily focuses on projects upwards of USD 1–2 million (Middler 2008). Indeed, this 'rural finance gap' in access to finance is perceived for capital needs between USD 10,000 and USD 1 million. Figure 1 illustrates this problem.

Considering that, historically, private equity investment in the agricultural sector has been limited, it is not surprising that little research has examined the effects of private equity on rural labour and employment. Yet, evidence from finance literature points to considerable controversy regarding the general employment impacts of private equity (Davis *et al.* 2008). While unions and other workers' groups contend that private equity firms destroy jobs, private equity associations and other groups have released several recent studies that claim positive effects of private equity on employment (e.g. Private Equity Council 2008, British Venture Capital Association 2006, Kearney 2007, Taylor and Bryant 2007). Within academia, studies are no less conflicting. An empirical analysis by Davis and colleagues (2008) found that gross job destruction is substantially greater in firms acquired by private equity transactions as compared to control (non-private equity-backed) firms. Yet, economists Shapiro and

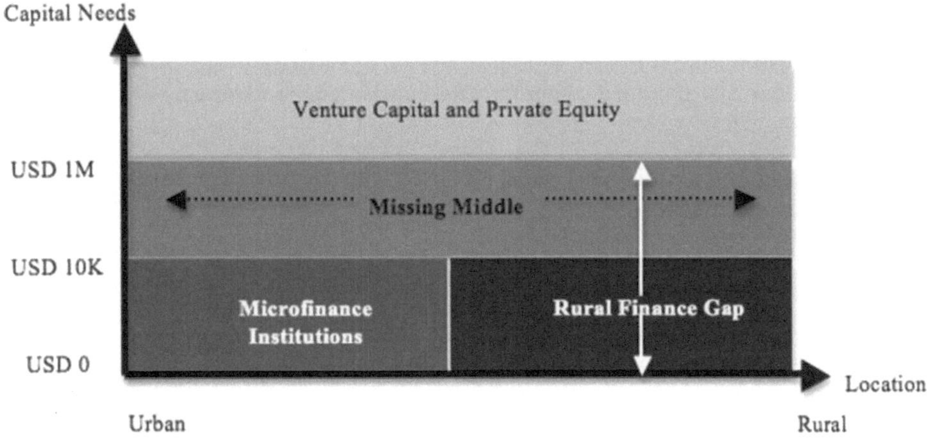

Figure 1. The rural finance gap and the 'missing middle'.
Source: Middler (2008).

Pham (2009) found that following a buyout there was a period of initial job losses but that, overall, private equity investments actually increased the number of jobs.

In the face of inconsistent findings, it is pertinent to turn to the inherent aim of private equity finance. With the goal of buying, investing, and growing companies ultimately to increase profit margins and maximize returns, private equity is, first and foremost, concerned with efficiency and competitiveness. As expressed by Mathis (2008), job creation is merely one measure of success: more jobs do not always translate into stronger, more competitive companies (14). Indeed, other measures of success include increased productivity or increased efficiency or output per unit of labour. This reality is detrimental to rural labour. The growth and competitiveness sought by private equity ventures require creative destruction – that is, continuous knowledge creation for labour and continuous innovation for companies. This often creates new opportunities for workers with higher value added skills – but less often for rural workers generally considered to be inefficient by the global capitalist system. Indeed, the alternative to increased efficiency is higher costs (due to the subsidization of competitively unproductive labour and/or capital), which eventually lead to a loss of competitiveness for the firm. According to Mathis (2008, 10), 'the primary objective of public and private companies is to increase shareholder value, not to increase employment, which is a public policy of government concern'. Thus, as existing agribusiness firms and commercial farms are acquired, private equity management will eliminate inefficiencies, including workers with inadequate working skills, and replace them with more knowledgeable workers or more efficient technology.

Interestingly, however, the discourse surrounding the investment philosophies of some private equity firms is one of active engagement with local communities. Under the premise of 'socially responsible investing', many Africa-focused funds claim to promote local job creation and sustainable development. EmVest's Africa Agriland Fund, for example, highlights the developmental benefits of its investments. Its website states,

> EmVest is establishing farming hubs throughout the region, working with national and local authorities to develop large-scale agriculture on a commercial basis, within a

> socially responsible framework, at a consistently high level of execution... Social responsibility is also a key tenet of [EmVest's] strategy and brings economic uplift to communities through commercially viable, first world practices, and by leveraging Africa's relative advantage of a labour force keen to work and seeking employment, not aid (Emergent 2011).

Similarly, Vital Capital Fund operates under the socially responsible notion of 'Impact Investing'. Its website states,

> Vital was born out of a vision to enhance the quality of life of communities and families in developing African nations, while simultaneously delivering attractive returns to investors – 'doing well' *and* 'doing good' (Vital 2011).

Chayton Capital, which invests throughout Africa, also acknowledges,

> Chayton is committed to establishing relationships in the region in order to ensure an effective transfer of skills, and sees significant opportunities to give back to the local communities and to train the workforce in effective, large-scale farming techniques (Chayton 2011).

While most of these funds are still too young to adequately evaluate the truth of such claims, recent Oakland Institute fieldwork in Mozambique sheds some light on existing realities. EmVest, in its Limpopo project in Matuba, Mozambique, plans to develop 2,000 hectares (ha) of virgin land into arable land using irrigation. It will then use the arable land to produce food crops and generate rates of return aimed at 25 percent per annum (Oakland Institute 2011b). EmVest's land lease is for a period of 50 years, and the firm has promised job creation with 'majority employment from the local community' (EmVest 2010). The Institute's fieldwork in 2010 and 2011, however, revealed that EmVest is, in reality, creating only a limited number of jobs through its Limpopo project. The project's employment headcount, as reported in May 2011, reveals that 355 positions had been created as of December 2010, and that these had been reduced to only 232 positions as of May 2011 (139 seasonal and 93 permanent) (Oakland Institute 2011b). Of these, the agricultural field workers comprise only 97 of the current positions (85 of which are seasonal), suggesting little long-term employment potential for local famers. Furthermore, it is important to note that, according to its Business Proposal, EmVest plans to spend a yearly average of USD 20,700 on wages and salaries, a mere 2.8 percent of the company's average yearly net income earned on production (Daniel 2011b).

Considering this small number of jobs created by a firm claiming to promote high standards of social responsibility, the ability or desire of private equity firms to generate lasting development impacts is questionable. Indeed, employment generation is quite juxtaposed to EmVest's overall investment strategy – and that of private equity in general – namely, that it aims to increase production yields 'based on the introduction of modern farming techniques and technologies. . .while agglomerating farms to increase efficiency and generate economies of scale' (Emergent 2011). As Li (2011, 294) points out, whether capital is tiny or large in scale, it always seeks subsidies (favorable regulations, free land and water, externalizable costs), and pushes down on the price of labour. Furthermore, large, plantation-style agriculture promoted by large-scale land investments is low in its labour requirements, particularly for the types of crops that EmVest plans to produce. As compared to perennials such as oil palm and rubber, food crops, including grains and soy, produce far fewer jobs per hectare. On average, the number of projected jobs is only

10 per 1000 ha for grains and 18 for soy, compared with around 150 for sugar and 300–400 for oil palm (World Bank 2010, 39). The labour-displacing 'mechanizability' of grains and soy production is precisely what gives these mega-farms their comparative advantage, and precisely what makes this farming method attractive to private equity capital.

4. Private equity-backed land deals and development finance: converging discourses

If we are to fully characterize the existing power relations involving private equity-based land deals, the role of development finance institutions (DFIs) cannot be ignored. Defined as multilateral institutions providing long-term finance for private sector enterprises in developing and reforming economies, development finance institutions facilitate private equity investors both in terms of direct funding, but also as they work to create attractive investment climates in the countries where investors seek to gain access to land. Indeed, a key factor distinguishing emerging markets-focused private equity funds, as compared to funds primarily operating in developed countries, is that the former receive funding from DFIs and similar multilateral agencies (including USAID and Overseas Private Investment Corporation) (Lerner and Schoar 2003).

The World Bank Group (WBG), in particular, has acted to promote foreign investment in developing countries. The world's largest development finance institution, WBG is comprised of five agencies,[7] each of which plays a distinct role in the Bank's general mission. The Bank's role within the global land grab has been characterized as conspicuous and multi-pronged, as its distinctly market-led approach to development has helped to create the conditions necessary for the proliferation of private sector investment in emerging markets by working directly, in varying capacities, with all stakeholders – from individual investors, to investment funds, to host country governments (Oakland Institute 2011c).

Within this greater investment promotion strategy, WBG is working to foment the influx of private equity capital into emerging markets. World Bank Group president, Robert B. Zoellick, believes private equity companies have a strong developmental impact due to their combined expertise and capital. He stated,

> Private equity is a cornerstone of our push to encourage growth and development led by the private sector. It is fundamental to building businesses, creating jobs, widening opportunities and establishing a virtuous upward spiral... Innovative financial instruments...enable the Bank Group's shareholders and donors to get the most development bang for their buck. And they generate very good returns for investors (World Bank Group 2010).

[7]The five agencies comprising the World Bank Group include the International Finance Corporation (IFC), the private sector arm of the World Bank Group, the Multilateral Investment Guarantee Agency (MIGA), which assists foreign investors by providing insurance against political investment risk; the International Development Association (IDA) and the International Bank for Reconstruction and Development (IBRD) which promote institutional, legal, and regulatory reform by providing technical assistance and policy advice to client governments, the former in conjunction with concessional financing, and the latter with debt financing; and the International Center for Settlement of Investment Disputes (ICSID) provides facilities for the conciliation and arbitration of investment disputes between member countries and individual investors.

WBG agencies directly assist private equity funds in a number of ways. The International Finance Corporation (IFC), the World Bank's private sector lending arm, has stakes in a number of agriculture-investing private equity funds, including Emerging Capital Partners' AIG African Infrastructure Fund, Citadel Capital's MENA Joint Investment Fund, Advanced Finance and Investment Group's Atlantic Coast Regional Fund (ACRF), and Agri-Vie's Agribusiness Private Equity Fund (see Table 1 for fund details). Zoellick has stated, 'We are looking at investments all across the [agribusiness] value chain: property rights, seeds, irrigation, fertilizers, basic technology advancements, financial services, harvesting, storage, getting goods to markets, and processing' (World Bank Group 2010).

Furthermore, the Bank's Multilateral Investment Guarantee Agency (MIGA) guarantees foreign direct investments against a number of risks, including currency inconvertibility and transfer restrictions, expropriation, war or civil disturbance, breach of contract, and the non-honoring of sovereign financial obligations (MIGA 2010). Yet, in addition to its traditional practice of underwriting individual projects, MIGA offers 'master contracts of guarantee' specifically for private equity fund investments. A master contract provides up-front pricing to the general partners of the fund for a specific period. The fund managers may use this contract to raise funds from institutional investors who are interested in taking the commercial risks associated with these investments but are concerned about noncommercial (political) risks (MIGA 2011). MIGA currently has master contracts with three private equity funds that invest in sub-Saharan Africa, including the African Development Corporation and the Sierra Leone Investment Fund LLC & ManoCap Soros Fund LLC, the latter of which focuses on small agribusiness companies in Sierra Leone (MIGA 2011). MIGA also signed a USD 50 million contract with Chayton Atlas Investments, an investment holding company within Chayton Atlas Agricultural Company; Chayton Agricultural is a private equity fund with extensive regional farming experience focused on investing in agribusiness in countries in the Southern African Development Community (SADC).

While private equity funds are currently making significant returns in emerging markets, this was not always the case. Despite its proliferation in the 1990s, the growth of emerging market private equity funds slowed drastically by the end of the decade, largely due to the inadequacy of regulatory and legal frameworks in emerging market countries, which failed to provide the necessary investor protection, accounting standards, corporate governance, and exit potential desired by private equity firms (Leeds and Sunderland 2003). World Bank Group has played an undeniable role in improving the private equity 'market', stepping in to provide technical assistance and advisory services in those areas that proved to be obstacles. Indeed, where access to markets has been bottlenecked by legal or regulatory inefficiencies, WBG – primarily through IFC and its partner organization, the Foreign Investment Advisory Service (FIAS) – has offered Technical Assistance and Advisory Services (TAAS) to developing country governments to overcome these barriers (Daniel 2010).[8] After two decades of advisory services, in addition to its

[8]TAAS comprises specific projects and initiatives designed to improve client governments' investment climates, and is therefore particularly relevant to the land grab trend; that is, to create the conditions necessary to attract foreign investment and to facilitate the investment process for investors. Such activities include investment legislation reforms, the reduction of administrative and institutional barriers to investment, the development of investment promotion agencies (IPAs) in these countries, and provision of policy assistance to

work on land tenure and titling, WBG has been successful in improving the investment climates of most African countries, having created conducive legal environments such that large-scale investments may more freely flow to emerging markets. Indeed, the substantial increase in foreign direct investment (FDI) and private equity capital to emerging markets in recent years is a testament to this.

Another key move on the part of WBG to support the mobility of private equity capital was the IFC-sponsored creation of the Emerging Markets Private Equity Association (EMPEA), a global membership association whose mission is to catalyze private equity and venture capital investment in emerging markets. EMPEA's 280 members include the leading institutional investors and private equity and venture capital fund managers across developing and developed markets (EMPEA 2011). IFC holds an annual Global Private Equity Conference, which is now the premier event for investors in the sector. Not surprisingly, WBG's interests in promoting access to land are in many ways directly aligned with those of private equity funds, owing to the fact that EMPEA's Board of Advisors includes several fund managers whose portfolios include major land deals. Hisham El-Khazindar, a current member of EMPEA's Advisory Council, is the Managing Director and co-founder of Citadel Capital, one of Africa's largest private equity funds. With USD 8.7 billion in assets currently under management, Citadel has acquired hundreds of thousands of hectares for agribusiness ventures in Kenya, Uganda, Tanzania, and South Sudan (Ombok 2011). Also on the Advisory Council is Shemara Wikramanayake, head of Macquarie Funds Group, a Sydney-based asset management business. In March 2010, Macquarie Agricultural Funds Management started up the Macquarie Crop Fund to 'acquire or lease grain and oilseed properties located in geographically diverse regions of Australia and Brazil' (Stock and Land 2010). Furthermore, former IFC executive George Manyere is currently the managing partner and chief investment officer of Brainworks Capital, a USD 20 million fund targeting Zimbabwe's agriculture, financial services, and mining sectors, pursuing 30 percent returns on its investments (Brainworks 2011).

Finally in one of its latest moves to support private equity in emerging markets, IFC launched a USD 500 million fund in 2010, which provides an exit option for emerging markets limited partner investors, thereby improving liquidity. Indeed, one of the most vexing aspects of private equity investing in developing nations has been the difficulty of exit (Lerner and Schoar 2003). While the gains of private equity investors in the developed world are largely linked to those of the market for initial public offerings (IPOs), private equity investors in developing countries cannot readily rely on these offerings. Even in 'hot markets' where large foreign capital inflows are occurring, new firms typically do not attract significant institutional holdings and have much less liquidity (Lerner and Schoar 2003). Consequently, private equity investors in developing countries tend to rely on the sale to portfolio firms to strategic investors. This can be problematic, however, when the number of potential buyers is small.

IFC's new fund specifically addresses this problem. The IFC fund will be the first 'secondaries fund' dedicated to emerging markets, seeking to improve the liquidity of

---

governments on regarding tax, customs, and land laws. Technical assistance and advisory activities may be linked to a specific investment project, or, increasingly, to broader goals such as improving the 'legislative environment' for a specific industry. See Daniel (2010).

private equity markets by providing exit opportunities to investors. IFC appointed an undisclosed private equity manager, and fundraising reportedly began at the end of June 2010 (AllBusiness 2010). The fund targets both emerging and frontier markets – including Africa – with IFC as a cornerstone investor. According to Mark Alloway, IFC's head of business development for the financial sector in Western Europe, a thriving secondaries market will provide limited partners, including those from developed nations, with a 'way out' if they need to get out of a fund. This option, he says, should relieve some of the concern developing markets limited partners have in gaining exposure to the emerging markets in general (AllBusiness 2010).

The huge support to emerging markets-based private equity from WBG's various agencies is reflective of WBG's overall development strategy, that increased investment is unequivocally positive and necessary for economic development and poverty reduction. This logic has been criticized by those who point to the World Bank's tendency to simplify the development problem. Li (2011, 292) argues that World Bank repeatedly 'boils down' complex political economic problems driven by unequal power and 'parses them into components that can be addressed by technical means'. She then addresses two assumptions made by the World Bank reflective of this tendency. The first is that capital is the key to economic growth, from which the greater good – poverty reduction – will eventually follow (p. 293). Interestingly, this is the same development logic increasingly posed by private equity groups, who, under the guise of 'social responsibility,' enter African land markets and so benevolently transfer their capital, skills, and knowledge to awaiting Africans.

The second assumption Li identifies is that the nations of the global South will one day experience an agrarian transition similar to the one that occurred in Europe, characterized by the shift from farm to factory, country to town, or from subsistence production to high value commodity production or wage labour on large farms (p. 293). As Li and others point out,[9] WBG's task of transforming the investment climates and regulatory frameworks of developing countries is, in essence, a recommendation that rural smallholders who are unable to compete in high value production, should exit agriculture for the sake of efficiency and for the sake of development. Indeed, WBG's role in encouraging private equity in emerging market is the promotion of increased capital and increased efficiency – and ultimately, the 'transfer of land to the most productive users' (World Bank, 2008, 9). This World Bank 'transition language' that Li points to (e.g. that exit from agriculture is presented as a matter of overcoming 'a deep inertia in people's occupational transformation' [World Bank 2008, 26) mirrors the creative destruction logic which drives the private equity market: grow, innovate, and generate returns – or be outcompeted.

Civil society groups are alarmed at this convergence of action and discourse. WBG's allocation of public resources into opaque private equity funds, without allowing for proper oversight, seems diametrically juxtaposed to governance efforts aimed at promoting transparency, disclosure and responsible investing. For many, IFC's increased allocation of funds to private equity vehicles, as well as the creation of its own private equity investment vehicles, represents a disturbing trend of moving

[9]For similar critiques, see Kiely (2009), Araghi (2009), Watts (2009).

development aid through nontransparent financial intermediaries. While the World Bank claims that placing money in such funds will permit the leveraging of additional funding, it is unsafe to assume that these funds will act in the public interest (Eurodad 2011).

4.1. Private equity as development?

Private equity-backed land deals are a new, little understood form of land grab, and they are growing at a significant rate. Indeed, the private sector investors now recognize that the investment returns in the sector make business sense in terms of risk versus reward (FAO 2010). In particular, the immense volume of private equity capital being directed at farmland in emerging markets has important implications for agrarian societies, which I have sought to analyze, albeit in this very limited exploration. A fundamental question underlying the current debate on the global land grab is whether development benefits resulting from land deals are realistic and realizable. The consideration of private equity-backed land investments, in terms of its inherent aims and distinguishing characteristics, expands this fundamental question in new ways.

Private equity investment is, at best, a limited tool for spurring economic development in developing countries. Rather, it is a financial vehicle providing capital to target companies to nurture expansion, new product development, or restructuring of a company's operations and management. It has been argued that private equity, as a development mechanism, is 'fundamentally neutral'; that it is not the mechanism itself, but rather the actors who undertake private equity investments who have the ability to invest responsibly or irresponsibly (Crutcher 2010). Yet, could private equity fund managers be any more clear as to where their priorities fall? As McNellis (2009) tells us, private equity fund managers, in overwhelming numbers, are 'ideologically free' in the sense that the social ramifications of their investments are a secondary or tertiary consideration; the only factor that counts, he contends, is sufficient return versus the estimated risk. McNellis explains, 'fund managers have the mandate to invest in almost any asset class in any location provided the return potential makes investment sense'. It is into such hands that the World Bank Group places funds while claiming to further its mission of global poverty reduction.

References

AAC. 2011. African Agricultural Capital. Available at: http://www.aac.co.ke/web, (accessed 6 April 2012).

AllBusiness. 2010. IFC to set up $500m secondaries fund. LexisNexis. 17 May.

Araghi, F. 2009. The invisible hand and the visible foot: peasants, dispossession and globalization. *In*: A.H. Akram-Lodhi and C. Kay, eds. *Peasants and globalization: political economy, rural transformation and the agrarian question*. London: Routledge, pp. 111–47.

Bachrach, P. and M.S. Baratz. 1962. Two faces of power. *American Political Science Review*, 56(4), 947–952.

Borras, S. Jr. and J.C. Franco. 2010. From threat to opportunity? Problems with the idea of a 'code of conduct' for land grabbing. *Yale Human Rights and Development Law Journal*, 13(2), 507–523.

Brainworks. 2011. Zimbabwe's Brainworks launches $20m fund. Private Equity Africa. 1 April.

Braun, J.V. and R. Meinzen-Dick. 2009. *'Land-grabbing' by foreign investors in developing countries: risks and opportunities*. IRPRI Policy Brief. Washington, DC: International Food Policy Research Institute.

British Venture Capital Association (BVCA). 2006. The economic impact of private equity in the UK. London.

Chayton. 2011. Investment Approach. Available from: http://www.chaytoncapital.com/html/products.html [accessed 22 August 2011].

Citadel Capital. 2011. Wafra. Available at: http://citadelcapital.com/current-investments/sabina-agriculture, (accessed 6 April 2012).

Coller Capital and EMPEA. 2011. *Emerging market private equity survey*, 7th edition.

Collier, P. 2006. International political economy: some African applications. Centre for the Study of African Economies, Department of Economics, Oxford University. May. Available at: http://users.ox.ac.uk/~econpco/research/pdfs/AERCPlenary-InternationalPoliticalEconomy-AfricanApps.pdf (accessed 6 April 2012).

Cotula L., S. Vermeulen, R. Leonard and J. Keeley. 2009. *Land grab or development opportunity? Agricultural investment and international land deals in Africa*. London and Rome: FAO, IIED and IFAD.

Crutcher, J. 2010. The private sector turn: private equity, financial intermediaries and what they mean for development. "Private Equity as Development?". London, UK. 22 November.

Daniel, S. 2010. *(Mis)Investment in agriculture: the role of the international finance corporation in global land grabs*. Oakland, CA: The Oakland Institute. December.

Daniel, S. 2011a. *The myth of economic development: understanding land investment deals in Africa*. Oakland, CA: The Oakland Institute.

Daniel, S. 2011b. *The myth of job creations: understanding land investment deals in Africa*. Oakland, CA: The Oakland Institute. December.

Davies, G. 2011. Farmland as an asset class: the focus of private equity firms in Africa. Presentation at 'Global Land Grabbing Conference,' Institute of Development Studies, University of Sussex, UK. 6–8 April 2011.

Davis, S., J. Haltiwanger, R. Jarmin, J. Lerner and J. Miranda. 2008. *Private equity and employment*. Washington, DC: Center for Economic Studies.

Deininger, K. 2011. Challenges posed by the new wave of farmland investment. *Journal of Peasant Studies*, 38(2), 217–247.

de Sa'Pinto, M. 2011. Geldof's Africa private equity fund nears $200 mln. *Reuters*, 29 August.

De Schutter, O. 2011. How not to think of land grabbing: three critiques of large-scale investments in farmland. *Journal of Peasant Studies*, 38(2), 249–279.

Diem, G. 2002. *The information deficiency problem of private equity fund-of-funds: a risk and monitoring management perspective*. Birmingham, UK: University of Birmingham.

Emergent. 2011. Africa AgriLand Fund. Available from: https://www.emergentasset.com/?func=PageAfricanLandFund [accessed 22 August 2011].

Emerging Capital Partners. 2011. ECP Capital Fund III PCC. Available at: http://www.ecpinvestments.com/fund.xml?id=1012&p=1012&d= (accessed 6 April 2012).

EMPEA. 2011. About EMPEA. Available from: http://www.empea.net/Main-Menu-Category/about.aspx [accessed 13 July 2011].

EmVest. 2010. Investment Proposal and Update. October.

EmVest. 2011. African Agricultural Land Fund. *Quarterly Update*. Issue 9, Q1 2011.

Eurodad. 2011. Civil society organisations sound alarm over World Bank plan to set up $1 billion private equity fund. 6 April.

FAO. 2010. *Agricultural investment funds for developing countries*. Rome: FAO.

Gillam, C. 2010. Funds flow toward farmland as experts eye deals. *Reuters*, 7 May.

GRET. 2011. Agriculture in Africa: Stakes and Public Policies. Available from: http://www.gret.org/publications/ouvrages/infoomc/en/F01en.html [accessed 23August 2011].

Gulf Daily News. 2008. UK firm plans to launch Africa Agriculture Fund. 21 September. Available at: http://www.gulf-daily-news.com/NewsDetails.aspx?storyid=229488 (accessed 6 April 2012).

IFAD. 2009. *The growing demand for land: risks and opportunities for smallholder farmers*. Rome: IFAD.

Kaplan, S. N. and P. Strömberg. 2008. Leveraged buyouts and private equity. *Social Science Research Network*. June.

Kearney, A.T. 2007. Creating new jobs and value with private equity. Available at: http://www.atkearney.com/index.php/Publications/creating-new-jobs-and-value-with-private-equity-ea.html (accessed 6 April 2012).

Kiely, R. 2009. The globalization of manufacturing production. Warrenite fantasies and uneven and unequal realities. *In*: A.H. Akram-Lodhi and C. Kay, eds. *Peasants and globalization: political economy, rural transformation and the agrarian question*. London: Routledge.

Kolesnikova, M. 2011. Grantham says farmland will outperform all global assets. *Bloomberg*, August 10.

Leeds, R. and J. Sunderland. 2003. Private equity investing in emerging markets. *Journal of Applied Corporate Finance*, 15(4), 111–119.

Lerner, J. and A. Schoar. 2003. Private equity in the developing world: the determinants of transaction structures. Working papers 4468–04, Massachusetts Institute of Technology (MIT), Sloan School of Management.

Li, T.M. 2011. Centering labour in the land grab debate. *Journal of Peasant Studies*, 38 (2), 281–298.

Lukes, S. 1974. *Power: a radical view*. London: Macmillan Press.

Makhene, M. 2009. Alternative growth: the impact of emerging market private equity on economic development. *Neumann Business Review*, Spring 2009, 17–47.

Mathis, F.J. 2008. The competitive impact of global private equity: creative destruction and innovation. Thunderbird School of Global Management. Thunderbird Global Private Equity Center. April. Available at: http://www.pegcc.org/wordpress/wp-content/uploads/thunderbird-study.pdf (accessed 6 April 2012).

McNellis, P. 2009. Foreign investment in developing country agriculture – the emerging role of private sector finance. FAO Commodity and Trade Policy Research Working Paper, No. 28.

Middler, B. 2008. Closing the gap: reaching the missing middle and rural poor through value chain finance. *Enterprise Development and Microfinance*, 19, 301–316.

MIGA. 2010. MIGA: bolstering private equity investment. Available from: www.miga.org/documents/privateequity.pdf [accessed 13 July 2011].

MIGA. 2011. Private equity investment guarantees – managing risk in Africa, Available from: http://www.miga.org/guarantees/index_sv.cfm?stid=1757 [accessed 3 August 2011].

Mosse, D. 2004. Power relations and poverty reduction. In: R. Alsop, ed. *Power, rights, and poverty: concepts and connections*. Washington, DC: The International Bank for Reconstruction and Development/The World Bank and the Department for International Development.

Mullen, K. 2010. African agricultural finance under the spotlight. *Reuters*, 24 August.

Oakland Institute. 2011a. *Deciphering Emergent's investments in Africa. understanding land investment deals in Africa*. Oakland, CA: The Oakland Institute.

Oakland Institute. 2011b. *EmVest asset management in Matuba, Mozambique. Understanding land investment deals in Africa*. Oakland, CA: The Oakland Institute.

Oakland Institute. 2011c, forthcoming. *The role of the World Bank Group in agricultural investment and land acquisition*. Oakland, CA: The Oakland Institute.

OECD. 2008. Private equity: an eye for investment under African skies? *Policy Insights*, 60, April.

Ombok, E. 2011. Citadel capital of Egypt plans to invest more in East Africa. *Bloomberg*. 1 August.

Pharos. 2011. Pharos Global Agriculture Fund LP. Available at: http://www.pharosfund.com/fund_agriculture.html (accessed 6 April 2012).

Phatisa. 2010. African Agricultural Fund First Closes On US$ 135 Million. 29 November. Available at: http://www.phatisa.com/images/file/Announcement%20Final%20-%20AAF%20First%20close%20(concise).pdf (accessed 6 April 2012).

Porter, M.E. 1990. The competitive advantage of nations. *Harvard Business Review*, March 1990, 73–93.

Preqin. 2010. Preqin special report: the private equity market in Africa. October. Available at: http://www.mandaportal.com/getattachment/b0760ba3-04be-4157-80b4-83d923afec98/Preqin-Special-Report-The-Private-Equity-Market- (accessed 6 April 2012).

Private Equity Africa. 2011a. SilverStreet raises $198m from PKA and OPIC. 11 July. Available at: http://www.privateequityafrica.com/funds/silverstreet-gets-198m-from-pka-and-opic (accessed 6 April 2012).
Private Equity Africa. 2011b. Vital I reaches $250m first close. 4 May. Available at: http://www.privateequityafrica.com/funds/vital-launches-500m-africa-fund (accessed 6 April 2012).
Private Equity Africa. 2011c. Carlyle to launch Africa fund. 10 March. Available at: http://www.privateequityafrica.com/funds/carlyle-to-launch-750m-africa-fund (accessed 6 April 2012).
Private Equity Africa. 2010a. Agri-Vie final close exceeds target. 19 November. Available at: http://www.privateequityafrica.com/funds/agri-vie-final-close-exceeds-target (accessed 6 April 2012).
Private Equity Africa. 2010b. Catalyst to launch $100m fund. 4 November. Available at: http://www.privateequityafrica.com/countries/catalyst-to-launch-100m-fund (accessed 6 April 2012).
Private Equity Africa. 2010c. Symbiotics launches Sub-Saharan Africa SME fund. Available at: http://www.privateequityafrica.com/funds/symbiotics-launches-sub-saharan-africa-sme-fund (accessed 6 April 2012).
Private Equity Council. 2008. American jobs and the impact of private equity transactions. Available at: http://www.pegcc.org/wordpress/wp-content/uploads/pec-jobs-study-01-17-08.pdf (accessed 6 April 2012).
Protess, B. and E.M. Rusli. 2011. S.E.C. approves new reporting requirements for hedge funds. *New York Times*, June 22.
RAI. 2011. Principles for responsible agricultural investment (RAI) that respects rights, livelihoods and resources: introduction. Available from: http://www.responsibleagroinvestment.org/rai/node/256 [accessed 23 August 2011].
Reuters. 2010a. African agricultural finance under the spotlight. 24 August.
Reuters. 2010b. Funds flow toward farmland as experts eye deals. 7 May.
Reuters. 2011. Geldof's Africa private equity fund nears $200 mln. 29 August.
Schumpeter, J. 1975. *Capitalism, socialism, and democracy*. Harper.
Shapiro, R.J. and N.D. Pham. 2009. The impact of private equity acquisitions and operations on capital spending, sales, productivity, and employment. SONECON. January. Available at: http://www.pegcc.org/wordpress/wp-content/uploads/pec_economic-impact-study-01-23-09-final.pdf (accessed 6 April 2012).
Stock and Land. 2010. Macquarie interest in grain business elevated. 19 March. Available at: http://sl.farmonline.com.au/news/nationalrural/agribusiness-and-general/finance/macquarie-interest-in-grain-business-elevated/1780950.aspx (accessed 6 April 2012).
Taylor, A. and C. Bryant. 2007. Private equity deals that cement business growth. *Financial Times*, 2 April.
Taylor, S. 2009. Labour markets in Africa: multiple challenges, limited opportunities. *Current History*, 108(718), 214–220.
Trade Invest Africa. 2012. Private equity deal to spur growth in African agribusiness. 4 April. Available at: http://www.tradeinvestafrica.com/feature_articles/1223804.htm (accessed 9 April 2012).
Uellenberg, A. 2009. Foreign direct investment in land in Madagascar. Eschborn: GTZ on behalf of the German Federal Ministry for economic Cooperation and Development. Available at: http://www2.gtz.de/wbf/4tDx9kw63gma/gtz2010-0063en-foreign-direct-investment-madagascar.pdf (accessed 6 April 2012).
Vital. 2011. Vital capital fund. Available from: http://www.vital-capital.com [accessed 29 August 2011].
Walsh, K. 2008. Fund targets African agriculture. *The Sunday Times*, 31 August.
Watts, M. 2009. The southern question: agrarian questions of labour and capital. *In*: A.H. Akram-Lodhi and C. Kay, eds. *Peasants and globalization: political economy, rural transformation and the agrarian question*. London: Routledge.
World Bank. 2008. World development report: agriculture for development. Available at: http://siteresources.worldbank.org/INTWDR2008/Resources/WDR_00_book.pdf (accessed 6 April 2012).

World Bank. 2010. Rising global interest in farmland: can it yield sustainable and equitable benefits? September. Available at: http://siteresources.worldbank.org/INTARD/Resources/ESW_Sept7_final_final.pdf (accessed 6 April 2012).
World Bank Group. 2010. Returning to growth: private equity in emerging markets. Speech by Robert B. Zoellick, President, World Bank Group. IFC's 12th Annual Global Private Equity Conference. 12 May.

Shepard Daniel is a Fellow at the Oakland Institute. She writes on topics of international food security and land policy.

Global resource grabs, agribusiness concentration and the smallholder: two West African case studies

Kojo Sebastian Amanor

This paper places land grabbing within the context of developments within agribusiness within the last 30 years, tracing the various trajectories of increasing competition and concentration and pressures on commodity prices that have resulted in increasing dispossession of smallholders and a move in some agri-food chains towards large estate production. The paper explores the ways in which contemporary agricultural policies and neoliberal market reforms reflect these developments and examines recent framing of land policies in Africa in the context of the development of agrarian capital and agribusiness. Competitiveness results in dispossession of less successful smallholders from below by commercial smallholders, and from above by large estates vertically integrated into agribusiness marketing chains. This is illustrated with examples from the cocoa sector in Côte d'Ivoire and pineapples in Ghana.

Introduction

An important and significant body of literature now exists on land grabbing within Africa. This either draws attention to the specific global crises in the 2006–2008 period leading to international investments in biofuels or large-scale food production in Africa for export (Borras *et al.* 2010) or places land grabbing within the context of the development of liberalisation policies that created land markets and new avenues for international investments in large-scale land acquisitions (Zoomers 2010). This paper draws upon recent literature on concentration in agribusiness and supermarket food value chains, and financialisation within agriculture to gain a deeper understanding of the dynamics, processes and pressures of agrarian accumulation and the relationship between these trends and land grabbing. It examines the impacts of recent transformations in agrarian accumulation upon smallholder production, scales and economies of production, and on livelihoods and access to land. It argues that increasing competitiveness and concentration within agriculture creates pressures on smallholders that ultimately result in dispossession. Dispossession arises in two ways. Firstly, from below, in which the increasing commercialisation of smallholder production and its integration into global agri-food chains results in the expansion of more successful commercial smallholders at the expense of less commercial producers, or in the movement of producers with options into other

---

Thanks to Ian Scoones for useful comments on an earlier draft and to the reviewers of this paper.

sectors leaving those best able to absorb declining margins of profit to fend for themselves. However, there may be further displacements of producers as markets impose regulatory entry barriers on the remaining producers leading to further exits. Secondly, increasing competitiveness and declining farmgate prices enable large-scale estates to gain an increasing foothold through technically advanced production and the organisation of logistical chains and economies of scale to capture production and displace struggling smallholders. Thus, the rise of investment in large-scale agriculture is closely connected with increasing competitiveness in agriculture and increasing concentration in agri-food chains.

Liberalisation policies have facilitated the globalisation of agribusiness. The increasing divestiture of state organisations and freedom from regulation of financial services has enabled agribusiness firms to move into agricultural sectors in new ways, gaining control over marketing structures through buying into former parastatal organisations, or gaining control over processing through the purchase of companies. Agribusiness corporations can expand operations through horizontal integration, hedging risks by investing in diversified sectors and through takeovers and mergers, or gaining control of whole sectors through vertical integration of large-scale production with marketing and financial services. The relative position of smallholders in various agri-food chains is determined by the relative profitability of different modes of organisation, rather than the relative efficiency of smallholder agriculture in contrast to large-scale agriculture, or a commitment of agribusiness to equity through corporate social responsibility (Zoomers 2010, Borras and Franco 2012). Some of the literature on land grabbing portrays it a distinctly new trajectory for accumulation, beyond the norms of transnational agribusiness:

> The food crisis coupled with the broader financial crisis has turned control over land into an important new magnet for private investors. We're not talking about typical transnational agribusiness operations, where Cargill might invest in a soya bean crushing plant in Mato Grosso in Brazil. We're talking about a new interest in acquiring control over farmland itself. There are two main players here: the food industry and, much more significantly, the finance industry. (GRAIN 2008, 7)

In contrast, this paper argues that many agribusiness companies have been at the forefront of large-scale land acquisitions for many years, and that the dynamic towards land grabbing arises internally from the logic of agribusiness accumulation rather than from exogenous developments.

The dominant practices which have evolved over the last 30 years in agribusiness are increasingly characterised by intense competition, monopoly control, aggressive takeover bids, price fluctuations and marketing wars, financialisation and hedge fund portfolio investments, new technology and logistic controls, and increased scale of production (Newman 2009, Lapavitsas 2009). All of these factors transform the land question and contribute to the insecurity of smallholders. While large-scale agribusiness estate production may not have been prevalent in Africa in the last 30 years, the organisation of large-scale production in Latin America and southeast Asia ultimately impacts on African producers within competing agri-food sectors, creating pressures for the reorganisation of agricultural production by transnational agribusiness moving into new markets. The aims of this paper are to trace the contours of recent developments in agribusiness in different sectors within established agri-food chains (rather than in new sectors such as agrofuels) and the impacts of these developments on smallholder livelihoods and access to land.

To understand recent trends in agrarian accumulation, this paper first outlines changing theories on the productivity of smallholders in Africa. It shows how these theories reflect processes of concentration, vertical and horizontal integration within agriculture, and the incorporation or dispossession of various socially differentiated categories of smallholder agriculture. These changes reflect strategies of accumulation based on either the tight integration of smallholders into agribusiness production, or the displacement of the least competitive agriculture by commercial smallholders and large estates. An important bridge between the land grabbing and agribusiness literatures can be found in the debate on the relative efficiency of smallholder and large commercial production in Africa and the implications for land reform. The contours of this debate are explored in relation to changing patterns of accumulation and the rise of increasing competitiveness and concentration in agriculture. While equity and efficiency were originally conceptualised as moving hand in hand, contemporary policy formulations make way for the dispossession of less successful farmers and explore pathways for their exit from agriculture or transformation into labourers. The increasing competitiveness of global food chains creates a hostile environment for smallholder agriculture. The second section examines the consequences of increasing competitiveness, monopolisation, price volatility of agricultural commodities, and agricultural financialisation on rural livelihoods and the security of smallholders. The third section uses case studies of cocoa farmers in Côte d'Ivoire and pineapple farmers in Ghana to further explore the impact of competition, concentration and monopoly control over markets and food chains on farmer livelihoods and their access to land. These examples are chosen to illustrate different trajectories of production under agribusiness governance, including the continued dominance of smallholder production in cocoa and the displacement of smallholders by large estates in pineapple. They show that control and expropriation can occur at different levels from land to marketing institutions, inputs, knowledge and intellectual property rights, and quality control (as in standards and grading systems). Thus, access to land in itself does not guarantee security for smallholders, and control over land is merely one resource among many through which transnational agriculture can control production and accumulation.

Smallholders, competitiveness and the inverse relationship

Debates about land reform have been structured around theories of productivity growth in farming, rural poverty and labour. At the core of these debates are controversies about the relative efficiency of large-scale and smallholder agriculture (Bernstein 2004). These provide key insights into the processes of capital accumulation in agriculture.

Land reform in the immediate post-war period, particularly in southeast Asia and Latin America, was justified in terms of equity and productivity concerns. Productivity issues were raised in the context of an inverse relationship (IR) between farm size and productivity, in which it was argued that smallholders were more productive than large estates in terms of output per area of land (Berry and Cline, 1979, Binswanger *et al.* 1995, Lipton 2010). In this framework large estates were associated with archaic social relations that constrained the development of a more dynamic agriculture based on efficient utilisation of land. Barraclough (1973) argued that the prevalence of the *latifundio* in Latin America prevented the development of smallholder agriculture because it created scarcity of land and distorted land markets. Land reform would create new opportunities for growth and expansion through a process of forced capitalisation that discouraged extensive production, non-capitalist tenure forms, and speculation in landholdings. Either idle lands would be expropriated or landowners would be forced to break them up into smaller productive estates, facilitating the development of a more productive capitalist agriculture (De Janvry 1981). The thesis was supported by several case studies showing the IR between farm size and productivity. The IR formed a theoretical basis for justifying state-led land reform with ceilings on holding size.

Within most of Africa radical land reform was absent during the post-war and early independence period. Land was assumed to be readily available and the major policies to increase production encouraged agricultural modernisation through state subsidisation of large estate commercial agriculture and through state direct investment in agricultural enterprises (Noronha 1985, Basset 1993, Okoth-Ogendo 1993). Land administration was largely concerned with the consolidation of fragmented holdings and promotion of large commercial farms and state farms.

The IR entered into African agrarian debates in the late 1970s and early 1980s as part of the restructuring of agriculture, in the context of a critical theory of the 'urban bias' in state-led developmentalism (Lipton 1977), and as a justification for liberalisation policies that called for the end of government support for large-scale nationalist commercial agriculture. This was seen as producing policy distortions to the emergence of efficient smallholder agriculture, by subsidising large-scale agriculture and facilitating the expropriation of smallholders by large-scale commercial farmers. Lack of clearly defined and officially recognised customary rights in land produced tenure insecurity and facilitated the expropriation of the land of small farmers held under customary tenure. By the late 1980s the IR began to form a central tenet for market-based land reform, as part of the justification for decentralisation of land administration from state authorities to customary institutions, and the promotion of private markets in land. This involved a process of strengthening and harmonising customary and individual titles to land, promoting individual peasant rights to land, regulating transactions in land within the large sections of the population in which customary land rights were prevalent, and developing techniques for mapping these rights at the community level (Lavigne Delville 2000).

The opening up of markets following structural adjustment in Africa resulted in the rapid expansion of foreign investment in mining, timber, wildlife and tourist sectors in Africa and divested public sectors, including utilities and agricultural services (Gibbon and Olukoshi 1996, Gibbon 1999, Zoomers 2010). Since the majority of land in Africa lay in the customary sector, foreign investment in land required clearly defined processes of demarcating land in the customary sector that could be converted into private property. This could be achieved through a zonal system delineating customary land boundaries, which enabled other lands to be transacted by the state with foreign investors, as occurs in Ethiopia (Rahmato 2009); or a process of empowering traditional rulers to transact land with investors in collaboration with the state, as in Ghana (Amanor 2009) and Mozambique (Lahiff 2003,Dinerman 2006); or of elected village councils, as in Tanzania (Alden Wiley 2003, Manji 2006, Chachage 2010).

Since the 1980s there has been a rapid expansion of policy-oriented research on institutionalising customary land tenure arrangements into formal land administration, working within frameworks of promoting market-based land reform or of social and livelihood protection. However, the recognition of the 'customary' tends to reinvent customary land as a fungible bundle of property rights that can easily be converted, transferred, and transformed from user rights in land into private property. It allows the dominant political coalitions at the local level to redefine customary tenure in line with their narrow group interests and appropriate the land of the poor and marginalised. While this recognition of fungible rights in land can create security of ownership for individuals, this is only one possible outcome. For those with weaker rights or rights that are less easily converted into fungible property, the establishment of new customary rights to own and transact land can result in eviction from the land. The excluded may include shifting cultivators, pastoralists, the rural poor, youth and women. Lavigne Delville argues that claims based on ancestral rights to land can result in the expropriation of land users: 'In some cases, the intervention of the Rural Land Plan in Ivory Coast has enabled local people to claim ownership of land that has been made available to groups of incomers several generations ago' (Lavigne Delville 2000, 118). Similarly, the decentralisation of land administration in the Fouta Djallon in Guinea led to the former Fulbe overlords of the Rimaube cultivators claiming back lands which had been redistributed to their former slaves following independence (Boiro 1996, Diop 2007). In the Ashanti Region of Ghana, Ubink (2008) describes how chiefs are using their newly revamped powers in land administration to appropriate lands from smallholders in peri-urban areas and sell them to property developers. In eastern Congo Hunde chiefs have sold large parts of the forest to aspiring commercial farmers (Vlassenroot and Huggins 2005).

There is a growing awareness among social scientists working within a political economy framework of the ways in which notions of an undifferentiated community may have adverse effects on the rights of the rural poor with the implementation of community-based land administration. However, within policy circles a framework of social differentiation or 'heterogeneity' (World Bank 2008) related to efficiency is also used to justify the promotion of commercial smallholder agendas, linking smallholders into global value chains.

During the 1990s the IR was reformulated to recognise new contractual relations between agribusiness and smallholder farmers (Lipton 1993, Binswanger and Deininger 1993). While the productivity advantages of smallholder agriculture continued to be recognised, this was now incorporated into a framework of scale and linkages. The efficiency of the family farm in production was constrained by problems of scale, which made it difficult to introduce technical innovations. Banks and marketing companies preferred to transact with larger entities rather than deal with many small production units. The strengths of large-scale agriculture lay in its ability to take up advanced technology that was beyond the reach of small farmers, such as farm and irrigation machinery, processing, transport and marketing facilities; to gain access to credit and risk-insurance; and to introduce efficient logistic management. In the smallholder sector these constraints could be overcome by developing 'power compatible' approaches (Lipton 1993) based on linkages between smallholders and larger agricultural enterprises, as found in contract farming (Lipton 1993, Binswanger and Deininger 1993).

From this juncture, the IR was developed into a theory of commercial smallholders by the World Bank, in which the productivity gains of smallholders applied only to those with access to market information and integration into input and industrial food processing and horticultural produce markets. Increasingly, the smallholder sector was characterised by bifurcation into two sectors. The first consisted of highly competitive commercial smallholders networked into agribusiness and supermarket chains, and contracts with large estates. They constantly upgraded their skills in response to market changes, and new entry barriers based on grades, standards, quality controls and supply logistics. The second consisted of smallholders who struggled to meet barriers of entry in the more profitable export crop sectors and who languished in producing low-value basic staples, or became transformed into labourers (World Bank 2008). Under these conditions of increasing social differentiation, theories of equity and efficiency no longer converge along the IR, and the way is paved for the less successful farmers to give way to more commercial production. The freedom of the market dictates that land must be distributed to the more efficient. The *World Development Report* of 2008 states: 'Secure and unambiguous property rights also allow markets to transfer land to more productive uses and users' (World Bank 2008, 138), and: 'Preparing people to migrate out of agriculture is the flipside of the economy's structural transformation as agriculture grows' (World Bank 2008, 248).

From initially proffering support for pro-poor land reform to facilitate the emergence of agrarian capitalism, the World Bank (2008) has transformed the IR into a conceptual device that justifies the expropriation of the rural poor. The dominant policy vision of the future of agriculture is not of an egalitarian structure of autonomous smallholders with secure rights in small family farms. It is rather of an increasingly competitive but leaner commercial smallholder sector, which is expanding to displace less competitive smallholders and convert their land to more profitable land uses. These commercial smallholders are integrated into corporate food chains, combining networks of large estates, smallholders, suppliers, transporters, financial investments and hedge funds, and transnational food traders. This ushers in a gradual process of dispossession from below, in which the pressures of market competitiveness lead to the exit of independent farm production by the rural poor, and the increasing concentration of farm property.

In recent re-workings of the IR, the World Bank is more ambivalent about the significance of the IR for land redistribution. A recent formulation suggests that while commercial smallholders may be more efficient in areas characterised by well developed agricultural infrastructures (which integrate farmers into agribusiness value chains), in areas with a deficit of infrastructure and poor integration of smallholders into markets large estates may be the most viable form of commercial agriculture, which then develop the requisite infrastructure for agricultural modernisation (World Bank 2009).

The IR has also featured in debates about land reform in southern Africa (Lipton 1993, Binswanger and Deininger 1993, Bernstein 2004, Sender and Johnston 2004, Moyo and Yeros 2005, Scoones *et al.* 2010) and the relative merits of the land reform programmes in South Africa and Zimbabwe. What is most interesting in the context of this paper are recent offshoots of this debate which examine the reasons why redistribution of land to smallholders within a market-based framework has had such little impact in South Africa. Hall (2007) argues that the main tensions in South African policy result from market-led land reform processes within the policy context of the liberalisation of agriculture. This has forced agriculture to become more competitive and operate without support services, resulting in many farm bankruptcies. Increasing competiveness leads to concentration and increasing exploitation of labour. This is reflected in declining numbers of commercial farms, increasing farm debt, and exploitative labour relations reflected in the shedding of formal labour and increased hiring of cheap, informal and contract labour (Aliber *et al.* 2007). The numbers of commercial farmers dropped dramatically from 60,000 from 1990s to 45,000 in 2002 (Jacobs 2009). Hall (2007) argues that the highly competitive nature of commercial agriculture creates barriers for new African entrants into farming, which makes it difficult for them to be economically viable.

Conversely, this insight can be extended to the world of contemporary agribusiness, characterised by increasing competitiveness, barriers of entry, and concentration. This environment erodes the security of smallholder livelihoods and creates pressures for a redistribution of land *from* the poor to more competitive sectors. Since, without ceilings, markets cannot differentiate between smallholders and large estates, increasing concentration in agriculture will create conditions that destabilise the land rights and access to land of the rural poor, and facilitate land allocations to commercial sectors, including large corporations that are able to take advantage of economies of scale to expand their production and gain increasing control over the market. The expansion of global markets integrates smallholders into market oligopolies, subjects them to increasing competition, and ensures the exit of those unable to compete, and the appropriation of their land by sectors better integrated into the technology treadmill of global capitalist agri-food chains.

Nevertheless, trends towards larger estate production are not uniform and vary between different agricultural value chains. Smallholder agriculture continues to prevail in many sectors of the global agricultural economy, such as in coffee and cocoa, although a few transnational companies control international trade and processing in these food chains. Where farm operations are dominated by simple manual labour and intensive tasks, such as in selective picking of products that do not ripen uniformly, or in laborious hand drying methods, smallholder production can be competitive (Watts 1994). However, where such processes are subject to uniformity and can be mechanised, large-scale production can gain advantages in logistics and scale of production. Where smallholders remain competitive they often operate with low margins of return and receive a small proportion of the profits in the commodity value chain (Daviron and Ponte 2005). Smallholder production in these food chains becomes subject to increasing competitiveness and many producers may be driven out of production by low profit margins. Thus, profitability and the potentials of exploiting cheap labour are more pressing factors in agrarian production than abstracted notions of efficiency. In agrarian policy frameworks notions of efficiency and equity have responded to the increasing competiveness of agribusiness domination of global agriculture, and include considerations for paving the way for the exit of non-competitive smallholders. The framework of contemporary land reform is characterised by a structural tension between pro-market capitalism in which transnational monopolies dominate the agri-food sector, and pro-poor policies that seek to protect the livelihoods and land interests of smallholders from the ravages of the state, while facilitating their incorporation into the market.

The rise of transnational agri-food systems

The world economic crisis of the 1970s and the resulting neoliberal policy ascriptions created conditions for the structural transformation of agribusiness and its emergence as a global industry (McMichael and Friedmann 2007). The deregulation of financial and currency markets enabled agribusiness to rapidly expand its operations throughout the world. This was further facilitated by the abandonment of multilateral price controls and commodity agreements and the liberalisation of commodity markets through structural adjustment policies introduced in developing countries. These developments encouraged financial speculation and hedging in commodities. During the late 1970s, commodities attracted financial portfolio investments. Investments in commodities provide a good hedging strategy since there tends to be a negative correlation between commodities and other assets. Commodity prices also reflect inflation, offering less insecure investments in times of rapid inflation. As a consequence of increasing investment in commodities, prices of agricultural commodities became increasingly volatile and distorted by speculative activities (Newman 2009). Large agribusiness corporations began to develop their own financial operations, gaining new sources of rents and profits by hedging and speculating against price volatility.

The 1980s were also characterised by major developments in food processing and information technologies, the expansion of cheap freight, and the emergence of internet-based financial markets. With increasing price volatility and aggressive competition, corporate agribusiness expanded the scale of its operation throughout the world and diversified production as hedging strategies. This resulted in a reconfiguration of the agri-food system, with increasing concentration of production in the hands of a few food traders and branders through takeovers and mergers, and the rise of a handful of transnational supermarket chains controlling fresh food marketing chains. Since 2000 there has also been further expansion of financial investments in commodity markets following the dot-com crisis and the 2006–2008 fuel and food crises. Thus, the 2006–2008 food crisis represents the culmination of a 30-year history of financial speculation and investment in agricultural commodities and natural resources, rather than the beginning of the financialisation of the agricultural sector and land (Newman 2009, Lapavitsas 2009).

Major agribusiness companies have been able to move into various specialised food processing and input manufacturing sectors through acquisitions. For example, Cargill, originally a Mid-West USA trader in grains, diversified into livestock feed, livestock, seeds, and fertiliser (Broehl 1998). By the early 1990s Cargill had operations and subsidiaries in over 60 countries. In 2005 it acquired a two-thirds share in the Mosaic Company, the largest potash fertiliser producer in the world, with plants in 10 countries and operation in 40 countries. During the 1990s and 2000s Cargill had become one of the three major cocoa grinding processors in the world, through the acquisition of private sector and state grinding facilities in Cote D'Ivoire and Ghana (Losch 2002). In 2000 Cargill acquired Esteve to form ECOM, the third largest coffee export company in the world (Newman 2009). Through mergers and acquisitions and hedging on agricultural commodities Cargill has expanded into an increasing range of products, and built upon its experiences in processing and supply chain logistics, to develop new investments in energy, ethanol products, futures markets and financial services. This includes hedge funds, such as the Black River Assets Management Fund (BRAMF) with assets of USD 6 billion. BRAMF has been a major investor in land deals, acquiring large-scale dairy farms in Asia and aquatic farms in South America (Gilliam 2010).

From the 1980s the agri-food industry has been characterised by increasing concentration through both vertical and horizontal takeovers and mergers in which a few transnational companies dominate the markets (Watts 1994). In the USA, many food markets are dominated by four firms that control over 40 percent of retailing. This indicates the prevalence of market monopolies (Heffernan and Hendrickson 2002). Moreover, firms such as Cargill and ConAgra dominate meat, grains and food packing industries, reflecting a very high degree of concentration (Heffernan and Hendrickson 2002). Supermarket chains now control a considerable part of the food trade, and process their own brands. In the UK the dominant four supermarket chains control about 80 percent of food sales. In the USA the top five grocery stores control over 50 percent of food retailing. Firms such as Carrefour and Wal-Mart operate in global markets (Fox and Vorley 2004).

The dominance of world agriculture by monopoly firms has led to the demise of wholesale markets where farmers and intermediary buyers can negotiate prices of small quantities of produce in developed countries. Buying is mainly through forward contracts with large suppliers, and rigorous controls over logistics and quality of production result in entry barriers. Prices are no longer stable and farmers and suppliers are often forced to absorb the costs of price-cutting promotional wars, or the costs of new technologies (Young 2004, Fox and Vorley 2004). International traders are able to access supplies from a variety of locations and find new sources of cheaper production, which results in less competitive producers and regions becoming increasingly sidelined and unable to find markets. The liberalisation of agricultural markets has tended to erode the bargaining power of smallholders, strengthened demand-driven marketing chains and buyer control of commodity markets, and eroded the proportion of value of commodities accruing to farmers. This tends to undermine the standards of living of smallholders and results in the rapid movement of production to areas most able to withstand lower prices. This promotes increased social differentiation. It also leads to the emergence of large estates that can absorb price volatility through technical innovations and expansion of scales of production.

Agribusiness and smallholders

This section examines two contrasting case studies of the impacts of agribusiness food chains on different categories of smallholders in West Africa, and the complex relationship between land, livelihood, competitiveness and food chain governance. In the instance of pineapple cultivation, this involves the undermining of smallholder production in Ghana by vertically integrated chains of large estates owned by transnational companies of US origins operating in Costa Rica, and the subsequent re-organisation of pineapple production in Ghana along large-scale estates. These estates are increasingly dominated by transnational capital including subsidiaries of major Costa Rican producers. In the case of cocoa production in Côte d'Ivoire, farm production continues to be the preserve of smallholder farmers, but transnational corporations have gained control of former state marketing and processing monopolies, and effective control over world processing and trading of cocoa. This domination arose through price wars which destabilised cocoa farm production throughout the world, resulting in only three countries in which significant production remains. Within Côte d'Ivoire, there has been a marked transformation in production, with larger planters exiting, and production becoming dominated by small-scale migrants who are able to withstand diminishing profit margins and the traumas of price volatility. The immiseration caused by this process has resulted in severe civil conflicts.

Ivoirian cocoa

Before the 1990s cocoa grinding and product marketing were carried out by the same firm, which sourced particular beans from around the world and blended them into distinct brands in the North. New processing techniques enabled grinders to achieve a wider range of new products from similar beans, which has enabled new economies of scale. Most manufacturers have now become dependent upon specialised grinders for their supplies. Liberalisation during the 1980s resulted in increasing price volatility, intense competition, takeovers and concentration within cocoa markets. During the 2000s the number of cocoa grinding companies operating in Europe declined from 40 to nine. Four corporations now dominate cocoa grinding: Cargill, Archer Daniel Midland (ADM), Barré Callebaut and Nestlé (Losch 2002, Fold 2002). By the early 1980s five countries – Côte d'Ivoire, Ghana, Malaysia, Indonesia and Brazil – produced over 90 percent of cocoa output, of which Côte d'Ivoire produced 40 percent. With heightened market wars and price volatility, production in Brazil and Malaysia has rapidly declined leaving three countries controlling over 80 percent of world production. The major grinders now control cocoa at the source, transforming grinding from at activity largely concentrated in Europe and Northern America to one located in the main centres of production.

Before the introduction of liberalisation the Ivoirian cocoa sector was characterised by about 600,000 farmers, 800 buying agents, three grinding enterprises, one chocolate factory and 50 exporting companies (Losch 2002). A state marketing board controlled Ivoirian cocoa, but worked with private export firms who were regulated and given export quotas. As Losch comments:

> The system was underwritten by the strong class alliances [that] cemented together the political class and private business, with advantages accorded by the state to companies in exchange for the latter's 'understanding' of political shareholding. On the other hand electoral support by growers for Houphouët-Boigny's party was cemented by the income security that price stabilization provided growers, and by their ease of access to land and (immigrant) labour. The two forms of alliance are the source of the original politico-economic complex central to understanding the trajectory of the Côte d'Ivoire (2002, 210).

Large migrant Ivoirian cocoa farmers from the Baole region dominated farm production. They had strong political representation in the ruling party (Léonard and Oswald 1997). The state facilitated the acquisition of land by migrant Baole farmers and attracted networks of migrant labour by enabling non-Ivoirian migrants to gain rights to land through cultivation from the early 1970s, particularly in the new frontier areas of the southwest. The rapid expansion of the frontier brought in revenues for the state through its control over cocoa marketing. These revenues were used to forge the political and economic alliance and to support development initiatives. Given low levels of capital accumulation among the farmers in the southwest (the main area with reserves of forest land for cocoa cultivation) the chiefs and elders in this area were encouraged to release frontier land to migrants. Customary institutions were adapted for the allocation of land to migrants (Chauveau 2006, Chauveau and Colin 2010). Sahelian migrants came to fill the lowest ranks of labourers and smallholder cultivators in the southwest, and the new generation of rural indigenes were educated and looked to the urban areas for their livelihoods. In this phase of development, the expansion of cocoa was based on an alliance of the state political elite with a class of wealthy planters with capital to invest in new frontier area and cocoa trading firms. Poor Sahelian migrants formed the main labouring class, and networks of labouring people were attracted to frontier areas by offering them some access to land. The local rural chiefs and elders were co-opted into this process by enabling them to gain revenues from transactions in land and access to state-funded development projects, and the displaced youth were provided with education, and absorbed into the urban areas.

During the 1980s and 1990s a series of price wars occurred as a result of increasing attempts by transnational corporations to capture world cocoa trade in the face of increasing production gluts and volatile prices. Intense international competition led to the collapse of cocoa prices in 1986. The Ivoirian state attempted to halt the slide in cocoa prices by withdrawing from the world market, halting Ivoirian shipments between July 1986 and October 1989. This strategy failed, since a large percentage of cocoa stocks was traded in future markets, and the Ivoirian economy was too heavily dependent upon cocoa (Losch 2002). As a consequence the Ivoirian state became bankrupt, and with insufficient reserves to purchase the 1989 crop it was forced to adopt IMF-imposed austerity measures that resulted in the divestiture of state cocoa trading and processing. The 1980s and 1990s were characterised by increasing concentration in the global cocoa industry as large transnational food traders launched aggressive takeovers and mergers. A large part of this global reconfiguration was played out in the Côte d'Ivoire, involving the development of trading alliances to build up grinding capacities to add value to Ivoirian cocoa and offset the large supplies in a world market glut. The end result was the emergence of three new major international grinding companies – Cargill, ADM, and Barré Callebaut – all with grinding plants in Côte d'Ivoire (Fold 2002, Losch 2002).

These intense market wars resulted in the collapse of farmgate prices and increasing insecurity for farmers. The aggressive takeover of Ivoirian state cocoa agencies by transnational corporations, and the neoliberal policy environment imposed by international financial institutions (IFIs) as a basis for financial aid, destabilised the Ivoirian cocoa industry. It undermined the complex social contract and accommodations between the political regime and various class strata that had held together the state. The collapse in cocoa prices coincided with the decline of the cocoa frontier and the need to replant old plantations in old frontier areas. The cost of this was significantly higher than that of establishing new plantations in new frontier forest areas. Ruf (2001, 293) estimates that it costs twice as much to establish a plantation on grassy fallows as it does in new forest areas. This placed intensive pressures on cocoa planters, and the large Baule cocoa planters began to retreat from new cocoa development as their standards of living declined. They found it increasingly difficult to maintain their social status and networks of social expenditure from declining cocoa revenues and increasing social expenditure resulting from the introduction of structural adjustment measures. Within this new competitive environment, the migrant Burkinabe became the dominant cocoa farmers. They worked small plots of land, used more inputs per area of land, utilised more labour in weeding, were accustomed to lower standards of living, invested less in the education of their children, and could draw upon mobile networks of Burkinabe labourers (Léonard and Oswald 1997, Léonard 1997, Ruf 2001).

The rise of the Burkinabe cocoa farmer coincided with allegations of labour exploitation and use of 'child slave labour' in the Côte d'Ivoire, and the enactment of the Elgin-Harkin Protocol in the USA which called for the chocolate industry to guarantee that its products were 'child labour' and 'slave labour' free (Off 2006). This led to the introduction of new entry barriers. With the decline of the Baole planters Burkinabe migrants became increasingly politically vulnerable. The decline of national cocoa revenues resulted in increasing urban economic recession and unemployment. The dominance of migrant producers in the Ivoirian industry was also politically unacceptable within the Ivoirian state, which sought to deflect the blame for the instability of the industry onto vulnerable Sahelian migrants. With few prospects in the urban centres, many young people from the southwest returned home, where they turned their frustrations upon migrants, taking up xenophobic slogans of *Ivoirité* and ejecting migrants from the land (Chauveau 2006, Kouame 2010). Attempts to harmonise customary land law with statutory rights, and recognition of formal customary rights in land in the 1998 Rural Land Law, tended to exacerbate social, ethnic and nationalistic conflicts, particularly in relation to the problem of formalising the lands that had been granted to migrants on the basis of tiller rights to land (Chauveau 2006, 2009, Kouame 2010). The attempts to readdress the imbalance in the land question through a framework of communitarian land reform failed to address the reality of the existence of a heterogeneous, socially differentiated rural population, and the dependence of the industry on Sahelian migrant labour.

Ultimately, the resource grab in the Ivoirian case lay at the level of the takeover of state monopoly institutions, which gave transnational corporations control over the whole cocoa value chain and access to the sizeable rents originally captured by the state. While it is possible to see the changing productive relations in Ivoirian cocoa as an example of the movement of land to the more efficient smaller producers (Ruf 2001), the organisation of frontier production was influenced by state interventions and political alliances that maintained an intricate social contract between migrant farmers with capital, migrant labour networks with access to small plots of land, and cocoa trading interests. When this unravelled as a result of monopoly transnational takeover of the industry and increasing price volatility, the farmers and labourers had to increasingly absorb the costs of production in the context of the decline of the frontier and rising costs of production. International accusations of unacceptable labour relations resulted in the imposition of new entry barriers and forms of cocoa chain governance that further facilitated increased monopoly control over the industry. Although concentration in cocoa has not led to large estates expropriating the poor, transnational food traders have captured increasing value from cocoa, while smallholder producers have seen their incomes decline, often below socially acceptable levels. The price squeeze has forced many farmers out of cocoa worldwide, including substantial sections of cocoa farmers in Brazil and Malaysia.

The rise and fall of smallholder pineapple cultivation in Ghana

During the 1980s Ghanaian business investors involved in import-export trade, looking for new sources of foreign exchange, pioneered export pineapples. They took advantage of the proximity of the European markets and cheap freight prices to airfreight pineapples to Europe. The dominant variety produced in Ghana, Smooth Cayenne, was favourably received in European markets and preferred to products from Central America (Whitfield 2010). Pineapple production developed in close proximity to Accra, in the Nsawam area. By the end of the 1990s pineapple was largely a smallholder crop with between 50 to 70 companies exporting pineapples, of which only two companies exported more than 2,000 tons. About 40 percent of export pineapples originated from smallholders (Whitfield 2010).

By the early 2000s Ghana exports of pineapple accounted for 11 percent of the European market, but this fell to 3 percent in 2005, as Ghanaian pineapple had to compete with the resurgence of Costa Rican production (Whitfield 2010). This was the result of the introduction of a new variety by Del Monte, MD2, which stored well, had good taste attributes, and was resistant to parasites and rot.

During the 1970s leading pineapple producing companies relocated from Hawaii to Costa Rica, where they were able to gain access to large cheap agrarian estates, cheap migrant labour from Nicaragua unencumbered by strict labour laws, and relaxed environmental laws (ILRF 2008). Del Monte established the Pineapple Development Corporation in 1978 in Costa Rica and set up large industrial plantations.

Throughout the 1980s and 1990s Del Monte struggled to produce high quality pineapples until it began experimenting with the MD2 variety (Whitfield 2010). MD2 is highly sensitive to cultural practices and requires large investments in land preparation, fertilisers, pesticides and water management, and post harvest management and storage. Without careful technical management MD2 produces small inferior fruits and low yields. With its investment in large mechanised estates, Del Monte was ideally placed to develop large-scale production of MD2 on estates of several thousand hectares. The ability of Del Monte to produce large volumes of high quality pineapples at low costs gave it privileged access to European markets, leading to a rapid switch in demand from West African pineapples (Whitfield 2010, Fold and Gough 2007). Del Monte also launched an effective lobbying campaign for the dominant supermarket chains to stock MD2, and introduced innovations in downstream packaging and logistics to integrate production into supermarket chains (Fold 2008).

Del Monte attempted to introduce barriers of entry for the competition by claiming controversial proprietary rights over MD2. This was eventually legally rejected, but this enabled Del Monte to gain patenting rights until 2003, and establish a dominant role in the production of MD2 over competing agribusiness companies. As a consequence, it became difficult for West African cultivators to get access to MD2 and to work out suitable agronomic practices. The large capital investments required for successful cultivation of MD2 precluded the participation of the majority of Ghanaian producers, heralding the collapse of the smallholder pineapple industry. Since then, attempts to rehabilitate the Ghanaian pineapple industry have been based on the introduction of MD2 on large estates controlled by transnational sourcing companies, such as Golden Exotics Ltd (Compagnie Fruitière), a subsidiary of Dole – a major competitor of Del Monte in Coast Rica (Fold and Gough 2008, Fold 2008).

Large transnational companies largely control the pineapple industry with large estates vertically integrated with trading operations that have considerable control over packing, transportation, proprietary technology and influence over retail outlets in Europe and North America. Having reorganised production in Central America, these companies have now moved into the Philippines and West Africa, extending their monopoly control through subsidiaries. Yet these are also companies that wish to project themselves as ethical. Dole has become the largest organic company in the world through takeovers, and is also a major player in fair trade initiatives. Thus, the rise of large-scale pineapple production in West Africa is not the result of dubious and unethical land deals between governments and private corporations, but arises from intensive global market competition, which leads to the exit of smaller producers and market pressures for new scales of production related to technical innovations and logistics.

Conclusion

These two case studies illustrate different trajectories in contemporary agrarian accumulation and the relationships between smallholders and transnational corporations. Competition in the pineapple industry resulted in the decline of smallholder production in Ghana and the consolidation of vertical integration by transnational corporations. The collapse of smallholder production has made way for the emergence of large estates and the reorganisation of production. In contrast, increasing competition within the cocoa industry has not as yet been accompanied by the emergence of vertically integrated cocoa estates, although it has resulted in significant pressures on the smallholder sector and considerable immiseration. The major focus of transnational corporations within the Ivoirian cocoa sector has been on gaining control over state marketing and processing facilities, which has given them considerable leverage in controlling world cocoa grinding. This control has been further enhanced as the three major grinders moved into Ghana in the early 2000s and acquired grinding facilities. In Ghana they have not displaced the state monopoly over cocoa but work hand in hand with the state to grind cocoa in export free zones. Through these arrangements the three dominant transnational grinding firms have been able to gain a global monopoly control over cocoa grinding and considerable power through the relations with the state over the governance of cocoa production. Smallholder production is likely to be subject to considerable reorganisation and entry barriers as the industry adjusts to the decline of new frontier zones throughout the world, and the need to increase investments to produce high quality cocoa under the control of a highly concentrated industry. Although cocoa production is still dominated by independent peasant production, the transnational corporations dominating the industry are not bound by a universal ethical code of preserving smallholder production. Cocoa traders, such as Cargill, have investments in many other sectors in other countries characterised by large-scale acquisitions of land and vertical integration of production.

In both case studies the market has been characterised by concentration and monopoly behaviour that has undermined smallholder production. A significant number of agrarian markets are now controlled by a handful of transnational companies, who can source production from farmers throughout the world. Moreover a small proportion of these farmers, located in small regions, can saturate global markets for food commodities. Smallholders occupy an increasingly perilous position in global agri-food chains, which cannot be addressed by merely ensuring them security of rights to own land.

In some policy circles there are attempts to frame land grabbing as an aberrant trend to liberal market reforms and mainstream agribusiness. This deflects from a critical investigation of concentration within agribusiness and its impact on smallholders. By attempting to differentiate land grabbing from concentration, the fiction can be maintained that mainstream agribusiness activities and land grabbing are unrelated. It has been argued that privatisation is distinct from concentration. Privatisation refers to the appropriation of customary and communitarian resources by the private sector, while concentration refers to competition over privately owned resources and their further appropriation within commercial and private sectors (Land Tenure and Technical Development Committee 2010). This distinction enables the large areas of estate agriculture in Latin America, South Africa, Indonesia and Malaysia to be removed from the equation of land grabbing since they are based on historical processes of appropriation rather than contemporary processes. This is an artificial and unproductive distinction, since it prevents an understanding of the processes that create pressures for further concentration and the emergence of large estates throughout the world, as is evident in the impact of the rise of transnational production of pineapples in Costa Rica on Ghana smallholders.

Resource grabs occur in land, other natural resources, crop genetic resources, intellectual property rights, in the privatisation of public institutions, and in the rise of financial services and hedge funds. Financial investments and speculation in agriculture emerged in the crisis of the 1970s with the rapid expansion of futures markets and development of hedging and speculation as a central part of strategies of agribusiness companies. This gave rise to highly volatile commodity markets and increasing concentration through takeovers, mergers and price wars. This frequently undermined the livelihoods, standards of living and autonomy of smallholder farmers whose share of the value of agricultural production plummeted (Kaplinsky 2004, Daviron and Ponte 2005). The insecurity of smallholder existence in Africa needs to be rooted in the rise of concentration in export and horticultural sectors. The understanding of contemporary processes of agriculture accumulation and organisation needs to be located in larger political economy issues of the real concentration of wealth and power in agribusiness and its complex impact on smallholder production. Land appropriation interacts with other forms of appropriation, financial speculation in commodities, and entry barriers to erode smallholder livelihoods. Concentration and monopoly control can be established at a number of levels including trading, processing and production. The erosion of farm profits can result in a race to the bottom, in which only farmers willing to accept the lowest standards of living remain. This may result in lower quality of production and unsavoury production relations that threaten commodity branding, and political and social tensions that disrupt production, as in Ivoirian cocoa production. This may result in attempts to restructure the industry and introduce entry barriers that favour more commercial smallholders, and lead to the demise of less competitive smallholders. Competition and low farmgate prices may also lead to the fallout of commercial smallholders and the rise of highly capitalised large estates that are able to introduce further economies of scale, as in the Ghanaian pineapple sector. Under these conditions the expansion of land markets and proprietor rights in land facilitates the exit of less competitive farmers. The creation of land markets operates in synergy with several other forms of commodification to undermine the viability of smallholder agriculture, erode autonomous rural livelihoods, ruin regional forms of agriculture, and increase the insecurity of smallholder agriculture brought about by the expansion of transnational agribusiness. A more nuanced understanding of these issues can be gained by contextualising them within the history of the emergence and expansion of transnational agribusiness from the 1970s.

References

Alden Wily, L. 2003. *Community-based land management: questions and answers about Tanzania's new Village Act, 1999*. Drylands Programme issue paper no 120. London: IIED.

Aliber, M., Baipheti, M. and P. Jacobs. 2007. Agricultural employment scenarios. *In:* R. Hall, ed. *Another countryside: policy options for land and agrarian reform in South Africa*. Cape Town: Institute for Poverty, Land and Agrarian Studies, School of Government, University of the Western Cape, pp. 133–63.

Amanor, K.S. 2009. Securing land rights in Ghana. *In:* J.M. Ubink, A.J. Hoekema and W.J. Assies, eds. *Legalising land rights: local practice, state responses and tenure security in Africa, Asia and Latin America*. Leiden: Leiden University Press, pp. 97–132.

Barraclough, S. 1973. *Agrarian structure in Latin America*. Lexington, MA: Lexington Books.

Basset, T.J. 1993. Introduction: the land question and agricultural transformation in sub-Saharan Africa. *In:* T.J. Basset and D.E. Crummey, eds. *Land in African agrarian systems*. Madison, Wisconsin: Wisconsin University Press, pp. 3–34.

Bernstein, H. 2004. Changing before our very eyes: agrarian questions and the politics of land in capitalism today. *Journal of Agrarian Change*, 4(1&2), 190–225.

Berry R.A. and W. Cline. 1979. *Agrarian structures and productivity in developing countries*. Baltimore, MD: John Hopkins University Press.

Binswanger, H.P. and K. Deininger. 1993. South African land policy: the legacy of history and current options. *World Development*, 21(9), 1451–75.

Binswanger, H.P., K. Deininger and G. Feder. 1995. Power distortions, revolt and reform in agricultural land relations. *In:* J. Behrman and T.N Srinivasan, eds. *Handbook of development economics, Volume 3*. Dordrecht: Elsevier Science, pp. 2661–939.

Boiro, I. 1996. The Guinean land tenure system: a methodological approach to sustainable resource management. Case study Kollangui-Pita (Fouta Djallon). Managing land tenure and resource access in West Africa. *In*: Proceedings of a regional workshop held at Gorée, Senegal. London: IIED.

Borras, S.M. Jr. and J.C. Franco. 2012. Global land grabbing and trajectories of agricultural change: a preliminary analysis. *Journal of Agrarian Change*, 12(1), 34–59.

Borras, S.M Jr., P.M. McMichael and I. Scoones. 2010. The politics of biofuels, land and agrarian change: editors' introduction. *The Journal of Peasant Studies*, 37(4), 575–92.

Broehl, W.G. Jr. 1998. *Cargill: going global*. Hanover, NH: University Press of New England.

Chachage, C. 2010. Land acquisition and accumulation in Tanzania: the case of Morogoro, Iringa and Pwani regions. Morogoro: Pelum Tanzania. Available from: http://www.commercialpressuresonland.org/research-papers/land-acquisition-and accumulation-tanzania [Accessed 10 October 2011].

Chauveau, J.-P. 2006. How does an institution evolve? Land, politics, intergenerational relations and the institution of the *tutorat* amongst autochthones and immigrants (Gban region, Côte d'Ivoire). *In:* R. Kuba and C. Lenz, eds. *Land and the politics of belonging in West Africa*. Leiden and Boston: Brill, pp. 213–40.

Chauveau, J.-P. 2009. La loi de 1998 sur le domaine rural dans l'histoire des politiques foncières en Côte d'Ivoire: la politique des transferts de droits entre 'autochtones' et 'étrangers' en zone forestière. *In:* J.-P. Colin, P.-Y. Le Meur and É. Léonard, eds. *Les politiques enregistrement des droits fonciers: du cadre aux pratiques locales*. Paris: Karthala, pp. 105–40.

Chauveau, J.-P. and J.-P. Colin. 2010. Customary transfers and land sales in Côte d'Ivoire: revisiting the embeddedness issue. *Africa*, 80(1), 81–103.

Daviron, B. and P. Gibbon. 2002. Global commodity chains and African export agriculture. *Journal of Agrarian Change*, 2(2), 137–61.

Daviron, B. and S. Ponte. 2005. *The coffee paradox: global markets, commodity trade and the elusive promise of development*. London: Zed Books.

De Janvry, A. 1981. *The agrarian question and reformism in Latin America*. Baltimore: Johns Hopkins University Press.

Dinerman, A. 2006. *Revolution, counter-revolution and revisionism in postcolonial Africa: the case of Mozambique 1975–1994*. Abingdon, Oxford and New York: Routledge.

Diop, M. 2007. *Réformes foncières et gestion des resources naturelles en Guinée: enjeux de patrimonialité et de propriété dans le Timbi au Fouta Djalon*. Paris: Karthala.

Fold, N. 2002. Lead firms and competition in 'bi-polar' commodity chains: grinders and branders in the global cocoa-chocolate industry. *Journal of Agrarian Change*, 2(2), 228–47.

Fold, N. 2008. Transnational sourcing practices in Ghana's perennial crops sectors. *Journal of Agrarian Change*, 8(1), 94–122.

Fold, N. and K. Gough. 2008. From smallholders to transnationals: the impact of changing consumer preferences in the EU on Ghana's pineapple sector. *Geoforum*, 39, 1687–97.

Fox, T. and B. Vorley. 2004. *Stakeholder accountability in the UK supermarket sector: final report of the 'Race to the Top' project*. London: IIED.

Gibbon, P. 1999. *Privatisation and foreign direct investment in mainland Tanzania, 1992–1998*. Copenhagen: Centre for Development Research.

Gibbon, P. and A.O. Olukoshi. 1996. *Structural adjustment and socio-economic change in sub-Saharan Africa: some conceptual, methodological and research issues*. Uppsala: Nordic Africa Institute.

Gilliam C. 2010. Funds flow towards farmlands as experts eye deal. *Reuters edition US*, 7 May. Available from: http://www.reuters.com/article/2010/05/07/aginvesting-idUSN07249981 20100507 [Accessed 10 October 2011].

GRAIN 2008. *Seized! The 2008 Landgrab for food and financial security*. GRAIN Briefing Paper. Available from: http://www.grain.org/briefings/?id=212 [Assessed 10 October 2011].

Hall, R. 2007. Transforming rural South Africa? Taking stock of land reform. *In:* L. Ntsebeza and R. Hall, eds. *The land question in South Africa: the challenge of transformation and redistribution*. Cape Town: HSRC Press, pp. 87–106.

Heffernan, W.D. and M.K. Hendrickson. 2002. Multi-national concentrated food processing and marketing systems and the farm crisis. Paper presented at the Annual Meeting of the American Association for the Advancement of Science Symposium on Science and sustainability: the farm crisis: how the heck did we get here, February 14–19, Boston, MA.

ILRF 2008. *The sour taste of pineapples: how an expanding export industry undermines workers and their communities*. Washington, DC: International Labor Rights Forum.

Jacobs P. 2009. Agricultural market reform and the rural poor in South Africa. PLAAS poverty workshop paper, Cape Town. Available from: www.hsrc.ac.za/Document-3216.phtml [Accessed 10 October 2011].

Kaplinksy, R. 2004. Competitions, policy and global coffee and cocoa chains. IDS, Brighton. Available from: http://www.ids.ac.uk/ids/global/pdfs/RKaplinskycocoacoffee05.pdf [Accessed 10 December 2010].

Kouame, G. 2010. Intra-family and socio-political dimension of land markets and land conflicts: the case of Abure, Côte d'Ivoire. *Africa*, 80(1), 126–46.

Lahiff, E. 2003. The politics of land reform in Southern Africa. *Sustainable livelihoods in Southern Africa*. Research Paper No. 19, March, PLAAS, Cape Town. Available from: http://www.ids.ac.uk/FA0BB47C-5056-8171-7BE8FD1457B52069 [Accessed 10 October 2011].

Land Tenure and Technical Development Committee 2010. *Large-scale land appropriations: analysis of the phenomenon and proposed guidelines for future action*. Nogent Sur Marne: AGTER.

Lapavitsas, C. 2009. Financialisation, or the search for profits in the sphere of circulation. Research on Money and Finance Discussion Paper No. 10, London: School of Oriental and African Studies. Available from: www.researchonmoneyandfinance.org [Accessed 10 October 2011].

Lavigne Delville, P. 2000. Harmonising formal law and customary land rights in French-speaking West Africa. *In:* C. Toulmin and J. Quan, eds. *Evolving land rights, policy and tenure in Africa*. London: IIED, NRI and DFID, pp. 97–122.

Léonard, E. 1997. Crise écologique, crise économique, crise d'un modèle d'exploitation agricole: ajustements et recomposition sociales sur les anciens fronts pionniers Ivoiriens. *In:* B. Contamin and H. Memel-Foté, eds. *Le modèle Ivoirien en question: crises, ajustements, recompositions*. Paris: KATHALA and L'ORSTOM.

Léonard, E. and M. Oswald. 1997. Cocoa smallholders facing a double structural adjustment in Côte D'Ivoire: responses to a predicted crisis. *In:* F. Ruf and P.S. Siswoputranto, eds. *Cocoa cycles: the economies of cocoa supply*. Cambridge: Woodhead Publishing Ltd, pp. 125–50.

Lipton, M. 1977. *Why poor people stay poor: a study of urban bias in world development*. London: Temple Smith.

Lipton, M. 1993. Land reform as commenced business: the evidence against stopping. *World Development*, 21(4), 641–57.

Lipton, M. 2010. *Land reform in developing countries: property rights and property wrongs*. London and New York: Routledge.

Losch, B. 2002. Global restructuring and liberalization: Côte d'Ivoire and the end of the cocoa market? *Journal of Agrarian Change*, 2(2), 206–27.

Manji, A. 2006. *The politics of land reform in Africa from communal tenure to free markets*. London: Zed Books.

McMichael, P. and H. Friedmann. 2007. Situating the 'retail revolution'. *In:* D. Burch and G. Lawrence, eds. *Supermarkets and agri-food supply chains: transformations in the production and consumption of food*. Cheltenham: Edward Elgar, pp. 291–319.

Moyo, S. and P. Yeros. 2005. Land occupations and land reforms in Zimbabwe: towards the national democratic revolution. *In:* S. Moyo and P. Yeros, eds. *Reclaiming the land: the resurgence of rural movements in Africa, Asia and Latin America*. London: Zed Books, pp. 165–208.

Newman, S.E 2009. Financialization and changes in the social relations along commodity chains: the case of coffee. *Review of Radical Political Economics*, 41(4), 539–59.

Noronha R. 1985. *A review of the literature on land tenure systems in sub-Saharan Africa*. Washington, DC: World Bank, Agricultural Development Department.

Off, C. 2006. *Bitter chocolate: investigating the dark side of the world's most seductive sweet*. Toronto: Random House Canada.

Okoth-Ogendo, H.W.O. 1993. Agrarian reform in sub-Saharan Africa: an assessment of state responses to the African Agrarian crisis and their implications for agricultural development. *In:* T.J. Basset and D.E. Crummey, eds. *Land in African Agrarian Systems*. Madison, Wisconsin: Wisconsin University Press, p. 247–73.

Rahmato, D. 2009. Land rights and tenure security: rural land registration in Ethiopia. *In:* J.M. Ubink, A.J. Hoekema and W.J. Assies, eds. *Legalising land rights: local practice, state responses and tenure security in Africa, Asia and Latin America*. Leiden: Leiden University Press, pp. 33–58.

Ruf, F. 2001. Tree crops as deforestation and reforestation agents: the case of cocoa in Côte d'Ivoire and Sulawesi. *In:* A. Angelsen and D. Kaimowitz, eds. *Agricultural technologies and tropical deforestation*. Wallingford: CAB International, pp. 291–316.

Scoones, I., N. Maorongwe, B. Mavedzenge, J. Mahenehene, F. Murimbarimba and C. Sukume. 2010. *Zimbabwe's land reform: myths and realities*. London: James Currey, Harare: Weaver Press and Sunnyside, South Africa: Jacuma Media.

Sender, J. and D. Johnston. 2004. Searching for a weapon of mass production in rural Africa: unconvincing arguments for land reform. *Journal of Agrarian Change*, 4(1&2), 142–64.

Ubink, J.M. 2008. Negotiated or negated? The rhetoric and reality of customary tenure in an Ashanti village in Ghana. *Africa*, 78(2), 264–87.

Vlassenroot, K. and C. Huggins. 2005. Land, migration and conflict in eastern DRC. *In:* C. Huggins and J. Clover, eds. *From the ground up: land rights, conflict and peace in sub-Saharan Africa*. Nairobi: African Centre for Technology Studies and Pretoria: Institute for Security Studies, pp. 115–94.

Watts, M. 1994. Life under contract: contract farming, agrarian restructuring, and flexible accumulation. *In:* P. Little and M. Watts, eds. *Living under contract: contract farming and agrarian transformation in sub-Saharan Africa*. Madison: University of Wisconsin Press, pp. 21–77.

Whitfield, L. 2010. *Developing technological capabilities in agro-industry: Ghana's experience with fresh pineapple exports in comparative perspective*. DIIS Working Paper 2010: 28, Copenhagen: DIIS.

World Bank 2008. *World development report 2008: agriculture for development*. Washington, DC: World Bank.

World Bank 2009. *Awakening Africa's sleeping giant: prospects for commercial agriculture in the Guinea savannah zone and beyond.* Washington, DC: World Bank.
Young, W. 2004. *Sold out: the true cost of supermarket shopping.* London: Vision Paperbacks.
Zoomers, A. 2010. Globalisation and the foreignisation of space: seven processes driving the current global land grab. *Journal of Peasant Studies*, 37(2), 429–47.

Kojo Sebastian Amanor is an Associate Professor at the Institute of African Studies, University of Ghana. His main research interests and publications are on agrarian political economy, land, seeds, agribusiness, the agricultural-forest interface and smallholders, and community development.

Looking back to see forward: the legal niceties of land theft in land rushes

Liz Alden Wily

This paper aims to make a modest contribution to an overdue need to locate the current land rush in its historical context, less as a new phenomenon than as a surge in the continuing capture of ordinary people's rights and assets by capital-led and class-creating social transformation. It aims to do so by looking back to earlier land rushes, and particularly to those which have bearing upon sub-Saharan Africa, the site of most large-scale involuntary land loss today. In particular, the paper focuses upon a central tool of land rushes, property law. The core argument made is that land rushes past and present have relied upon legal manipulations which deny that local indigenous ('customary') tenures deliver property rights, thereby legalizing the theft of the lands of the poor or subject peoples. Even prior to capitalist transformation this feudal-derived machination was an instrument of aligned class privilege and power, later elaborated to justify mass land and resource capture through colonialism. Now it is routinely embedded in the legal canons of elite-aligned agrarian governance as the means of retaining control over the land resources which rural communities presume are their own.

Introduction

Although the global land rush is the context for this paper, it is not discussed in detail. The global land rush (or land grab) refers to acceleration since (especially) 2008 in the formal transfer of lands in agrarian states from the peasant farming and pastoral sector into the hands of large-scale land producers and/or speculators. They are both local and international and include not just private sector actors but foreign states or their parastatals, adding force to the trend (Deininger *et al.* 2010, JPS 2011 *passim*, Anseeuw *et al.* 2011).

This paper does not delve into the extent, character, or projection of this trend, around which a mass of research and opinion now exists. Instead, for the purposes of this paper, it takes as given that first, a land rush is indeed occurring; that the rush is already or potentially detrimental to the land rights of thousands of land-dependent households; that the surrender of lands is significantly involuntary on the part of those deprived of lands they believe to be their own, or at least they are

---

The author is very grateful to Ian Scoones for his helpful comments on the first draft of this paper and to four anonymous reviewers of a subsequent draft.

ill-informed or even misled as to the benefits of surrendering these lands; and that so far, most of the lands being taken or proposed for land takings are in sub-Saharan Africa.

This paper also takes as given that the primary agent in often industrial-scale land transfers are governments. This is because the source of lands being sold or mainly leased are lands or land types which governments claim as their own, and/or as rightfully theirs to dispose of. This means, as I have argued elsewhere (Alden Wily 2011c, 2011f), that the current land rush amply demonstrates the use of perfectly legal means of dispossession or reallocation of lands involving significant loss of access for rural poor and not a little physical displacement in practice. In this context, the 'land grabbers' are not the beneficiary investors but governments of the people affected, and for as yet uncertain public good. Land transfers are undertaken strictly within the terms of domestic property laws: laws made by host governments and their predecessors. Broadly, these laws are designed to render untitled (but traditionally occupied and used) lands as unowned, and the state, by default, their legal owner.

This is not a minor scenario. For, to over-generalise only slightly, most of the land area of the agrarian world, the target of the land rush, is untitled. The rights to land of millions of rural households are accordingly under threat should the trend continue. This particularly affects sub-Saharan Africa. There, by the start of the twenty-first century, around 90 percent of the region was the *de jure* or *de facto* property of governments (Alden Wily 2011e). The vulnerability of nearly half a billion Africans to state-led involuntary loss of land, resources and livelihood is presently significant.

How, this paper asks, has this situation come about? The overarching explanation clearly lies in the ever-expanding and now globalised capitalist transformation; this has, *inter alia*, consistently left in its wake mass dispossession of agrarian poor in service of engineered elite capture, commercialized land use, and other developments proclaimed necessary for growth. And yet, the plethora of critiques upon the contemporary land rush is still primarily descriptive and ahistorical. Location of the rush in a social transformation context, and structural analysis of its process, is now beginning to be seen, for example, in the work of Murray Li (2011) on labour relations, Patnaik and Moyo on agrarian crisis (2011), the helpful reminder by Bernstein (2010) of how class formation is integral to capitalist transformation and the polarised accumulation which marks it, and in the recent work of Borras and Franco (2012). This last add significantly to our understanding of distinctions between land ownership and land use transitions which are 'part of everyday dynamics of accumulation, differentiation and dispossession/displacement' and those which are 'land grab-induced' (p. 46). Further pursuit of such analyses by scholars now seems urgent. This is not least as advocacy for containment of what appears to be an unstoppable trend of land loss by poor rural communities, and irrevocable given the legally binding ways in which it is being effected.

The aim of this paper is to proffer insights into these legal dimensions. For this, it is argued, help from the past is needed to better understand how these aspects have come about. It is for this reason that this paper pays less attention to the current land rush than to those which have established the legal manipulations which continue to make rushes possible. I choose to begin the story with Ireland, some centuries past.

Looking back

1607 in Ireland

The English have been taking Irish lands since 1169 but mainly around Dublin ('The Pale') and where feudal land relations have been satisfactorily established among better-off Gaelic-Irish and the Catholic Church. This leaves most of the island still governed by indigenous law (*Brehon* law). First written down in the seventh century, this customary law is sophisticated and administered by trained traditional magistrates (*Brithem* or *Brehons*) (Ginnell 1894). Even English Normans secure properties through Brehon norms when it suits, not least for their practical inheritance rules.

Now the revitalised Elizabethan conquest of Ireland (1603) makes it necessary to find a cheap but legal way of acquiring vast but occupied lands beyond the Pale. (Legality is necessary in the minds of the new legal elites who have such a great hand in governing the new United Kingdom of Wales, England, Scotland and Ireland, and who have been busy for some decades bringing disparate courts and laws under a single superior modern 'common' law (Dorsett 2002). Accordingly, through manipulated cases of *Gavelkind* (1605) and *Tanistry* (1606, 1608), the courts helpfully rule that Brehon land law does not deliver property in a manner compatible with English sovereignty and feudal common law.

Noblemen, including traditional elites, scramble to prove their title under the individualising norms of this new English common law in order to avoid compulsory surrender of their lands, helping to seal the demise of communitarian law and governance (Morrissey 2004). The majority of poor Irish fail in this endeavour. For centuries, clan-owned lands have been carefully (if inequitably) sub-divided into family homesteads, larger lands for clan leaders and officials – including areas they must devote to feeding the destitute – communal lands available for shifting cultivation under usufruct, and highland and bog commons regulated for grazing, hunting and peat fuel collection. These now begin to be systematically allocated to new waves of English and Scottish settlers (an estimated 100,000 between 1603 and 1641). The Irish become their increasingly exploited tenants. 'Mercenary individualistic attitudes to land ownership' and the 'infiltration of opportunistic colonial investors' see a market in land develop (Morrissey 2004, 12). Seeds of local land rebellion are sown, first surfacing in 1641.

As to legal stratagem, justifications brought before the Court of King's Bench in the *Case of Tanistry* in 1608 include: the impossibility of the customary law of the *Irishry* (sic) surviving the establishment of the *superior* common law of the conquerors; that as a Christian kingdom Ireland must follow the more Christian (i.e. Protestant) law of England; and that the Irish will only benefit by coming under the 'just and honourable common law of England' (Dorsett 2002, 9–13, citing Sir John Davies, who later became the Attorney-General for Ireland). Additionally, laws must serve public good, says Davies. Public good requires 'certain ownership of land'. Customary law which allows the extended family to elect the most suitable heir (not necessarily the eldest son) cannot be certain enough to encourage investment in the land. Indeed such *tanistry* is 'the true cause of barbarism and desolation which is in all of the Irish counties where the custom is in use'. Most of all, Davies argues that such customs are 'prejudicial to the profit and prerogative' of the King who, if allowing clan tenure to continue, deprives himself of the (lucrative) benefits and powers of being the ultimate owner (Dorsett 2002, 9–13).

1823 in America

Chief Justice Marshall of the new U.S. Supreme Court (U.S. Supreme Court, 1823) rules that while Indian natives were rightfully *in possession* of 43,000 square miles of disputed land in a case he engineers to be brought before the Supreme Court (he has personal interests in the results), they did not have the right to sell this to developers (*Johnson and Graham's Lessee v. William M'Intosh,* U.S. 1823). However, Marshall cannot deny Indian land interests entirely; after all, it has been routine for Europeans since the seventeenth century to acquire land by direct purchase from the natives, indicating acknowledgement of Indian ownership. Further, a number of states have new native protection policies which have to be taken into account. Still, he finds ways to diminish native rights to 'a mere right of usufruct and habitation, without power of alienation' (U.S. 1823, 569).

Establishing the right of discovery

Marshall's main argument is that by virtue of conquest, the British Crown not only established political suzerainty over the territories but became the owner of all lands within them ('Discovery is the foundation of title, in European nations, and this overlooks all proprietary rights in the natives', p. 567). Therefore, only the Crown or its successors may sell or grant lands in ways which English common law regards as real property. Marshall draws deeply upon the very feudal notions of tenure which newly independent America had supposedly reviled ('All titles must be derived from the Crown', p. 570), and more speciously, upon opinions delivered in England on other matters in 1772 and 1774, which prove to him that English common law supersedes local law (pp. 568, 572).

Establishing the notion of vacant lands

Yet the Chief Justice has more strings to his bow, for Indians are 'savages' and the force used to acquire title by conquest 'finds some excuse in the character and habits of the people' (p. 589). Marshall also claims that Indians fled from bloody wars ahead of colonial agriculturalists, leaving the soil 'no longer occupied' (p. 591). Therefore the country was uninhabited, and 'according to the theory of the British constitution, all vacant lands are vested in the crown' (p. 595). Finally, Marshall argues that even prior to the arrival of Europeans Indians were in fact no more than occupants of the Americas. As they were not cultivators, they could not have acquired 'proprietary interest in the vast tracts of territory which they wandered over... they had no individual rights to land' (p. 570).

Establishing that possession does not necessarily mean ownership

Nevertheless, Marshall concludes, 'It has never been contended that the Indian title amounted to nothing. Their *right of possession* has never been questioned. The claim of government extends to the complete ultimate title...' (p. 603). Possession, or occupancy, and title and ownership are thus carefully separated, enabling colonizers thereafter to ignore the fact that the continents and islands they capture are in any sense owned.

1845 in England

The villagers of Otmoor, Oxfordshire, as described by Linebaugh (2010) have lost the fight to keep their commons including their smallholdings as have hundreds of other communities across the realm. In fact, feudal land law in England (and the rest of Europe) has dictated for some centuries (since 1285) that only those grantees of the king, i.e. the land lords, own the land. Local populations hold no more than use rights. Enclosures and concentration of ownership in the hands of the minority has been steady, with a peak in the fifteenth and sixteenth centuries, and now reoccurring with a vengeance. The legal realities of permissive occupancy over farms and 20 million hectares of pastures, meadows, moors, and other commons now comes harshly into focus with industrialisation, private capital hungry for lands and the financial killings which may be made from selling, in particular, unfarmed commons to railways and factories. Parliament, made up of wealthy (land) lords, is on their side, passing law after law since 1773 to legalise privatisation of this or that commonage. *The General Inclosure Act 1845* (sic) administers the *coup de grâce*, unifying and accelerating the process.

Of course these private gains are declared as being in the public interest. Public interest was, Eric Hobsbawn writes, 'no more than the forces of profit-pursuing private enterprise' which sought 'to turn land into a commodity', 'to pass this land into the ownership of a class of men impelled by reason; i.e. enlightened self-interest and profit', and 'to transform the great mass of the rural population into freely mobile wage-workers' (1962, 184). The parliamentary enclosures were, in a word, writes the equally famous historian, E.P. Thompson,

> ... a plain enough case of class robbery, played according to the fair rules of property and law laid down by a parliament of property-owners and lawyers.... But what was 'perfectly proper' in terms of capitalist property-relations involved ... a rupture of the traditional integument of village custom and of right; and the social violence of enclosure consisted precisely in the drastic, total imposition upon the village of capitalist property-definitions.... (1991, 237–239).

Better-off commoners could be bought, and 'whose eyes might glitter at the short-term prospect of outright proprietorship – even the four or five acres which enclosure might bring – but the cottager without any proprietary rights by enclosure lost all' (Thompson 1991, 241).

Resistance and relief

As the majority poor resisted, customary land arrangements were seen as dangerous centres of indiscipline, and dispossession more rigorously applied. Meanwhile the jobs and opportunities promised by the politicians to evicted labour and the growing urban working class were never enough; the regime of poor houses during the 19th century would eventually become the national welfare system. Additionally, rural population growth more than compensated for migration from village to town. Landowners themselves began to 'regret the lost commons – the cow, the geese, the turfs – which had enabled the poor to subsist without coming to the parish... But it was too late to reverse a general process: no common was ever brought back though many more were enclosed...'

(Thompson 1991, 245). The centrality of land loss did not dissipate. While the groundswell of rural grievance gave rise to urban revolts and unionism, 'it always came back to access to the land' (pp. 253–254).

Looking to Africa

The first land grab: 1885 to 1915

In 1885 plenipotentiaries of European Powers (as they refer to themselves) agree to establish respective 'spheres of economic influence' in Africa, making the key entry points of the Niger and Congo Rivers free trade zones. As newly industrially emergent, Germany hosts the diplomatic meet. Stress levels are high. Europe is in economic crisis (the Great Depression 1873–1896) and a vast export drive is seen as solution (Hobsbawn 1987, 66). Factory owners need new markets for textiles and other manufactures lying unsold. They have their eyes on China and Africa. Fabulously wealthy entrepreneurs, 'with vast accumulations of capital burning holes in (their) owners' pockets' are also seeking new investment opportunities (Hobsbawn 1975, 112 as cited by Kuponen 1993). The new working classes, having lost their lands and livelihoods, and now dependent upon (failing) factory jobs, are 'distressed' (restive). Additionally, population growth continues to surge. More locales for emigration are needed.

Trade intentions are not one-way. Europe needs more rubber, oil, tin and copper to expand industry; gold and diamonds to satisfy the rising number of bourgeois consumers; and palm oil to feed soap manufacturing(one of the few expanding industries). More grain and wheat is also needed as well as the 'colonial goods' which European populations have come to depend upon: sugar, coffee, tea, cocoa, and tropical fruits (Hobsbawn 1987, 64). If expanding consumerism is one pillar of the economic scramble, then plantations are another and metropolitan traders and financiers are the third (Hobsbawn 1987, 64).

The issue of free trade is so important that the Powers end the conference with an international trade law, the *General Act of the Berlin Conference on West Africa, 26 February 1885* (Berlin 1885). African land rights are not mentioned in this law. The suggestion during the Conference by the American plenipotentiary that the American Government '... would gladly adhere to a principle which should aim at the voluntary consent of the natives whose country is taken possession of...' is quietly ignored (Huberich 1947, 1207). It is merely agreed in the law that each Signatory Power will notify the others if it 'takes possession of a tract of land on the coasts of the African continent outside of its present possession' (Article 34). They also pledge to protect 'existing rights', meaning those already established by Europeans (Article 35). In any event, Africans will gain from the 'blessings of civilization', notably Christianization (Article 6).

In practice, by 1890, establishing enterprise and markets in Africa is not working out so well; clearly political conquest is needed to secure supremacy. Free trade goes out the window as each Signatory Power becomes rampantly protectionist. By 1895 the economic scramble for Africa has segued into a political scramble with the creation of colonies to protect new markets and to tap the extraordinary wealth of resources and cheap labour that expansionism into hinterlands is revealing.

Capturing not just territory but all rights to land therein

The problem in the 1890s is once again how to acquire such massive lands and resources cheaply – and legally, now necessary to pacify public opinion at home, politicised by anti-slavery and humanist movements. As in America, there are awkward precedents to overcome. For many decades, (some) traders, profiteers and missionaries have been buying lands (albeit for petty prices) from coastal chiefs to establish trading posts, ports, missionary enclaves, and latterly, anti-slavery monitoring posts. Companies backed by European governments have been doing the same. The British Royal Niger Company alone has several hundred land contracts made with West African chiefs guaranteeing access to land for oil palm plantations. There is also the 1844 Bond (bilateral investment treaty) to consider, agreed by 'sovereigns of equal power' on the Gold Coast and the British Crown (Amanor 2008a). These conditions make it difficult to pretend that Africa is entirely unowned.

Luckily, the old feudal land laws of Europe along with the above Marshall Ruling of 1823 come to the rescue. They offer a clutch of routes to legalise dispossession at scale, refined through practice in the Americas, Asia and Oceania. The 'right of discovery' assures the new colonizers of Africa root ownership of the soil. It also opens the way for European law, including its property norms, to be applied. Africans unwittingly assist; many claim that only God can own the soil and that ownership of rights to the land is vested in the community, an inter-generational entity which exists in perpetuity. While firm in this temporal property right (and routinely fighting wars over boundaries) some chiefs admit the land itself cannot be sold outright and should not even be lent or leased without the consent of communities. This lack of fungibility and tendency to communalism further proves to colonizers that Africans do not own their lands (Alden Wily 2007). Nevertheless, it is practical to admit native *possession*, at least where Africans have settled and permanently farm. It would not, in any event, be wise to make it difficult for 'the natives' to feed themselves. A policy of 'peaceful native occupancy' permeates policy making, meaning natives will not be disturbed unless necessary (Alden Wily 2011a). 'Necessary' means when their lands are required for the more important needs of non-natives.

Singling out the commons

Colonizers can afford to be generous. Respect for peaceful occupancy over village and farm lands leaves nine-tenths of the continent available, the several billion hectares of unfarmed lands owned and used by respective tribes and villages for off-farm purposes such as hunting, grazing and wood collection, and purposely impermanent shifting cultivation. For have not Adam Smith, Locke, Mills, and others established long ago that 'property' only comes into existence by man's labour and thus involves clearance and cultivation? The 'wastes' or commons of feudal tenure (the land left over as unrequired by the (land) lord) are neatly applied to the continent by each Signatory Power. Counterpart constructs quickly evolve to define lands which are 'effectively occupied' (homesteads and farms) and lucrative presumed 'unowned and vacant lands'; *wastelands* to the British, *Herrenlos* to the German Empire, and *terres sans maîtres* in Francophone Africa. These fall directly to new colonial administrations.

Through such means dispossession of Africans is more or less total – in new colonial *law*. In practice, the ability of Europeans (and Afro-American freed slaves in Liberia) to settle and develop more than several million hectares in each new polity is constrained (British South Africa and German Southwest Africa aside). Africans continue to occupy and use lands which they do not legally own.

Resistance and compromise

Enforcing mass dispossession also proves not so easy. First, Africans are not as compliant as hoped where their lands are concerned. Protests and rebellions occur up and down the continent in response to land takings (Iliffe 1972, Clover 2005, Magaziner 2007). Resistance is most sustained in Southwest Africa where only angry public response at home in Berlin puts a stop to the long slow genocide of Herero and Nama defending their lands (1890–1908) (Werner 1993).

Second, those who use civil servants rather than soldiers to colonize Africa (i.e. the British and less fully, the French) are forced to compromise. The British find themselves settling for protectorate arrangements in some areas, acknowledging Arab estates along the Sudan Nile as absolute property (1899) and rewarding the King of the Buganda for his military support with legal title for himself and his extended family (thereafter turning his subjects into tenants) (Mugambwa 1987).

Constraints from home

Nor is European law entirely helpful to the colonial mission. The French Civil Code, operating in French coastal enclaves since the 1850s, guarantees privileges to 'assimilated' Africans including the right to hold property. Several of its articles are supportive of customary rights (e.g. Article 713). Such provisions require curtailment, effected in laws of 1891 (for Gabon and Congo), 1900 and 1906. English common law is also clear that protectorate arrangements do not allow colonizers to take and alienate native lands (McAuslan 2006). This drives British determination to convert protectorate agreements into full colonies, not achieved in Ghana until 1895, 1906 in Southern Nigeria, and 1920 in Kenya. Barotseland in modern-day Zambia and all of Bechuanaland, Basutoland and Swaziland avoid this fate, setting treatment of customary land rights on a novel path. The Basuto King issues his own Laws of Lerotholi in 1903 (Juma 2011). The Tswana also secure recognition that customary law governs their land relations in Bechuanaland (Botswana), other than in areas occupied by unwary San (Bushmen) whose lands automatically become Crown property (Schapera 1970).

Playing the colonizers at their own game

More frustrating for colonizers is the capacity of certain world-wise local leaders to use the colonizers' own laws against them. Having long maintained missions in European capitals to service their slaving, then commodity, interests, Gold Coast chiefs and elites are far from unknowing of European intentions. They rebuff British efforts in 1894, and again 1897 and 1910, to turn their gold-rich forest lands into Crown Lands, aware how this would limit the money they themselves may make by leasing lands to fast-arriving gold mining companies. They form an Aborigines Rights Protection Society and send a deputation to the Privy Council in England. They argue that 'no land is unowned' in the Gold Coast and that customary tenure should form the basis on which the modern colonial state should be built (Amanor 2008a). Various colonial researches back them up. Their success in London means that 70 percent of Ghana will remain the private collective property of communities, but which chiefly elites will steadily reshape as vested in themselves, less as custodians than as absolute owners.

African leaders in Lagos also use the English courts to recoup their rights. In 1912 and finally in 1921 the colonial government of Southern Nigeria will be forced to live with the decision of the Privy Council in London that King Docemo of the Benin Empire had not actually *sold* Lagos Island to the United Africa Company in 1861 (bought by the British Government) but only ceded his political sovereignty, 'leaving native land ownership unimpaired'.[1] However, the British are still amply able to limit those rights by virtue of denying that unfarmed lands are owned and therefore do not require compensation when taken.

The common cause of dispossession

Despite these difficulties, the colonial land rush satisfactorily engineers the dispossession needed to support European enterprise. The mechanisms are quite uniform, which is unsurprising, given the shared agenda of the Powers and the federated nature of especially French colonial governance wherein laws are handed down to its colonies to adopt from focal points in Port Louis or Brazzaville. The shared objectives of these laws between 1890 and 1918 are straightforward: to establish the ultimate title and controlling authority of the new colonial state; to restrict the tendency of settlers and profiteers to make land deals themselves; to reject or regularize European acquisitions already made; to meet the demands of new arrivals for entitlements which will stand up in the courts at home, and, as time passes, to clamp down on the tendency of Africans to protest at land takings.[2]

As an example, early provisions in Sudan (1899) which recognize that most of Sudan is subject to rights vested in tribes, sections and villages are refined in 1901, 1903 and 1905 to make it more bluntly clear that these lands are Government property 'until the contrary is proven'. Faced with persisting and rapacious land grabbing by settlers in Tanganyika (Iliffe 1972, 8), the German administration reduces all land grants to 25-year leaseholds until the holder has shown clearance of at least half the land (James 1971 as cited in Wily 1988, 45). Following the swift capture of most of the pastoral area of Southwest Africa into eight private concessions by 1893 (Werner 1993), Germany is also the first to delimit areas where

[1]Privy Council, 1921. The issue had already come before the Supreme Court of the Colony of Southern Rhodesia in 1912–1914, not decided upon until 1919, at which point native title was not acknowledged as existing (*Re Southern Rhodesia* (1919) A.C. 211,233–4).

[2]Such provisions are clear in the land laws enacted for the German territory of Tanganyika, Ruanda and Burundi (1891) and Cameroon (1896); laws enacted throughout the 13 French possessions (such as in Senegal in 1900, 1904 and 1906); laws enacted in the Belgium Congo Free State (1885, 1886, 1906), and Anglophone laws such as enacted in British Kenya (1901, 1902, 1915). See Alden Wily (2011b) for details.

Europeans may *not* acquire land, such as in its 1896 Ordinance for Cameroon. The argument is made that as well as arousing local unrest, private deals with natives challenge the 'profit and prerogative' of the colonizing state as the sole authority which may take people's land (and for free). Initially these off-limits areas around the continent are expansive, except in British South Africa where the Land Act of 1913 designates only 7 percent of the country as native areas, leaving inhabitants in the remainder with the choice of becoming wage labourers on what were their own lands or squeezing themselves into the reserves. At a lesser scale, this choice is replayed wherever Europeans are granted title.

A second land rush: 1919–1939

Following the First World War (1914–1918) the state-led scramble for Africa settles down and more orderly land takings ensue. Profiteers become 'investors'. The system for land grants is better organized with new legislation enacted in almost all territories (Alden Wily 2011b). Mercantile companies in French territories (individual settlers are discouraged except in one part of the Ivory Coast), American companies in Liberia (including Firestone, granted rights to up to one million acres), and both corporate and individual grantees in Anglophone Africa lead the expansion of plantation agriculture for sisal, cotton, rubber and oil palm. Mega-schemes are developed, especially in Francophone Africa, such as through the *Office du Niger*, set up in 1932, to green the desert in what is modern-day Chad, Burkina Faso and Mali. White settlement in most eastern and southern Africa states multiplies (nine-fold in Nyasaland (Malawi) in three years alone (Kanyongolo 2008)). Without the barriers of reserves, Africans in Lusophone territories probably fare worst; having already endured the arrival before 1900 of 10,000 criminals shipped to find land for themselves in Angola, they are now confronted with even larger waves of arrivals in the 1920s and in circumstances where the new fascist regime in Lisbon (1926) does not feel bound by pre-war support for 'peaceful native occupancy' (Clover 2005).

Challenges from without

At the same time, global opinion is decreasingly tolerant of land takings in Africa. Following gleeful declaration of Tanganyika as Crown property (and certainly not 'one whit less British nor our tenure one whit less permanent' (Tanganyika League 1927 as cited by Wily 1988, 51)), the League of Nations coldly reminds Britain (and Belgium and France) that they have been ceded former German territories as trustees, not owners, and have agreed under international law to protect native land rights (Article 6 of the Mandatory Agreement). The Governor of Tanganyika gets around this by enacting the *Land Tenure Ordinance, 1923*, which respects African rights to 'use and enjoy' the King's land. He then launches 'a certain amount of white settlement to develop the country's resources' (1926) involving a million more hectares (Wily 1988, 56). Still hounded by the League, Britain grudgingly deems these rights to be similar to the titled Rights of Occupancy awarded settlers (1928), but only so long as *use* is visible and active. This safely leaves the Governor with millions of hectares of unfarmed but community-owned lands to play with. Unfortunately, both Europeans and Africans prefer to cultivate the same not-so-expansive fertile lands, and from which the latter are forcibly removed.

Finding the need to squeeze African rights further

As post-war investment and settler pressure for lands mount, virtually all territories review native tenure policies, limiting peaceful native occupancy further 'in the interest of development and growth' (1920 Nyasaland Land Commission as cited by Kanyangolo 2008, 89). Sudan, for one, clarifies in new laws of 1925 and 1939 that native land rights include only rights to cultivate, pasture livestock, and collect forest produce. It also makes clear for the avoidance of doubt that these are merely use rights (Alden Wily 2010). A new law promulgated throughout Francophone territories in 1925 does much the same.

Anglophone and Francophone land laws differ in one notable respect. The latter, bound by the Civil Code, permit (educated) Africans to acquire titles for their house plots and farms, although subject to more onerous conditions than required of Frenchmen (Alden Wily 2011a). The British avoid such provisions, claiming these may increase already apparent polarisation among Africans (Kanyinga *et al.* 2008, Sender and Smith 1986). In any event, the French do not mean to be so lax; by 1935 the French colonies are all enacting another law confirming that lands not covered by title and not exploited or occupied for more than ten years belong to the State (Hessling 2009). As economic depression bites in the 1930s, capture of African lands for European plantations gives way to rampantly coerced commodity production for export by African peasant farmers (Bryceson 1980). This is particularly productive for French companies in Francophone Africa which have systems for promoting and exploiting already well-established peasant commodity production (Amanor 2008b).

Indirect rule lends a helping hand

French Direct Rule (*Indigenat*) and British Indirect Rule play their role in both subordinating African rights and polarising African society, by co-opting or creating chiefs as their representatives charged with collecting taxes, organizing forced labour and applying rules of tenure as directed or permitted. Traditional territories are themselves redefined for administrative convenience. For example, in 1932 the Nuba in central Sudan find their plains, crucial to shifting cultivation, handed over to Arab pastoralists and 'more qualified' Arab leaders appointed to govern them (Suliman 1999). Native Councils in Ghana (1928) centralise powers of village chiefs into the hands of paramount chiefs, sealing the subordination of family and village rights to these aristocrats. Paramount Chiefs will spend the next decades rent-seeking from timber and mining companies and from more humble cocoa farmers moving in the from the north (Amanor 2008a).

Elites are so well developed in highland Kenya that by 1939, as officials of Native Authorities or educated protégés of Christian missions, they begin to buy tractors and carve out large farms at the expense of poorer families. The poor in turn are forced to labour on even larger white-owned tea plantations (Kanyinga *et al.* 2008). Kikuyu elites form an association to lobby for issue of private land titles such as held by white farmers and landowning companies like Del Monte and Brooke Bond. Poorer Kikuyu support the East Africa Association, a resistance movement inspired by Gandhi's Indian nationalism and Marcus Garvey's Black Nationalism. Meanwhile, land shortage caused by alienation triggers migrations into the Rift Valley, laying the seeds for later inter-tribal violence (Kanyinga *et al.* 2008).

Chiefs from the Liberian hinterland are also active but with a more equitable end result. This is prompted by a Supreme Court ruling in Monrovia in 1920 which overturns a court decision of 1916 which ruled that Monrovia's jurisdiction in the hinterland was limited to political matters (Alden Wily 2007). Chief Justice Dossen now rules that international law (i.e. the Berlin Agreement) never required that the willing consent of uncivilised people be sought to take their lands 'as through this they gained civilization'. Therefore it is quite in order for hinterland areas to be allocated to coastal elites and to rising numbers of international mining, timber and agricultural investors (Alden Wily 2007, 77–80).

Hinterland chiefs meet in Suehn in 1923 to argue (inter alia) for protection against expanding land losses. This is granted in Hinterland Regulations (1926, revised in 1936 and 1949). These acknowledge that tribal title exists with or without registration. Chiefdoms are also permitted to secure fee simple Aboriginal Title Deeds for their respective territories, should they be willing to pay the cost of surveys. This opportunity is taken up by five wealthy chiefdoms before 1945. This doubly secures (they think) one million acres as community-owned lands (Alden Wily 2007).

A third land rush: 1945–1955

The end of the Second World War (1939–1945) brings with it a new foreign land rush. Commercial farming expands dramatically. In Namibia settler ranches expand to 5,214 by 1954 and absorb over a third of the country (Werner 1993). In Tanganyika, the British launch the Groundnut Scheme (1946) funded partly by multi-nationals, administered by the British Ministry of Food, and implemented by former British army officers promised land at the end of the war. This will fail within a decade but in the meantime absorbs the lion's share of credit, inputs and subsidies (Bryceson 1980, Raikes 1975 as cited in Wily 1988, 57). A surge in issue of concessions for timber and mining also occurs across the continent. Thousands of families, colony to colony, are evicted. Where natives resist, legal ways are found to circumvent their claims.

Thus, when Meru elders in Tanganyika present their grievances to the UN, the Government in Dar es Salaam passes a new law improving local consultation but not requiring consent before eviction (1950), and enacts the *Public Preserved Areas Act, 1954* to limit conditions in which compensation is payable. Attempts to reassess the status of customary tenure in the courts fail in a landmark case (1953–1957) which rules customary land rights to be no more than lawful possession (not ownership) (*Mtoro Bin Mwamba v. Attorney-General*, High Court, Tanzania; 1953–1955). In Liberia, under Tubman who establishes an Open Door Policy for foreign (American) investment, provisions for hinterland communities to secure collective titles are re-worded in 1956 to reduce title to possession (lawful occupancy and use). By now, much of the country's forest lands are also under foreign concession (Alden Wily 2007).

Impatience with African tenure

By the mid-1950s, communal tenure is marked down as an impediment to agricultural growth and land ownership, in general advisedly to be 'removed entirely from the sphere of customary law' (Phillips 1955, 58). The East African Royal Commission (1953–1955) leads the way, arguing that Africans need indefeasible title as individuals to farm commercially and a free land market of freeholders to allow 'energetic and rich Africans to acquire more land and bad ones less land thus creating landed and landless classes... a normal step in the evolution of a country' (Swynnerton 1954). The same positions are advocated in Francophone studies of 1958–1959 (Hesseling 2009). Elites need no encouragement. Independence from European states is around the corner, nurtured by rising resentment of colonial theft of lands. The scene is set for liberation of African land rights.

Independence in Africa: 1960–1990

This liberation does not eventuate, at least not for poor majorities. By 1975, 47 African territories declare independence from Europe, and Ethiopians overthrow the local feudal Menelik dynasty. Liberation of customary land tenure is not on the agenda. Instead, late colonial exhortation that this be done away with becomes formal policy in state after state. Landholding by any other than individuals is especially reviled. Clan, family and other forms of communal ownership are to be replaced with state-issued entitlements through voluntary or systematic adjudication, titling and registration (Bruce and Migot-Adholla 1994). Model adjudication and registration laws are quickly proffered by new donors (the former colonizers) and by now thriving UN agencies, which have a lot to say about land reform across the globe in this era (Deininger and Binswanger 1999, Cox *et al.* 2003). The World Bank is especially active and by 1975 is producing papers which make privatization and concentration of landholding through conversionary titling its advised policy (World Bank 1975).

The new state land grab in the name of progress

In the case of Africa, titling reforms will have to derive from a platform of state tenure, for the much stronger political agenda is nationalization. This sweeps the continent in a swath of new laws which leave much of the continent even more trenchantly the property of the state, in practice governments. This is the case in no fewer than 30 states by 1980, to be added to later by Eritrea, Rwanda and Zimbabwe (Alden Wily 2011c). In the process, anachronistic situations where the odd king or chief has retained privileges as land controllers are done away with, such as in Malawi (1965), Zambia (1970) and Lesotho (1979). Even in Ghana, Ashanti chiefs find forests separated from the land, new law giving the state complete control over logging activity and revenues (1969). This separation of trees from the land is a trick which Gabon (1963), Cameroon (1964, 1974) and most other Francophone administrations also engineer, as will Charles Taylor as late as 2000 in Liberia.

Nationalization of land ownership is accompanied by rigorous centralization in land administration. The powers of even newly reformed native authorities across the continent are drawn upwards into the hands of all-powerful land ministries. Much of this is undertaken in the new spirit of African national socialism. In countries where this is most developed (Ghana under Nkrumah, Guinea under Sekou Toure, Senegal under Leopold Senghor, Tanzania under Julius Nyerere, later in Ethiopia and Mozambique, and in milder versions in Zambia and Sudan), forceful new policies are applied which have a pro-poor appearance. Farm landlessness is actually outlawed in some of these states. Collective farming is mobilised to make use of tractors possible, scattered homesteads are aggregated into more service-accessible settlements, and cooperatives are created country-wide to facilitate inputs and marketing. Residual feudal landlordism is also legally extinguished in Uganda, Burundi, Tanzania, Cape Verde and Nigeria. Through all this customary landholders remain legal users, not owners.

Nevertheless, exceptions or loopholes are routinely left in place, catering to the burgeoning African urban elites and rural 'kulak' sector (Wily 1988), and where colonial focus on progressive farmers during the 1950s has triggered much polarization and amassing of lands by absentee landlords. To service such elites, freehold tenure is often left intact in enclaves but reduced to long leaseholds for foreigners (Alden Wily 2011b).

While majority rural populations await the promised award of statutory entitlements for at least their dwellings and farms, they have even less security than in the past. This is executed in laws which variously remove valuable investment zones from communities (as in the 'pioneer' zones in the law of 1964 in Senegal); abolish clan tenure in Somalia (1975), laying the basis for violent conflict lasting to the present; vest rural lands in local governments but with key land powers given to the central state (Kenya 1963, Nigeria 1978); suborn customary tenure to *Shari'a*, meaning documentary evidence of rights is required for rights to be upheld (Mauritania in 1967); declare all unregistered land to be Government property (Sudan 1970); or reduce the entire customary sector to 'tenants at sufferance' (Gabon in 1963, DRC in 1966 and 1973, and Uganda in 1975).

Reinforcing the notion of vacant lands

Only in Botswana does Independence in 1966 bring with it acknowledgement that tribal land is owned, not merely occupied (1968). However, this is in a manner which leaves the larger part of these domains, the grazing zones, vulnerable to the privatised ranching schemes which the new government also favours (1975). Similar commercialised private farming schemes mushroom around the continent. These include large-scale land allocations to burgeoning numbers of agricultural parastatals or to revitalized agencies like the Office du Niger. Projects range from the cotton and peanut schemes of the Niger Basin, the sorghum and sesame schemes of Sudan, to the elite-led ranching schemes of Kenya, and the wheat schemes of Tanzania, governments persuaded by donors and international agencies that large-scale mechanised farming and ranching schemes are the route to growth (Golan 1990, Niamur Fuller *et al.* 1994, Galaty 1999, Suliman 1999). Millions of hectares of untilled lands are appropriated from communities in the process. These include several million hectares taken from Nuba and Funj communities in central Sudan. Their lands are given to the Mechanised Farming Authority to allocate to officials, politicians, military men, local traders, and Egyptian banks and other entrepreneurs in the region with the capacity to farm commercially (Komey 2008). USAID and the World Bank are lead supporters.

Such appropriation is perfectly legal (or where there are doubts, made so, such as by the introduction of the Unregistered Lands Act 1970 in Sudan at the bidding of the World Bank which removes any doubt that untitled lands belong to Government). Almost without exception, post-independence land laws confirm that untilled lands are therefore lands without owners or even lawful occupants (*terres sans maîtres*). This is most explicitly the case in Francophone Africa, where only Togo (1974) departs from this colonial norm. As the years proceed, interference in unregistered lands and communal lands multiplies in a host of states, and where the line between the personal interests of those in authority and public interest blurs. Later, the facts of wilful dispossession and irregular allocations of untitled or state lands in the 1970s–1980s will be most systematically documented in Kenya (Klopp 2000, Republic of Kenya 2004). Issue of concessions for oil, mining and timber extraction also soar during this era. This displaces thousands of customary landholding from six Congo Basin States alone (Sunderlin *et al.* 2008). Extension of game reserves and declaration of Hunting Areas in East and Southern Africa has similar effects (Nelson 2010).

Creating a vacuum of tenure: no rights and no remedy

In all this, the fact that Africans own their land already, the provisions of sustained colonial-derived laws notwithstanding, is conveniently forgotten. Indigenous tenure regimes in general and collective forms of customary ownership in particular, continue to be located as obstructions to individual-centric growth and development up until the 1990s. As is now so well known, Gareth Hardin, confusing collective group-owned property with open access regimes, adds his penny's worth to destructively good effect (1968).

In the event, the privatised titling which from the 1950s is supposed to meet majority needs for at least their homesteads to be secured in 'modern' ways, fails to materialize other than in Kenya (Bruce and Migot-Adholla 1994). Even there the project will reach a natural limit by 1990, its farm-centred provisions irrelevant to the agro-pastoral and pastoral zones which extend over 60 percent of the country. Nor are the results of farm titling satisfactory. In the first instance, rural communities find this does not include their collective properties, the forests, grazing lands and wetlands they keep for common use. These are variously handed over to wealthy individuals in the community who have the capacity to extend farming into these areas, given to local authorities to manage or dispose of, or declared to be national forest or wildlife reserves (Alden Wily 2006). Even on the farm, family tenure is extinguished in favour of entitlement in the name of family heads, generating a plethora of disputes. In some areas the majority do not even bother to collect their title deeds and continue to regulate transfers and inheritance by customary norms (Hunt 2005). The sanctity of title deeds is brought into question as the registry itself is corrupted (Republic of Kenya 2010). Titling is increasingly left up to individuals to acquire and defend, largely confined to elites who can afford the costs and know how to circumvent the risks. Polarization and landlessness in rural areas reaches such extremes that national policy papers simply observe this as incalculable (Republic of Kenya 2010).

By 1990 it is clear that titling has been a failed project. First, areas actually titled since 1960 are limited, amounting to probably no more than 10 percent of the continental land area, and most of this located in the (former, and not so former) white farming areas of Namibia, Zimbabwe and South Africa (Deininger 2003). This leaves more than a billion hectares of land on the sub-continent under contradictory status as customary property as perceived by still-operating customary norms, and as national or public property, under statutory law (Alden Wily 2011e). Statutory law takes precedence; whether they are aware of the facts or not, millions of Africans are no more than tenants of state, occupying and using lands formally belonging to the state.

The practical reality around the continent is not much better. Through population growth (nine-fold over the twentieth century) as well as through land grabbing by local and national elites unleashed in the vacuum of legal support for customary rights, most rural families in 1990 are eking out a living on less and less farmland per capita. Distribution of cultivable land in the smallholder sector is skewed, with a Gini Coefficient comparable to feudal ratios in the 1960s and 1970s (Jayne *et al.* 2005). When the large estate sector is included, the inequities are even greater (Jayne *et al.* 2005). Off-farm resources, where they remain under the control of communities, begin to be noted as making the difference between pauperisation and survival. Woodlands and forests are especially valuable, often doubling the livelihood of the poor (Emerton 2010).

With hindsight, none of this is surprising. Class formation and land commoditisation had grown apace since 1945. Politicians and civil servants who assume power in the 1960s constituted economic elites, often closely tied to traditional forms of leadership (Chabal and Daloz 1999). They bring with them to power strongly centralist inclinations, delivered in one-party and often tribal-aligned administrations. Forms of unaccountable neopatrimonialism, clientelism or more developed class-bound social relations abound (Erdmann and Engel 2006). North *et al.* (2009) later describe these norms as the predominant and even natural political order around the world; personal and commercial interests and authority are so closely intermeshed that there is little incentive for elites and politicians to surrender to regulation of their behaviour, including in land dealings. Moreover, the African administrations of 1960–1990 have enjoyed substantial international support, egged on by, and themselves convinced of, the rightness and inevitability of land market-led development as the way forward (Deininger and Binswanger 1999). All that is missing by 1990 is to make land in Africa much more freely available for sale and for foreigners to again be able to join in the bonanza.

Looking to the present

Land reforms of the 1990s provide these market-friendly conditions. For while it is political democratization (at times following years of bitter social conflict and civil war) which allows reformism across many sectors in the 1990s, it is the more pedestrian demands of private sectors to freely acquire unlimited lands which first triggers tenure reform outside Southern Africa and, overall, which wins the day in the terms of new land laws which emerge (Alden Wily 2011d). This objective is forcibly reinforced by World Bank-led structural adjustment conditions for failing economies which require indebted governments to work harder to free up the market in land, re-launch titling programmes, open their doors to foreign land ownership, and remove land ceilings, development conditions and other rules which impede emergence of large commercial farms. By 2005 most African states have modified their land laws towards the above, and additionally, enacted foreign investment-friendly promotion laws modelled by international agencies.

Recognising Africa is owned – at last

Nevertheless, in the process of reviewing land tenure and administration, most African states are forced to consider the status of customary tenure as affecting the rural majority, if only to encourage these holders to make their lands more freely available. In the process, a handful of states significantly depart from the norms of colonial and post-colonial tenure in place. This is most apparent in the new land laws of Mozambique (1997), Uganda (1998), Tanzania (1999), South Africa (1996, 2004), and then a decade later in Benin (2007), Burkina Faso (2009) and Southern Sudan (2009) (Alden Wily 2011c, 2011e).

Broadly, these laws finally do away with the embarrassing situation in which majority rural populations have remained, despite independence, what Judge Nyalali in Tanzania describes in 1994 as tantamount to 'squatters on their own land' (*Civil Appeal No. 31, Court of Appeal of Tanzania*). Constitutional and/or land law provisions are enacted in these countries which for the first time endow customary rights with legal force as real property, not mere rights of occupation and use of government owned and controlled lands. This is made so irrespective of whether or not these interests are formally titled and registered, or owned by individuals, families or communities. The last opens the way for communities to secure thousands of hectares of their most vulnerable assets – their unfarmed commons – as their legal, collectively owned private property.

However, the procedures for any level of formalization are weak with a retained focus upon titling of individual houses and farms, and/or without the localised institutional empowerment needed to support such decision-making in a lawful manner. Only in Tanzania, with the advantage of longstanding formalised local government at village level including regulatory powers, does new land law formally devolve primary authority over land matters to the community. The Tanzanian Village Land Act, 1999, establishes a regime whereby every village has the right to determine its own landholding norms as well as to register collective, family and individual properties in its own Village Land Register. Even without such procedures (or titles) the law assures the members of some 12,000 villages that they are the legal owners of all lands within their respective domains. This enables communities to secure some 70 percent of the total country area as declared Village Land. Yet, at the same time, the state maintains ultimate title to all land in Tanzania, and a sister law (*Land Act, 1999*) makes it relatively easy for the state to claim lands within the Village Land sector which are not actively used. Still, the new investment promotion centre will have to find itself working hard over the next decade to persuade villagers to voluntarily surrender lands for private investment. This initiative will not meet with great success, so that by 2009 the President is announcing that the law will be amended to make more village-owned land available to investors (Government of Tanzania 2009).

A handful of other states, among which Benin, Madagascar, and Namibia must be counted, make less striking changes to their laws, leaving important obstacles to mass tenure security in place (Alden Wily 2011c). The commonest handicap to emerge is to retain the colonial and post-colonial norm of limiting the right to secure tenure to house and farm plots, excluding what usually much more substantial unfarmed land areas. Another limitation is to subject acknowledgement of customary interests to the same expensive survey-dependent and conversionary mechanisms that have long existed for the settler, then the statutory landholding sector, the case for example in Benin.

In practice, Madagascar, Rwanda and Ethiopia are alone in re-launching mass titling. These programmes again only cover homesteads, state title is retained over most of the much larger communal assets of each community. This is most detrimental in Rwanda where its Organic Land Law 2005 first extinguishes all customary rights. These countries, like almost all other African states, are tangibly loath to surrender the 'profit and prerogative' which continues to be gained by regarding these lands as 'empty of owners' and available for state grant or sale in the interest of public purpose. Public purpose itself is explicitly expanded in most land laws during the 1990–2010 era to include private investments which support economic growth as in the public interest (Alden Wily 2011e).

The current land rush

The new land rush, the subject of this volume, enters into this environment, and as will be clear by now, is greatly facilitated by the above conditions. The rush is global, noticeably foreign-driven, and involves sales or mainly leases to upwards of 200 million hectares since 2000, and especially since 2008 (Anseeuw *et al.* 2011). As remarked earlier, most of the large deals (in fact two thirds of them) are for lands in poverty-stricken and investment-hungry sub-Saharan Africa. Acquisitions of many thousands of hectares dominate reported deals but are known to be underlain by myriad and less well-reported smaller allocations, often taken up by local investors or speculators (Deininger *et al.* 2010). The oil and food market crises of 2007–2008 are widely acknowledged as triggers, compounded by financial crisis and economic recession, especially in Europe and America. This raises the speculative stakes enormously; the land rush provides new investment opportunities for fabulously wealthy entrepreneurs forced to withdraw from failing mortgage and related markets, and who Hobsbawn might again describe as having 'vast amounts of accumulated capital burning holes in their owners' pockets'.

The land acquirers themselves are more global and diverse than their 1880s counterparts or the European settlers and companies who followed them through the twentieth century. As well as prominently involving China and India and some Middle Eastern states along with traditional European and American companies, South Africa shows signs of being a major player in Africa, with negotiations for land underway in more than 20 states (Hall 2011). China may prove its main competitor, not least in the land and resource-rich Congo Basin where Chinese companies have acquired 10 percent of Gabon's forests under concession and are reportedly negotiating multi-billion dollar deals for large tracts of land in Cameroon and DRC (Putzel *et al.* 2011). Recent analysis suggests the main declared purpose to be for industrial-level production of jatropha, sugar cane and especially oil palm to feed the lucrative biofuel industry, followed by intentions to use leased lands for cereal, horticultural, floricultural and livestock feed production market (Anseeuw *et al.* 2011). Acquisition of forested or forest-able lands to earn anticipated carbon credits is also on the agenda, as are lands rich in timber, water, oil and minerals. Exploration concessions for oil and minerals are especially vast, such as in the 1.1 million square kilometres under oil concession in Sudan (Global Witness 2009, 15). Concession data is unevenly included in compilations of data on large-scale leases. This may be because issue of such grants has a longer history and acceptance in the public domain, is considered less avoidable, and in practice causes fewer evictions for as long as the grants remain exploratory.

Extending spheres of economic influence

As of old, the land rush does not stand alone. Local banks, communications, infrastructural projects, tourism ventures and local industry are also rapidly being bought up by foreign capital. Nor are these developments a one-way street. Extending competitive 'spheres of economic influence' in the form of capture of population-rich new markets for home manufactures is just as active as large-scale land acquisitions. This is illustrated in the largely foreign capital buy-in and buy-up of Special Economic Zones (SEZ), most advanced in India but multiplying in Africa, including eight proposed very large 'Shenzhen' for development by Chinese interests (Brautigam 2011). Should these take off, they will provide tariff-free entry and finishing hubs for Chinese goods at scale and locales for Chinese producers and labourers seeking to escape the saturation of home markets. In these respects, the new land rush is hardly different from the first scramble for Africa, founded upon, we will recall, the pillars of growth in mass consumption, 'plantations', and metropolitan traders and financiers (Hobsbawn 1987, 64).

The state as land grabber

Few hectares being made available derive from the existing private sector. On the contrary, as remarked early in this paper, sales and long-term leases are being issued by governments. The reasons for this will be clear: that, with exceptions, majority lands, no matter how anciently possessed by ordinary Africans as communities or households, are in law owned by the state in most African states. This suits investors well, assured that the deal will hold when their lessor is the host state, especially where thousands of hectares are involved.

It deserves note that in the countries where communities are recognized as lawful owners of their land, communities may still be easily persuaded by the promise of jobs to sell those lands. Such cases are often engineered by local elites.[3] In practice, the meeting of promises by investors and state agents is proving elusive thus far, especially with respect to jobs, and cause for rising protest (Anseeuw *et al.* 2011).

Laying out the welcome mat

As shown above, governments, both colonial and post-colonial, have for the most part actively welcomed commercial land investors and structured legislation accordingly. If anything has changed in the current gifting of their people's lands, it is in the warmth of the welcome. Indeed, as suggested above, a great deal of time and thought has gone into laying the welcome mat over the last decade. This includes investor-friendly tenure legislation, investment promotion centres to help investors get land and move swiftly through licensing procedures, a clutch of simplified investor-orientated template treaties and contracts to work from, and the whole backed up by a mesh of supporting international guarantees and insurance schemes for those willing to participate (Heri *et al.* 2011). The land itself is cheap; premiums and annual rents minor (around USD 0.50 to 1.00 an acre in Tanzania and Cameroon); lease terms of up to 99 years, or of shorter, but renewable terms; virtually total tax and duty free exemption for the first decade of production; the

[3]Cases of this have been documented in Sierra Leone (Da Via 2011), Kenya (FIAN 2010), Southern Sudan (Deng 2011), Ethiopia (Fisseha 2011), Ghana (Schoenveld *et al.* 2010) and Rwanda (Veldman and Lankhorst 2011).

right to export most if not all income and production; the right to employ mainly foreign labour and to access low interest loans on the basis of their new title deeds (Tanzania Investment Centre 2010).

The enticements are in fact so numerous that the benefits to host governments are opaque. Aside from likely personal rewards to facilitators and signatories, and the strategic conviction that such investments offer the new path to growth (especially in the declining bilateral aid environment), the answer must lie in anticipated technology transfer, prompts to infrastructure development, the pickings and leavings of products not fit for export, and most of all, the promise of jobs. As suggested above, these are not yet forthcoming. One is reminded that even the much more expansive industrialisation of the nineteenth century enclosures could not produce the number of jobs required and more often than not meant the poorhouse for the most disadvantaged and poor.

The law as facilitator

Bilateral investment treaties, of which nearly 5000 have been signed between north and south states over the last decade, provide the governing framework for these developments. Like the land contracts which they shelter, these are drafted with a keen eye on international trade law, protecting investors (Perez *et al.* 2011). International human rights law, through which communities might secure a hearing for their rights, is a good deal less precise or effective, delivered in conventions and declarations to which few states outside Latin America bind themselves and in Africa lacking a widely subscribed-to regional court whose rulings must be implemented (Alden Wily 2011c). Thus far, despite the immense public rhetoric around the need to protect the interests of local populations, few moves have been made to bring international law up to speed with much more energy being invested in refinement of investment procedures and protection (Da Via 2011, Daniel 2011) and devising of entirely voluntary guidelines. Nor has the land rush prompted land reforms. If anything, it has slowed down reforms begun in the 1990 almost to stopping point.

Nevertheless, the accumulated niceties of not disturbing lands which are settled or permanently farmed still largely apply. Both host lessor governments and investors prefer to keep away from settlements and farms, if only to avoid the potential conflicts and (usually petty) costs for compensation this may entail. Looking beyond the village and the farm for lands also allows for unbroken landscapes for investors to clear and cultivate, and ideally, where the land is forested, yielding ideal soils for oil palm and other biofuel crops. Therefore, while preferring to be near ports and airports, and with good roads for ease of marketing and export, today's investors, like their predecessors, are ultimately being forced more and more into hinterlands. Initially, this may expose lands which governments have already long ago taken from communities to create (now failed) state farms, wildlife and (now degraded) forest reserves, refugee settlements, and other zones where indigenous settlement is (comparatively) more sparse (Alden Wily 2011f). If such areas are not available, there always remain the untilled forest/woodlands and rangelands of rural communities. The fact is ignored that these lands are of crucial livelihood importance to the three quarters of rural households who still fall below the international poverty line (Emerton 2010, Population Reference Bureau 2005).

Fortunately for investors, these communally-owned and -used lands, by custom, are still deemed in most state laws to be unowned, vacant and available. Yet, these are also the very assets upon which off-farm community-based enterprise could be built without dispossession and taking communities well above the poverty line, as examples of conservancies in Namibia and Kenya demonstrate (Nelson 2010) or small to medium forest-based enterprise as increasingly evidenced in Congo Basin and other states by Molnar *et al.* (2011).

But African administrations and aligned private sectors have another agenda, and one which revitalizes the traditional shortcut to wealth, privatization and enclosure of land and resources by and for elites. Moreover, African governments may fairly protest that they are only doing what has been so specifically urged upon them since the 1950s: to bring large areas of their countries under privatised, European-derived forms of entitlement, and for the purposes of investors. This is now occurring at speed, soon to double the estimated area of rural lands which were privately titled in 1990. It is unfortunate, and provocative that the majority of their citizens are excluded in this revitalized enterprise.

References

Alden Wily, L. 2006. Formalizing the informal: is there a way to secure human potential through land entitlement? A review of changing land administration in Africa. *In*: B. Guha-khasnobis, R. Kanbur and E. Ostrom, eds. *Linking the formal and informal economy concepts and policies*. Oxford: Oxford University Press.

Alden Wily, L. 2007. *So who owns the forest? An investigation into forest ownership and customary land tenure in Liberia*. Moreton-on-Marsh: SDI Liberia and FERN.

Alden Wily, L. 2010. Making peace impossible? Failure to honour the land obligations of the Comprehensive Peace Agreement in Sudan. Available from: http://www.oxfam.org.uk/resources/learning/landrights/downloads/making_peace_impossible_land_obligations_in_central_sudan.pdf [Accessed 18 December 2011].

Alden Wily, L. 2011a. *Whose land is it? The status of customary land tenure in Cameroon*. Moreton-on-Marsh: Centre for Environment and Development, FERN, The Rainforest Foundation UK.

Alden Wily, L. 2011b. Putting 20th century land policies in perspective. *In*: L. Alden Wily, *Rights to resources in crisis. Reviewing the status of customary land tenure in Africa*. Brief No. 2. Washington, DC: Rights and Resources Group.

Alden Wily, L. 2011c. 'The law is to blame'. Taking a hard look at the vulnerable status of customary land rights in Africa. *Development and Change* 42(3), 733–57.

Alden Wily, L. 2011d. Land reform in Africa: a reappraisal. *In*: L. Alden Wily, *Rights to resources in crisis. Reviewing the status of customary land tenure in Africa*. Brief No. 3. Washington, DC: Rights and Resources Group.

Alden Wily, L. 2011e. The status of customary land tenure today. *In*: L. Alden Wily, *Rights to resources in crisis. Reviewing the status of customary land tenure in Africa*. Brief No. 4. Washington, DC: Rights and Resources Group.

Alden Wily, L. 2011f. *The tragedy of public lands: understanding the fate of the commons under global commercial pressure*. Rome: International Land Coalition.

Amanor, K. 2008a. The changing face of customary land tenure. *In*: J. Ubink and K. Amanor, eds. *Contesting land and custom in Ghana: state, chief and the citizen*. Leiden: Leiden University Press, pp. 55–78.

Amanor, K. 2008b. Sustainable development, corporate accumulation and community expropriation: land and natural resources in West Africa. *In*: K. Amanor and S. Moyo, eds. *Land and sustainable development in Africa*. London and New York: ZED Books, pp. 126–58.

Anseeuw, W., Alden Wily, L., Cotula, L. and M. Taylor. 2011. *Land rights and the rush for land: findings of the Global Commercial Pressures on Land Research Project*. Rome: International Land Coalition.

Berlin 1885. General Act of the Berlin Conference on West Africa, 26 February 1885. Available at: http://africanhistory.about.com/od/eracolonialism/l/bl/BerlinAct1885.htm

Bernstein, H. 2010. *Class dynamics of agrarian change*. Agrarian and Peasant Studies Series. Halifax and Sterling: Rernwood Publishing and Kumarian Press.

Borras, J. and J. Franco, 2012. Global land grabbing and trajectories of agrarian change: a preliminary analysis. *Journal of Agrarian Change*, 12 (1), 34–59.

Brautigam, D. 2011. African shenzhen: China's special economic zones in Africa. *Journal of Modern African Studies*, 49 (1), 27–54.

Bruce, J. and S. Migot-Adholla, eds. 1994. *Searching for land tenure security in Africa*. Washington, DC: The World Bank.

Bryceson, D. 1980. Changes in peasant food production and food supply in relation to the historical development of commodity production in pre-colonial and colonial Tanganyika. *Journal of Peasant Studies*, 7 (3), 281–309.

Chabal, P. and J.-P. Daloz, 1999. *Africa works: disorder as political instrument*. Oxford: James Currey; Bloomington and Indianapolis, IN: Indiana University Press, in association with the International Institute.

Clover, J. 2005. Land reform in Angola: establishing the ground rules. *In*: J. Clover and C. Huggins, eds. *From the ground up: land rights, conflict and peace in sub-Saharan Africa*. Nairobi: African Centre for Technology Studies, pp. 347–80.

Cox, M. *et al.* 2003. FAO in agrarian reform. *Land Reform, Land Settlement and Cooperatives*. Volume 2003/2, 13–30.

Daniel, S. 2011. The Role of the international finance corporation in promoting agricultural investment and large-scale land acquisitions. Paper presented at the International Conference on Global Land Grabbing, sponsored by the Land Deals Policy Initiative, 6–8 April 2011, University of Sussex, Brighton, UK.

Da Via, E. 2011. The politics of "win-win" narratives: land grab as development opportunity? Paper presented at the International Conference on Global Land Grabbing, sponsored by the Land Deals Policy Initiative, 6–8 April 2011, University of Sussex, Brighton, UK.

Deininger, K. 2003. *Land policies for growth and poverty reduction*. Washington, DC: World Bank Policy Research Report.

Deininger, K. and K. Binswanger, 1999. The evolution of the World Bank's land policy. *Research Observer* 14 (2), 247–76.

Deininger, K. *et al.* 2010. Rising global interest in farmland: can it yield sustainable and equitable benefits? Washington, DC: The World Bank.

Deng, D. 2011 'Land belongs to the community': demystifying the 'global land grab' in Southern Sudan. Paper presented at the International Conference on Global Land Grabbing, sponsored by the Land Deals Policy Initiative, 6–8 April 2011, University of Sussex, Brighton, UK.

Dorsett, S. 2002. 'Since time immemorial': a story of common law jurisdiction, native title and the case of tanistry. *Melbourne University Law Review*, 3, 1–26. Available from: http://www.austlii.edu.au/journals/MULR/2002/3.html [Accessed 4 August 2011].

Emerton L. 2010. *Quantifying the impacts of barriers to pro-poor forest management. livelihoods and landscapes strategy*. Gland: IUCN Markets and Incentives Discussion Paper.

Erdmann, G. and U. Engel, 2006. *Neo-patrimonialism revisited – beyond a catch-all concept*. GIGA Research Paper No. 16, Hamburg: German Institute of Global and Area Studies.

FIAN, 2010. *Land grabbing in Kenya and Mozambique: a report on two research missions – and a human rights analysis of land grabbing*. Heidelberg, Germany: FIAN International Secretariat.

Fisseha, M. 2011. *A case study of the Bechera agricultural development project, Ethiopia*. Rome: International Land Coalition.

Galaty, J. 1999. *The rhetoric of rights: construing Maasai land claims*. ALARM Working Paper No. 7. Kampala: Centre for Basic Research.

Ginnell, L. 1894. The Land in the brehon laws, a legal handbook. Available from: http://www.aughty.org/pdf/clan_ceiles_land.pdf [Accessed 4 August 2011].

Global Witness, 2009. Fuelling mistrust: the need for transparency in Sudan's oil industry. Global Witness Limited. London.

Golan, E. 1990. *Land tenure reform in Senegal: an economic study from the peanut basin*. Madison, WI: Land Tenure Center, Research Paper 1001.

Government of Tanzania. 2009. *Kilimo Kwanza Resolution*. Tanzania National Business Council. Dar es Salaam.

Hall, R. 2011. The next Great Trek? South African commercial farmers move north. Paper presented at the International Conference on Global Land Grabbing, sponsored by the Land Deals Policy Initiative, 6–8 April 2011, University of Sussex, Brighton, UK.

Hardin, G. 1968. The tragedy of the commons. *Science*, 162, 1243–8.

Heri, S. *et al.* 2011. *International instruments influencing the rights of people facing investments in agricultural land*. Rome: International Land Coalition.

Hesseling, G. 2009. Land reform in Senegal: l'histoire se repete? *In*: J. Ubink *et al.* (eds.) *Legalising land rights*. Leiden: Leiden University Press, pp. 243–70.

Hobsbawn, E. 1962. *The age of revolution. Europe 1789–1848*. London: Abacus.

Hobsbawn, E. 1987. *The age of empire 1875–1914*. London: Abacus.

Huberich, C.H. 1947. *The political and legislative history of Liberia, Volume II: constitution and laws of the Republic 1847–1944*. New York: Central Book Company Inc.

Hunt, D. 2005. Some outstanding issues in the debate on external promotion of land privatization. *Development Policy Review*, 23(2), 199–231.

Iliffe, J. 1972. *Tanganyika under German rule, 1905–1912*. Cambridge, UK: Cambridge University Press.

Jayne, T., D. Mather and E. Mghenyi, 2005. Smallholder farming in difficult circumstances: policy issues for Africa. Paper presented at the Conference on The Future of Small Farms, June 26–29, 2005, Wye, UK.

(JPS) Journal of Peasant Studies (The), 2011. Vo. 36, No. 1, January 2009.

Juma, L. 2011. The laws of Lerotholi: role and status of codified rules of customary in the Kingdom of Lesotho. *Pace International Law Review*, 23(1), 92–145.

Kanyinga, K., O. Lumumba and K. Amanor, 2008. The struggle for sustainable land management and democratic development in Kenya: a history of greed and grievance. *In*: K. Amanor and S. Moyo eds. *Land and sustainable development in Africa*. London and New York: ZED Books, pp. 100–26.

Kanyongolo, F., 2008. Law, land and sustainable development in Malawi. *In*: K. Amanor and S. Moyo, eds. *Land and sustainable development in Africa*. London and New York: ZED Books, pp. 82–99.

Klopp, J. 2000. Pilfering the public: the problem of land grabbing in contemporary Kenya. *Africa Today*, 47(1), 7–26.

Komey, G. 2008. The denied land rights of the indigenous peoples and their endangered livelihood and survival: the case of the Nuba of the Sudan. *Ethnic and Racial Studies* 31(5), 991–1008.

Kuponen, J, 1993. The partition of Africa: a scramble for a mirage? *Nordic Journal of African Studies*, 2(1), 117–35.

Linebaugh, P. 2010. Enclosures from the bottom up. *Radical History Review*, 108, 11–27.

Magaziner, D. 2007. Removing the blinders and adjusting the view: a case study from early colonial Sierra Leone. *History in Africa*, 34, 169–88.

McAuslan, P. 2006. Property and empire. Paper presented to the Sixth Biennial Conference, 21–23 March 2006, The Centre for Property Law, School of Law, University of Reading, UK.

Molnar, A. *et al.* 2011. *Community-based forest management the extent and potential scope of community and smallholder forest management and enterprises*. Washington, DC: Rights and Resources Initiative.

Morrissey, J. 2004. Kilnamanagh and the frontier: surviving the new English of the early seventeenth century, Department of Geography, National University of Ireland, Galway. Available from: http://www.courts.ie/Courts.ie/library3.nsf/pagecurrent/3CBAE4FEB856E917B80256 [Accessed 4 August 2011].

Mugambwa, J., 1987. The legal aspects of the 1900 Buganda Agreement revisited. *Journal of Legal Pluralism*, 1987(25/26), 243–74.

Murray Li, T. 2011. Centering labour in the land grab debate. *The Journal of Peasant Studies* 38(2), 281–98.

Nelson, F., ed. 2010. *Community rights, conservation & contested land the politics of natural resource governance in Africa*. London and Washington, DC: Earthscan.

Niamir-Fuller *et al.* 1994. Barabaig displacement from Hanang district to the Usangu plains: changes in natural resource management and pastoral production in Tanzania. London: Institute for International Development.

North, D., J. Wallis and B. Weingast, 2009. *Violence and social order: a conceptual framework for interpreting recorded human history*. Cambridge, MA: Cambridge University Press.

Patnaik, U. and S. Moyo, 2011. *The agrarian question in the neoliberal era: primitive accumulation and the peasantry*. Oxford: Pambazuka Press.

Perez, J., M. Gistelinck and D. Karbala, 2011. *Sleeping lions international investment treaties, state-investor disputes and access to food, land and water*. The Hague: Oxfam Discussion Paper.

Phillips, A. 1955. The future of customary law in Africa. *Journal of African Administration*, VII(4), 151–9.

Population Reference Bureau, 2005. PRB World Population Data Sheet. Available from: http://www.prb.org [Accessed on 21 January 2012].

Privy Council, 1921. Amodu Tijani v. The Secretary, Southern Provinces. The Judicial Committee, His Majesty's Privy Council, London.

Putzel, L. *et al.* 2011. *Chinese trade and investment and the forests of the Congo Basin*. Working Paper 67 Bogor: CIFOR.

Republic of Kenya, 2004. *Report of the Commission of Inquiry into Illegal/Irregular Allocation of Public Land*. Nairobi: Government Printer.

Republic of Kenya, 2010. *National land policy*. Nairobi: Government Printer.

Schapera, I. 1970. *Tribal innovators, Tswana chiefs and social change 1795–1940*. London School of Economics Monographs on Social Anthropology No. 43. London: The Athlone Press.

Schoenveld, G., L. German and E. Nutakor. 2010. Towards sustainable biofuel development: assessing the local impacts of large-scale foreign land acquisitions in Ghana. Paper presented at The World Bank Annual Conference on Land Policy & Administration, Washington, DC, 26–27 April 2010.

Sender, J. and S. Smith, 1986. *The development of capitalism in Africa*. London and New York: Methuen.

Suliman, M. 1999. The Nuba Mountains of Sudan: resource access, violent conflict, and identity. *In*: D. Buckles, ed. *Cultivating peace: conflict and collaboration in natural resource management*. Washington, DC: International Development Research Centre, World Bank Institute, pp. 205–20.

Sunderlin, W., Hatcher, J. and M. Liddle, 2008. *From exclusion to ownership? Challenges and opportunities in advancing forest tenure reform*. Washington DC: Rights and Resources Initiative.

Swynnerton. J. 1954. *A plan to intensify the development of African agriculture in Kenya*. Nairobi: Government Printer.

Tanzania Investment Centre, 2010. *Procedures for obtaining TIC Certificates of Incentives*. Dar es Salaam: President's Office.

Thompson, E.P. 1991. *The making of the English working class*. Penguin Books. First published by Victor Gollanz 1963.

U.S. Supreme Court, 1983. Johnson & Graham's Lessee v. McIntosh, 21 U.S. 8 Wheat. 543 (1823). http://supreme.justia.com/cases/federal/us/21/543/case.html [Accessed January 21 2012].

Veldman, M. and M. Lankhorst, 2011. *Socio-economic impact of commercial exploitation of Rwandan marshes. A case study of sugar cane production in rural Kigali*. Rome: International Land Coalition.

Werner, W. 1993. A brief history of land dispossession in Namibia. *Journal of Southern African Studies*, 19(1), 135–46.

Wily, L. 1988. The political economy of African land tenure: a case study from Tanzania. PhD Dissertation, *Development* No. 2, School of Development Studies, University of East Anglia.

World Bank. 1975. *Land reform: a sector policy paper*. The World Bank. Washington, DC.

Liz Alden Wily, political economist, is an independent scholar and technical adviser who works with agrarian governments, donors and INGOs on rural land tenure issues. She is a fellow of the Rights and Resources Initiative, a global consortium of agencies, and is an affiliated fellow of the Van Vollenhoven Institute of Leiden Law School.

New investment, old challenges. Land deals and the water constraint in African agriculture

Philip Woodhouse

Foreign investment in agricultural land acquisition in sub-Saharan Africa has been viewed primarily as driven by a set of linked 'crises': in financial capital markets, in security of energy and food supply, and in global environmental governance. This paper argues that a focus on the 'buyers' of land risks overlooking the dynamics that operate on the side of the land 'sellers'. Accordingly, the first part of the paper argues that it is important to view the current 'land grab' as the latest stage in a longer historical process of competition for control of land and other natural resources by different 'domestic' economic and political actors within African countries. While such struggles are often characterised as the 'state versus the peasantry', with the state acting on behalf of 'urban elites', the paper argues that processes of accumulation and associated enclosure of natural resources need to be examined more critically in specific contexts if the role and impact of foreign capital investment are to be understood. The second part of the paper seeks to identify the ways in which questions of scale (in the sense of greater capital intensity) can be considered to be constraints to the development of African agriculture. Particularly, it considers the extent to which the production models most frequently mentioned in connection with foreign investment (large-scale mechanised farms and small-scale outgrower contract farming) respond to current productivity constraints. The paper argues that current debates about foreign investment in agricultural land underplay the importance of water resources needed to overcome production risks associated with irregular rainfall. Bringing the water dimension of land deals more clearly into focus is necessary if the scope for positive and negative impacts of new investment on existing land users is to be fully understood. The paper concludes by considering the implications of such challenges in the current context of foreign investment in agriculture in Africa.

Introduction: the 'land grab' as a crisis narrative

The characterisation of the recent rapid growth of foreign investment in agricultural land as a 'land grab' is consistent with the widespread perception that it is a manifestation of 'a context of ecological, food and energy crises' (de Schutter 2011, 250). This suggests a sense of crisis that is pervasive, and Borras *et al.* (2011, 209) also state 'in many cases buyers and investors are simply preparing for the next global crisis'. This generalised perception of crisis is understandable in the midst of

The author wishes to thank Jun Borras and three anonymous referees for their very helpful and constructive comments on an earlier draft of this paper.

the worst economic recession in 70 years and will undoubtedly influence the behaviour of market actors. However, a discussion of the consequences of such market transactions demands also an understanding of the different facets of crisis to which they are a response. In particular, it is important to distinguish more proximate 'crises' from those that are rooted in a perception of longer-term problems of 'sustainability'. In this paper, however, I wish to question the usefulness of 'crisis' as a way of approaching the analysis of international land deals. I do this by considering the logic of foreign investment in agricultural land within processes of agrarian change within sub-Saharan Africa, where the majority of land deals have been located. Africa has, of course, been the subject of a series of crisis narratives – humanitarian, economic and ecological – since independence from colonial rule, and in that sense the 'land grab' represents a certain continuity. However, whereas previous African crises were located in narratives about African society, the 'land grab' has so far focused attention mainly on the 'crises' motivating foreign 'buyers' of land, which I will first review briefly in order to ask: what is the nature of the crisis?

It seems reasonable to start by identifying the land deals as a response to an abrupt change in the behaviour of food commodity markets. The 2007–2008 food price spike ended a three-decade-long period of stable, but declining, agricultural commodity prices. Further, unlike the previous such episode in the 1970s, prices failed to fall back to their pre-spike levels and, after only two years at relatively lower levels, in late 2010, returned to close to the peak values seen in 2008. Agricultural commodity markets over the past five years have thus been characterised by both increasing and volatile prices (HPLE 2011). For these events to be understood as a trigger for foreign investment in agricultural land in developing countries, however, the key observation is that 'the global food crisis of 2007–2008 convinced many governments and private commodity buyers that international markets . . .could not be trusted to provide a stable supply of food commodities. . .' (de Schutter 2011, 251). The growth in land deals are thus most clearly a response to a crisis of confidence in global food markets.

The question of why the behaviour of food commodity markets should have changed has been reviewed by the UN Committee on World Food Security. They note the Commodity Futures Modernization Act passed by the US Congress in 2000 allowed trading of financial derivatives based on commodities, notably the commodity index funds, reported to have increased fivefold in value in the three years prior to 2008, when they accounted for investments of US$250billion (HPLE 2011, 25–26). However, the move of finance into agricultural commodity markets was itself promoted by a series of related crisis narratives. The chief of these was a concern with energy security arising from increasing oil prices after 2003 and interpreted as signalling an inability of new discoveries of oil reserves to meet the growing demand (the 'peak oil' narrative). This prompted government subsidies for biofuel production in both the EU and the US, amounting to US$10 billion in 2006 (FAO 2009a) and US$8 billion in 2009 (IEA 2010) and generating a biofuel industry that consumes about 40 percent of US maize production and 66 percent of EU vegetable oil production (HPLE 2011, 32). The biofuel boom has been identified as a major factor causing food commodity prices to rise (FAO 2009a), but it has, in turn, been reinforced by other crisis narratives, notably that of climate change. This argues that the existing agricultural systems are threatened both by their dependence on increasingly costly fossil energy sources (particularly for nitrogen fertilizer), and by their vulnerability to changes in temperature and rainfall that will result from continued emissions of greenhouse gases. Of the latter about 12 percent of the global total are attributed to agriculture itself (FAO 2009b), rising to 30 percent if changes in land use and deforestation are included (IPCC 2007). This analysis is nested within a broader narrative of an impending crisis in modern agriculture, which draws on well-documented negative impacts of input-intensive commercial agriculture on biodiversity and water quality as evidence for underlying unsustainability and imminent crisis (Weis 2010, Moore 2010). Thus a series of interlocking narratives have suggested both a crisis in the ability of existing agricultural systems to supply enough food, while also identifying agriculture as a solution to a crisis of energy security. While these broader perceptions of crisis have reinforced more immediate effects of price volatility of food commodity markets, it is important to note that policy responses, of which the land deals are one, have displayed a collapse in confidence in the efficiency of food and energy markets as the basis of reliable supply.

The UN Committee on World Food Security (CFS) observes that not all food markets were affected by the 2007–2008 food price spike. In countries such as India and China, government food stocks were released and exports banned to counteract rises in domestic food prices. The CFS has proposed international coordination of food stocks held by governments, as well as tighter regulation of speculation in commodity markets (HPLE 2011). Whether such measures will be able to dispel the wider sense of agricultural crisis remains to be seen, but it is arguable that, at least in terms of understanding the impacts of the 'land grab' on rural poverty, viewing land deals primarily in terms of a conjuncture of crises in financial markets, energy and food supply, or governance of global trade, misses an important part of the story. In particular, this focus on the 'buyers' in land deals risks overlooking the dynamics that operate on the side of the land 'sellers'. This paper attempts to redress the balance. It does so by first sketching the enduring quest for agricultural modernisation by governments in sub-Saharan Africa and its significance for their current role as willing 'sellers' in land deals. The paper then argues that alternative visions for agriculture have been restricted by official narratives of rural society that tend to characterise rural society in terms of undifferentiated 'smallholders', and thus often fail to recognise the highly differentiated nature of rural communities generated by the dynamics of labour markets and local patterns of agricultural investment, particularly in the management of water for agriculture. The paper then considers the potential impacts of two models of foreign investment contemplated in land deals: large-scale production and contract farming for small-scale producers. It argues that water remains a key blind spot in assessing land deal impacts, not only because the need for irrigation to guarantee agricultural productivity is often only implicit in project proposals but also because the new water infrastructure required by many – if not all – foreign investments in agricultural land will have potential effects – both positive and negative – far beyond the boundaries of the land included in the deals.

The long quest for agricultural modernisation in sub-Saharan Africa

Rural poverty in Africa has been blamed on inefficient 'backward' farming methods since colonial times, and the search for a response through the introduction of modern farming methods has a similarly long history. In areas where impoverishment was most acute, notably in the colonies with significant European settlement and alienation of land, the period from the 1930s to the 1960s saw the wholesale reorganisation of African farming through a broadly consistent set of 'scientific' land use planning measures variously termed the Swynnerton Plan in Kenya, the Native Land Husbandry Act in Rhodesia and Betterment Planning in South Africa, and echoed in the *picadas* (rural roads) established to enable surveillance of forced cultivation of cotton in Mozambique. The principles that informed these schemes were: first, to separate and consolidate zones of cultivation, residence and grazing; second, to implement more productive technology, notably soil and water conservation and, where possible, mechanised (oxen or tractor) ploughing on arable areas, and improved grazing management; and third, to facilitate better access to markets for agricultural produce via government-controlled marketing agencies or cooperatives. In more arid areas, such as the Sudan and the Sahel, the modernisation project involved resettling farmers within areas served by large-scale irrigation infrastructure, such as the Gezira, the Office du Niger and the lower valley of the Senegal River. From the point of view of water management in agriculture, a sharp line was drawn between 'rainfed' and 'irrigated' agriculture. The modernisation project was underpinned by a basic assumption of government support for, and control of, input supply and output marketing, in large part reflecting colonial governments' perceptions of a strategic need to secure food production during successive crises of economic recession, wartime and postwar reconstruction that constituted the 1930s–1960s.

These assumptions of the developmental state continued to govern agricultural development under post-independence administrations, whether they focused on supporting a 'green revolution' among small-scale producers through subsidised supply of fertilizer, seed and pesticides and state-run marketing agencies, or embarked on state-owned large-scale mechanised production, as in Mozambique and Ethiopia. Large-scale commercial plantations, with or without small-scale 'outgrowers' remained the model under both colonial and post-independence administrations for crops such as sugar and tea that require immediate processing. Large-scale European-owned farms in some instances, notably Zimbabwe, also provided working models for modernisation goals. By the late 1970s, the pervasive failure to raise productivity in African agriculture meant the basic assumptions of state support for (and control of) agricultural production had become both the subject of critique from a variety of political perspectives (Shivji 1976, Lipton 1977, Bates 1981, Berg *et al.* 1982, Wuyts 1985), and also the object of reforms, most commonly referred to as 'structural adjustment programmes' (SAPs), imposed on many African governments by foreign funding agencies as a condition for loans to finance public budget deficits. Reforms followed a consistent formula, including an immediate period of retrenchment of government employees and divestiture of government ownership in agriculture and agricultural services. The ensuing two decades involved low rates of public investment in agriculture (6–7 percent of agricultural GDP in Africa compared to 20 percent commonly encountered in developed economies) and a collapse in public expenditure on agricultural research. The same period saw a decline of about 50 percent in overseas development aid to agriculture between 1985 and 2000 (HPLE 2011, 29–30). Despite this SAP reduction in public funding for services to agriculture, the World Development Report 2008 (World Bank 2007) makes clear that in sub-Saharan Africa, commercial agencies of input supply and output marketing have failed to provide an alternative. Further, it argues that the weakness of commercial input supply in Africa is due to the absence of 'complementary investment' in roads, irrigation, marketing infrastructure and financial services, which underpin dynamic commercial inputs in Asia and Latin America (World Bank 2007, 150).

This raises the question of where such complementary investment is to come from, other than public funds. More broadly, the failure of two decades of market deregulation to deliver a significant growth in aggregate agricultural output in much of sub-Saharan Africa suggests a need to reverse SAPs and re-instate government intervention as an element of increasing agricultural productivity. For many African governments, however, the impossibility of making such investments without recourse to foreign funding is an inescapable part of the context for land deals with foreign investors. Here it needs to be clear that it is not necessary to impute motives to specific governments or officials in relation to specific land deals. The issue here is to map a context in which a long-standing narrative of the failure of aggregate agricultural growth (i.e. at the level of the national economy) is understood in terms of a failure to modernise, whether in terms of small-scale 'green revolution' (high input use) farming or in terms of large-scale mechanised cultivation. Where public funds with which to pursue modernisation have been choked off by SAPs and subsequent restrictions on government expenditure, the appearance of foreign investors with proposals to establish 'modern' systems of production will hardly be unwelcome. The point here is not to defend the appropriateness of the modernisation narrative, with its consequences for designation of 'unused' or 'underutilised' areas under 'backward' and inefficient agriculture (de Schutter 2011, Li 2011), in understanding or improving agriculture in specific African contexts. Rather, it is to argue that its continuing prevalence among officials of both African governments and development funding agencies establishes a clear presumption in favour of seeking private investment in agriculture. As de Schutter (2011) argues, this reflects an absence of a credible alternative to 'modernisation' with which to guide agricultural investment. One reason why such an alternative may be difficult to identify is that the discussion of agricultural development in Africa is often bound by social models that do not engage with actually existing dynamics of social and agrarian change. In the next section, I seek to identify aspects that appear to be important in analysing the impacts of particular patterns of agricultural investment.

Dynamics of accumulation and agrarian change

A discussion of the political economy of foreign investment in agricultural land in Africa typically counterposes the interests of investors, allied with government officials and/or 'local elites' with interests vested in the land deal, against existing local land users or communities. It does indeed appear often the case that 'foreign' investment involves local private investment partners (*cf.* Deininger 2011, 218). Also, government officials will inevitably be involved in authorising and registering the terms of the investment that usually involve a leasehold contract, often with a central government agency if the area involved is large. However, it is less often the case that existing land users are characterised as other than a homogeneous social group, uniformly affected by the proposed investment. In de Schutter's (2011) terms, they are likely to be small-scale farmers who hold customary land rights through a history of generations of residence and land use in the area. I wish to argue that such a picture misses important elements of African rural economy that have a bearing on both the potential nature and outcome of foreign investment. In particular, rural people are engaged in labour markets, but in highly differentiated ways, with implications for both labour and capital investment in agriculture. On one hand, labour markets may mean a loss of 'household' labour from small-scale farms. On the other, alternative sources of income may explain the survival of otherwise 'non-viable' farms (see Cousins and Scoones 2010, for a critical discussion of 'viability') and, in the most positive scenarios, provide a source of investment to improve farm productivity. This linkage between 'off-farm' income and farming investment establishes an important capacity for private investment in agriculture, irrespective of any formal 'development' project or programme to promote it. This, in turn, raises the possibility of entrepreneurial small-scale agriculture becoming responsive to market opportunities and investing to achieve significant productivity growth. Finally, rural people are mobile, seeking better income opportunities not only in industrial and urban employment but also in other rural areas, which has consequences for the nature of rural 'communities'. I shall briefly elaborate on each of these in turn.

Labour markets, off-farm income and agricultural investment

There is a paradox between the often-stated importance of agriculture as source of employment – typically above 60 percent of the working population in sub-Saharan African countries – and the low share of GDP – no more than a third, and typically a fifth – generated by agriculture (World Bank 2007, 340–341). The paradox is 'resolved' in the form of very low earnings from agricultural labour, and resort to non-agricultural work. Ellis (2000) suggested that non-farm income contributes 30–50 percent of rural income in much of sub-Saharan Africa, increasing to more than 80 percent in southern Africa. It needs to be acknowledged that this characteristic of rural incomes in Africa is not new, having historical antecedents in patterns of labour migration at least since the early twentieth century (Adams 1977, First 1983), but is now commonplace and documented by an extensive literature (Reardon 1997, 1998, Ellis 2000, Barrett *et al.* 2001, Ellis and Freeman 2005).

While this may be interpreted as a 'de-agrarianisation' of the rural economy (Bryceson 1999, 2002), it is also possible to consider it as integral to a re-ordering of the relationships of labour and capital within African farming in two important ways. First, the engagement of labour from rural households in employment outside the household establishes an opportunity cost for labour on the family farm. Second, the level of income earned away from the family farm will determine whether 'off-farm income' will compensate for the labour withdrawn from family farming. In making such assessments, it is important to note that the timing of labour use in agriculture is critical. Thus, removal of labour may have critical impacts on farm productivity at certain, relatively short periods in the year, such as planting, weeding or dry season clearing of fallowed land. Yet removal of that labour for a 12-month period may not generate earnings sufficient to compensate the loss in farm productivity.

Conversely, more remunerative off-farm employment may generate flows of capital that can not only hire replacement labour but also invest in other inputs and equipment. Thus, off-farm income has been widely observed to constitute not only the principal source of agricultural investment in sub-Saharan Africa (Collier and Lal 1986, Bonnevie 1987, Reardon 1998, Lanjouw and Lanjouw 2001) but also a major mechanism for generating disparities of wealth within rural communities as a result of differences between individual households' capacity to achieve high earnings in off-farm work. This process is exemplified in Murton's (1999) detailed quantitative study of household investment in terrace construction for soil and water conservation in Machakos, Kenya. The corollary of the above observation is that local private investment in raising agricultural productivity can and does take place, although the capacity to do so is very unevenly distributed among the rural population. This remains the case, even where the technology for raising productivity is labour intensive. This is clear not only from the Machakos terracing case, cited earlier, but also from the successful experience in the Sahel of promoting 'indigenous' labour-intensive methods (*tassa* and *zwai* pits) to bring degraded land back into production (Hassan 1996), or mulch application in Burkina Faso (Slingerland and Masdewel 1996), or green-manure and composting technologies in Kenya (Place *et al.* 2007). In these instances, higher off-farm income translates into greater ability to 'command' (hire) labour. The other means by which rural households may invest more labour to increase production is through control of land, by rental or sharecropping arrangements that involve major labour and possibly capital investments by tenants, as in the establishment of tree crop plantations (Amanor 2010, Kouame 2010), or in clearing forested areas for cultivation (Woodhouse *et al.* 2000). From the point of view of analysing the impacts of land deals, the above method suggests rural communities will contain individuals with widely differing stakes in existing patterns of land and water use, raising the possibility that new forms of investment may offer opportunities to some while damaging the livelihoods of others. The presumption of a range of potentially conflicting impacts on members of rural communities must therefore underpin the discussion of land deals.

Market-responsive smallholder production

Investment in increased agricultural productivity can only be made in response to market demand, and there is evidence of investment in, and intensification of, agriculture where market access is good. This is most consistently evident in peri-urban zones of irrigated fruit and vegetable production, but also in rural areas where major transport corridors provide good access to urban markets, as in the case of small-scale irrigated horticultural production in Kajiado district, Kenya, on the main road between Mombasa and Nairobi (Southgate and Hulme 2000), or the several thousand hectares of furrow irrigation developed over the past two decades in the Penhalonga Hills, Mozambique, near the main road between Beira and Harare (Bolding *et al.* 2010).

The majority of such local investments of capital and labour are to meet growing urban demand for high-value, more perishable products. Such investments will not necessarily preclude food staples such as rice and maize, but the 'return on investment' is primarily generated by high-value crops. It is striking that improvement in water management is central to such strategies of agricultural development. As reviewed elsewhere (Woodhouse and Ganho 2011), water management forms an integral part of indigenous agricultural production systems throughout semi-arid and sub-humid rainfall zones that cover the majority of sub-Saharan Africa because of the strong seasonality of rainfall and the spatial variability of water distribution within the landscape. Whereas historic cultivation tended to emphasise adaptation to such variability by determining the timing and location of cultivation, for example, through planting following a receding flood or in wet valley bottoms during the dry season, investment of capital and/or labour has allowed greater degrees of alteration of water flows through interception, diversion or storage to push back the spatial and temporal limits to cultivation, and hence expand the range of crops that may be grown.

Such water management constitutes investment of labour, capital and knowledge, and in many cases, involves visible technology such as terraces, weirs, irrigation furrows and small pumpsets and boreholes. However, in many parts of rural Africa, its dispersed nature and its integration with rainfed cultivation, rather than separation in discrete blocks of 'irrigated farming', has tended to make it less visible to officials and technical staff trained to differentiate 'irrigated' and 'rainfed' agriculture as separate production systems. Further, even where informal irrigation more closely approximates water control used in formal irrigation schemes (e.g. diversion of streams into canals), it tends to vary in extent from one year to the next according to water availability (Lankford 2004). As a consequence, although official statistics have since the 1980s recognised some 2 million ha of 'informal irrigation' in sub-Saharan Africa (excluding Madagascar) alongside 6 million ha of formal irrigation (FAO 2005), there can be little confidence that these reflect the significance of water management in African agriculture. This 'invisibility' of existing water management is further increased by the dearth of studies on how 'informal' water management strategies fit within broader African agricultural systems or livelihoods (Woodhouse 2012). These considerations highlight the importance of understanding existing production systems in terms of their use of resources such as water. In doing so, it is critical to appreciate the variation in water availability both spatially and temporally, suggesting particular attention to the way cultivation on different sites and in different seasons may be interdependent within an overall production system. Unless the distinct, interlocking elements of existing land use are identified, there is a risk that significant aspects of existing water management will be rendered invisible.

Mobility of rural population

Discussion of migration of people in sub-Saharan Africa has tended to focus on circular patterns of migration, particularly between agricultural and mining areas, or migration from rural areas to towns and cities. Much less attention has been paid to migration between rural areas. Yet the key role of migrants recurs in empirical studies of agricultural development. The development of plantations of cocoa in West Africa (Hill 1963), the growth of commercial groundnut production in the Gambia (Robertson 1987), and the more recent growth of commercial pineapple growing in Cote d'Ivoire (Kouame 2010) all featured large inflows of migrants who cultivated under a variety of rental and sharecropping arrangements with existing landholders. Similarly, case studies of exploitation of water resources for agriculture in Mali (Woodhouse *et al.* 2000) and Kenya (Southgate and Hulme 2000), and Mozambique (Bolding *et al.* 2010) all highlight the role of migrants, as suppliers of not only labour but also farming expertise. In some instances of rapid agricultural development (e.g. in parts of Kajiado District in Kenya's Maasailand) migrants had become the majority of the rural population. Such population inflows into areas with favourable agricultural conditions may be associated with growing competition for land and increasingly commoditised transactions in land in the form of rental, sharecropping or sale agreements (Woodhouse 2003, Benjaminsen and Sjaastad 2004, Peters and Kambewa 2007). Here it is important to underline that such tendencies towards commoditised access to land do not require formal land titles but may be observed within customary land tenure systems (Chimhowu and Woodhouse 2006). The shifting terms of land access are a manifestation of broader change in the rural society in which they are embedded (Colin and Woodhouse 2010), and Pauline Peters argues that evidence of growing competition for land 'calls into serious question the image of relatively open, negotiable, and adaptive customary systems of landholding and land use and, instead, reveals processes of exclusion, deepening social divisions, and class formation' (Peters 2009, 1319).

In summarising the above observations, it is important to note that I have focussed on instances of agrarian change that have not been initiated or mediated by externally funded development 'projects'. Such processes of rural change, although evidently susceptible to modification by political and economic initiatives of the state, are present and active whether or not development agencies intervene. Thus foreign investment projects cannot be assumed to take place within rural areas made up of smallholders dependent on household labour and non-commodified access to land via customary tenure. A more appropriate set of assumptions would embrace the likelihood that rural communities will contain households with significant engagement in the rural and non-rural labour market, and highly differentiated according to wealth and rights to land. Where the area is well-connected to markets, local investment in intensive agriculture, particularly involving water management, may be expected, and a significant part of the population may be migrants from elsewhere, whose access to land may be defined by sharecropping, rental or other monetary arrangements with local holders of customary land rights. Such assumptions would have to be tested in specific rural contexts, but they imply that any simple counterposing of 'state' and 'urban elites' to 'communities' is unhelpful in identifying the impacts of foreign agricultural investment. In the following section, I consider how models of agricultural production under foreign investment land deals are likely to respond to the productivity constraint posed by risks of low and/or uncertain rainfall.

Investment, productivity and the water constraint in African agriculture

One of two models of farming structure feature most commonly in discussion of land deals: large-scale mechanised production units, and small-scale 'outgrowers' supplying a central processing and/or marketing facility. Both models promise improvement in market access, a key constraint evidenced by FAO's (2009a) finding in Kenya, Zambia and Mozambique that less than half of maize-growing households regularly sell their crop. This is partly a reflection of low levels of marketable surplus, but also of the significant disadvantages of smaller producers, such as lack of adequate storage or financial resources to enable them to delay sales after harvest, ensuring that they receive the lowest prices from traders. Yet similar dynamics of bargaining power may apply in outgrower schemes as in other markets in which small-scale producers have little choice but to sell to the first buyer they encounter, raising questions about the regulation of the terms of outgrower contracts (Little and Watts 1994).

If the need for investment to achieve economies of scale in marketing is widely accepted, the desirability of increases in scale of agricultural production is highly contested. On one hand, advocates of small-scale production units argue that small farms are more labour intensive and hence more 'efficient', producing greater yield per unit area as well as greater employment (Lipton 1977, Griffin *et al.* 2002). On the other hand, critics argue that higher land productivity on smaller farms may be illusory, reflecting denser population in more fertile areas (Dyer 2004), or that it arises not from greater efficiency, but from greater self-exploitation, of family labour (Byres 2004). This latter point is illustrated by a World Bank study comparing production costs in Brazil, Thailand and four African countries, which observed,

> The competitiveness of Africa's producers at the farm level derives mainly from very low returns to labor (reflecting the absence of alternative employment opportunities in rural areas) and limited use of purchased inputs (which results in significant soil nutrient mining). .... at current productivity levels and farm size, agriculture is economically impoverishing and technically unsustainable. (World Bank 2009, 5)

The significance of labour markets, highlighted earlier, is also suggested in the same study, for data from Nigeria:

> ... when family labour is assigned a value equivalent to 60% of the wage rate for hired labour, small-scale cassava production with traditional labour-intensive production practices and resulting low yields looks much less attractive compared with medium and large-scale production. (World Bank 2009, 81)

It hardly needs saying that (over)-exploitation of family labour is heavily conditioned by intra-family relationships of age and gender (Jones 1986, Kouame 2010).

Employment effects of large-scale and small-scale units of production are also disputed. Much writing on the subject has argued that more employment (and hence wider reduction in poverty) is generated by supporting the productivity growth of small farms (Mellor 1976, Lipton 1977, de Schutter 2011). However, recent empirical work in Africa gives grounds to question this view. Introduction of higher-quality standards for vegetable exports to the EU from the *Niayes* area of Senegal (Maertens and Swinnnen 2009) caused production to be re-structured away from smallholders towards larger-scale production units integrated with processing and packing facilities. However, extreme poverty fell by half in the population studied, since the larger units employed many people who had no access to land, effectively too poor to be smallholders. Different, but not inconsistent, results from a study of rural labour markets in Mozambique (Cramer *et al.* 2008) showed that wage labour was both prevalent and important to livelihoods, particularly among poorer women, and that wage rates were higher on larger farms than on smaller ones. This does not provide conclusive support for the proposition that larger farms will generate more employment than small-scale cultivation (Sender and Johnson 2004), but it certainly underlines the importance of understanding rural communities as differentiated by wealth and access to land, so that smallholders farming their own land may not be the 'poorest of the poor'.

It is also evident that debate about the employment impact of large-scale farms must differentiate between the type of agriculture being practiced. In large-scale grain farming, of the kind practiced in Brazil and Argentina, there seems little doubt that there are few jobs (albeit highly paid ones). In contrast, commercial fruit and vegetable production requires relatively high labour inputs, not only in field operations but also in processing and packing, typically generating large numbers of jobs. The question, then, is whether these jobs can constitute 'decent work'. Tania Li, in a previous JPS Forum on the Land Grab (Li 2011) provides evidence that large-scale investment in oil palm plantations in South-East Asia operate on the principle of recruiting from an impoverished labour pool with little access to land of their own, and hence more willing to accept low wages. She argues that, since commercial considerations do not lead companies to pay anything other than poverty wages, state intervention is needed to regulate employment conditions if immiseration and poverty are to be avoided. I return to consider the implications of this argument in the concluding section of the paper.

The preceding paragraphs suggest that simple arguments in favour of large or small production units will not suffice to predict impacts of land deals on labour productivity or employment. In practice, such impacts are likely to be highly context-specific. However, a focus on labour productivity in agriculture shifts the farm size/productivity debate in important ways. It indicates that the levels of investment ('scale'of production) need to be differentiated from the 'size' of land area farmed. For example, the comparison of small-scale cultivation arising from land reform in Brazil with underutilised *latifundia* landholdings will not necessarily produce the same conclusions as the comparison of the land reform farms with highly capitalised intensive grain cultivation on large-scale mechanised farms on the *cerrado*. Nor should it be assumed, moreover, that productivity is dependent on land ownership. Deininger and Byerlee (2011) observe that the largest-scale agricultural producers in Argentina own neither land nor farm machinery but are constituted by farm management companies that lease land from landowners and contract machinery services from specialist operators. Conversely, however, it needs to be recognised that increasing farm scale also confronts constraints. Operating costs of farm machinery will be substantially higher in sub-Saharan Africa, as supply chains for spare parts are longer and equipment is more vulnerable to spend time out of service (World Bank 2009, 85). Even on commercial mechanised farms in the 1980s Zimbabwe, where operators had long experience of local conditions, farm machinery costs were reckoned to be over 20 percent higher than for similar farms in the US (Stanning 1983). The evidence reviewed above suggests that scale of production unit in foreign investment projects cannot be used as a simple indicator of productivity or employment impacts. I now turn to consider questions of scale in relation to investment in water management.

As stated earlier, water management to reduce risk can be seen to play a central role in African agriculture, not only in the structuring of long-established patterns of land use but also in the way local producers make new investments of labour and capital to improve agricultural productivity. Scale is significant, in that water management can seldom be achieved in a way that is entirely contained within a single small-scale unit. Generally, interception, diversion or storage of water creates downstream effects or may place demands on upstream land users. It thus places particular priority on larger-scale intervention and investment, even if this is a matter of many coordinated small-scale actions (e.g. terracing, irrigation furrows, drainage ditches) on neighbouring farm units. The remainder of this section focuses on the implications of large-scale agricultural investment for developing water management.

I have argued earlier that, although water management has often been a focus of farmers' own investments in agricultural intensification this may often not register within formal planning systems. The dispersal of water management investments among small-scale units, their often seasonal nature, and their deployment of techniques unfamiliar to those trained to see water management only in terms of formal irrigation projects, means that they are not easily visible to administrative and planning agencies. This problem may be compounded by the way small-scale water use has been addressed within water legislation, including – paradoxically – by its relatively favourable treatment within recent reforms. For example, South Africa's 1998 National Water Act stipulates that water allocation is to be governed through the issue of water use permits to individual water users. However, in certain areas (usually ex-'bantustan' territories reserved for African settlement), the requirement for individual water licences has been waived in favour of a 'general authorisation' for small-scale water use. Similarly, under Mozambique's water law of 1991, all water use must be registered and licensed except 'common use' (*usos comuns*), which has priority over all other water use. Common use is defined as water used by rural households for domestic needs, watering livestock and irrigation of an area of up to one hectare, as long as this does not use mechanical devices. While such uses have priority over commercial use, for example in large-scale agriculture, the fact that they are not registered or quantified means that they are invisible to the planning process. This has prompted van Koppen *et al.* (2009, 28) to comment 'small-scale water uses are tolerated, as long as any new water uptake remains insignificant and ... as long as the marginalised remain marginalised, there is no problem'.

This potential blind spot for existing water use is matched by the lack of explicit mention of water requirements in many land deals. A few land deal contracts in the public domain concern proposed projects that deal explicitly with expansion of irrigation in semi-arid areas, such as the agreement by Libya to finance the expansion of irrigation infrastructure to allow cultivation of a further 100,000 ha in Mali's Office du Niger scheme (GRAIN 2009, Cotula 2011, 36). Otherwise, it is striking how little mention is made of water management in discussion of agricultural land deals in Africa. Mann and Smaller (2010, 6) are among the few commentators who draw attention to this, arguing access to water resources is a key goal of land deals: '... a critical motivation in the current trend towards large-scale land acquisitions is the water factor.'

The World Bank (2010) discusses land deals only in terms of rainfed agriculture, as does an earlier report (World Bank 2009) emphasising the large potential for expansion of cultivation in Africa's sub-humid 'guinea savannah'. However, the latter does concede

> Irrigation may not be as critical in the Guinea Savannah as in other more arid production environments, but the potential contribution of irrigation to African agriculture should not be underestimated. Commercial farmers in southern Africa have long known that even a single pre-planting irrigation can make an enormous difference in enabling timely planting and ensuring that crops get off to a vigorous start, which can significantly affect eventual yields and reduce risks. (World Bank 2009, 119)

This acknowledgement signals a weakness in conventional approaches to agricultural water management, namely the sharp division between 'rainfed' and 'irrigated' agriculture. In practice, agriculture in much of Africa's savannas may require something in between. The potential benefits of some intermediary form of water management cannot be judged by commonly cited aggregate indicators, such as total annual rainfall, or length of growing period. This is because, as stated earlier, the unreliability of rainfall extends to drought periods during the growing season, making supplementary irrigation or some other form of water management essential to avoid water shortages at critical moments of crop development. The IIASA study (Fischer *et al.* 2002) on which the World Bank (2010) draws for its assessments of amounts of 'available land' for agricultural expansion includes an assessment of the impact of irrigation on both land 'availability' and potential production. In the case of maize, irrigation is estimated to raise potential harvests by 49 percent and 39 percent above that of high-input rainfed production in East and West African regions respectively. For southern Africa, the equivalent figure is a three-fold increase (332 percent) in maize harvests. For South America the potential increase was estimated at only 29 percent (Fischer *et al.* 2002,Table 5.19). It does not follow that irrigation of maize in sub-Saharan Africa is always a financially rational strategy, but these estimates underline the point that agricultural potential is likely to be highly influenced by water management.

De Schutter (2011) has observed that foreign investors' interest is focused on fertile and high-rainfall, irrigated or irrigable areas, and it is unlikely that commercial investments would accept the risks arising from the vagaries of African rainfall. A review of crops proposed for cultivation in land deals in Africa (Woodhouse and Ganho 2011) strongly indicates that these will need irrigation, even where this has not been stated in the published information. This is because projects involve crops with high year-round water requirements (sugar cane, bananas), or because they involve dry-season cultivation (wheat, vegetable, fruit), or because the projected yields are only obtainable under optimum (including moisture) conditions (e.g. jatropha, *cf.* Schut *et al.* 2010). Large-scale land deals, therefore, imply large-scale use of water resources. This need not be negative in a region routinely observed to suffer from underutilisation of its water resources (FAO 2005). Investment in dams for water storage can make more water available through the dry season, with positive impacts on agricultural output beyond the immediate project land. For example, a 3000-ha commercial banana plantation established in 2010 in northern Mozambique is to be irrigated using water from a 60-million cubic metre capacity reservoir behind a dam constructed upstream on the Monapo river. The significance of this water storage can be appreciated by comparing it to the capacity of 4 million cubic metres for the dam which supplies the nearby provincial capital, Nampula, with a population of a little less than half a million people. As a consequence of the new dam, the dry season river flow between the reservoir and the banana plantation will be higher and more regular than previously, offering a potential new resource for other local economic activity. One possible use for such a resource would be to provide supplemental irrigation to protect small-scale producers from the worst effects of dought periods. To do so would require additional infrastructure but it could radically widen the potential distribution of benefits from the scheme. Whether such potential wider benefits will be realised will depend on the recognition of the important 'public good' benefits that such investment may provide. In particular, larger investments offer possibilities of water management on a larger scale and for many more people than existing small-scale investment has thus far achieved. To say this is not to underestimate the political challenges such a strategy would confront, not least as a consequence of increasing competition over land that had become more productive as a result of improved access to water (Woodhouse 2003).

It is evident that such a strategy would also need to avoid damage to the myriad forms of existing water management by small-scale agricultural producers and people engaged in fishing, livestock-rearing and other water-using enterprises. It was, indeed, the failure to identify and avoid such impacts that was the hallmark of failed state-sponsored water infrastructure projects in the 1970s (Diemer and Vincent 1992). Here, as in the assessment of water management needs of crops in African savannas , there is a need to question whether conventional approaches to measuring land and water use are adequate. Just as I observed earlier the problem of 'visibility' of existing patterns of water use, it is important to assess potential impacts at the times and places where they are most critical. This is exemplified by the Malibya project, noted earlier, to extend the area of the existing Office du Niger irrigation scheme in Mali. GRAIN (2009) reported that the project funders were negotiating for the project to have priority access to irrigation when river flows are lowest. However, to do so would potentially jeopardise existing producers' most valuable crop (cool season fruit and vegetables), thus undermining the financial viability of the system as a whole. Bues (2011) has documented the imposition of priority water access for foreign-owned floriculture farms on canal systems in Ethiopia, disrupting institutional arrangements for water sharing among the existing small-scale irrigators. Van der Zaag *et al.* (2010) have also shown how the impact of a 30,000-ha irrigated project on the Limpopo, in Mozambique, could only be fully assessed by estimating the probability of water flow in the river during the driest months of the year. In each of these cases, the critical impacts of projects on water availability during the time of greatest scarcity can be assessed neither simply by the land area occupied nor by total rainfall or total water consumption. Only an appreciation of the existing water use, particularly during the dry season, and likely annual variability in water supply, would enable an assessment of impact.

To identify opportunities for broad-based improvement in water availability, whether or not it includes large-scale units of agricultural production, and avoid damage to the range of existing water users clearly goes beyond the remit of investors in the land deals that have formed part of the current 'land grab'. Most obviously, it demands a role for the state in defining and enforcing a public interest as part of land deal contracts.

FDI and agricultural modernisation

In this paper I have argued that the focus in 'land grab' discussion on international investors, and the crisis narratives that motivate their investments, risks distracting attention from the dynamic character of the African rural societies in which such land deals will be implemented. In particular, an acknowledgement of rural areas as connected to markets, and particularly labour markets – both local and distant – leads to an understanding of African rural communities as significantly differentiated in terms of wealth and land rights. Such differences may be translated in various ways to differences of political identity and power, defined in a variety of terms such as ethnicity, lineage, gender, age or length of settlement in the area. I have highlighted the range of ingenuity and skill with which African agricultural producers manage land and water, expanding production, investing in new technology or moving to new areas where market opportunities warrant it. Rather than a tradition-bound society of small farmers, it may be more useful to envisage African rural communities as being made up of a variety of more or less mobile populations, probing the frontiers of the political economy in which they live. Yet, as Peters (2009) sets out in detail, these are processes creating winners and losers, and greater inequality within rural societies, with the impoverishment of some linked to the enrichment of others. Narratives of a homogeneous rural society of 'smallholder farmers' allow policymakers to ignore such processes. Equally, it is important to acknowledge that winners and losers will also be created even where 'public goods' can be maximised from large-scale investment in agricultural water management.

In calling for a re-thinking of land-grabbing, de Schutter (2011) questions the assumption that large-scale, export-oriented agriculture is the inevitable direction which agriculture must take in developing countries, with only the capacity to regulate contract design and implementation as concerns for successful social outcomes. While accepting this point, in this paper I have suggested a need to go further. An understanding of rural communities as differentiated, mobile and changing, as outlined here, raises questions about what other assumptions might be being made about patterns of rural development that, in the *absence* of large-scale investment, would permit 'access to land and water...democratized for the local farming communities' (de Schutter 2011, 250). In this paper I have argued that the advantages of small-scale farming, whether with respect to employment, productivity or poverty reduction are by no means established. The type of crop, effects on labour productivity and wage rates as well as the generation of public goods are all aspects of large-scale investment that need context-specific evaluation. The question of improved water management is not only central to the challenge to improve the productivity of agricultural work but also highlights the social and political dimensions of any interventions to address it. Moreover, even where small-scale production is preferred, it is likely that, if improved labour productivity and reduced production risks through better water control are to be achieved for more than a minority of better-off smallholders, then large-scale investment by the state or state-regulated private investment will be needed. While this suggests a central role of state intervention, it is important to avoid reducing this to a matter of technical or financial capacity. The dynamics of increasing inequality in African rural society outlined earlier demand, above all, political leadership in defining what values and goals shall set the parameters for investment. This will need to negotiate *inter alia* the role that markets, customary hierarchy, gender, age and ethnicity play in determining the rights of individuals to property and income, as well as the public accountability to be expected of government officials. Without such political leadership, it seems likely that modernising investment, whether in small-scale or large-scale production, will exacerbate the tensions already present within rural societies.

References

Adams, A. 1977. *Le long voyage des gens du fleuve*. Paris: Maspero.

Amanor, K. 2010. Land sales, family values and commodification: the dynamics of agrarian change in southeastern Ghana. *Africa*, 80(1), 104–125.

Barrett, C., T. Reardon and P. Webb. 2001. Nonfarm income diversification and household livelihood strategies in rural Africa: concepts, dynamics, and policy implications. *Food Policy*, 26, 315–331.

Bates, R. 1981. *Markets and states in tropical Africa*. Berkeley: University of California Press.

Benjaminsen, T. and E. Sjaastad. 2002. Race for the prize: land transactions and rent appropriation in the Malian cotton zone. *European Journal of Development Research*, 14(2), 129–152.

Berg, E., K. Amoako and R. Guesten. 1982. *Accelerated development in Sub-Saharan Africa: an agenda for action*. Washington DC: The World Bank.

Bolding, A., N. Post Uiterweer and J. Schippers. 2010. The fluid nature of hydraulic property: a case study of Mukudu, Maira and Penha Longa irrigation furrows in the upper Revue river, Manica District. *In*: P. van der Zaag, ed. *What role of law in promoting and protecting the productive uses of water by smallholder farmers in Mozambique?* CGIAR Challenge Programme on Water and Food, pp. 105–136. Delft: Water Resources Section, Delft University of Technology.

Bonnevie, H. 1987. *Migration and malformation. Case studies from Zimbabwe.* Copenhagen: Centre for Development Research.

Borras, S., R. Hall, I. Scoones, B. White and W. Wolford. 2011. Towards a better understanding of global land grabbing: an editorial introduction. *Journal of Peasant Studies*, 38(2), 209–216.

Bryceson, D. 1999. African rural labour, income diversification and livelihood approaches: a long-term development perspective. *Review of African Political Economy*, 80, 171–189.

Bryceson, D. 2002. The scramble in Africa: reorienting rural livelihoods. *World Development*, 30(5), 725–739.

Bues, A. 2011. Agriculture foreign direct investment and water rights: an institutional analysis from Ethiopia. Presented at the Conference on Global Land Grabbing 6–8 April, 2011, Institute of Development Studies, University of Sussex, UK.

Byres, T. J., ed. 2004. Introduction: contextualising and interrogating the GKI case for redistributive land reform. *Journal of Agrarian Change*, 4(1–2), 1–16.

Chimhowu A. and P. Woodhouse. 2006. Customary vs private property rights? Dynamics and trajectories of vernacular land markets in Sub-Saharan Africa. *Journal of Agrarian Change*, 6(3), 346–371.

Colin, J-P. and P. Woodhouse. 2010. Introduction: interpreting land markets in Africa. *Africa*, 80(1), 1–13.

Collier, P. and D. Lal. 1986. *Labor and poverty in Kenya 1900–1980*. Oxford: Clarendon Press.

Cotula, L. 2011. *Land deals in Africa: what is in the contracts?* London: International Institute for Environment and Development.

Cousins, B. and I. Scoones. 2010. Contested paradigms of 'viability' in redistributive land reform: perspectives from southern Africa. *Journal of Peasant Studies*, 37(1), 31–66.

Cramer, C., C. Oya and J. Sender. 2008. Lifting the blinkers: a new view of power diversity and poverty in Mozambican labour markets. *Journal of Modern African Studies*, 46(3), 361–392.

Deininger, K. 2011. Challenges posed by the new wave of farmland investment. *Journal of Peasant Studies*, 38(2), 217–247.

Deininger, K. and D. Byerlee. 2011. The rise of large farms in land abundant countries. *Do they have a future? Policy Research Working Paper 5588*. Washington DC: The World Bank.

Diemer, G. and L. Vincent. 1992. Irrigation in Africa: the failure of collective memory and collective understanding. *Development Policy Review*, 10, 131–154.

De Schutter, O. 2011. How not to think of land-grabbing: three critiques of large-scale investments in farmland. *Journal of Peasant Studies*, 38(2), 249–279.

Dyer, G. 2004. Redistributive land reform: no April rose. The poverty of Berry and Cline and GKI on the inverse relationship. *Journal of Agrarian Change*, 4(1–2), 45–72.

Ellis, F. 2000. *Rural livelihoods and diversity in developing countries*. Oxford: Oxford University Press.

Ellis, F. and A. Freeman, eds. 2005. Rural livelihoods and poverty reduction. London: Routledge.

FAO, 2005. *Irrigation in Africa in figures – Aquastat Survey 2005*. Rome: Food and Agriculture Organisation of the UN.

FAO, 2009a. *The state of agricultural commodity markets 2009*. Rome: Food and Agriculture Organisation of the United Nations.

FAO, 2009b. *Low greenhouse gas agriculture. Mitigation and adaptation potential of sustainable farming systems*. Rome: Food and Agriculture Organisation of the United Nations.

Fischer, G., H. van Velthuizen, M. Shah and F. Nachtergaele. 2002. *Global agro-ecological assessment for agriculture in the 21st century: Methodology and results*. Laxenberg, Austria: International Institute for Applied Systems Analysis.

First, R. 1983. *Black gold*. Brighton: Harvester.

GRAIN, 2009. Rice land grabs undermine food sovereignty in Africa. Available from: http://www.mail-archive.com/sustainablelorgbiofuel@sustainablelists.org/msg73860 [Accessed 15 April 2010].

Griffin, K., A. Khan and A. Ickowitz. 2002. Poverty and the distribution of land. *Journal of Agrarian Change*, 2(3): 279–330.

Hassan, A. 1996. Improved traditional planting pits in the Tahoua Department (Niger): an example of rapid adoption by farmers. *In*: C. Reij, I. Scoones, and C. Toulmin, eds. *Sustaining the soil. Indigenous soil and water conservation in Africa*. London: Earthscan, pp. 56–61.
Hill, P. 1963. *The migrant cocoa farmers of southern Ghana*. Cambridge: Cambridge University Press.
HPLE, 2011. *Price volatility and food security. A report by the High Level Panel of Experts on Food Security and Nutrition. July 2011*. Rome: United Nations Committee on World Food Security.
IEA, 2010. *World Energy Outlook 2010*. Paris: International Energy Agency.
IPCC, 2007. *Climate change 2007: The physical science basis.Contribution of Working Group I to the Fourth Assessment Report of the Intergovernmental Panel on Climate Change*. Cambridge: Cambridge University Press.
Jones, C. 1986. Intra-household bargaining in response to the introduction of new crops: a case study from North Cameroon. *In*: J. L. Moock, ed. *Understanding Africa's rural households and farming systems*. Boulder, CO: Westview Press, pp. 105–123.
Kouame, G. 2010. Land markets and land conflicts: intra-family and socio-political dimensions in the Abure case, Cote d'Ivoire. *Africa*, 80(1), 126–146.
Lanjouw, J. and P. Lanjouw. 2001. The rural non-farm sector: issues and evidence from developing countries. *Agricultural Economics*, 26, 1–23.
Lankford, B. 2004. Resource-centred thinking in river basins; should we revoke the crop water requirement approach to irrigation planning? *Agricultural Water Management*, 68, 33–46.
Li, T. M. 2011. Centering labor in the land grab debate. *Journal of Peasant Studies*, 38(2), 281–298.
Lipton, M. 1977. *Why poor people stay poor*. London: Temple Smith.
Little, P. and M. Watts. 1994. *Life under contract: contract farming, agrarian restructuring and flexible accumulation*. Madison: University of Wisconsin Press.
Maertens, M. and J. Swinnen. 2009. Trade, standards and poverty: evidence from Senegal. *World Development*, 37(1), 161–178.
Mann, H. and C. Smaller. 2010. *Foreign land purchases for agriculture: what impact on sustainable development? Sustainable Development Innovation Brief No. 8*. New York: United Nations Department of Economic and Social Affairs.
Mellor, J. 1976. *The new economics of growth*. Ithaca, New York: Cornell University Press.
Moore, J.W. 2010. The End of the Road? Agricultural Revolutions in the Capitalist World-Ecology, 1450–2010. *Journal of Agrarian Change*, 10(2), 389–413.
Murton, J. 1999. Population growth and poverty in Machakos District, Kenya. *Geographical Journal*, 165(1), 37–46.
Peters, P. 2009. Challenges in land tenure and land reform in Africa: anthropological contributions. *World Development*, 37(8), 1317–1325.
Peters P. and D. Kambewa. 2007. Social conflicts over access and transfer of land in Malawi and their implications for new land policy. *Journal of Modern African Studies*, 45, 447–72.
Place, F., M. Adato, P. Hebinck and M. Omosa. 2007. Impacts of agroforestry-based soil fertility replenishment practices on the poor in Western Kenya. *In*: M. Adato and R. Meinzen-Dick, eds. *Agricultural research, livelihoods and poverty*. Baltimore: Johns Hopkins University Press, pp. 149–197.
Reardon, T. 1997. Using evidence of household income diversification to inform study of the rural non-farm labor market in Africa. *World Development*, 25(5), 735–747.
Reardon, T. 1998. Rural non-farm income in developing countries. *In*: *The state of food and agriculture 1998*. Rome: Food and Agriculture Organization of the United Nations, pp. 283–356.
Robertson A. 1987. *The dynamics of productive relationships. African share contracts in comparative perspective*. Cambridge: Cambridge University Press.
Schut, M., M. Slingerland and A. Locke. 2010. Biofuel developments in Mozambique. Update and analysis of policy, potential and reality. *Energy Policy*, 38, 5151–5165.
Sender, J. and D. Johnson. 2004. Searching for a weapon of mass production in rural Africa: Unconvincing arguments for land reform. *Journal of Agrarian Change*, 4(1–2), 142–164.
Shivji, I. 1976. *Class struggles in Tanzania*. London: Heinemann.

Slingerland, M. and M. Masdewel. 1996. Mulching on the central plateau of Burkina Faso. *In*: C. Reij, I. Scoones and C. Toulmin, eds. *Sustaining the soil. Indigenous soil and water conservation in Africa*. London: Earthscan, pp. 85–89.

Southgate, C. and D. Hulme. 2000. Uncommon property: the scramble for wetland in southern Kenya. In: P. Woodhouse, H. Bernstein and D. Hulme, eds. *African enclosures? Social dynamics of wetlands in drylands*. Oxford: James Currey, pp. 73–118.

Stanning, J. 1983. Production costs of farming in Zimbabwe. *Zimbabwe Agricultural Journal*, 80, 57–60.

van der Zaag, P., D. Juizo, A. Vilanculos, A. Bolding and N. Post Uiterweer. 2010. Does the Limpopo river basin have sufficient water for massive irrigation in the plains of Mozambique?. *In*: P. van der Zaag, ed. *What role of law in promoting and protecting the productive uses of water by smallholder farmers in Mozambique?* CGIAR Challenge Programme on Water and Food, pp. 41–49. Delft: Water Resources Section, Delft University of Technology.

van Koppen, B., H. Sally, M. Aliber, B. Cousins and B. Tapela. 2009. Water resources management, rural redress and agrarian reform. Development Planning Division Working Paper Series No. 7. Halfway House: Development Bank of Southern Africa.

Weis, T. 2010. The Accelerating Biophysical Contradictions of Industrial Capitalist Agriculture. *Journal of Agrarian Change*, 10(3), 315–341.

Woodhouse, P. 2003. African enclosures: a default mode of development. *World Development*, 31(10), 1719–1733.

Woodhouse, P. 2012. Water in African agronomy. *In*: J. Sumberg and J. Thompson, eds. *Contested agronomy: agricultural research in a changing world*. London: Earthscan, pp. 102–115.

Woodhouse, P. and A-S. Ganho. 2011. Is water the hidden agenda of agricultural land acquisition in sub-Saharan Africa? International Conference on Global Land Grabbing 6–8 April 2011, Institute of Development Studies, University of Sussex, UK.

Woodhouse, P., P. Trench and M. Tessougué. 2000. A very decentralised development. *In*: P. Woodhouse, H. Bernstein and D. Hulme, eds. *African enclosures? Social dynamics of wetlands in drylands*. Oxford: James Currey, pp. 29–72.

World Bank, 2007. *World Development Report 2008: agriculture for development*. Washington DC: World Bank.

World Bank, 2009. *Awakening Africa's sleeping giant. Prospects for commercial agriculture in the Guinea Savannah Zone and beyond*. Washington DC: The World Bank.

World Bank, 2010. *Rising global interest in farmland. Can it yield sustainable and equitable benefits?* Washington DC: The World Bank.

Wuyts, M. 1985. Money, planning and rural transformation in Mozambique. *Journal of Development Studies*, 22(1), 180–207.

Philip Woodhouse has worked on questions of land and water management for more than 30 years, initially based in Mozambique and subsequently in francophone West Africa and South Africa. He is a Senior Lecturer in the School of Environment and Development at the University of Manchester, UK. His current research is focused on livelihood impacts of large-scale farming in South Africa and Mozambique. His publications include *African enclosures? Social dynamics of wetlands in drylands* (James Currey, 2000), and *Water resources and development* (Routledge, 2011).

Patterns of agrarian transformation in Ethiopia: State-mediated commercialisation and the 'land grab'

Tom Lavers

Much of the literature on the 'land grab' has thus far focused on the international drivers of foreign agricultural investment, with far less attention paid to the roles of developing country states and domestic political economy in changing forms of agrarian production. This paper analyses how global and domestic processes combine to produce patterns of agrarian transformation in Ethiopia, one of the main targets of foreign agricultural investment. The paper presents a typology of changes in land use and examines in detail three case studies of investments in Ethiopia drawn from this typology. The paper concludes that the most dramatic changes are taking place in lowland, peripheral regions where large-scale, capital-intensive farms employing wage labour pose a serious risk to pastoralists whose 'use' of land is contested by the state. Although the government has been careful to avoid mass displacement of settled smallholders, there are also important changes taking place in highland areas, with the government encouraging investments that combine the resources of investors with the labour and land of smallholders. These investments have resulted in exposure to new forms of market risk.

Introduction

Reports of large-scale foreign investments in Africa have understandably raised concerns of a 'land grab' that could result in the mass displacement of African peasantries (GRAIN 2008, Baxter 2010, Global Campaign for Agrarian Reform 2010, Anseeuw *et al.* 2011). The vast scale of many investments and the speed at which change is occurring presents an urgent need for empirical research examining the impacts of investment on existing systems of agricultural production and assessment of the new forms of production established. Here a critical issue is whether business models exist that can benefit local populations, as well as providing an economic return to investors, or whether these projects truly constitute 'land grabs' – the extraction of resources by external actors at the expense of local populations. Existing research has identified crises of finance, fuel, food and climate as important drivers of recent investment processes (GRAIN 2008, Cotula *et al.* 2009, McMichael 2010, Zoomers 2010, De Schutter 2011). In comparison, little

---

This research was carried out as part of a PhD supported by a studentship from the Economic and Social Research Council (ESRC), and this paper was written with financial support from the Land Deal Politics Initiative (LPDI). The support of both organisations is gratefully acknowledged. In addition, I would like to thank Addis Ababa University for facilitating my fieldwork in Ethiopia; Amsalu Darge and Tolasa Mamuye for their interpretation skills during the case studies; and Shea McClanahan, Ruth Hall and two anonymous reviewers for their comments on a draft of this paper.

research thus far has examined how these international processes interact with domestic political economy, what role the state plays in agricultural investment and its role in influencing changing agrarian relations.

This paper examines these issues in Ethiopia, a country which has for many years pursued a smallholder-focused development strategy, but which has nevertheless started to promote large-scale agricultural investment and has become one of the main hotspots for the 'land grab'. Existing research on the 'land grab' in Ethiopia has, on the whole, been descriptive, rather than analytical, and has paid limited attention to the domestic political economy of investment processes. For example, Weissleder (2009) presents a picture of increasing trends in agricultural Foreign Direct Investment in Ethiopia, while global reports such as those by the International Institute for the Environment and Development (IIED) and the World Bank investigating the land grab present data on investment trends in Ethiopia to support their global analyses (Cotula *et al.* 2009, Deininger and Byerlee 2010). In both cases, detailed analysis of how these trends relate to Ethiopian politics is beyond their remit. Tamrat (2010), meanwhile, provides a useful summary of the legal framework governing investment in Ethiopia, without analysing the implementation of these laws or assessing the government's motivations. In addition, several recent studies have focused on the impacts of specific investments (Fisseha 2011, Guillozet and Bliss 2011, Shete 2011), but provide only brief details of the broader context in which the cases are situated. Recent papers by Rahmato (2011) and The Oakland Institute (2011) both go further by situating specific projects within their broader context. Nevertheless, both stop short of a systematic analysis of the role of the state in the promotion of investment and the impact on agrarian transformation.

This paper goes beyond the existing literature to examine the impact of investment on changing land use and the impacts on systems of production. In doing so, the analysis relates these emerging patterns of agrarian transformation to domestic political economy and, in particular, the key role of the state in mediating investment processes. In Ethiopia, commercial pressures resulting from international drivers are filtered through the state as it attempts to promote its own developmental objectives and manage the competing interests of dominant groups in society. The result, rather than mass displacement and proletarianisation, is a varied pattern of changing land use. Those most likely to be negatively affected by investment are pastoralists and shifting cultivators in lowlands whose 'use' of land is contested by the state. While the political importance of the smallholder majority means that mass displacement of the peasantry is unlikely, new models of agricultural production that seek to combine increased investment with smallholder farms nevertheless imply new risks.

The analysis that follows draws on inventories of investments obtained from federal and regional governments in Ethiopia, as well as semi-structured interviews with government officials, investors operating in Ethiopia and local people affected by investments. These interviews were conducted during fieldwork between October 2009 and September 2010 and are supplemented by secondary sources, which provide additional information on particular investment projects. The analysis utilises three case studies to illustrate particular paths of changing land use and to assess the impact of the projects on local people differentiated by class, ethnicity and gender.

Following this introduction, section two outlines the framework used to analyse agricultural investment and its impact on land use and systems of agricultural production. Section three provides an overview of the forms of production and dominant land uses in Ethiopia prior to recent investment processes. Section four then outlines the emerging pattern of agrarian transformation resulting from investment, providing case studies of common transitions and highlighting areas of continuity. Section five analyses the political economy of this changing land use, explaining the patterns based on international drivers in the context of domestic political economy. Section six concludes.

The role of agricultural investment in agrarian transformation

The central concern of this paper is the role of the state in influencing patterns of agrarian transformation in Ethiopia. Key to this analysis is the identification of the existing systems of agricultural production that are being transformed by investment processes and those that are not, how this pattern relates to key political interests, the way in which these transformations are discursively justified by the state and the impacts of the changes on local people, differentiated by class, ethnicity, gender and age.

The main impacts on people living in the vicinity of investments stem from changing patterns of land use and changing systems of agricultural production (Borras and Franco 2010). This paper therefore examines the changing social context of production (Crouch and de Janvry 1980) resulting from agricultural investment. These systems of production are in part influenced by the characteristics of the crops grown but also vitally by political, historical and social factors influencing the system of production (Crouch and de Janvry 1980, Little and Watts 1994, White and Dasgupta 2010). In the Ethiopian context, it is necessary to move beyond Crouch and de Janvry's (1980) distinction between peasant and capitalist production. First, the analysis differentiates in the non-capitalist sector between settled smallholders and those involved in shifting agriculture and agro-pastoralism. Key factors here include the nature of the land rights of these groups and their political importance to the state. Second, the paper acknowledges that capitalist production, and consequently its impacts on local people, takes different forms. The paper draws a distinction between plantations directly managed by investors and farmed using wage labour, and different types of outgrower schemes. A key concern of the 'land grab' debate is the extent to which plantations create employment and consequently address problems of rural unemployment (Li 2011). In turn, employment creation is related to the relative emphasis on labour-intensive versus mechanised forms of production. Outgrower schemes are arrangements where investors negotiate agreements with smallholders to grow a specific crop on their land and to sell exclusively to the investor, usually at a pre-agreed price. The 'outgrowers' are often supplied with technology and expertise to aid their production. As such, outgrower schemes are a means by which smallholders are incorporated into capitalist (and frequently export-oriented) agricultural production chains (Little and Watts 1994). While the landless must earn their living as wage labourers, outgrowers retain control of their land, albeit with a diminished 'bundle of rights' (Schlager and Ostrom 1992). Nonetheless, wage labourers and outgrowers are similar in that they are both participants in the monetary economy and are exposed to the inherent risks of the market (Scott 1976).

According to some critics, the land grab is part of a broader neo-liberal project of globalisation, which privileges capital mobility and has unleashed processes that are creating a capitalist agrarian transformation in developing countries, threatening global de-peasantisation and proletarianisation (Araghi 2009, McMichael 2010). One potential danger of the 'land grab' is therefore the mass displacement of peasantries by large-scale, capital-intensive agricultural production and the creation of a class of landless labourers. Nevertheless, the state has always played a key role in influencing the outcome of agrarian transformations (Byres 1991). Indeed, in Ethiopia, in addition to its economic importance, land has also been central to political power (Rahmato 2009). As a result, past economic pressures for commercialisation have been mediated by the political priorities of the state, leading to varied patterns of transformation (Joireman 2000). A key question of this paper is, in the context of the international drivers of the land grab, what role does the Ethiopian state play in mediating changes in land use, transformations of existing systems of production and the establishment of new ones? To answer this question, the research must focus on the interests of state organisations and their attempts to balance their own economic and political priorities with those of dominant groups within domestic society – including those based on class and ethnicity – and international influences (Migdal 2001) – including the states and companies driving the 'land grab'. Within countries, variations in these political relations between dominant groups and the state are likely to result in regional variation in processes of agrarian transition (Byres 1991).

The prior social context of production

Agricultural production in Ethiopia continues to be dominated by the radical land reform conducted by the *Derg* following the revolution which overthrew the Imperial regime in 1974. This land reform nationalised all land, destroyed the landholding elite and redistributed usufruct rights to smallholders, creating a smallholder majority in the highlands. At the time of the revolution, there were only a few capitalist farms, mostly foreign companies producing sesame in Humera, and cotton and sugar on the banks of the Awash (Clapham 1988, Zewde 2008, Rahmato 2009). With the revolution, these holdings were nationalised and turned into state farms (Clapham 1988).

Many lowland areas, where the dominant systems of production were agro-pastoralism and shifting cultivation, had never been fully incorporated into the Imperial state following their conquest at the end of the nineteenth century but had been subject to the extraction of resources through tribute, taxes and raids for slaves and ivory (Donham 2002, Garretson 2002, Feyissa 2006). Under the *Derg*, forced resettlement campaigns moved people from the heavily populated highlands to parts of these less densely populated lowlands, where the government claimed there were vast expanses of fertile 'virgin' lands, largely ignoring existing land uses (Pankhurst 1992, 1997). Nevertheless, many pastoral areas remained largely beyond the day-to-day control of the state (James *et al.* 2002).

In 1991, the *Derg* was militarily defeated by the Tigrayan People's Liberation Front (TPLF), an ethnic-based insurgency organised along Maoist principles, which prioritised living and working with the peasantry in order to gain their support (Young 1997). On seizing power at the national level, the TPLF formed the Ethiopian Peoples' Revolutionary Democratic Front (EPRDF), a coalition of ethnic parties that were also founded by the TPLF, and established a federal system based on ethnic regional administrations. Largely to secure international legitimacy and development assistance, the EPRDF claimed to have switched to a market-based development strategy, although in practice the government retains close control over the economy (Vaughan and Tronvoll 2003, Clapham 2006). Similarly, despite claims of democratisation, space for political opposition has reduced substantially in recent years, leaving Ethiopia as an effective one party state, with the EPRDF and its affiliate parties winning 99 percent of seats at the 2010 federal and 2008 local elections and controlling all levels of government (Aalen and Tronvoll 2009, Tronvoll 2009).

On coming to power, the EPRDF retained many aspects of the *Derg*'s land tenure system. Indeed, in 1993, the EPRDF made smallholder agriculture the central pillar of its Agricultural Development-Led Industrialisation (ADLI) strategy, arguing that labour-intensive smallholder agriculture, alongside improved inputs that complement rather than replace labour, offered the potential of egalitarian economic growth providing forward and backward linkages from agriculture to stimulate industrialisation (MoFED 2003). The state ownership of land is key to this strategy, preventing the distress sale of land when farmers face shocks and thereby ensuring that land remains in the hands of smallholders.

These social and economic objectives also overlap with key political interests. First, for the EPRDF, which began as a peasant-based movement, the protection of the peasant majority from displacement constitutes a populist policy that resonates with its political base. Second, by prioritising equality in landholdings, the government has prevented the accumulation of land by large landholders (Rahmato 2009) and the emergence of independent economic actors that might be able to translate economic power into political influence. Third, the land policy limits migration, which is considered by the government to be the 'source of economic, political and social instabilities' (MoFED 2002, 56), through restrictions on land transactions. In particular concerns relate to the fear that migration across ethno-regional boundaries would lead to ethnic conflict (MoFED 2002, 56), while the lack of industrial development raises concerns that urban migration would result in high rates of urban unemployment leading to social and political unrest. As a result of these factors, it has been argued that state ownership of land brings the state and the peasantry into direct contact, while the government's role in the distribution of usufruct rights means that the peasantry is dependent on and therefore subordinate to the state, leaving little opportunity for political mobilisation, and as such is a key means of maintaining state authority (Rahmato 2009). Land and agriculture are therefore central issues in Ethiopian political economy, while any changes to land tenure and agricultural production resulting from investment are vitally important to both the government's development objectives and its political authority.

The ADLI strategy focuses on the majority involved in settled agriculture. In contrast, the minority subsisting from pastoralism and shifting cultivation remain politically marginal, while the government sees no long-term alternative to sedentarisation (MoFED 2003, 2010). In the words of Dr Aberra Deresa, State Minister for Agriculture and Rural Development, 'we are not really appreciating pastoralists remaining as they are. We have to improve their livelihood by creating job opportunities. Pastoralism, as it is, is not sustainable. We want to change the environment' (Butler 2010).

Thus far ADLI has failed to meet its socioeconomic objectives. In particular, there have been only limited improvements in agricultural productivity and the marketed surplus remains small due to the shortage of land, a failure to increase the supply of improved inputs and volatile output markets, which make investment in inputs risky for smallholders (Rahmato 2003, Dercon and Vargas Hill 2009). High cereal prices and limited production of crops used as industrial inputs highlight the failure of the smallholder sector to contribute to industrialisation through forward linkages, while agricultural exports remain limited and reliant on coffee. Meanwhile, the end to land redistribution in the 1990s, alongside rapid population growth, has resulted in land shortages (Rahmato 1997, 2003) and a degree of differentiation within the peasantry, dividing independent smallholders from the landless and land-short, who are forced to rely on sharecropping and land rental or work as wage labourers (Habtu 1997). The problems in the smallholder sector are clearly evident in the reliance of more than seven million chronically food-insecure people on the food- and cash-for-work Productive Safety Net Programme (PSNP), while each year several million others facing shocks rely on emergency assistance to survive (MoARD 2009). The latter figure reached 4.5 million in 2011 during the East African famine (FAO 2011). The donor-financed PSNP now constitutes a key means of household reproduction in parts of the highlands, in effect supporting government policies that in many parts of the country offer little potential for transforming smallholder productivity or solving the problem of food insecurity.

Lavers (2012) shows that ADLI's failings have prompted the government to make the transition to a more trade-based development strategy, encouraging foreign investment as part of a broader push for agricultural commercialisation. The government has used generous tax holidays and low land lease costs as incentives to promote investment and consequently expects the main benefit of agricultural investment to be increased foreign exchange earnings, which will enable the import of machinery to advance industrialisation and food to address problems of food insecurity. However, with most agricultural investment projects still at relatively early stages of development, the actual impact on foreign exchange remains to be seen. Nevertheless, in terms of agricultural production, this change in strategy is resulting in uneven impacts in different parts of the country. Indeed, past government policy has given rise to four broad categories of land, each of which has been affected differently by the promotion of investment. First, smallholders, predominately in the highlands, farm 11.5 million of the 11.8 million hectares used for grain production (CSA 2010) with less than one hectare per household, on average. Second, within each highland village, despite encroachment of individual plots due to population pressure, there remain a few hectares of communal land used for grazing or firewood collection. Third, primarily in highland areas, there are a number of state farms left over from the *Derg* era. The final 'category', from the perspective of the state, constitutes the majority of the lowlands, comprising a great variety of pastoral and agro-pastoral lands, land used for shifting cultivation and forests. Before the government started to promote agricultural investment during the last decade, there was virtually no capitalist agricultural production.

The spatial distribution of agricultural investment

Although government data do not contain information on the previous use of investment land, in Ethiopia, variations in land use broadly overlap with population density. For example, rural areas in the central highlands with high population density are almost always in smallholder-dominated areas, while low population densities in lowland border areas signal pasture, areas used by shifting cultivators and forests (see Figure 1). Consequently, important insights into the types of land leased to investors can be obtained by comparing investment locations with population density. Figure 2 shows the locations of land already leased to investors, where data are available. Comparison with Figure 1 reveals that the vast majority of investment land is located in sparsely populated border regions. Although smaller amounts of land have been

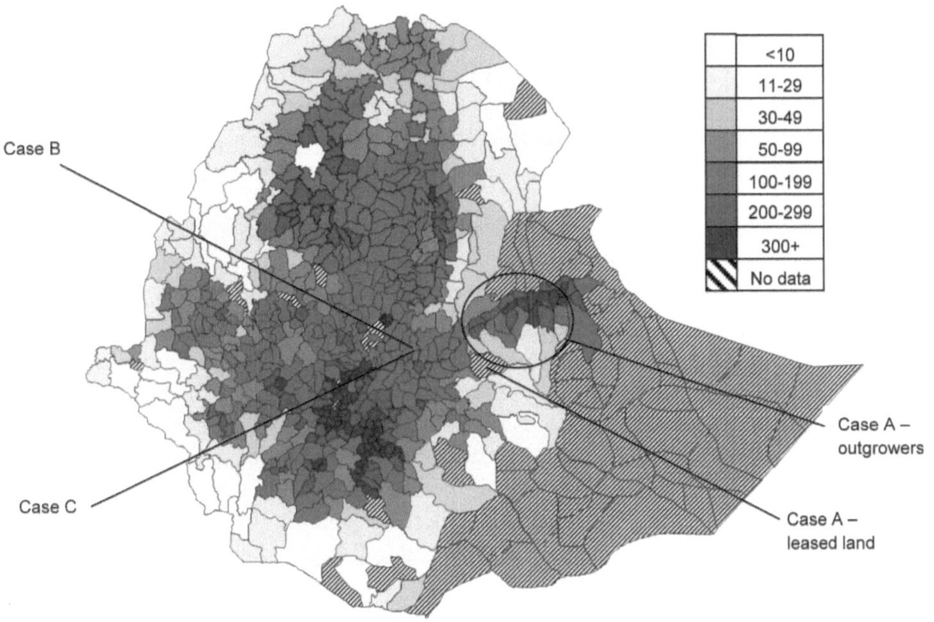

Figure 1. Population density (people per sq. km) by *wereda*.
Source: CSA, UNOCHA.

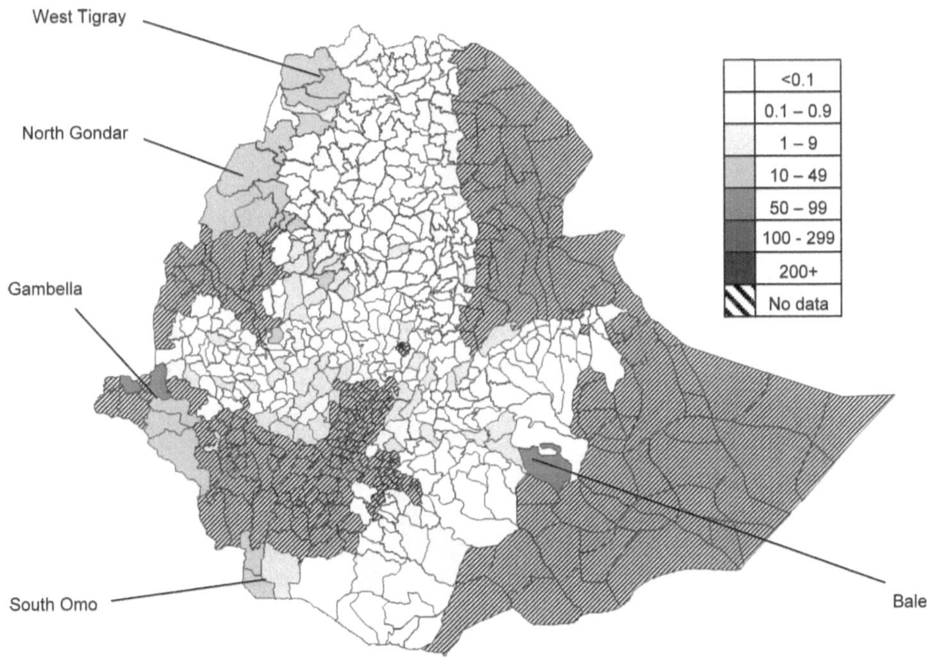

Figure 2. Land leased to investors (in thousand ha) by *wereda*.
Source: Own calculations based on data from CSA, Amhara Investment Agency, EIA, MoARD (2011), SNNPR Investment Promotion Agency.
Note: Data for Oromiya do not distinguish between pre-implementation and active projects. Data for Tigray are only reported by zone, so totals have been equally divided between the relevant *wereda*.

leased in highland, densely populated areas, the main sites for investment are West Tigray, North Gondar, Gambella, South Omo and Bale.

To promote investment, the federal government has instructed regions to identify 'banks' of suitable land for investment and has identified 3.7 million hectares of 'unused' land so far (respondent AE, see annex). Figure 3 presents the location of land in these land banks for all regions except Oromiya and Somali, for which sub-regional breakdowns are not available. It shows a similar pattern to Figure 2, with all main targets for investment situated in lowland areas – nearly 2 million hectares in Gambella and Benishangul-Gumuz, 400,000 hectares in Afar as well as substantial amounts in northwest Amhara and South Omo. Land identified for investment constitutes a remarkable 42 percent of the surface area of Gambella and 14 percent of Benishangul-Gumuz (based on data in CSA 2009). Many of the investment hotspots are near major rivers that have potential for irrigation, including the Awash in eastern Oromiya and Afar, the Omo in Southern Nations, Nationalities and Peoples Region (SNNPR), the Baro in Gambella and the Blue Nile in Benishangul-Gumuz. In addition to these lowland areas, there are also significant amounts of land identified for investment in some densely populated highland *wereda* (districts)[1] in East and West Gojjam zones in Amhara.

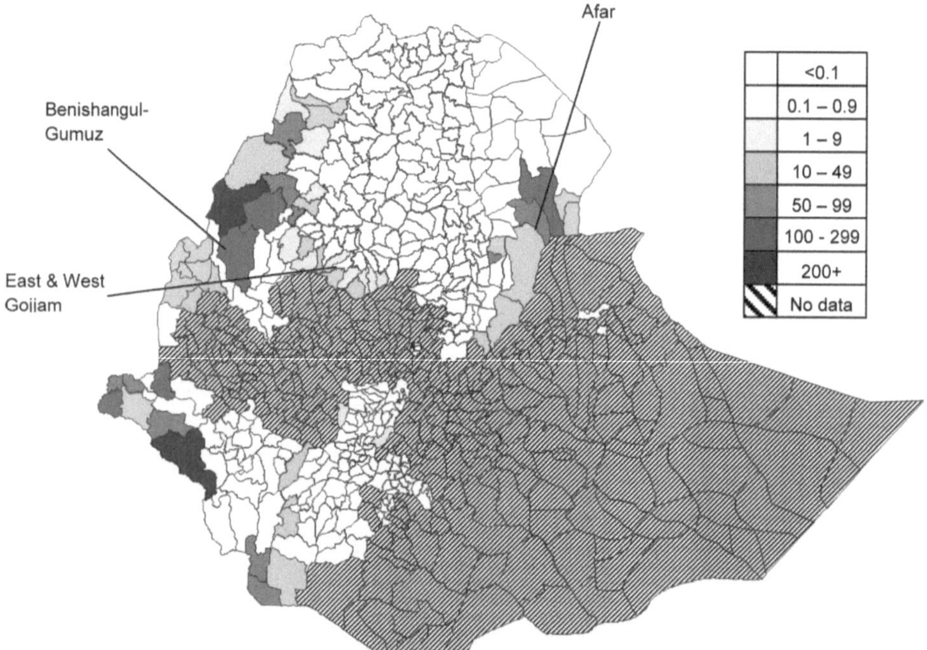

Figure 3. Land identified for investment (in thousand ha) by *wereda*.
Source: Own calculations based on data from CSA, AISD/MoARD, Amhara Investment Agency, Oromiya Investment Commission, Tigray BoEPLAUA.
Note: Data for Asosa zone, Benishangul-Gumuz are only reported by zone, so the total has been equally divided between the relevant *wereda*.

[1]Below the federal government, the hierarchy of government administrations is as follows (from largest to smallest): region, zone, *wereda* and *kebele*.

As well as identifying this 'available' land, the government has provided investors with incentives to establish projects in remote areas. For example, projects in Gambella, Benishangul and South Omo are eligible for an additional tax holiday on top of those used to encourage export-oriented investments (FDRE 2003, para. 4). Furthermore, areas not previously under cultivation are prioritised by exempting investors from land use fees for up to five years if they use improved seeds and irrigation (ONRG 2001, para. 2). After these holidays expire, land fees are also set to encourage investment in target areas. For example, remote land in Tigray is leased for 40 *birr*[2] per hectare compared to 100 *birr* for more accessible land (TNRG 2000 (Ethiopian Calendar)), while in SNNPR, prices range from 30 to 117 *birr* per hectare (SNNPR Investment Agency 2008).

Deconstructing the discourse of 'unused' land

The government claims that land identified for future investment and the vast majority already leased to investors was previously 'unused'. This land can therefore be leased to investors to expand the country's total cultivated area without damaging existing production and livelihoods. As the Prime Minister, Meles Zenawi states, 'We have three million hectares of unutilized land. This land is not used by anybody. This land should be developed' (cited in ProKerala News 2011). Nevertheless, the categorisation of all this land as 'unused' is highly questionable. The identification of 'unused' land is conducted from the state's perspective (Scott 1998, Borras and Franco 2010). 'Unused' land may therefore include some land which is objectively unused for any human purpose, as well as land used by people for purposes which are considered insufficiently productive or are invisible to the state.

In discussions with government officials and in government policy statements, the terms 'unutilized' (Meles Zenawi, cited in ProKerala News 2011), 'empty' (interview with respondent D), 'unsettled' (MoFED 2002, 53) and 'uncultivated' (Esayas Kebede, head of the Agricultural Investment Support Directorate, cited in McClure 2009, and respondent C) are used interchangeably to describe the land identified for future investment. These descriptions betray the construction of land utilisation from the highland perspective of settled crop cultivation, which views pastoralism as unsustainable (MoFED 2003). This construction of 'unused' land is also reflected in government land use statistics. According to studies cited by the government to minimise the significance of investment in comparison with the agricultural sector as a whole (Rice 2010, Vidal 2010, Kebede 2011), approximately 75 million hectares in Ethiopia are suitable for cultivation, while only 14 million hectares are currently cultivated, leaving about 60 million hectares 'free'.[3] Clearly, this identification of 'cultivation' with 'use' dismisses the importance of other land uses.

So what criteria have actually been applied to identify 'unused' land? A process of land registration has been conducted during the last decade in highland regions with the objective of increasing tenure security (Adenew and Abdi 2005, Deininger

[2]US$1 was approximately 16.5 *birr* in September 2010.

[3]I have been unable to trace the origin of these estimates, which do not refer to specific sources. Similar, claims of vast areas of fertile 'virgin' land in the lowlands were used by the *Derg* to justify resettlement programmes (Pankhurst 1992) and probably contributed to the construction of 'unused' land. They also differ considerably from those provided by the World Bank in their recent report on foreign agricultural investment, which suggest that less than 5 million hectares of currently *uncultivated* land are suitable for cultivation (World Bank 2010, 110).

et al. 2008). Although the government can displace farmers with land certificates in the 'public interest', these land certificates do at least serve as the basis for landholders to claim monetary compensation (see below). Nevertheless, land registration has not been attempted in Afar, Gambella, Benishangul-Gumuz or remote parts of SNNPR (respondents H) due to limited state capacity in these areas and the logistical challenges of registering land used by pastoralists or shifting cultivators. As a result, one of the key problems facing these groups is an inability to prove that they have historically occupied or used the land in question.

While administrations in Gambella and Benishangul-Gumuz, which are among the weakest of the regional governments, have rapidly identified vast tracts of land for investment which they confidently claim to be 'unused', respondents in the Amhara Bureau of Environmental Protection, Land Administration and Use (BoEPLAU), widely recognised as the most capable, stated that the rapid identification of large amounts of suitable investment land was beyond their capacity (respondent A). Consequently, Amhara annually identifies comparatively small tracts of land for investment. Similarly, although the Oromiya Land Use and Environmental Protection department (LUEP) identified a land bank for investment, a senior respondent in the department admitted that the data were unreliable (respondent B). Serious questions therefore must be raised regarding the criteria used to identify 'unused' land. In many cases it is impossible to tell with certainty whether land in these areas is truly unused or merely unregistered.

There are numerous examples where the label 'unused' is contestable. Case A, discussed below, illustrates one instance where an investor in Oromiya leased a plot of land used by pastoralists for grazing. Meanwhile, in Gambella, despite low population density, access to land was the subject of conflict even prior to the arrival of investors (Feyissa 2005). Reports now suggest that local people are unhappy that land they previously used has been leased to investors (Daniel and Mittal 2010, Vidal 2010, The Oakland Institute 2011). In addition, a report by the IIED finds that the classification of all investment land as 'wasteland' is highly questionable and claims that at least some of the land leased to investors in Benishangul-Gumuz was previously used for shifting cultivation and grazing (Cotula *et al.* 2009). Finally, land allocated to investors in West Tigray may be sparsely populated but the same area has been the destination for resettlement from the highlands (respondent I) and these areas have already seen tensions between 'indigenous' pastoralists and new arrivals, according to resettled farmers (respondents J). Misclassifications of 'unused' land therefore stem both from inadequate land surveys as well as the conceptualisation of land use from the perspective of settled agriculture, which dismisses the use of land by other livelihoods.

Investment in the highlands

Although most investment land lies in sparsely populated lowlands, Figure 2 shows that some land has been leased in heavily populated areas that were, or could be, cultivated by smallholders. As such, the smallholder and investor sectors are not entirely separate, as frequently claimed by government officials. First, several state farms have been leased to investors in recent years. These include: 3,000 hectares in Bale, Oromiya leased to the Government of Djibouti to grow grain (Zenebe 2010); and an unspecified amount of land in Hadiya and Kembata, SNNPR leased to flower farms (respondent D).

A second category of land in highland areas is communal grazing land. In recent years, local governments have leased plots of communal land to mostly domestic, but also some foreign, investors. These plots are usually relatively small (usually 5–10 hectares), although one project established by an Indian investor, Karuturi, on communal grazing land in Bako, Oromiya, is exceptionally large and probably unique in covering 10,000 hectares (Rahmato 2011, Fisseha 2011). Like land in pastoral areas, most communal land is unregistered and is considered by government officials to be a government, not a community, resource.[4] Based on my own case research, leases of communal lands to investors were subject to discussion between local government officials and local people at community meetings. Nevertheless, a minority of respondents stated that these meetings were merely a formality and in practice they did not have the power to reject the projects. While investors pay a land lease fee to local government for communal land, they provide no financial compensation to the community, instead making informal promises that they will contribute to communities by building schools or clinics, and providing employment. Nonetheless, the decision to allocate such land ignores the valuable role of communal land and property resources in local livelihoods (Platteau 2005).

Perhaps the most controversial category of investment land is that already cultivated by smallholders. Despite economic challenges in the smallholder sector, it retains considerable political importance, and the government is keen to avoid displacement. Indeed, restrictions on land sales and rental are explicitly designed to prevent the transfer of land directly from smallholders to investors and limit smallholder displacement (respondent B). Nevertheless, in certain cases, government assessments of agricultural potential have concluded that labour-intensive agriculture has been insufficient to take full advantage of available agricultural resources. Investors potentially offer the capacity to develop resources, in particular water for irrigation, beyond that available to smallholders. Nonetheless, the question, addressed below, remains whether there are investment models which can provide local people with secure opportunities for inclusion in this process.

Agricultural investment and the new social context of production

This section uses three case studies of investments in Oromiya to examine the impacts of changing land use, particularly focusing on issues of employment and displacement. These case studies constitute three examples drawn from the range of possible trajectories from prior to subsequent land use. Table 1 presents a typology of changing land use and the principal changes involved in systems of production. Drawing on the limited existing academic literature and newspaper articles on agricultural investments in Ethiopia, the table also identifies examples of particular trajectories of changing land use, where they exist. Oromiya, where all three in-depth cases are situated, is one of the regions with the most land leased (Lavers 2012). Nevertheless, the important role of the regional government in the administration of investment processes may limit the relevance of the case findings for other regions in Ethiopia. Similarly, given the diversity of trajectories, the small number of cases

[4]Respondent C questioned the categorisation of any land as communal land: 'communal land means it belongs to the community, land in Ethiopia is government land'.

Table 1. A typology of changing land use.

From To	Investor plantation	Outgrower scheme
'Unused' land	Type 1 – Creates employment, threatens pastoralist livelihoods – ***e.g. case A (Oromiya), Karuturi (Gambella), Saudi Star (Gambella)***	Type 5 – Combines investment with re/settlement, threatens pastoralist livelihoods – ***No examples***
Communal land	Type 2 – Loss of communal resources (for all), gain of employment (for a few) – ***e.g. case C (Oromiya), Karuturi (Oromiya)***	Type 6 – Loss of communal resources (for all), gain of land (for a few) – ***No examples***
State farms	Type 3 – State employees to private employees – ***e.g. Govt of Djibouti (Oromiya), flower farms (SNNPR)***	Type 7 – Transformation of wage labourers into smallholders – ***No examples***
Individual holdings	Type 4 – Transformation of smallholders into wage labourers – ***e.g. flower farms (Oromiya)***	Type 8 – Smallholders inserted into monetary economy – ***e.g. cases A and B (Oromiya)***

examined in this study and the limited available empirical data available on other projects, any general conclusions drawn from this analysis must be somewhat tentative.

I have found no examples of investments in which uncultivated land (type 5 in Table 1), communal land (type 6) or state farms (type 7) have been leased to investors using outgrowers, a change which, plausibly, could combine investment with smallholder expansion to address land shortages. Instead, investments on 'unused' land tend to be large, mostly foreign-owned investments with capital-intensive production employing some wage labourers (type 1). Case A is an example of a foreign investor growing bio-fuel crops and part of the project is on 'unused' land in West Hararghe and Arssi (for location see Figure 1). In contrast, investments on communal lands are usually small and often involve domestic investors employing wage labourers (type 2). Case C examines a number of small, domestic private investors growing wage foods, in particular vegetables, on communal land in a smallholder-dominated area in Arssi.

The only examples of outgrower schemes have occurred on individual holdings (type 8). The majority of the land covered by the bio-fuel project discussed in case A uses outgrowers, as does the entirety of case B. The latter is an expansion of an Ethiopian state-owned sugarcane plantation in East Shewa, which has incorporated land previously farmed by smallholders into an outgrower scheme to produce sugar for domestic consumption and bio-ethanol. Although it would appear that the proportion of investors using outgrowers is small, with most preferring to manage their own land, as a result of the vast extent of the outgrower scheme in case A, a large proportion of land under active projects does actually use outgrowers.

Wereda officials have the right to expropriate smallholders' land 'where it believes that it should be used for a better development project' (FDRE 2005, para. 3(1)), giving local government enormous power and leaving smallholders with little possibility of appeal. The principal example of the displacement of smallholders to make way for investor plantations (type 4) is the expanding flower industry, in particular in central Oromiya. The technologically advanced nature of production and the controlled conditions required in floriculture are considered to make such projects unsuitable for outgrower schemes. Nevertheless, production on flower farms is labour-intensive, so although smallholders lose their land, government requirements for investors to employ local people are expected to limit the impact on migration. On the whole, displaced smallholders, the majority of whom now have certificates for their landholdings, do seem to receive the legally required compensation of ten times the average annual income over the previous five years (FDRE 2005, para. 8(1)). However, it is questionable whether this is sufficient given that farmers are not allowed to buy replacement land.

Case study A: A foreign private investor producing castor

This case study examined one of the largest investments in Ethiopia. The project, established by Israeli managers with finance from European investors, covered approximately 140,000 hectares. The project was established in 2006 to grow castor for bio-diesel, cosmetics and paints (respondents G, K). Demand for castor has risen in recent years due to high fuel prices and EU requirements for bio-fuel use based on environmental criteria.

In 2007 the company leased 8,000 hectares in three *wereda* in East Hararghe. This land was classified as 'unused' but was identified by the investors using only satellite images rather than on-site assessments (respondent L). Indeed, when they arrived at the site, the investors found that much of the land was already cultivated by smallholders, while the rest was used by pastoralists (respondent G). Government officials expected the investors to negotiate with local people in smallholder areas, but told the investors that since the pastoralists were not 'settled', the land belonged to the government and they could easily take their animals to graze 'elsewhere' (respondent G). Depending on the previous land use, different production systems were implemented: on pastoral lands the investors farmed the land directly with machinery and some wage labourers; and on cultivated land outgrower schemes using labour-intensive techniques were negotiated with elders acting on behalf of the local people (respondent M).

On the basis of this land, and a commitment from the government for up to 200,000 hectares, the managers went to look for finance (respondent G). At that time, before the financial crisis and with considerable interest in renewable energy, they found finance from European investment funds easy to obtain, raising €17 million, way beyond the €5 million for which they were hoping (respondent F). Encouraged by this investment and optimistic about the potential of outgrower schemes, the investors planned a vast expansion. They convinced the government to support their plans because the type of investment and the production system fit its development strategy – it is an agricultural project that provides industrial inputs and the resulting fuel can either be exported or can substitute for imports addressing foreign exchange shortages. Additionally, the production avoided smallholder displacement and, in a food insecure area, offered the potential for 'graduation' of PSNP recipients from government assistance (respondent N). Based on support for the project by high-level government officials, the zone, *wereda* and *kebele* administrations played a vital role in helping the investors to expand the project (respondent L). Community meetings were held in each *kebele*, *kebele* and traditional leaders were paid incentives by the investors to convince people to join the project (respondent M) and *kebele* committees signed contracts with the investors on behalf of the farmers.[5]

This rapid expansion was so effective that by 2008, the project covered 72,000 hectares under the production of an estimated 84,000 (respondent G) to 124,000 (respondent M) smallholders[6] in 240 *kebele* in East and West Hararghe. Most of these *kebele* were classified as chronically food insecure and a substantial proportion of farmers, who farm maize, sorghum and some cash crops like *chat* and coffee, qualified for assistance from the PSNP (respondent L). The *kebele* signed three-year contracts at a fixed price of 50 *birr* per quintal for the castor produced. The investors calculated this price to ensure a substantial increase in smallholder incomes based on estimated price and yield for sorghum, and the seed supplier's estimates of castor yield (respondents L, M). Outgrowers were required to switch up to half their land from cereals to castor, retaining some land for food production, and were supplied with seeds and fertiliser, the cost of which was deducted from their payment when seeds were sold to the company (respondent M). Although the company advised the farmers, the smallholders managed production themselves. In line with government policy of using agricultural growth to contribute to industrialisation, the company built a bio-diesel processing factory with a 30 million *birr* loan from an Ethiopian bank, as well as creating employment by hiring 5,000 mostly Ethiopian staff, including agronomists, accountants and supervisors (respondent M). The company also leased an additional 60,000 hectares of 'unused' land in West Hararghe, to be farmed as a plantation, though the outgrower scheme was initially prioritised (respondents P).

The project quickly ran into problems. Castor yields were massively overestimated as the seeds had not been tested across the range of soil and rainfall conditions covered by the expansion (respondent M). In addition, by the time the company bought the seeds produced, sorghum prices in local markets had tripled and, given fixed castor prices, farmers were unhappy with their payment (respondents M, O). This news spread to neighbouring *kebele* and productivity deteriorated further as farmers neglected castor to focus on other crops (respondents M, O). Consequently, instead of the 70 quintals per hectare that the investors had expected, average yield was only 3–4 quintals per hectare and the factory took just three days to process all the seeds bought (respondent M). Facing failure, the managers fled the country in April 2009 and, according to the new manager, stole the remaining money (respondent F). They left massive debts and no money to pay wages or buy the remaining seeds from the outgrowers (respondents M, F, O).

The failure of the project created a major political problem for government, which was heavily involved in its establishment. Having switched from subsistence crops, with little or no income as a replacement, the farmers lost up to half their annual production and held the government to be partly responsible. Additionally, the pesticides used on the castor plants killed their bees, which had provided

[5]According to the contract, the chairman and deputy chairman should have been granted power of attorney by the *kebele* members, though this does not appear to have been the case according to respondents O.

[6]This number is the total of (mostly male) landholders. A conservative estimate of family size would suggest that at least 500,000 people were involved in the project.

important extra income (respondents O, F). Consequently, some had to sell cattle to buy food and others increased reliance on the PSNP (respondents O).

The investment company, now under new management, is in the process of restarting castor production. Nevertheless they have abandoned outgrower schemes since they believe that an inability to supervise production effectively makes such schemes unprofitable in practice (respondent F). Instead they plan to use mechanised farming, and consequently relatively little hired labour, on the 60,000 hectares of land in West Hararghe, which has never been cleared, with the possibility of another 100,000 hectares in Arssi and Bale promised by the regional government (respondent F). All this land is classified as 'unused' as it is not cultivated, although it was clearly being used by pastoralists to graze camels and cattle when I visited. Unfortunately one important constraint of the study on which this paper is based was its focus on areas of settled agricultural production and, consequently, an absence of interviews with pastoralists affected by investment processes. Nevertheless, when asked about the pastoralists on the land, both the investor and government officials indicated that they would be moved elsewhere, with one official stating that their presence 'is no problem at all because we shall make our pastoralists settle' (respondents L, M). The only concession is that the farm is not allowed to extend as far as the river, as the water is used by pastoralists in the dry season (respondent M).

The project is exceptional both in its size and the extent of its failure. Nonetheless, the case does have findings of general relevance. First, it demonstrates how investment is a government-owned process. Key aspects of the project resonated with the government's development strategy, convincing officials to take huge risks with the welfare of smallholders. The promotion of investment is based on the premise that the government can manage investors and direct them to specific purposes, nevertheless this case raises important questions regarding its capability to do so. Second, the government has only limited information on land use in some remote areas that are the target of investment. In particular, the classification of land use disregards pastoralists who are seen as merely passing through land, rather than 'using' it and therefore can easily move on to other 'unused' areas. Finally, while the government makes efforts to prevent smallholder displacement, the outgrower schemes entail inherent risks, which need to be taken into consideration. In this case, smallholders became exposed to fluctuations in international and national prices for fuel and food.

Case study B: State-owned sugar production using outgrowers

Recent years have seen demand for increased sugar production in Ethiopia resulting from rising domestic demand, the need for bio-ethanol production to reduce expensive oil imports and a preferential EU trade agreement for sugar imports from developing countries (van Berkum *et al.* 2005). These factors came to a head in 2009–2010, resulting in a severe domestic sugar shortage and a ban on exports. The state Ethiopian Sugar Development Agency (ESDA), manages domestic sugar supply and four factories. As part of its sugar development strategy, all these factories, including Wonji-Shoa, the subject of this case study, are being expanded to increase production of sugar and bio-ethanol (ESDA 2010). Ironically, the Wonji-Shoa sugar estate was itself created as part of a previous 'land grab', when Haile Selassie's government promoted foreign investment on the Awash River in the 1950s and 1960s. A Dutch company established the plantation by leasing land from the government and displacing the Jille Oromo pastoralists who had previously used it for seasonal grazing (Zewde 2008). The plantation was subsequently nationalised under the *Derg*.

The first phase of the current Wonji-Shoa expansion involves 600 hectares in East Shewa, and 2,600 hectares and a new factory in Arssi to be built by an Indian company. The second phase will involve another 6,000 hectares nearby in East Shewa. The land for the first phase was previously held by smallholders with 0.5–3.0 hectares. With no irrigation infrastructure, farmers depended upon inconsistent rainfall for cereal and pulse production for self-consumption. The site in East Shewa was not classified as food insecure, whereas the area in Arssi remains food insecure and receives the PSNP. Nonetheless, due to proximity to the Awash River and the paved road to Addis Ababa and Djibouti, the area has considerable economic potential.

The proposed expansion generated debate between government agencies regarding the system of production. Statements in 2006 by the ESDA suggest that a private plantation may have been considered (Ethical Sugar 2006), while Wonji-Shoa wanted to displace the smallholders and manage the plantation directly (respondents Q, R, S). Nonetheless, the final agreement between federal and regional governments attempted to combine state investment, the commercial requirements of a secure sugarcane supply to justify investment in a new factory and the political imperative of avoiding displacement (respondents R, S).

The project is a 'dividend scheme', a type of outgrower arrangement in which land is managed as a block by the investor, while landholders work as labourers but play no role in managing production (Li 2011). The smallholders were formed into cooperatives and for two years while the land was left fallow and an additional two years before the first crop was ready for harvest, Wonji-Shoa paid members a monthly allowance of 229 *birr* per hectare. Wage labourers farm the land, with cooperative members given priority when jobs are allocated. The cooperatives are required to repay the production costs – wages, inputs and technical assistance provided by Wonji-Shoa – and are potentially liable for the substantial cost of constructing the irrigation system, though a final decision has not yet been taken (respondents Q). Meanwhile, the *wereda* farmers' union, elected by the cooperatives to negotiate the sugarcane price, has been pressurised into accepting low prices, set at three-year intervals, by the *wereda* administration and the cooperatives office (respondents T, U).

The initial results have so far been negative. The maintenance payments, set in 2006, but not adjusted for inflation, quickly became insufficient given rapid food price increases (respondents V, W, X). In addition, the first sugarcane harvest sold in March 2010 at the agreed price did not cover production costs and the cooperatives were paid nothing (respondents Q). Despite the understandable discontentment of the members, they are unable to extricate themselves from the scheme as the cooperatives are tied to an indefinite agreement to supply sugar exclusively to Wonji-Shoa (respondents Q). Indeed, the land that cooperative members still believe belongs to them is actually being registered to the cooperatives, not individual farmers (respondent R). Consequently, the only exit option the farmers have would be to leave the cooperative, losing 'their' land without compensation.

The sugar factory and government officials claim that the cooperative members made a free choice to join the project (respondents Q, R, S). Nonetheless, several cooperative members say they opposed the arrangement and were forced to join by the government (respondents T, W). Whatever the case, it is clear that the farmers did not have sufficient information to make an informed decision. Even by the time of fieldwork in March 2010, the cooperatives were unsure of the debt to be repaid, had no oversight regarding the running costs paid to Wonji-Shoa and therefore what constituted a fair price for the sugarcane. In theory, the interests of farmers are represented by the cooperative, although these organisations, which are regulated by the *wereda* cooperatives office (respondent Y), are widely seen as state organisations and a means of disciplining outgrowers (respondents Z, W).

The impact of the plantation is differentiated along class, generational and gender lines. Only landholders are members of the cooperatives and receive priority access to day labour (respondents R, S). Nonetheless, all the work done by local people is unskilled and relatively poorly paid, with 'skilled' workers brought in to operate machinery (respondents R, S). According to respondents, there are no training schemes to replace skilled workers with local staff. This class differentiation, stemming from land rights, reinforces generational and gender divisions. As in most of Ethiopia, land shortage has meant that most older men have land and consequently became cooperative members while young people are landless and seek wage labour. Although land registration in Ethiopia has received some praise by gender assessments for issuing certificates to both husband and wife (Teklu 2005), the formation of the cooperatives pre-dated registration and consequently only the landholder – usually a man – was allowed to join. There are no attempts to rectify this situation with the result that the few female members are widows of former male members. Women also tend to be allocated low status and poorly paid jobs, especially weeding, because they are not thought able to do jobs which involve lifting and would not be respected as team leaders, giving instructions to men (respondents W).

Like case study A, the Wonji-Shoa project constitutes an attempt to combine the anti-displacement focus of ADLI with investment to promote commercialisation. Nonetheless, the need for a secure sugar supply to justify investment in the factory, which might otherwise favour a directly managed plantation, requires the government to discipline the cooperatives through coercion and monopoly conditions in the sugar sector. The case again demonstrates the risks to outgrowers. First, outgrowers are exposed to variation in the cost of living and agricultural inputs. Second, the outgrowers have little bargaining power in negotiations with the buyers of their produce as they lack market information and there are no alternative buyers.

Case study C: Small investors on communal land

Case study C is situated in a food insecure area dominated by smallholder crop production and is included in the PSNP. Nevertheless, as a result of the limited success of government and NGO projects which aimed to increase smallholder productivity and address food insecurity, the government has begun to lease plots of up to 30 hectares of communal grazing land with potential for irrigation to small-scale investors growing vegetables for sale in Addis Ababa.

The site is situated on the Awash River in Arssi. The area experiences regular droughts, although most local people lack the resources to invest in motorised pumps to use the river for irrigation, relying instead on rainfall for agriculture. In the past, a number of production systems have been established by local government, NGOs and local people in an attempt to utilise river water for irrigation. These irrigation schemes have met with at best partial success primarily due to the limited capacity of local people to deal with shocks such as motor failures and price fluctuations. The largest scheme covers 60 hectares divided into 0.25 hectare plots per household. It was established by a foreign NGO nearly 30 years ago and is now managed by the *kebele* but has had repeated breaks in production, often for years at a time, due to flood damage to the motor in the 1990s and, in 2010, stolen machinery. Each time limited resources in the local community have made it difficult to pay for repairs.

Another scheme established in 2001 covers 30 hectares in 0.5 hectare plots. It is managed by a cooperative established by another foreign NGO (respondents AA), which created a fund to pay for running costs and maintenance. However, this money has since been exhausted, partly due to rising fuel prices (respondents AA) and the members now need to replenish the fund. Although crops such as tomatoes and onions can be extremely profitable, they require significant investment in inputs and are considered risky due to variable market prices (respondents AB). Given the unreliable nature of production on non-irrigated land, cooperative members therefore prefer to grow maize on irrigated land, providing a secure source of subsistence but little surplus. Consequently, cooperative members are concerned that they will be unable to contribute to the fund, placing future production at risk (respondents AB).

In the early 2000s, the *wereda* land administration decided to lease communal grazing land adjacent to these schemes to investors in the expectation that they would have the resources to farm high value cash crops (respondent AC). There are four investors in the village, three Ethiopian and one Australian. The largest has 30 hectares and the others approximately 10 hectares each. They grow vegetables like tomatoes, onions and chillies, which are taken to market in Addis Ababa, though it is rumoured that the Australian intends to export his crop (respondent E). The community received no monetary compensation for the loss of grazing land since communal land is considered a government rather than a community resource. Instead, the investors promised to contribute to the community by building infrastructure such as a raft across the river and a mill and to bring electricity to the village. So far, however, only one investor has kept his promise.

The main impact on the local community has been to provide wage labour, in particular during harvest time. Cultivation is for the most part labour-intensive and one of the domestic investors hires 25 people on his investment of 10 hectares and many more at harvest time (respondent E). Although investors in Oromiya are required to hire all unskilled workers from the local community, investors complain that local people are lazy and demand high wages, and have brought many workers from other areas (respondent E). Several workers complained that the wages paid to labourers have not risen in line with food prices (respondents AD), nonetheless, day labour on investment land has become an important supplement to the incomes of many local people who struggle to meet their subsistence requirements from their own agricultural production. In particular many of those employed are young and landless.

There have been few direct negative impacts of the investments. Even though investors have occupied communal land, grazing land remains relatively plentiful and no respondents stated that the projects had damaged their ability to keep livestock. Instead, the main source of differentiation in the area is access to irrigable land and the financial resources to maintain irrigated production. As such, the key issue related to the case study is the counterfactual – whether other systems of production might have achieved more substantial and broader based benefits to the community. According to ADLI, rather than leasing land to investors, the best solution would have been to distribute communal land to local landless along similar lines to the NGO irrigation schemes described earlier. However, these group irrigation schemes have made limited contributions to commercialisation or even reliable production. The experience with the existing schemes appears to have convinced the *wereda* that investors implementing 'modern' agriculture offer better prospects of increasing productivity.

The political economy of changing land use

A common theme in discussions with government officials and in policy documents is that the state sees its role as bringing together capital – either its own resources or directing the activities of foreign and domestic investors – with land and labour in the most productive combinations in the national interest. Although many of the international drivers of the land grab have been influential in the projects discussed earlier, the state's role in ensuring preferred combinations of factors of production is evident in Table 1. For the time being, the political importance of the smallholder sector remains a key factor in government decision making. Cases A and B show that the government has gone to considerable lengths to avoid the displacement of smallholders. Indeed, for the foreseeable future it seems highly unlikely that investment will be allowed to displace the peasantry, as has been a concern of some 'land grab' articles (e.g. GRAIN 2008). On the whole, investors have only been allowed access to land in smallholder areas when they have convinced the government that they have the potential to achieve a major increase in productivity, either by developing water resources for irrigation or based on knowledge of high value crop production such as flower farms or castor.

Indeed, flower farms are the main example of investments that have actually displaced smallholders as they require land close to airports for quick transportation to European markets and production is not considered to be suited to outgrower arrangements. Nevertheless, they have been permitted by the government because they are labour-intensive projects that are expected to employ more labourers than had been supported by previous smallholder agriculture, limiting any impact on migration.[7] On the whole, however, the government's preferred option is for investors in smallholder areas to establish outgrower schemes in the expectation that investor capital can be combined with the land and labour of smallholders. Such outgrower schemes have been suggested by some observers as a means by which agricultural projects could produce 'win-win' situations as a means of including smallholders in projects which draw on the resources of investors to boost

[7]An additional concern, which is not addressed in the present study, regards the environmental impact of agricultural investment projects. Many of the investments are based on the intensive use of chemical fertilisers and pesticides, which, if not carefully managed, could pollute the local environment. Indeed, the limited research that exists on this topic in Ethiopia suggests that there are significant weaknesses in government environmental legislation and monitoring of investment projects (Getu 2010, The Oakland Institute 2011). Meanwhile studies have concluded that health and safety guidelines relating to the use of pesticides are frequently not adhered to in many flower farms, posing health risks for farmworkers (Haddis 2007), and contributing to pollution of water bodies used for drinking water by local communities and their livestock (Tadele 2009).

productivity (Cotula *et al.* 2009, von Braun and Meinzen-Dick 2009). Nevertheless, for such schemes to benefit the outgrowers themselves, there must be a degree of balance in negotiating power between the parties to the agreement (Little and Watts 1994). This requires that the outgrowers have the capacity to decline an agreement if it is not in their interests and, as Li (2011) shows, may necessitate the intervention of the state to support outgrowers who are often in a weak bargaining position with respect to investors. In the cases examined, state organisations were heavily involved in the establishment of outgrower schemes, although the state did not intervene to safeguard the interests of the outgrowers but to support the investors and ensure the submission of the outgrowers. In the Ethiopian context, agricultural investment is a government policy, which requires the allocation of labour resources by the government, leaving little room for outgrowers as active participants. The cases therefore suggest that past research by Rahmato (2009), who claims that the land policy facilitates the political control of the state over the peasantry, remains relevant in emerging systems of agricultural production, where opposition to government-sponsored investment projects is seen as political opposition to the government.

In contrast to its approach to the smallholder sector the EPRDF does not consider pastoralism and shifting cultivation to be sustainable or efficient land uses, and consequently has begun leasing vast tracts of land in lowland areas to investors. These recent investment processes continue trends under previous governments. For example, studies of pastoralists in the Awash River Basin, which covers parts of Afar, Oromiya and Somali regions, show that pastoralists' access to grazing land has been eroded over many years as a result of commercial sugar plantations established in the 1950s and subsequently expanded, and the establishment of the Awash National Park (Beyene 2009, Elias and Abdi 2010, Müller-Mahn *et al.* 2010, Behnke and Kervan 2011). Indeed one study concluded that these irrigation projects, which restricted pastoralists' access to key water resources, damaged their ability to manage drought risks and were a major cause of the famine in 1973 (Kloos 1982).

At the same time as recent large-scale land leases, the government has started to implement its plans for the 'voluntary resettlement' of pastoralists in villages, which will supposedly provide improved access to services (MoFED 2010). The resettlement programme is at an early stage, which makes it impossible to draw firm conclusions, but while the government claims that resettlement and investment are entirely unrelated, the two strategies are, at the very least, compatible and the danger remains that resettlement will be used to clear land for investors (Rahmato 2011). The differential impact that state-mediated investment is having on different forms of livelihood in many cases also overlaps with ethnic divisions. In particular, heavily targeted regions such as Gambella and Benishangul-Gumuz are home to ethnic groups that practice pastoralism and shifting cultivation and which have historically been both physically remote and politically marginal to state power. Despite the creation of an ethnic federal system, which is intended to protect the rights of minority ethnic groups, recent processes of agricultural investment seem likely to continue past patterns of exploitation of the resources of minority ethnic groups for the benefit of the centre.

The system of production adopted in these sparsely populated areas represents a choice of development strategy that diverges from the principles of ADLI, with important consequences for equality and rural social protection. In principle, smallholder agriculture or investments using outgrowers could be established on state farms, communal land and potentially even land in lowland areas. Instead, investors using 'modern' technology have been prioritised over smallholders. While political imperative prevents the displacement of the existing smallholder sector, the economic arguments outlined in ADLI in favour of expanding smallholder agriculture appear to have lost their appeal within the government, with investors rather than smallholders now given priority in new land allocations. The result is an emerging dualism between the existing smallholders, who have tiny landholdings and are prevented from expanding through purchase or long-term lease, and investors, who operate on a much larger scale and are increasingly seen by the government as the engine of agricultural transformation.

The result of this pattern of transformation is that the impact of agricultural investment on employment is likely to be relatively limited. In the highlands, investment projects utilising outgrowers rely primarily on household labour, as was the case with the previous agricultural production. In some cases, the labour-intensive nature of flower production may create some additional demand for labour compared with previous production, easing local unemployment problems. As an indication, Mano *et al.* (2011) estimate that flower farms employ some 20,000–33,000 people, although the paper makes no comparison with the labour requirements of the previous smallholder production. Many of these jobs are taken by local people although there has certainly been a degree of labour migration to flower farms around Addis Ababa and the Rift Valley, in particular from other parts of Oromiya. Nonetheless, any additional employment created by the change in production is dwarfed by the enormous problems of land shortage and unemployment in rural Ethiopia. In lowland areas, meanwhile, the capital-intensive nature of production means that there are likely to be relatively few job opportunities created. In the absence of detailed case studies, it remains unclear who will benefit from this limited number of jobs – local pastoralists, highland migrants or foreigners.

Nonetheless, by preventing large-scale displacement, the government has ensured that smallholder production will continue to constitute a key part of the livelihoods of the majority of the population and will constitute the dominant land use for the foreseeable future. Although the government has suggested that one benefit of investment will be the transfer of technology from investors to smallholders, one of the main reasons for encouraging investors is that they have resources that smallholders lack and are therefore able to make large investments, in particular in irrigation infrastructure. Given the challenges facing smallholders attempting to develop and maintain irrigation infrastructure, as illustrated in case study C, it is therefore illogical to expect this technology to transfer from large-scale investors to independent smallholders. Similar arguments apply to the mechanised and energy-intensive farming techniques envisaged by many agricultural investors, which are unlikely to be viable on the tiny plots of most smallholders.[8] There therefore remains an urgent need to address the problems that have constrained the productivity of smallholder agriculture including the supply of improved seeds and fertilisers, price incentives for produce and protection from price shocks for inputs and outputs, as well as off-farm employment to reduce the pressure on land.

The promotion of investment and the emerging dualism in agricultural production has potentially important consequences for Ethiopia's political

[8]This conclusion is supported by past research which suggests that wealthier farmers have been better placed to take advantage of green revolution technologies (Harriss 1987, Bernstein 2010).

economy. In effect, the government is enabling the creation of a class of large landholders – both domestic and foreign – which has not existed in Ethiopia since before the 1974 revolution. Thus far, what evidence exists would suggest that co-operation between investors, and the identification and pursuit of common interests remain limited. Certainly fieldwork showed little sign of linkages between investors and anecdotal evidence of other projects would suggest that this is probably a common state of affairs. In addition, one study of the flower sector concluded that co-operation between foreign and domestic investors was minimal and limited to basic issues such as transportation (Tiruwha Melese and Helmsing 2010). Interviews conducted for this project meanwhile suggested that many domestic investors actually resent foreigners who, they perceive, are given preferential treatment in the allocation of land by the government. However, depending on the growth of the sector and its contribution to the Ethiopian economy, this capitalist class could in the future become a powerful influence on government policymaking. For the time being, the numerical dominance of the smallholder sector is such that it will continue to constitute a central concern to the government and investors will be largely confined to remote and politically marginal areas. However, in the future, capitalist farmers could demand greater access to land in highland areas and even greater privatisation of land tenure, constituting an important challenge to the smallholder sector.

Meanwhile, many studies examining the trends in agricultural investments have suggested that investment deals might be more inclusive and make a greater contribution to poverty reduction if accompanied by initiatives to improve land governance and promote land registration, introduce voluntary guidelines for investors, improve transparency around land deals and promote advocacy and legal support for those at risk from investments (e.g. Cotula *et al.* 2009, Deininger and Byerlee 2010). Nevertheless these recommendations appear to assume the prior existence of the basic elements of a liberal democracy. It would seem implausible that such initiatives could be implemented and achieve the desired results in the political context of Ethiopia where investment is a policy promoted by the government and space for any opposition to government policy is extremely restricted.

Conclusions

In addition to the international drivers of the land grab, in Ethiopia the state plays a key role in promoting agricultural investment. Indeed, the Ethiopian government is attempting to manage the resources of capital – from foreign as well as domestic investors – land and labour, in an attempt to achieve combinations which meet its political and economic objectives. Rather than widespread dispossession of the peasantry, state mediation has meant that commercialisation pressures have resulted in a selective transformation in systems of agricultural production.

Those most seriously affected by investment are likely to be those whose livelihoods are based on pastoralism and shifting cultivation in lowland, border regions and whose 'use' of the land is questioned by the government. These areas have already seen the establishment of vast capital-intensive projects, with even larger tracts of land identified for future investment. The livelihood distinction between those involved in settled agriculture, and pastoralism and shifting cultivation also overlaps with ethnic divisions and a history of exploitation of the periphery by the state (Lavers 2012).

Although smallholder areas, with the exception of those targeted for labour-intensive flower farms, have largely been spared the risk of displacement, commercialisation pressures have also resulted in new forms of agricultural production in highland areas. The government has leased land to investors using wage labourers on communal grazing land and former state farms, as well as attempting to combine increased investment with smallholder production through outgrower schemes. Despite the potential benefits of employment and increased earnings, these new forms of investment also involve risks to smallholders, namely the loss of grazing land and exposure to market risks for inputs, produce and food.

The impacts of agricultural investment in the years to come are likely to depend on a number of factors. First, future trends in the international processes that led to the current wave of foreign investment. Although many agricultural projects have been established in Ethiopia, a considerable amount of investment is based on speculation on volatile international commodity prices. Consequently, falls in food and fuel prices or the presence of more attractive investment opportunities elsewhere could mean that many proposed investments fail to materialise.

Second, the government's promotion of investment is based on the premise that investment can be managed both economically and politically. The case studies have raised doubts whether the government is able to manage investment in order to combine its objectives of increasing production with equitable growth and security for smallholders, or whether there is an inherent trade-off between these goals. Meanwhile, the promotion of foreign and domestic investment will create a class of commercial farmers with economic power and, consequently, political influence. Coming years will show the extent to which this class is able to influence government policy and whether the government will continue to be able to manage these actors.

References

Aalen, L. and K. Tronvoll. 2009. The 2008 Ethiopian local elections: The return of electoral authoritarianism. *African Affairs*, 108(430), 111–120.

Adenew, B. and F. Abdi. 2005. Research report 3: Land registration in Amhara Region, Ethiopia. London: IIED.

Anseeuw, W., A. Ducastel and J.-J. Gabas. 2011. The end of the African peasant? From investment funds and finance value chains to peasant related questions. Paper presented at the International Conference on Global Land Grabbing 6–8 April. IDS, University of Sussex: Land Deal Politics Initiative (LDPI).

Araghi, F. 2009. The invisible hand and the visible foot: peasants, dispossession and globalization. *In:* A. H. Akram-Lodhi and C. Kay, eds. *Peasants and globalization: Political economy, rural transformation and the agrarian question*. London: Routledge.

Baxter, J. 2010. Africa's land and family farms – up for grabs? *Seedling*, January, 13–16.

Behnke, R. and C. Kervan. 2011. Replacing pastoralism with irrigated agriculture in the Awash Valley, North-Eastern Ethiopia: Counting the costs. Paper presented at the International Conference on the Future of Pastoralism, 21–23 March. Addis Ababa.

Bernstein, H. 2010. *Class dynamics of agrarian change*. Halifax and Winnipeg: Initiatives in Critical Agrarian Studies/Fernwood Publishing.

Beyene, F. 2009. Property rights conflict, customary institutions and the state: The case of agro-pastoralists in Mieso District, Eastern Ethiopia. *The Journal of Modern African Studies*, 47(02), 213–239.

Borras, S.M. and J. Franco. 2010. Towards a broader view of the politics of global land grab: Rethinking land issues, reframing resistance. ICAS Working Paper Series, 001.

Butler, E. 2010. Land grab fears in Ethiopia. *BBC News*. Available at: http://www.bbc.co.uk/news/business-11991926 [Accessed 20 January 2011].

Byres, T.J. 1991. The agrarian question and differing forms of capitalist transition: An essay with reference to Asia. *In:* J. Breman and S. Mundle, eds. *Rural transformation in Asia*. Delhi: Oxford University Press.
Clapham, C. 1988. *Transformation and continuity in revolutionary Ethiopia*. Cambridge: Cambridge University Press.
Clapham, C. 2006. Ethiopian development: The politics of emulation. *Commonwealth and Comparative Politics*, 44, 137–150.
Cotula, L., *et al.* 2009. Land grab or development opportunity? Agricultural investment and international land deals in Africa. Rome: IIED, FAO, IFAD.
Crouch, L. and A. de Janvry. 1980. The class basis of agricultural growth. *Food Policy*, 5(1), 3–13.
CSA, 2009. *Statistical abstract: Section B – population*. Addis Ababa: Central Statistical Agency.
CSA, 2010. Report on area and production of crops (private peasant holdings, meher season) Agricultural Sample Survey 2009/10 (2002 E.C.). *Central Statistical Agency Statistical Bulletin*, 446.
Daniel, S. and A. Mittal. 2010. (Mis)investment in agriculture: The role of the International Finance Corporation in global land grabs. Oakland, CA: The Oakland Institute.
Deininger, K. and D. Byerlee. 2010. *Rising global interest in farmland: Can it yield sustainable and equitable benefits?* Washington, DC: World Bank Publications.
Deininger, K., *et al.* 2008. Rural land certification in Ethiopia: Process, initial impact, and implications for other African countries. *World Development*, 36(10), 1786–1812.
Dercon, S. and R. Vargas Hill. 2009. Growth from agriculture in Ethiopia: Identifying key constraints. Paper presented at the ESSP-II policy conference, 'Accelerating agricultural development, economic growth and poverty reduction in Ethiopia', 22–24 October. Hilton Hotel, Addis Ababa: IFPRI.
De Schutter, O. 2011. How not to think of land-grabbing: three critiques of large-scale investments in farmland. *Journal of Peasant Studies*, 38(2), 249–279.
Donham, D.L. 2002. Old Abyssinia and the new Ethiopian empire: themes in social history. *In:* D. L. Donham and W. James, eds. *The southern marches of Imperial Ethiopia*. London: James Currey, pp. 3–50.
Elias, E. and F. Abdi. 2010. Putting pastoralists on the policy agenda: Land alienation in southern Ethiopia. *IIED Gatekeeper*, 145.
ESDA, 2010. Five years (2010–2015) sugar industry sub-sector development plan. Addis Ababa: Ethiopian Sugar Development Agency.
Ethical Sugar, 2006. Ethiopia to boost sugar output, seeks investors. Available from: http://www.sucre-ethique.org/Ethiopia-to-boost-sugar-output.html [Accessed 20 January 2011].
FAO, 2011. Crisis in the Horn of Africa. Food and Agriculture Organization of the United Nations. Available from: http://www.fao.org/crisis/horn-africa/home/en/ [Accessed 9 August 2011].
FDRE, 2003. Federal Democratic Republic of Ethiopia (FDRE) – Investment incentives and investment areas reserved for domestic investors. Federal Negarit Gazeta, Council of Ministers Regulation No. 84/2003.
FDRE, 2005. Federal Democratic Republic of Ethiopia (FDRE) – Expropriation of land holdings for Public Purposes and Payment of Compensation Proclamation. Federal Negarit Gazeta, Proclamation No. 455/2005.
Feyissa, D. 2005. Land and the politics of identity. The case of Anywaa-Nuer relations in the Gambella region. *In:* S. Evers, M. Spierenburg, and H. Wels, eds. *Competing jurisdictions: settling land claims in Africa*. Leiden: Brill Academic Publishers, pp. 203–222.
Feyissa, D. 2006. The experience of Gambella Regional State. *In:* D. Turton, ed. *Ethnic federalism: The Ethiopian experience in comparative perspective*. London: James Currey, pp. 208–230.
Fisseha, M. 2011. *Commercial pressures on land in Ethiopia: A case study of the Bechera agricultural development project*. Rome: International Land Coalition/CIRAD.
Garretson, P.P. 2002. Vicious cycles: Ivory, slaves, and arms on the new Maji frontier. *In:* D. L. Donham and W. James, eds. *The southern marches of Imperial Ethiopia*. London: James Currey, pp. 196–218.
Getu, M. 2010. Ethiopian floriculture and its impact on the environment. *Mizan Law Review*, 3(2), 240–270.

Global Campaign for Agrarian Reform, 2010. *Why we oppose the principles for Responsible Agricultural Investment (RAI)*, Land Research Action Network, FIAN International, Focus on the Global South, La Via Campesina, Social Network for Justice and Human Rights.

GRAIN, 2008. Seized: The 2008 land grab for food and financial security. *GRAIN Briefing*, October 2008.

Guillozet, K. and J.C. Bliss. 2011. Household livelihoods and increasing foreign investment pressure in Ethiopia's natural forests. Land Deal Politics Initiative (LDPI) Working Paper No. 3.

Habtu, Y. 1997. Farmers without land: the return of landlessness to Rural Ethiopia. *In:* D. F. Bryceson and V. Jamal, eds. *Farewell to farms: de-agrarianisation and employment in Africa*. Ashgate Publishing Ltd., pp. 41–59.

Haddis, E. 2007. An assessment of the working conditions of the floriculture industry: The case of four flower farms in West Showa Zone, Oromiya Regional State. MA Thesis, Addis Ababa University.

Harriss, J. 1987. Capitalism and peasant production: the green revolution in India. *In:* T. Shanin, ed., *Peasants and peasant societies*. Oxford: Blackwell, pp. 227–236.

James, W., *et al.* eds. 2002. *Remapping Ethiopia: socialism and after*. London: James Currey.

Joireman, S.F. 2000. Property rights and political development: The state and land in Ethiopia and Eritrea, 1941–1974. Oxford: James Currey.

Kebede, B. 2011. Land deals in Ethiopia bring food self-sufficiency, and prosperity. *The Guardian*. Available from: http://www.guardian.co.uk/global-development/poverty-matters/2011/apr/04/ethiopia-land-deals-food-self-sufficiency [Accessed 6 June 2011].

Kloos, H. 1982. Development, drought, and famine in the Awash Valley of Ethiopia. *African Studies Review*, 25(4), 21–48.

Lavers, T. 2012. 'Land grab' as development strategy? The political economy of agricultural investment in Ethiopia. *Journal of Peasant Studies*, 39(1), pp. 105–132.

Li, T.M. 2011. Centering labor in the land grab debate. *Journal of Peasant Studies*, 38(2), 281–298.

Little, P.D. and M.J. Watts. eds. 1994. *Living under contract: contract farming and agrarian transformation in Sub-Saharan Africa*. Madison, WI: University of Wisconsin Press.

Mano, Y., *et al.* 2011. Local and personal networks in employment and the development of labor markets: Evidence from the cut flower industry in Ethiopia. *World Development*, 39(10), 1760–1770.

McClure, J. 2009. Ethiopia leases land for agriculture to earn foreign exchange. *Bloomberg*. Available from: http://www.bloomberg.com/apps/news?pid=newsarchive&sid=aJEzbUEy4xTE [Accessed 14 June 2011].

McMichael, P. 2010. Agrofuels in the food regime. *Journal of Peasant Studies*, 37(4), 609–629.

Migdal, J.S. 2001. *State in society: Studying how states and societies transform and constitute one another*. Cambridge: Cambridge University Press.

MoFED, 2002. *Ethiopia: sustainable development and poverty reduction program*. Addis Ababa: Ministry of Finance and Economic Development (MoFED), Federal Democratic Republic of Ethiopia.

MoFED, 2003. *Rural development policy and strategies*. Addis Ababa: Ministry of Finance and Economic Development (MoFED).

MoARD, 2009. *Food Security Programme, 2010–2014*. Addis Ababa: Ministry of Agriculture and Rural Development (MoARD).

MoFED, 2010. *Growth and Transformation Plan (GTP)*. Addis Ababa: Ministry of Finance and Economic Development (MoFED), Federal Democratic Republic of Ethiopia.

Müller-Mahn, D., S. Rettberg and G. Getachew. 2010. Pathways and dead ends of pastoral development among the Afar and Karrayu in Ethiopia. *European Journal of Development Research*, 22(5), 660–677.

ONRG, 2001. Oromia National Regional Government (ONRG) – Utilization of rural lands for investment activities in the Oromia region proclamation no. 3/1995 (amendment) proclamation no. 41/2001. Megeleta Oromia, Regulation No. 41/2001.

Pankhurst, A. 1992. *Resettlement and famine in Ethiopia: The villagers' experience*. Manchester: Manchester University Press.

Pankhurst, A. 1997. When the centre relocates the periphery: Resettlement during the Derg. In: Institute of Ethiopian Studies, ed. *ICES: International Conference of Ethiopian studies*, 2, 540–558.

Platteau, J.-P. 2005. The gradual erosion of the social security function of customary land tenure arrangements in lineage-based societies. *In:* S. Dercon, ed. *Insurance against poverty*. Oxford: Oxford University Press, pp. 247–280.

ProKerala News, 2011. No Indian land grab, says Ethiopian PM. Available from: http://www.prokerala.com/news/articles/a224066.html [Accessed 6 June 2011].

Rahmato, D. 1997. *Manufacturing poverty: rural policy and micro-agriculture*. Addis Ababa: Institute of Development Research, Addis Ababa University.

Rahmato, D. 2003. Poverty and agricultural involution. *In:* D. Rahmato, ed. *Some aspects of poverty in Ethiopia*. Addis Ababa: Forum for Social Studies, pp. 1–15.

Rahmato, D. 2009. *The peasant and the state: Studies in agrarian change in Ethiopia 1950s–2000s*. Addis Ababa: Addis Ababa University Press.

Rahmato, D. 2011. *Land to investors: Large-scale land transfers in Ethiopia*. Utrecht: IS Academie.

Rice, X. 2010. Ethiopia – country of the silver sickle – offers land dirt cheap to farming giants. *The Guardian*. Available from: http://www.guardian.co.uk/world/2010/jan/15/ethiopia-sells-land-farming-giants [Accessed 17 March 2010].

Schlager, E. and E. Ostrom. 1992. Property-rights regimes and natural resources: a conceptual analysis. *Land Economics*, 68(3), 249–262.

Scott, J.C. 1976. *The moral economy of the peasant: Rebellion and subsistence in Southeast Asia*. New Haven, CT; London: Yale University Press.

Scott, J.C. 1998. *Seeing like a state: How certain schemes to improve the human condition have failed*. London: Yale University Press.

Shete, M. 2011. Implications of land deals to livelihood security and natural resource management in Benshanguel Gumuz Regional State, Ethiopia. Paper presented at the International Conference on Global Land Grabbing 6–8 April. IDS, University of Sussex: Land Deal Politics Initiative (LDPI).

SNNPR Investment Agency, 2008. *The southern nations, nationalities and peoples' regional state resource potential and investment opportunities*. Awassa: SNNPR Bureau of Trade and Industry.

Tadele, M. 2009. Environmental impacts of floriculture industries on Lake Ziway: With particular reference to water quality. MSc Thesis, Addis Ababa University.

Tamrat, I. 2010. Governance of large scale agricultural investments in Africa: The case of Ethiopia. Paper presented at the World Bank Conference on Land Policy and Administration, 26–27 April. Washington, DC: World Bank.

Teklu, A. 2005. Research report 4: Land registration and women's land rights in Amhara Region, Ethiopia. London: IIED.

The Oakland Institute, 2011. Understanding land investment deals in Africa. Country report: Ethiopia. Oakland, CA: The Oakland Institute.

Tiruwha Melese, A. and A.H.J. Helmsing. 2010. Endogenisation or enclave formation? The development of the Ethiopian cut flower industry. *The Journal of Modern African Studies*, 48(01), 35–66.

TNRG, 2000. Tigray National Regional Government – Rural land administration and utilization revised regulation. Negarit Gazeta of Tigray, Regulation No. 48/2000.

Tronvoll, K. 2009. Ambiguous elections: The influence of non-electoral politics in Ethiopian democratisation. *Journal of Modern African Studies*, 47(3), 449–474.

van Berkum, S., P. Roza and F. van Tongeren. 2005. Impacts of the EU sugar policy reforms on developing countries. *Agricultural Economics Research Institute (LEI)*, 6(09).

Vaughan, S. and K. Tronvoll. 2003. *The culture of power in contemporary Ethiopian political life*. Stockholm: Swedish International Development Cooperation Agency.

Vidal, J. 2010. How food and water are driving a 21st-century African land grab. *The Observer*. Available from: http://www.guardian.co.uk/environment/2010/mar/07/food-water-africa-land-grab [Accessed 17 March 2010].

von Braun, J. and R. Meinzen-Dick. 2009. 'Land grabbing' by foreign investors in developing countries: Risks and opportunities. *IFPRI Policy Brief*, 13.

Weissleder, L. 2009. Foreign direct investment in the agricultural sector in Ethiopia. Ecofair Trade Dialogue Discussion Papers No. 12.

White, B. and A. Dasgupta. 2010. Agrofuels capitalism: a view from political economy. *Journal of Peasant Studies*, 37(4), 593–607.

World Bank, 2010. Rising global interest in farmland: Can it yield sustainable and equitable benefits?. Washington, DC: World Bank.

Young, J. 1997. Peasant revolution in Ethiopia: The Tigray People's Liberation Front, 1975–1991. Cambridge: Cambridge University Press.

Zenebe, W. 2010. Djibouti generates grain on 3,000ht gesture. *Addis Fortune*. Available from: http://www.addisfortune.com/Vol%2010%20No%20499%20Archive/Djibouti%20Generates%20Grain%20on%203,000ht%20Gesture.htm [Accessed 20 January 2011].

Zewde, B. 2008. Environment and capital: Notes for a history of the Wonji-Shoa sugar estate (1951–1974). *In:* Zewde, B. ed. *Society, state and history: selected essays*. Addis Ababa: Addis Ababa University Press, pp. 120–146.

Zoomers, A. 2010. Globalisation and the foreignisation of space: seven processes driving the current global land grab. *Journal of Peasant Studies*, 37(2), 429–447.

Tom Lavers is a PhD candidate in the Department of Social and Policy Sciences at the University of Bath. His research focuses on the political economy of social policy and agrarian change in Ethiopia, in particular examining the links between land tenure, food security and agricultural commercialisation.

Annex. Interview respondents.

Respondent	Position/Organisation	Place	Date conducted
A	Manager, Amhara BoEPLAU	Bahir Dar	15 Mar 2010
B	Manager, Oromiya Land Use and Environmental Protection Department	Addis Ababa	28 Apr 2010
C	Manager, Oromiya Regional Investment Commission	Addis Ababa	3 Feb 2010
D	Manager, SNNPR Investment Agency	Awassa	1 Mar 2010
E	Ethiopian investor, case C	East Shewa	19 Dec 2009
F	Current project head, case A	Addis Ababa	7 Sep 2010
G	Former project manager, case A	Addis Ababa	8 Sep 2010
H	Three managers in SNNPR Dept of Natural Resources and Environmental Protection	Awassa	23 Feb 2010
I	Tigray Environmental Protection, Land Administration and Use Adminstration	Mekele	29 Mar 2010
J	Three farmers resettled from East Tigray to West Tigray	East Tigray	Apr 2010
K	Castor seed buyer for Chinese Government	Addis Ababa	15 Sep 2010
L	A manager, East Hararghe Investment Commission	Harar	21 Sep 2010
M	Current project manager, case A	Hararghe	8 and 9 Sep 2010
N	Former representative of the Food Security Office, East Hararghe	Harar	21 Sep 2010
O	Ten outgrowers, case study A	Hararghe	19 and 20 Sep 2010
P	Two project supervisors, case A	Hararghe	19 Sep 2010
Q	Three project managers, Wonji-Shoa Sugar Factory, case B	Wenji	9 and 12 Mar 2010
R	Wereda land administration	Adama	8 and 12 Mar 2010
S	Wereda investment desk	Adama	8 and 12 Mar 2010
T	Management of sugar cooperative, case B	Adama wereda	13 Mar 2010
U	Government development agents, case B	Adama wereda	13 Mar 2010
V	Members of sugar cooperative, case B	Dodota wereda	17 Dec 2009
W	Members of sugar cooperative, case B	Adama wereda	10 Mar 2010
X	Members of sugar cooperative, case B	Adama wereda	13 Mar 2010
Y	Wereda cooperatives office	Adama	12 Mar 2010
Z	Kebele chairman, case B	Adama wereda	11 Mar 2010
AA	Two founding members of irrigation cooperative, case C	Dodota wereda	18 Dec 2009
AB	Six members of the NGO-established irrigation scheme	Dodota wereda	17, 18, 19, 20 Dec 2009
AC	Wereda land administration	Dera	17 Dec 2009
AD	Two female day labourers	Dodota wereda	20 Dec 2009
AE	Manager, Agricultural Investment Support Directorate, MoARD	Addis Ababa	28 Dec 2009

The next Great Trek? South African commercial farmers move north

Ruth Hall

This paper analyses the shifting role of South African farmers, agribusiness and capital elsewhere in the Southern African region and the rest of the continent. It explores recent trends in this expansion, and investigates the interests and agendas shaping such deals, and the ideologies and discourses of legitimation employed in favour of them. While for the past two decades small numbers of South African farmers have moved to Mozambique, Zambia and several other countries, this trend seems to be undergoing both a quantitative and a qualitative shift. Whereas in the past their migration was largely individual or in small groups, now it is being more centrally organised and coordinated, is more frequently taking the form of large concessions for newly formed consortia and agribusinesses, and is increasingly reliant on external financing through transnational partnerships. By early 2010, the commercial farmers' association Agri South Africa (AgriSA) was engaged in negotiations for land acquisitions with the governments of 22 African countries. This essay is the product of a scoping study to document and analyse major land acquisitions by South African farmers and agribusinesses, and the processes through which these have occurred and are occurring. It considers the changing character, scale and location of South African investments elsewhere in the region and the continent, and focuses specifically on the AgriSA-Congo deal (the largest deal concluded thus far), and acquisitions by the two South African sugar giants, Illovo and Tongaat-Hulett, for outgrower and estate expansion elsewhere in the region. The study addresses the degree to which South Africa is no longer merely exporting its farmers, but also its value chains, to the rest of the continent – and what this means for trajectories of agrarian change. It questions how we might understand the growing trend of 'intra-regional land grabbing' and, in the cases discussed, suggests that South African-based companies are acting as arteries of global capital.

1. Introduction

While for the past two decades small numbers of South African farmers have moved to Mozambique, Zambia and several other countries, this trend seems to be undergoing both a quantitative and a qualitative shift. Whereas in the past their

I am deeply grateful to the Future Agricultures Consortium, funded by the UK Department for International Development, for supporting this research, to my colleagues in the Land Deal Politics Initiative (LDPI) for advice on this paper, as well as to three anonymous reviewers whose comments were immensely helpful. All errors and omissions are mine.

migration was largely individual or in small groups, now it is being more centrally organised and coordinated and, as this paper shows, is more frequently taking the form of large concessions for newly formed consortia and agribusinesses, and is increasingly reliant on external financing through transnational partnerships.

This paper builds on a review of land deals in Southern Africa (Hall 2011) which noted among other trends the significant role of domestic and regional actors in what has often been considered 'foreign' land grabbing; the prevalence of deals for purposes other than food production (specifically biofuels, mining and forestry); and the ways in which what is being grabbed extends beyond the land, to water and other natural resources but also, as noted more generally by Li (2011), the labour with which to exploit these, through a range of institutional forms and business models. One conclusion of that paper was that South African businesses are not merely relocating elsewhere in the region, but are acquiring land as part of wider business strategies to extend South African value chains and to position themselves as intermediaries for other global interests. The actors and interests involved, then, are not merely those of 'farmers', but are of a wide range of private and public sector actors in 'host' countries and in South Africa, but also from further afield.

This paper picks up on these observations and analyses the shifting role of South African farmers, agribusiness and capital elsewhere in the Southern African region and the rest of the continent. It explores recent trends in this expansion, and investigates the interests and agendas shaping such deals, and the ideologies and discourses of legitimation employed in favour of them. It presents the first scoping phase of this work to document this expansion of South African farmers and agribusiness elsewhere in Africa, and reflects on three angles: (i) negotiations by commercial farmer association AgriSA to secure land leases for consortia of its members; (ii) the acquisition of land for estates and outgrower schemes by the major South African-based sugar companies; and (iii) the emerging models through which acquisitions are financed and new actors involved. The paper does not address the totality of South African land-based investments elsewhere on the continent, which include agricultural ventures in other sectors of farming, as well as deals in mining and coastal and safari tourism, among others.

2. Intra-regional land deals and the changing agro-food system

As the literature on transnational land deals expands, it is becoming apparent that several developing countries are not only the destination of large-scale land deals, but also the source of new investments. South-South land grabs challenge traditional conceptions in political economy, and raise questions as to whether these arise from or produce different dynamics than 'North-South' deals. This paper draws attention to some ways in which there are new and distinctive intra-regional dynamics of land grabbing in the present era. In this context, policy agendas of regional integration, and the development agendas of regional institutions, may be read as forms of legitimation of regional powers' expansionary interests.

These trends are evident elsewhere, in Latin America and East Asia. Mackey (2011, 1) shows that Brazil is 'not only a destination but also a driver of transnational land investments across the tropical world'. His analysis of the expansion of Brazilian soybean production in Bolivia draws attention to the legitimating discourses deployed by these intra-regional grabbers, and the ways in which private investors' strategies are located within and dependent upon the

Brazilian government's strategic ambitions and its policy, institutional and sometimes financial support. Similarly, Dwyer (2011) shows that Laos, having signed away 15 percent of its territory within the past decade, has been the target of 'land-hungry neighbours' – in this case, not only China but also Vietnam – rather than remote neo-colonialists. This is not merely due to corruption and collusion between deep-pocketed investors and greedy officials; rather, foreign investment by regional powers plays a role in the Laos government's efforts 'to govern and develop a rural hinterland where both natural resources and local populations are seen to be undeveloped' (Dwyer 2011, 4).

Further, Visser and Spoor (2011) demonstrate how the former Soviet Union is a major target for land deals, and also how ex-Soviet republics are involved in intra-regional grabbing, especially of their poorer neighbours. Their work suggests that, while global academic, policy and civil society attention has been turned to land deals in the developing world and Africa specifically, European, Gulf and Far East companies and investment funds have secured vast areas within the former Soviet Union (see also Visser *et al.* 2012). Among those that are part of this trend are Russian, Kazakh, Lithuanian and Estonian companies acquiring agricultural holdings in the far east of Russia, with some individual holdings exceeding half a million hectares (ha) (Visser and Spoor 2011).

Analysis of land deals, then, should focus not only on the traditional powers (Europe, North America), and the rising role of China, Korea and the Gulf states, but also on other emerging key actors. The global agro-food system is being recast by the rising influence, among others, of the BRICS countries (Brazil, Russia, India, China and South Africa), in contrast to previous food regimes anchored by empires on both sides of the Atlantic, in Western Europe and North America (Borras *et al.* forthcoming). In view of this, how do we understand this new range of actors, their interests and strategies, and the implications for new geopolitical configurations? By focusing on elements of the role of South Africa in Africa, this paper aims to contribute to the emerging literature that explores empirically how the interests of these regional and other international investors intersect in these deals. This should inform efforts to understand and to outline the contours of a new emerging 'agrofood regime' (McMichael 2009).

3. The restructuring of South African agricultural capital

South Africa's (still almost exclusively) white commercial farmers have over the past two decades experienced dramatic changes in their political and economic situation. A combination of pressures has put these farmers – once a primary political constituency of the National Party apartheid government – into new difficulties, both objective and subjective. The new pressures have been well documented and arise from the dismantling of an elaborate architecture of policy and institutional support for commercial farming: agricultural deregulation including the removal of direct and indirect subsidies, state-controlled marketing boards with floor prices and pan-territorial pricing, cheap credit and tax breaks; the rapid liberalisation of trade in agricultural products; and sharp increases in the prices of key farming inputs, particularly diesel and electricity (Bernstein 1996, Vink and Hall 2010, Williams *et al.* 1998). Further pressures adding to the actual and perceived difficulties of pursuing commercial farming include the introduction for the first time of basic labour rights for farm workers in the 1990s, and since then also minimum wage regulations, the extension of tenure rights to farm workers and their families (Atkinson 2007) and the placing of historical land claims to large areas of commercial farmland by former black occupiers, owners and tenants, in terms of the Restitution of Land Rights Act, 22 of 1994 (Walker *et al.* 2010).

These pressures have elicited a number of responses on the part of South Africa's white farmers. One response has been the decision by many to exit farming, sell their farms and invest in new careers in other sectors of the economy. Indeed, the total number of commercial farming units contracted dramatically from about 60,000 in 1996 to just under 40,000 by 2007; ownership of multiple farms suggests that these figures might even under-represent the scale of exit from commercial farming (DAFF 2009). A second response has been diversification into non-agricultural sectors, or into up- or downstream activities. A third response has been to either individually or collectively move out of South Africa, to locations elsewhere on the continent.

In the midst of these changes, the South African Agricultural Union (SAAU), formed in 1904 to represent white farmers, renamed itself Agri South Africa (AgriSA) in 1999 as a newly deracialised association of farmers' organisations, keen to recruit black membership and to build relations with the post-apartheid political dispensation and the ruling African National Congress (ANC). Chief negotiator for their commercial farmers is Theo de Jager, deputy president of AgriSA, who in recent years has led numerous delegations of farmers to meet with agriculture and foreign affairs ministers of other African countries offering land.

In contrast to AgriSA is the conservative Transvaal Agricultural Union (TAU) of South Africa – which refused to join this deracialised union – and which represents the agenda of white farmers battling to escape black rule, in part by betting on Georgia in the former USSR as the main new destination for its membership. TAU general manager Bennie van Zyl was invited on a three-month study tour, together with key advisors (TAU 2010), and reportedly concluded an agreement 'to send farmers who want to leave South Africa to Georgia' (Bloomberg 2011). This was backed up by a visit in February 2011 by 90 prospective new settlers to inspect land on offer (McGuinness 2011). The terms on which land is being offered by the government of Georgia have not been made known, though the South Africans involved suggest that:

> ... *what the ANC is prepared to expropriate, the government of the Republic of Georgia is willing to give away for free.* Is it possible that they know something about the character of the Boer-Afrikaner that the ANC just won't admit. When they [the ANC] lose the Boer, they'll lose the security of their food supply. When they lose that, what will they eat? (Phillips 2010, original emphasis).

News headlines and right-wing websites warn of an impending 'exodus' of 'thousands' of white 'boer' farmers; that 'the last of the white farmers are about to depart for greener pastures' (Phillips 2010). Many predict disaster for food production in South Africa if this small, skilled community decides to emulate its predecessors – who, in the 1830s and 40s, embarked on a 'Great Trek' from the Cape colony into the hinterland, and thence to establish the Boer republics of the Orange Free State and Transvaal. Harking back to this iconic moment in the contestation between Boer and British – over labour and the abolition of slavery – the implication is that these resilient farmers are once again without a secure homeland. Images abound of farmers in khaki shorts and Springbok rugby caps visiting snow-covered

Georgia in the former USSR, testing the soils of Libya, and imagining a new life in the Congo. Such images promote an understanding of racist farmers, resistant to change in a new South Africa and fearful of a future under ANC rule.

Elements of such a narrative may indeed be true in some cases. Yet, as this paper shows, the current migration is far more complex, largely signifying *not* a threatened 'boer' nation fleeing one regime to establish a new order elsewhere, but rather a web of interlocking strategies: diversification, building of political alliances, extension of value chains, sourcing of patronage, and the consolidation of a greater global market share (and days on the supermarket shelves of Europe). This is not to be understood (or written off) as the desperate knee-jerk reaction of alienated white Afrikaners unable to see a future for themselves in South Africa, but rather a realignment of interests at different scales – domestic, regional and global – which, as I show, is producing surprising new alliances.

4. Rebranding for the continent: 'AgriSA Africa'

In 2010, Agri South Africa formed an 'AgriSA Africa committee... with a view to securing funding for those farmers who wish to extend their farming operations into other African countries' (AgriSA 2010a). The committee was tasked with liasing with African governments regarding land acquisitions and farming support. By 2009, about 1000 farmers had signed up to be part of the expansion 'into Africa', according to AgriSA chief negotiator Theo de Jager (Hoffstatter 2009a). AgriSA has discursively constructed its land deals as development partnerships, and considers the Congo deal as a pilot project which would guide the format for further AgriSA to government deals (Pienaar cited in AgriSA 2010c). A resolution from the AgriSA congress in October 2010 declared that: 'Within [the] Southern African Development Community (SADC) and Africa it is essential to promote regional economies of scale in respect of primary and related development, i.e. joint projects must be developed' (AgriSA 2010b).

The South Africans have been actively networking with others negotiating for, and offering, land deals across the continent. A key such opportunity was the Conference on Large-Scale Farming in Africa, held in Cairo in April 2010. Here, AgriSA president Johannes Möller claimed that South African commercial farmers want to move into Africa 'as a result of scarcity of natural resources and land redistribution' at home (Derby 2009), echoing in more measured terms the claims of TAU that white farmers are being marginalised and threatened by economic transformation initiatives at home. At this event, South Africa and Zimbabwe were agreed to be the only countries on the continent not safe as destinations for agricultural investment, because of concerns about land reform programmes and labour regulation (*Farmers' Weekly* 2010).

South African farmers already have operations underway in Mozambique, Malawi, Botswana, Kenya and several other countries (Reuters 2010a). The terms vary quite substantially but, in the past year or two, diplomatic missions of various countries have made direct contact with AgriSA, offering preferential investment terms. Zambia, for instance, has offered to waive import duties and value added tax, invest in roads and extend power grids in order to attract South African farmers to its two new 150,000-ha 'farm blocks' available for colonization (Reuters 2010a). Among the recent South African farmland deals are some for sugarcane production by TSB (the sugar company formerly known as Transvaal Suiker Beperk) in Mali, sugarcane and palm oil in Madagascar, palm oil in Benin, jatropha in Mozambique, Tanzania and Zambia, and sugarcane in Zimbabwe and Namibia (ILC 2011). Of the 16 major land acquisitions by South African companies elsewhere in Africa, three are largely for food crop production, one is for carbon sequestration under the United Nations' Reducing Emissions from Deforestation and Forest Degradation (REDD) and the rest (12) are for agrofuels – jatropha for biodiesel, sugarcane for bioethanol, or a combination of other feedstocks like maize, soya and palm oil (ILC 2011). One high-profile deal that is now off is the proposed allocation of 35,000 ha of former state farms for grapes and olives in Libya (Hoffstatter 2009b, Reuters 2011). During 2010, further talks got underway with Angola, Democratic Republic of the Congo (DRC), Egypt, Gabon, Guinea, Cameroon, Mozambique, Namibia, Senegal, Sudan, Uganda and Zambia (Bloomberg 2011, Groenewald 2010, Reuters 2010a, Reuters 2011). By late 2010, 'AgriSA Africa' was in negotiations for land with the governments of 22 African countries (AgriSA 2010a, Reuters 2010c).

4.1. The Congo deal

The most significant recent deal offering African farmland to South African farmers has been concluded between AgriSA and the Republic of the Congo (Brazzaville, not DRC). In October 2009 the government of the Congo signed an agreement with AgriSA (AgriSA undated) in terms of which it allocated to a consortium of South African commercial farmers an initial area of 200,000 ha of former state farms, with the option of expanding to 10 million ha (Donnelly 2009). The land in question consists of state-owned farms, neglected for many years following civil conflict. The land lies in the southwest of the country, between the capital Brazzaville on the Congo River and the coastal city of Point-Noire to its west, and occupies much of the fertile Niari Valley in the departments of Niari and Bouenza (Roch Eulogue N'Zobo, Director, Observatoire congolais des droits de l'homme, Congo, personal communication, 26 March 2011).

Congo's Agriculture Minister claims that the country imports 95 percent of its food requirements – though this presumably refers only to marketed produce – and that the deal would stimulate agriculture as part of the national New Plan of Action (Fin24 2009). Because of its reliance on imports, food prices in domestic markets (particularly for fresh produce) are high by international standards, which makes production for the local market attractive to the prospective settlers – yet their entry will arguably affect those local farmers already supplying this market, a point to which I return below. Although initially mooted as a 99-year lease, it appears that a renewable 30-year lease was signed, in terms of which no rent is payable, and which contains guarantees regarding the tariff-free importation of agricultural inputs, and unlimited rights to export produce, and for these rights to be heritable (N'Zobo, personal communication, February 2010). Says de Jager, 'farmers would pay no import duties or tax on equipment, unlike in SA where we are heavily taxed in many ways'; in addition, farmers would be able to repatriate profits to any part of the world, and receive a five-year tax holiday (Sharife 2010).

Soon thereafter, on 8 May 2010, the governments of South Africa and Congo concluded a new bilateral investment treaty, expanding the terms of pre-existing cooperation agreements dating from 2003 and 2005. This suggests not that the two countries did not have bilateral relations, or agreed terms for these, but rather that the terms are changing. Next, *pro forma* 'Government to Farmer' contracts were drawn up by AgriSA and Congo authorities: 'The contract will apply more specifically to activities of individual farmers, the stipulation of boundaries and terms and conditions for farming activities. The agreement will help farmers... who establish businesses in the Republic of Congo, open bank accounts and start farming, with the protection of the said contracts and with all the benefits for the farmer as negotiated by AgriSA' (Swanepoel cited in AgriSA 2010c). The first contracts, for 88,000 ha, were concluded between the South African farmers and the Congolese Ministries of Agriculture, Livestock and Land at a ceremony at Point-Noire on 10 March 2011 (N'Zobo, personal communication, 26 March 2011). The next step by AgriSA was to develop an affiliation structure for prospective Congo farmers, now called 'AgriCongo', so that they have a local association of their own through which to voice their interests (Swanepoel cited in AgriSA 2010c).

Three ambiguities emerge from the many statements by the protagonists in this deal. These concern the South Africans' motivation for the deal; who is occupying this land currently, if anyone; and what types of production are envisaged, for which markets.

First, the deal in the Congo (as with negotiations elsewhere on the continent) have been accompanied by narratives that rest on a developmental discourse of pan Africanism and good neighbourliness – suggesting that South African farmers are fulfilling a duty, as mandated by their government, to bring about development in other African states. This focuses on the need for increased productivity and supply to meet national food demand, and the transfer of skills and technology to enable the growth of indigenous small- to medium-scale commercial farming. According to the Agricultural Business Chamber, explaining to the Johannesburg Stock Exchange the benefits of such deals with South African farmers, Africa can gain through 'technology transfer, employment creation, infrastructure development, GDP growth' and also increased local food supply for domestic markets (Maluleke 2009). At the same time, AgriSA (and to a greater degree, TAU) have emphasised 'push' factors: the need to escape government regulation and taxation and the threat of land expropriation. Less prominent in these narratives are the obvious economic advantages of the cheap deals and preferable terms on offer. Andre Botha, president of Agri Gauteng, a division of AgriSA, encompassed several arguments when he explained that:

> There are three main reasons we are in the Congo. The first is, of course, to diversify our businesses; the second is to assist local farmers to commercially develop their own land; the third reason is to assist the government of South Africa to fulfill the expectations of the world in stabilising the African continent through the exchange of skills and technology (Sharife 2010).

Second, the merits of the deal rest on what Cotula *et al.* (2009) call an 'idle land' or 'wasteland' discourse – the claim that the land is unoccupied and unused. AgriSA's expectation is that 'the farmers will move into the houses on the propert[ies]' (de Jager, cited in Sharife 2010). Despite such claims, both by the government of the Congo and by AgriSA itself which describes it as 'virgin land', there is evidence not only that people live there, but also that they have been farming on this land and consider themselves to have rights to remain there. A Congolese human rights organisation, the Observatoire Congolais des Droits d'Homme (OCDH), alleges that communities in the affected areas, which hold customary land rights, were not adequately consulted (N'Zobo, personal communication, February 2010). Though official sources claim that the land in this remote valley is vacant and unused, they have also promised that the local people would benefit from employment (Fin24 2009). OCDH reports that:

> we know that in Bouenza's Dihesse or Mantsoumba [which form part of the initial area leased to AgriSA], local communities have been using this land, which government abandoned decades ago. They had indeed planted manioc, groundnuts, etc. How come government does not provide technical and financial support to encourage this local initiative and increase yields? (N'Zobo, personal communication, 26 March 2011)

Third, there are ambiguities regarding the types of production and their destination markets. While it has been reported that the 17 South African farming syndicates will focus on 'producing grain and red meat for the local market and fruit for export' (Bloomberg 2011), AgriSA has claimed that its primary intention is to produce vegetables and poultry for the local market (Sharife 2010). A further motive, though, appears to be to expand its market share and shelf space in European supermarkets. Geographic diversification to a production base in Congo, with its two rainy seasons and earlier harvest, may just be the answer, as AgriSA's de Jager has acknowledged: 'The farmers will grow tropical fruit and hopefully give us an edge against farmers in South America where they traditionally beat us to the European markets by about a week' (de Jager, cited in Sharife 2010). Which production will predominate in practice remains to be seen; without restrictions on exports, the promise of boosting local food supplies for Congo will depend on its profitability compared with markets elsewhere.

Like so many instances of 'land grabbing', the deal has been characterised by a silence about those on the land – their rights, livelihoods and future – once the South Africans move onto these farms and 'move into the houses' that are still standing on them. But, as pointed out by de Schutter (2011, 258–9), even more obscured from view in many deals, and certainly in this one, are the potential impacts of the influx of large-scale commercial producers on *other* farmers elsewhere in the Congo. If indeed the South Africans focus on the domestic market, existing Congolese farmers, without external financial backing and without the tax breaks and other preferential terms offered to the South Africans, will compete with the new operators for domestic market share. On this, there is a total silence in all statements from both sides.

4.2. AgriSaMoz

Another current project of AgriSA Africa is the establishment of 'AgriSaMoz', an association of South African commercial farmers in Mozambique, launched in May 2011 at a function in Pretoria with senior government politicians and officials from both countries in attendance to affirm their support. According to AgriSA, the primary motivation for forming such an association is to enable individual commercial farmers to get expert advice (and political support) to secure land rights in their new country:

> A request received in November 2010 at an agricultural investment conference in Xai-Xai in the Gaza province of Mozambique to the effect that South African farmers should register their interest in writing with AgriSA, had attracted wide response... Although land is inexpensive in Mozambique, the establishment of land use rights is a lengthy and time-consuming process. The process commences when a prospective land user identifies the land and negotiates with the local traditional leader to register such a right (usually for 50 years) (AgriSA 2011).

AgriSA was also negotiating additional 50-year land leases in Mozambique in the past few months for land to grow sugar for ethanol, as well as tropical fruit and livestock. Authorities in Gaza province in the south of the country indicated that they were able to make available a million ha to AgriSA members (van Burick 2010). 'At this stage there are just over 800 [South African] farmers who have already established interests in Mozambique and seemingly another 800 are in the process of trying to tie up land, with a deal that is being closed with the Gaza province', said de Jager (Reuters 2011). A prominent exponent of these deals, and a personal friend of Zuma, is Charl Senekal – who faces a land restitution claim on his R160 million farm in South Africa's KwaZulu-Natal province – who declared at the Xai-Xai meeting:

> We are very anxious to provide and feed the Mozambican people with affordable food... Farmers wanting to diversify or who have lost their farms in [South Africa's land] redistribution have a new opportunity in Mozambique (Reuters 2010b).

The meeting in Xai-Xai was organised not by AgriSA but by South Africa's high commissioner in Mozambique, Dikgang Moopeloa, who explained the motive as being to create opportunities for South Africans, to strengthen diplomatic links with Mozambique, and to promote development and regional stability across the Southern African Development Community so as to 'stem the inflow of migrants to SA' (SAPA 2010). The meeting took place against the backdrop of protests against rising food prices, which left several people dead, shot by security forces in Maputo in September 2010.

5. The government connection: funding, political support and BITs

The major constraint on further deals within Africa is the absence of bilateral investment treaties (BITs) to secure investors' assets and the right to repatriate profits (Cotula and Vermeulen 2009). From a South African agribusiness point of view (a view shared with its partners in the Southern African Confederation of Agricultural Unions, representing largely the interests of commercial farmers), this is the precondition for implementation that has scuppered (or at least delayed) several in-principle agreements for allocation of farmland in other countries in the region (SACAU 2010).

In support of business expansion, the South African government recently signed and updated numerous BITs – in the past three years alone, with Angola, Cameroon, DRC, Gabon, Guinea, Ethiopia, Madagascar, Mauritania, Namibia, Sudan, Tanzania, Zambia, and Zimbabwe. These are mostly cited as 'Agreement[s] on Promotion and Reciprocal Protection of Investment' (plus related protocols), but there are many more memoranda of understanding on 'Cooperation in the Field of Agriculture' or 'Mining and Minerals Beneficiation' and agreements to avoid double taxation in source and host countries.

In recent years, the export of South Africa's agricultural skills became government policy, when Minister of Agriculture, Forestry and Fisheries Tina Joemat-Pettersson announced a R6 billion fund for supporting South African farmers, half of which would be spent on projects beyond South Africa's borders (*Farmers' Weekly* 2010). She was previously quoted as saying 'If we can't find opportunities for white South African farmers in this country, we must do it elsewhere in the continent' (Hoffstatter 2009a). This she characterised as 'an equal relationship between people of the African continent' (Joemat-Pettersson cited in Hoffstatter 2009a).

More generally, the mainstream discourse deployed by AgriSA and the South African government vacillates between pan-Africanism and claims of Boer exceptionalism. The government's interests appear to lie in part in regional expansion but also in appeasement of white farmer (and agribusiness) interests. Economically and politically nuanced strategies are at work, as these farmers are diversifying rather than leaving South Africa. None of the farmers signing up for land in the Congo intended to sell their South African operations and, of those previously considering taking up the Libyan deal (prior to political conflict in that country), most intended to remain based in South Africa, and employ managers (Libyan or South African) to manage their foreign operations (Hoffstatter 2009b, Sharife 2010).

6. The sugar

At the same time that South African farmers are launching into a wide variety of farming enterprises across the continent, South Africa's sugar industry is also expanding, not only establishing (or taking over existing) sugar estates, but also exporting its model of smallholder outgrower schemes – which some commentators consider the basis of potential 'win-win' solution to the problems of 'land grabbing'. Because the AgriSA initiatives are largely in horticulture, livestock and grains, and because of the size and rapidity of the sugar expansion, the latter deserves analysis in order to capture the distinct dynamics of these dual processes.

South Africa's two sugar giants, Illovo and Tongaat-Hulett, are the products of decades of apartheid-era state subsidies and trade protection. Both have rapidly reinvented themselves as regional businesses, and now operate in six countries each in Southern and East Africa. Sugar has long occupied a special place in South African agricultural policy, being the only commodity which – a decade after the closure of the last state marketing boards – still benefits from state-imposed price controls. Both companies have adjusted to new pressures and opportunities by reconfiguring their asset and production portfolios. Since the 1990s they have embarked on three modes of expansion: purchase or lease of state-owned estates (for example in Malawi, Mozambique and Zambia); outgrower schemes where they establish core estates and mills while also entering into supply contracts with neighbouring small-scale farmers; and contract farming only, to supply existing mills. Together, these companies now account for nearly half the sugar factories in Southern Africa (see Figure 1).

Both Illovo and Tongaat-Hulett's production volumes have expanded substantially year after year through the food and financial crises – their financial statements show no symptoms of the recent global recession. The main driver behind the sugar boom in the region is massive increase in global demand, due to a combination of major markets (such as India) becoming net importers, and rising demand for bioethanol buoyed by targets (notably in the European Union) for renewable fuel content (Biopact 2007, Richardson 2010). While McMichael (2010) among others notes the competition of 'food versus fuel', increasingly the food and fuel sectors are now fused together, with the South African sugar industry embodying both, by producing a fungible crop. Its fungibility – as food and as fuel – makes sugar a relatively safe hedge against fluctuations in demand. Added to this are the different institutional models for production (outgrower, estate and combinations) which provide options to adapt to local conditions, and vary the degree of fixed investment required. The experience of contract farming in sugar shows the adaptability of the industry to outgrower schemes to diverse contexts, and the ways in which contracts have been influenced by considerations of infrastructure, mechanisation and labour mobilisation – but have also responded to ideological contestation over the place of smallholder production in commercialising agriculture (Oya 2012).

While expanding into the region, the South African sugar industry is also lobbying hard at home for mandatory renewable content in national fuel supplies, for billions in state subsidies to enable it to meet this demand and for import tariffs to protect the industry from the competition of the global ethanol industry (Martin Eweg, Consultant, AgriCane, and former research specialist, South African Sugar Association, personal communication, 6 July 2010, RSA 2010, Williams 2011). Already, though, these sugar companies are diversifying far and wide, from the industry's traditional products of sugar and molasses into byproducts and derivatives like furfural, furfuryl alcohol, diacetyl, 2,3-pentanedione, natural methanol, ethyl alcohol and lactulose, which Illovo now sells to 81 countries. In addition, BioMass Sugar, 'a liquid organic fertiliser derived from sugar cane, is being sold to a growing number of international markets' (Illovo 2011). The next big step is to use the 'bagasse' biomass left over from crushed cane for 'co-generation' of electricity, not only for the sugar factories themselves but also to feed into national grids to shore up domestic energy supplies. This would be a boon for host governments but – other than Swaziland where this is happening already – is still largely in the realm of vague promises for the future.

Illovo, for which data is more readily available (and which therefore forms the focus of this discussion), now owns not only its South African operations, but is a

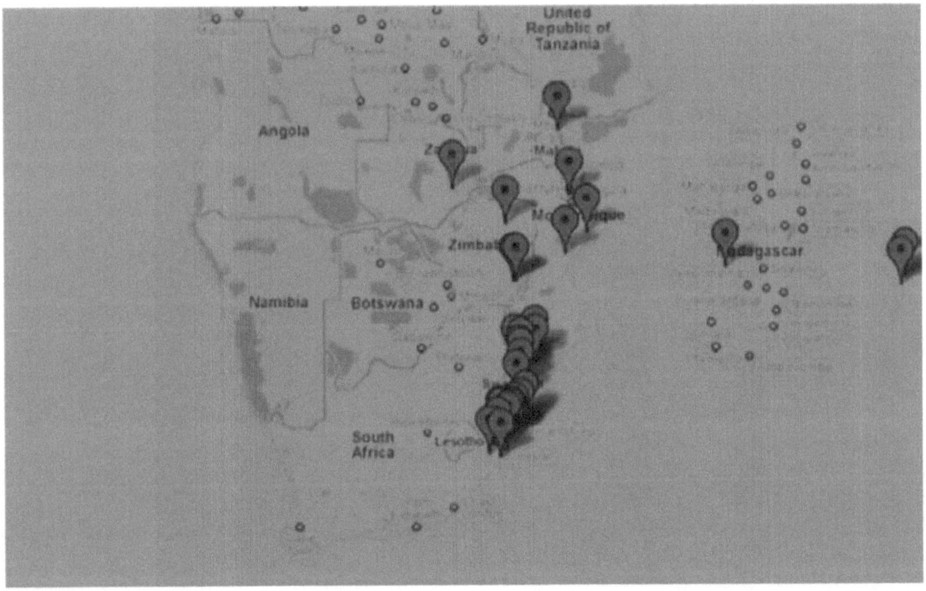

Figure 1. Map of sugar factories in Southern Africa.
Source: http://www.sugartech.co.za/factories/groupmap.php?regid=7

majority shareholder in five other 'Illovos' across the region (see Figure 2) at the same time that it, in turn, is majority-owned by British Sugar, which in turn is a wholly owned subsidiary of the UK-listed multinational Associated British Foods (ABF) (Richardson 2010). This expansion has been largely in partnerships with domestic investors or governments. Illovo's own description of its 'group strengths' emphasises its African (rather than South African) character, and its geographic diversity as key to its resilience, when it defines itself as an 'African sugar producer with African expertise in sugar cane agriculture, sugar manufacturing and the production of high-value downstream products' and having a 'wide geographic and climatic spread of core interests with good access to secure water supplies for irrigation' (Illovo 2011). In fact, Illovo's rise was on the back of cheap buy-ups of ailing government estates during the 1990s in Malawi, Mauritius, Swaziland and Zambia (Richardson 2010), on the basis of which it has been involved in a second-wave expansion in the past five years. Now making most of its profits outside South Africa, Illovo's sources of finance are also increasingly diverse, including through issuing of shares on domestic stock exchanges, as in Zambia. Rather than simply a matter of regional imperialism by South African agricultural capital, this reflects the globalised character of the capital behind the expansion, and the strategic location of South African actors to take advantage of it.

At the same time, production in South Africa is declining and rapidly being eclipsed by other countries (see Tables 1 and 2 below). The milling capacity (for sugar refining) is still highest in South Africa, whereas elsewhere increases in sugarcane output are largely for ethanol rather than sugar.

Equivalent information has not been made publicly available for Tongaat-Hulett. What is known is that through the mid-2000s they were expanding their Zimbabwean estates, aiming for 1 metric tonne per year at the Triangle and Hippo Valley plants, converting these into bioethanol plants, and were expanding in Mozambique at Xinavene and Mafambisse (Biopact 2007).

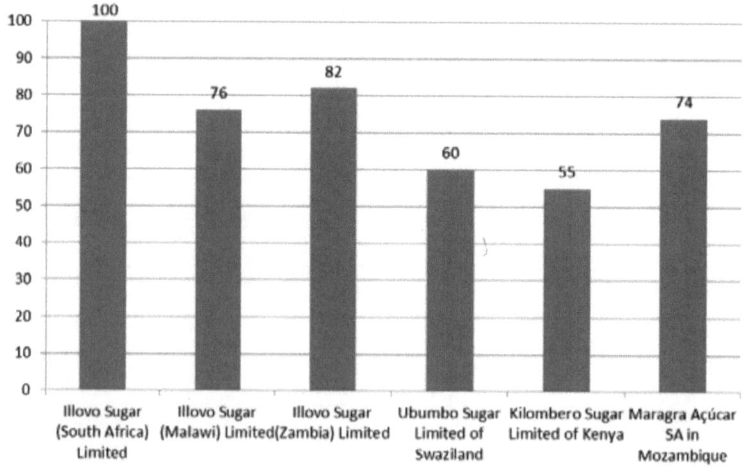

Figure 2. Illovo ownership of sugar companies in Southern Africa, 2010 (percent).
Source: Illovo 2010.

Table 1. Illovo production statistics: sugar cane (× 1000 tons).

Country	2010	2009
South Africa	361	464
Malawi	2 136	2 122
Zambia	1 705	722
Swaziland	804	741
Tanzania	614	555
Mozambique	488	450
Total	6108	5054

Table 2. Illovo production statistics: sugar (× 1000 tons).

Country	2010	2009
South Africa*	663∧	922
Malawi*	295	304
Zambia*	315	194
Swaziland*	211	210
Tanzania*	120	118
Mozambique	81	76
Total	1685	1824

* - Sugar cane supplied by the group's own agricultural operations and private growers
∧ - Excludes Pongola and Umfolzi [sic] – sold
Source: Illovo 2010.

Illovo data shows robust growth in regional operations and substantial dominance of the sugar industry in several countries, notably Malawi and Zambia (see Figures 3 and 4 below).

Rather than simply increasing output, one of the motives for South Africa's sugar expansion in the region appears to be improved access to international markets. Producing and exporting from Mozambique and Malawi, for instance, enables these South African companies to benefit from the EU's Everything But Arms agreement that offered preferential trade terms to least developed countries (LDCs) (Biopact 2007). Together, the South African, Malawian and Mozambican governments have jointly lobbied the EU and its Environment Committee to relax its proposed biodiversity requirements, arguing it was punitive to force LDCs to demonstrate that changes in land use 'should not disproportionately penalise countries rich in biodiversity with unjustified, wide-ranging restrictions on the sustainable use of their territories'. The strategic benefits of outsourced production extend beyond the region; de Jager has explained that 'The [South African] farmers in northern Sudan should be able to utilize a preferential sugar trade agreement between Arab nations' (Bloomberg 2011).

7. Multiple connections in the shifting agro-food regime

While Ben Richardson (2010) talks of 'Big Sugar' to describe the conglomeration of multinationals that dominate the industry, in the context of partnerships between the

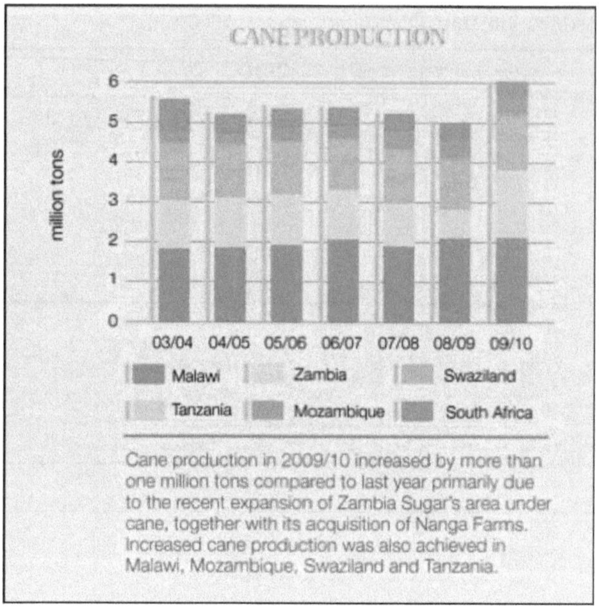

Figure 3. Illovo cane production by country and financial year (in million tons).
Source: Illovo 2010.
http://www.illovo.co.za/World_of_sugar/Sugar_Statistics/Illovo_Group.aspx

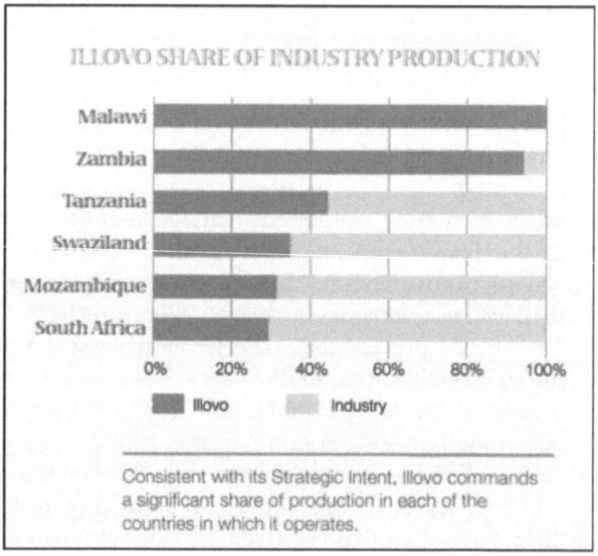

Figure 4. Illovo share of industry production (percentage by country) in 2010.
Source: Illovo 2010.
http://www.illovo.co.za/World_of_sugar/Sugar_Statistics/Illovo_Group.aspx

US and Brazil driving the expansion of the sugarcane plantation system for ethanol production, Gail Hollander (2010) speaks of 'assemblages' that are not only geographically diverse but also take on a distinctive regional character. In the case she examines, this is the 'hemispheric' project of building the ethanol industry in the Americas, as well as the more specific role of Miami as a conduit for information and networks for the multiple actors involved in this industry. Drawing on Hollander's notion of 'assemblages', I argue that, in the African context as in the Latin American, it is important to recognise that behind this expansion are the interests not only of sugar companies, or AgriSA, but a complex of commercial interests that span different sectors. In Southern Africa, many of those who stand to profit from large-scale land deals are not involved with farming at all.

7.1. Engineering, consulting and construction interests

Interests in the expansion of sugar production in the region extend beyond Illovo and Tongaat-Hulett, to a substantial industry involved in the construction and management of new sugar mills and related estates and outgrower schemes. Also worthy of mention are those involved with logistics, transport, cold storage, financial services, and so on.

A notable example of this wider set of interests in the sugar industry is PGBI consulting engineers. This South African company is one of the top recipients of contracts to build sugar mills and ethanol plants in Africa (Eweg, personal communication, 6 July 2010). It was also commissioned by the International Finance Corporation, part of the World Bank Group, to produce a guide for investors in the sugar industry on how to deal with land, social, and environmental issues to the satisfaction of financiers. PGBI works in Zimbabwe, Swaziland, Tanzania, Mozambique, Zambia, Mali, Ghana, Ethiopia, Kenya (for Mumias, the continent's largest sugar factory) and Sudan (for the Saudis and, soon, likely also for the South Africans) (PGBI 2010). As well as providing professional services to sugar companies, PGBI plays a number of other roles, working for their financiers and for governments, and is itself an investor – it is listed as a shareholder in the 10,000-ha allocation of land in Caprivi, Namibia, for sugarcane production (ILC 2011). The point is that beyond the sugar companies themselves, there are associated service and consulting industries, and there is enormous circulation of people between industry and consultancies to these, including those auditing conditions to the satisfaction of financiers and industry regulatory bodies, and much of this is far from public view. It is worth looking into how these interests work out in practice, yet the emerging literature on land grabbing focuses on the interests of investors themselves, largely obscuring these associated actors.

7.2. The banking connection

With the decline of South Africa's state-owned Land Bank, and the withdrawal of access to subsidised credit for white farmers as part of the wider deregulation process of the 1980s and 1990s, South African commercial banks have emerged as primary lenders to agriculture. Now three commercial banks are involved in financing AgriSA's continental expansion: Standard (at the forefront), ABSA and Standard Chartered (Reuters 2010a). At the same time, the 'South Africanness' of these banks – and the financial backing behind the farming expansion – is becoming more questionable. First, several South African banks have been acquiring substantial acquisitions of shares in banks in Africa; for instance Standard Bank recently acquired 90 percent of Uganda Commercial Bank, taking over its loan book and clients. Second, shareholding of these once 100 percent South African-owned banks is now diversifying. The Chinese stake in Standard Bank is growing, starting with the acquisition in 2007 of a 20 percent stake in this the oldest established South African bank, for USD 5.5 billion by the Industrial and Commercial Bank of China. This was followed by a club loan by 5 Chinese banks for USD 1 billion in 2009. These deals involved the Chinese acquiring existing shares, but also new shares – and so apparently funding new expansion. Incidentally, Standard Bank was at the Cairo conference, recruiting clientele looking to invest in African agriculture.

7.3. The China connection

More direct China-South Africa partnerships are under negotiation – not only for finance but also for more substantive cooperation (van der Lugt and Hamblin 2011). The multilateral policy context for these new partnerships is the Forum on China-Africa Cooperation (FOCAC) concluded between the African Union and China in 2006 and endowed with USD 35 billion to finance Chinese investment in Africa. Within this cooperative framework, one proposal under discussion is Chinese research and funding, through its parastatal institutions, for South African experimentation with genetically modified seed varieties, crop varieties and water management in its new landholdings elsewhere on the continent. Synergies of interests between Chinese biotechnology and agricultural research institutes and South African farmers are evident: 'They're aware South Africa has a research problem at the moment and they know they need farming skills for the land they're acquiring in Africa' (Christie 2010). The breakthrough in the partnership between AgriSA and China was at the Cairo conference; as AgriSA's lead negotiator for African land, de Jager, confirmed: 'They asked us to find land for them, so that if we all reach an agreement, they can immediately start with experimental work' (cited in Christie 2010). This was to be followed by AgriSA's visit to China to agree on 'joint investment ventures with Chinese public firms in agriculture on the African continent' (Shacinda 2010) – though it has not been possible to verify whether this visit took place and, if so, its outcomes. In such ways, agricultural investments through bi- (and multi)lateral agreements between South Africa and other African states are framed as regional and Pan-African integration – as opposed to land grabbing and Sino neo-imperialism.

7.4. The investment funds

A further way in which South African actors and interests are involved in large-scale land acquisitions elsewhere on the continent is through investment funds. Emergent Asset Management is a jointly registered UK/SA management firm, previously focused on hedge funds in international financial markets, and now specialising in farmland investments in Africa. Two of its flagship products are the 'African Agricultural Investment Fund' established in 2008, and the 'African Land Fund' established in 2010. The latter, through a partnership with Grainvest, one of the top five companies listed on the South African Futures Exchange (SAFEX), has formed Emvest Agricultural Corporation which constitutes the operational side of these two funds' investments in land and agriculture. Together, these funds offer a vehicle for South African, UK and other investors to diversify their investments into African agriculture, and by late 2010 had secured land holdings in Angola, Botswana, DRC, Kenya, Lesotho, Madagascar, Malawi, Mauritius, Mozambique, Namibia, South

Africa, Swaziland, Tanzania, Zambia and Zimbabwe (EAML 2010). Remarkable increases in land values and yields would be generated through 'modern farming techniques and technologies, while agglomerating farms to increase efficiency and generate economies of scale' (EAML 2010). The African AgriLand Fund:

> brings together the key themes of agriculture/food security, Africa, socially-responsible investing (SRI) and economic sustainability, purchasing and managing a wide spectrum of agricultural properties across the sub-Saharan region, with investments diversified both geographically and across agricultural sectors – including crops, biofuels, livestock, game farming and timber. Returns, based on those successfully achieved through a 4-year pilot project, are projected to be approximately 30 percent per annum over the Fund's 5-year term. (EAML 2010)

The European report on financial speculation and private equity funds as drivers of land deals – entitled 'The Vultures of Land Grabbing' – concludes that 'In short, private equity funds not only have a speculative business model, but also represent a conveyor belt for shareholder capitalism from the financial to the real economy' (Merian Research and CRBM 2010, 1). Susan Payne, head of EAML, was previously at JP Morgan and then was executive director at Goldman Sachs, which is itself one of the institutions at the forefront of this new model of hedging against international market fluctuations by investing in farmland. The Fund demands a minimum investment of 500,000 euros from private investors and 5 million euros from institutional investors, and claims to offer secure and diversified investment packages. Its selling point is explicitly as a vehicle to capitalise on cheap ('undervalued') African land.

> EmVest has been created as a dedicated agricultural operating and fund management company to execute the strategy. It is based in Pretoria, South Africa and is dedicated to the development and management of the Fund's assets. EmVest is establishing farming hubs throughout the region, working with national and local authorities to develop large scale agriculture on a commercial basis, within a socially responsible framework, at a consistently high level of execution... The Manager buys agricultural land and manages a wide spectrum of agricultural projects, spread across different commodities. It invests across 14 countries in sub-Saharan Africa, where land is substantially undervalued compared with arable real estate on other continents (EAML 2010).

Questions for further research must include precisely where and how they are investing, via whom, and what kinds of land uses and social relations are being produced through this influx of global capital. Importantly, to what extent is the capital focused on investment in durable production, or on speculative holding of land on a short-term basis? Spectacular promises of profits – as 30 percent returns to investors per annum – hint at the latter.

8. Conclusions

While the 'land grabbing' discourse has tended to focus broadly on 'foreign' actors (governments and companies) and, in the context of Africa, on actors from outside the continent, what this paper demonstrates is the rise of *intra-regional land grabbing*. Powerful narratives used to describe large-scale land acquisitions in Africa and elsewhere in the developing world emphasise historical precursors and suggest a return to old patterns of colonial enclaves and mercantilist colonialism. Yet growing evidence suggests new dynamics and different actors in the latest phase of 'land grabbing'. Widely acknowledged is the role of the Gulf States, moving away from reliance both on costly and intensive irrigated production at home, and on global commodity markets to meet their national food needs, and towards offshore production instead. But another such dynamic that merits further empirical investigation and theorisation is the intra-regional character of many such land acquisitions and investments. The rise of South-South deals have been led by emerging powers like the BRICs (now joined by South Africa, to form the 'BRICS' countries). In the context of 'rising global interest in farmland' – the euphemistic phrase deployed by the World Bank – it is not surprising that those regional powers that are well placed both politically and economically to act as initiators and brokers might be among the most significant beneficiaries of this new wave of transnational land leasing. South Africa is undoubtedly one such country.

This paper interrogates the discourse promoted by commercial farmers themselves that they are seeking greener pastures elsewhere in Africa because the new democratic government is driving them out of business – embarking on land reforms, threatening land expropriation, imposing labour regulation and minimum wages, and increasing taxes and exposure to competition from imports, while setting high prices for key inputs (in addition to labour) like diesel and water. There is a certain poignance in what has been described as the 'white tribe of Africa' – predominantly white (male) Afrikaner farmers – finding, or claiming, that they are not valued in a new South Africa, and seeing the expressed demand for their skills elsewhere as affirmation of their 'African-ness' and a sense of a place and role in the future of the continent. Andre Botha, a livestock and maize farmer who heads AgriSA's Gauteng branch, said it should *not* be seen as

> another Groot [Great] Trek. This is not running away – it is a calculated process of helping Africa take up its rightful place in the world. All over Africa the message we get is: we are looking for South African farmers, which is incredibly important for us. It means that our contribution to the future of this continent is valued (cited in Hoffstatter 2009b).

What exactly might South African 'farmers' be bringing to the rest of Africa, though, if (as in some cases so far) they get the land cheaply or for free, do not manage the farms themselves, and do not bring their own capital but raise this through international and political partnerships? One part of their answer lies in their social capital, as they build a platform as 'African farmers', in the eyes of their own government, and in the eyes of global investors to whom they look for finance. In this context, a narrative of Afrikaner farmers as hardy, innovative and enduring has been actively cultivated by South African farmers and agribusinesses, to convince African governments that, unlike European or Asian investors, these are 'expert commercial farmers [accustomed to operating] under tough African conditions' (Hoffstatter 2009b). None of these processes, though, can be understood outside of global changes.

In what McMichael (2009) terms the newly emerging global food regime, multiple centres of power are emerging; it is 'polycentric' (Borras *et al.* forthcoming). This direction of change, and its implications, should be fruitful subjects for further empirical investigation. As Dwyer observes, 'the transnationalization of corporate access to farmland will be a long-term process, and will necessarily involve a diverse array of coalitions, tactics and locations' (2011, 4–5). 'Land grabbing' by South African firms elsewhere in the region is not therefore merely a fascinating anomaly, but rather is the newest manifestation of these structural changes, and is symptomatic of the restructuring that is underway. This paper's dual focus on AgriSA and the sugar industry depicts, on the one hand, a case of consortia that link old farming capital (from South Africa and elsewhere) with new partners in new and less regulated locales, and a case of old agro-industries securing new market opportunities and protection through expansion both geographically and into new products and regimes of profitability, notably in agrofuels. In this changing context, it is not the national identity of the investors that proves most relevant for understanding what is really going on – which is why the depiction of these processes as another 'great trek' has missed the mark. Rather, what is of importance for deeper analysis is the articulation of regional and global capital. Further research on intra-regional land grabbing should focus on cooperation, consolidation and competition between 'capitals'.

Explicitly, the South African farmers are *not* going elsewhere on the continent to secure food supplies for South Africa. They are going to find cheap land, water, labour and more lenient tax conditions from which to export to whichever markets appear the most lucrative – though they may be constrained by deals that require some domestic sales, and by financiers that demand that some or all of their output be exported to the country financing these investments, like China. Though this trend takes the form of South African deals elsewhere on the continent, the phenomenon cannot be fully understood as 'South African imperialism' in Africa. The interests shaping it, the financing, and the destination markets extend far beyond South Africa. This is global capitalism.

References

AgriSA (South Africa). 2010a. *Newsletter*. Centurion: AgriSA. September.

AgriSA (South Africa). 2010b. *Newsletter*. Centurion: AgriSA. October.

AgriSA (South Africa). 2010c. *Newsletter*. Centurion: AgriSA. November.

AgriSA (South Africa). 2011. AgriSaMoz. Media Release. 24 January 2011.

AgriSA (South Africa). Undated. Settlement of Farmers in the Republic of Congo. Unpublished press release. Available from: http://www.agrisa.co.za/Dokumente/Afrika/KongoE.pdf [Accessed 30 March 2011].

Atkinson, D. 2007. *Going for broke: the fate of farm workers in arid South Africa*. Cape Town: Human Sciences Research Council Press.

Bernstein, Henry. 1996a. South Africa's Agrarian Question: Extreme and Exceptional? Journal of Peasant Studies, vol. 23, no. 2/3. January/April 1996, pp. 1–52 (Special Issue on the Agrarian Question in South Africa edited by Henry Bernstein).

Biopact. 2007. South African sugar producer expanding operations in Zimbabwe and Mozambique, eyeing ethanol and biomass cogeneration. 20 February 2007. Available from: http://news.mongabay.com/bioenergy/2007/02/south-african-sugar-producer-expanding.html [Accessed 19 November 2010].

Bloomberg. 2011. S. Africa farmers shun Libya, target Sudan bananas, Egypt grapes. 24 February 2011. http://farmlandgrab.org/post/view/18207 (accessed 30 March 2011).

Borras, S., J. Franco, S.Gomez, C. Kay and M. Spoor. 2012. Land grabbing in Latin America and the Caribbean. Journal of Peasant Studies 39(3&4).

Christie, S. 2010. China courts SA's commercial farmers. *Farmers' Weekly*, 1 October.

Cotula, L. and S. Vermeulen. 2009. Deal or no deal: the outlook for agricultural land investment in Africa. *International Affairs*, 85(6), 1233–47.

Cotula, L., S. Vermeulen, R. Leonard and J. Keeley. 2009. *Land grab or development opportunity? Agricultural development and international land deals in Africa*. London: International Institute for Environment and Development, Rome; Food and Agricultural Organization, Rome: International Fund for Agricultural Development.

DAFF (Department of Agriculture, Forestry and Fisheries). 2009. *Abstract of agricultural statistics*. Pretoria: DAFF.

Derby, R. 2009. South African farmers offered land in Angola, Uganda. *Bloomberg*, 9 October. Available from: http://www.bloomberg.com/apps/news?pid=newsarchive&sid=aXWZYn1ahnJ8 [Accessed 19 November 2010].

De Schutter, O. 2011. How not to think of land-grabbing: three critiques of large-scale investments in farmland. *Journal of Peasant Studies*, 38(2), 249–79.

Donnelly, L. 2009. SA farmers in new scramble for Africa. *Mail & Guardian*, 8 September.

Dwyer, M B. 2011. Building the politics machine: tools for resolving the global land grab. Paper presented at the International Conference on Global Land Grabbing, hosted by the Land Deal Politics Initiative, Institute of Development Studies, University of Sussex, 6–8 April 2011.

EAML (Emergent Asset Management Limited). 2010. African AgriLand Fund. https://www.emergentasset.com/?func=PageAfricanLandFund [Accessed 19 November 2010].

Farmers' Weekly. 2010. 'SA, Zim not safe for investments.' 9 May 2010. http://www.agriworldsa.com/article-archive/economic-environment/sa-zim-not-safe-for-investments [Accessed 14 May 2010].

Fin24. 2009. 'Huge Congo deal for SA farmers'. 20 October 2009. http://www.fin24.com/Business/Huge-Congo-deal-for-SA-farmers-20091020 [Accessed 12 September 2010].

Groenewald, Yolandi. 2010. Made in China, in Africa. Mail & Guardian, 30 May. Available from: http://www.mg.co.za/article/2009-05-30-made-in-china-africa [Accessed 31 May 2010].

Hall, R. 2011. Land grabbing in Southern Africa: the many faces of the investor rush. *Review of African Political Economy*. 128, 193–214.

Hoffstatter, S. 2009a. Government drive to set up white SA farmers in Africa in *Business Day*, 12 October. Available from: http://allafrica.com/stories/200910120009.html [Accessed 26 October 2009].

Hoffstatter, S. 2009b. Farmers sample what Libya offers. *Business Day*, 9 November. Available from: http://www.businessday.co.za/articles/Content.aspx?id=86387 [Accessed 19 November 2010].

Hollander, G. 2010. Power is sweet: sugarcane in the global ethanol assemblage. *Journal of Peasant Studies*, 37(4), 699–721.

ILC (International Land Coalition). 2011. Excel spreadsheet of South African investors (extracted from the Commercial Pressures on Land database). Updated to June 2010, obtained January 2011. Unpublished spreadsheet. Rome: ILC.

Illovo. 2010. Illovo group statistics. Available from: http://www.illovo.co.za/World_of_sugar/Sugar_Statistics/Illovo_Group.aspx [Accessed 19 November 2010].

Illovo. 2011. Illovo Sugar Limited. www.illovo.co.za [Accessed 30 March 2011].

Li, T.M. 2011. Centering labor in the land grab debate. *Journal of Peasant Studies*, 38(2), 281–98.

Mackey, L. 2011. Legitimating foreignization in Bolivia: Brazilian agriculture and the relations of conflict and consent in Santa Cruz, Bolivia. Paper presented at the International Conference on Global Land Grabbing, hosted by the Land Deal Politics Initiative, Institute of Development Studies, University of Sussex, 6–8 April 2011.

Maluleke, H. 2009. International investment in agriculture and the implications for Africa. Manager: International Trade Intelligence, Agribusiness Chamber. Presentation at the Johannesburg Stock Exchange seminar on 'The Farms Race: The rush for food security in Africa', 31 August 2009.

McGuinness, D. 2011. Following South African farmers to Georgia. BBC. Available from: http://www.bbc.co.uk/worldservice/worldagenda/2011/02/110228_worldagenda_georgia_farmers.shtml [Accessed 28 February 2011].

McMichael, P. 2009. A food regime genealogy. *Journal of Peasant Studies*, 36(1), 139–69.

McMichael, P. 2010. Agrofuels in the food regime. *Journal of Peasant Studies*, 37(4), 609–29.

Merian Research and CRBM (Campaign for the Reform of the World Bank). 2010. The vultures of land grabbing: the involvement of European financial companies in large-scale land acquisition abroad. No publication details. Available from: http://www.rinoceros.org/IMG/pdf/VULTURES-completo-2.pdf [Accessed 16 November 2010].

Oya, C. 2012. Contract farming in sub-Saharan Africa: a survey of approaches, debates and issues. *Journal of Agrarian Change*, 12(1), 1–33.

PGBI. 2010. PGBI Engineers and Constructors (Pty) Ltd. Available from: www.pgbi.co.za [Accessed 18 November 2010].

Phillips, J. 2010. Republic of Georgia to the rescue of South African farmers. *The Afrikaner Journal*. 29 October. Available from: http://afrikanerjournal.wordpress.com/2010/11/02/republic-of-georgia-to-the-rescue-of-south-african-farmers/ [Accessed 1 April 2011].

Reuters. 2010a. S. Africa firms launch $408 mln farmland fund. 24 March. Available from: http://farmlandgrab.org/11870 [Accessed 10 September 2010].
Reuters. 2010b. SA farmers eye Mozambique. 25 November. Available from: http://www.iol.co.za/business/international/sa-farmers-eye-mozambique-1.876615 [Accessed 2 April 2011].
Reuters. 2010c. South African farmers get more land deal offers in Africa. 17 September. Available from: http://www.timeslive.co.za/local/article663955.ece/South-African-farmers-get-more-land-deal-offers-in-Africa [Accessed 19 November 2010].
Reuters. 2011. S. African farmers seal Congo deal, wary of N. Africa. Available from: http://af.reuters.com/article/investingNews/idAFJOE71K09X20110221 [Accessed 30 March 2011].
Richardson, B. 2010. Big Sugar in southern Africa: rural development and the perverted potential of sugar/ethanol exports. *Journal of Peasant Studies*, 37(4), 917–38.
RSA (Republic of South Africa). 2010. 2010/11-2012/13 Industrial production action plan. economic sectors and employment cluster. February 2010.
SACAU (Southern African Confederation of Agricultural Unions). 2010. Land in Southern Africa: key issues for farmers. Report on the 2010 Conference, Misty Hills Conference Centre, Johannesburg, South Africa. 29–30 March.
SAPA (South Africa Press Agency). 2009. Huge Congo deal for SA farmers. 22 October.
SAPA (South African Press Agency). 2010. Xai-xai forum attracts farmers. *Business Day*, 26 November. Available from: http://www.businessday.co.za/articles/Content.aspx?id=127909 [Accessed 1 April 2011].
Shacinda, S. 2010. S Africa Farmers get land offers in Africa. *Reuters*, 9 July. Available from: http://farmlandgrab.org/14268 [Accessed 12 September 2010].
Sharife, K. 2010. The South Africa-Congo concession: exploitation or salvation? Pambazuka News, 7 January. Available from International Land Coalition: http://www.landcoalition.org/cpl-blog/?p=4172=1 [Accessed 19 November 2010].
TAU (Transvaal Agricultural Union). 2010. Georgia shows progress with dynamic agriculture project. TLU SA/TAU SA media release. 30 December. Available from: http://www.africancrisis.co.za/Article.php?ID=88384& [Accessed 1 April 2011].
Van Burick, N. 2010. Agri SA brei uit na Mosambiek. *Landbou.com*. 29 November. Available from: http://www.landbou.com/nuus/nuus-artikel/agri-sa-brei-uit-na-mosambiek [Accessed 2 April 2011].
Van der Lugt, S. and V. Hamblin (with M. Burgess and E. Schickerling). 2011. *Assessing China's role in foreign direct investment in Southern Africa*. Stellenbosch: Centre for Chinese Studies and Hong Kong: Oxfam Hong Kong.
Vink, N. and R. Hall. 2010. Agricultural and land reform in South Africa's first decade of democracy. *In*: S. Jones and R.W. Vivian, eds. *South Africa: economy and policy 1990–2000*. Manchester: Manchester University Press, pp. 644–70.
Visser, O. and M. Spoor. 2011. Land grabbing in post-Soviet Eurasia. *Journal of Peasant Studies*, 38(2), 299–323.
Visser, Oane, Max Spoor and Natalia Momonova. 2012. Land grabbing in Russia: The post-Soviet political economy of insecurity and dispossession. *Journal of Peasant Studies*, 39(3&4)
Walker, C., A. Bohlin, R. Hall and T. Kepe, eds. 2010. *Land, memory, reconstruction and justice: perspectives on land claims in South Africa*. Athens, OH: Ohio University Press and Scottsville: University of KwaZulu-Natal Press.
Williams, Francois, 2011. Sugar industry ready for green energy. Fin24. 6 November 2011. http://www.fin24.com/Companies/Industrial/Sugar-industry-ready-for-green-energy-20111106 (accessed 18 January 2012).
Williams, Gavin, Joachim Ewert, Johann Hamman and Nick Vink. 1998. Liberalizing Markets and Reforming Land in South Africa. *Journal of Contemporary African Studies*, vol. 16, no. 1, pp. 197–220.

Ruth Hall is Associate Professor at the Institute for Poverty, Land and Agrarian Studies (PLAAS) at the University of the Western Cape, South Africa

Land grabbing in Latin America and the Caribbean

Saturnino M. Borras Jr., Jennifer C. Franco, Sergio Gómez, Cristóbal Kay and Max Spoor

Land grabbing has gained momentum in Latin America and the Caribbean during the past decade. The phenomenon has taken different forms and character as compared to processes that occur in other regions of the world, especially Africa. It puts into question some of the assumptions in the emerging literature on land grabbing, suggesting these are too food-centered/too food crisis-centered, too land-centred, too centred on new global food regime players – China, South Korea, Gulf States and India – and too centred on Africa. There are four key mechanisms through which land grabbing in Latin American and the Caribbean has been carried out: food security initiatives, energy/fuel security ventures, other climate change mitigation strategies, and recent demands for resources from newer hubs of global capital. The hallmark of land grabbing in the region is its intra-regional character: the key investors are (Trans-)Latin American companies, often in alliance with international capital and the central state. Initial evidence suggests that recent land investments have consolidated the earlier trend away from (re)distributive land policies in most countries in the region, and are likely to result in widespread reconcentration of land and capital.

Introduction

In the initial surge of reports and studies on contemporary land grabbing there is a dominant assumption that the phenomenon has occurred because of the 2007–2008 food crisis, which in turn was largely caused by the emerging global biofuels complex (the initial 'food versus fuel' view – see White and Dasgupta 2010). The changes in the global agrofood system made some financially powerful countries (primarily China, South Korea, and the Gulf States) that could not produce sufficient food domestically feel insecure. They started to seek control over large tracts of lands overseas to secure food supply (GRAIN 2008). The principal target is Africa where vast empty lands are thought to be available, cheaply. It is generally assumed that 70 percent of all lands that were grabbed, estimated by Oxfam to be 227 million hectares (ha) in 2011, are in Africa (Oxfam 2011). (Inter)national public policy-making aimed at addressing some of the serious concerns in the current land rush (expulsion of peasants from their land, corrupt land deals, and so on) has been underway and is politically contested.

These assumptions have been increasingly challenged. Visser and Spoor (2011) identify the former Soviet Eurasia, and Borras and Franco (2011) Southeast Asia, as important regional sites of land grabs too. Levien (2011) raises the issue of domestic land grabs, Sommerville (2011) on the perspective from the Global North. Other challenges have been raised by Amanor (2012) on the role of Transnational Corporations (TNCs) and global commodity chains, Hall (2011) on pre-existing crop boom-bust cycles, McMichael (2012) on the location of land grabs in the restructuring of the global food regime, Woodhouse and Ganho (2011) and Kay and Franco (2012) on the water dimension of land grabbing, Hofman and Ho (2012) on an alternative view of Chinese land grabbers, Alden Wily (2012) on historical continuities, Li (2011) on centering labour in the debate, De Schutter (2011) on the right-to-food dimension, and Fairhead, Leach and Scoones (2012) on 'green grabbing' – land grabbing in the name of the environment. Peluso and Lund (2011) offer fresh analytical insights on new frontiers of land control more generally, while Cotula (2012) offers a useful comprehensive overview on key issues in current land grabs. In short, the current trajectory of scholarly thinking is to broaden the parameter of empirical and theoretical inquiry into land grabs. The emerging common thread is that there is a need to embed land grabs within our analysis of contemporary global capitalist development (Harvey 2003), in the specific context of the convergence of multiple crises: food, energy, climate change and finance capital (McMichael 2012, Sassen 2010).

We build on this emerging body of literature, and offer our contribution by bringing in fresh perspectives from Latin America and the Caribbean. We did a close reading of the FAO studies in 17 countries in the region, and came up with four main arguments. First, there are a number of key specific contexts that underpin current land grabs, and food production is just one of these. Others are energy/fuel (biofuels), other climate change mitigation strategies, and recent demands for resources destined to newly-emerged global hubs of capital. These are not unrelated contexts; a common thread among these is the dynamics of global capital reconfiguration and emerging accumulation imperatives and strategies. Financialization of these contextual processes is an emerging feature common to food, energy/fuel and other resource extraction. Hence, looking at food politics embedded within the dynamics of global capitalism (the Friedmann and McMichael 1989 formulation of 'food regime') is key, and logically links the former to other mechanisms of land grabbing. It broadens our unit of analysis, enabling us to go beyond analyzing the food security-induced land grabbing initiatives by China, India, South Korea and the Gulf States. In this context, we offer the concept of 'flex crops' – crops that have multiple and flexible uses – as a useful way of illustrating our point, i.e. in understanding the interconnection between and implications of the restructuring of agrofood, feed, and fuel complexes.

Second, land grabbing occurs in this region to an extent wider than previously assumed. When seen from a relational international perspective, it changes the relative shares of other regions, especially Africa, in global land grabbing. Third, refocusing our analysis on the key mechanisms of land grabbing (food, energy/fuel, and rising demands from newer hubs of global capital for natural resources), we see that the primary capital driving the land grab processes in this particular region seems to come from within the region itself: domestic capital and what others (e.g. Gomez 2011) call '(Trans-)Latina Companies'; in this paper, we will use the term '(Trans-)Latin American Companies' (TLCs). The actors in land grabbing under the spotlight in Africa – China, India, South Korea, Gulf

States – have marginal presence in the region in terms of agriculture FDI, at least for now. The hallmarks of land grabbing in the region are the 'intra-regional' character of the phenomenon that is comparatively more diverse, dynamic and larger in scale than the one observed by Hall (2012) in Africa with South African land grabbers, and the rise of flex crops. This provides us with some hints on the reconfiguration of global capital in general and the agrofood-feed-fuel complex more particularly.

Fourth, land grabbing has resulted in the expulsion of people from their lands – but not everywhere and not always. What we see in this region are pockets of expulsions, but the scale of them is relatively low. Incorporation of peasants, rural labourers and indigenous peoples into the emerging commercial farms, commodity chain, and industrial plantations are more common. Many of the land grab sites are in the agricultural frontiers and (former) cattle ranches that are relatively sparsely populated. In short, while land grabbing occurs in Latin America and the Caribbean within the same logic and processes of global capitalist development that underpinned land grabs elsewhere, it has taken different forms and character – and trajectories – in this region.

The limits of the original FAO study

Our paper is based on the Food and Agriculture Organization (FAO) research on the state and trends in land grabbing in 17 countries in Latin America and the Caribbean (Baumeister 2011, Carrera and Carrera Campos 2011, Donoso 2011a, b, Echenique 2011, Galeano 2011, Lavandier 2011, Martinez 2011, Murmis and Murmis 2011, Piñeiro 2011, Remy and de los Rios 2011, Robles 2011, Salinas 2011, Urioste 2011, Wilkinson *et al.* 2011, A. Williams 2011, P. Williams 2011; and the summary paper, Gómez 2011). The studies were concluded in late 2011. The common analytical framework of the studies is both wide and narrow. On the one hand it is wide because it looks into the broad processes of rural land and capital concentration in the context of neoliberal globalization, along the lines of argument by Amanor (2012). On the other hand, it is narrow because it looks into the phenomenon of land grabbing strictly based on three specific dimensions, namely: (i) significant extent of recent large-scale land acquisitions; (ii) involvement of foreign governments in these land deals; and (iii) negative impact of such renewed land investments on food security of the recipient country. This framing is firmly located within the initial salvo of reports and studies on land grabbing that we explained above. It is due to the wide framing (in the context mentioned above) that most of the studies were able to uncover, gather and assemble significant empirical material related to recent land dynamics in the region. At the same time, however, authors of the 17 studies have to focus their analyses and conclusions based on the definition and dimension of land grabbing cited above, arriving at a conclusion that land grabbing exists *only* in two countries in the region: Argentina and Brazil (in contrast, our re-interpretation of the data using broader analytical lenses led us to conclude that land grabbing occurs in at least 10 countries). Tables 1 and 2 provide a summary of the key findings of the 17 country studies based on the original FAO framework.

From Table 1, we highlight the following: First, across Latin America and the Caribbean there has been a significant increase in (foreign) investments in land and

Table 1. Land investments, land grabbing, and food security in selected countries.

Presence of recent large (foreign) investments in land			Presence of foreign 'land grabbing'		Negative impact on food security of investment recipient country	
High	Medium	Low To None	Yes	No	Yes	No
Argentina	Costa Rica	Trinidad & Tobago	Argentina	Bolivia		Argentina
Bolivia	Guatemala		Brazil	Chile		Bolivia
Brazil	Panama			Colombia		Brazil
Chile				Ecuador		Chile
Colombia				Paraguay		Colombia
Ecuador				Peru		Ecuador
Paraguay				Uruguay		Paraguay
Peru				Mexico		Peru
Uruguay				Costa Rica		Uruguay
Mexico				Guatemala		Mexico
Nicaragua				Nicaragua		Costa Rica
Dominican Republic				Panama		Guatemala
Guyana				Dominican Republic		Nicaragua
				Guyana		Panama
				Trinidad & Tobago		Dominican Republic
						Guyana
						Trinidad & Tobago

Source: 17 FAO country studies (see Annex 1) plus the summary paper of Gómez (2011). The categorization (high, medium low) is a matter of degree. It is a relative concept in which 300 hectares (ha) of land in Mendoza, Argentina may be equivalent in value with 20,000 ha in Colombia. Categorization is based on individual author's estimate of the scale of land acquisitions and/or scale of capital involved.

Table 2. Land and capital concentration, by country and sector.

Country	Sectors where recent significant (land & capital) concentration has occurred
Argentina	Soya, wheat, livestock, sugarcane, tobacco, fruit, conservation
Bolivia	Soya, livestock, forestry
Brazil	Soya, sugarcane, poultry, livestock, fruit, forestry
Chile	Fruit, dairy, wine, seeds, poultry, conservation
Colombia	Oil palm, sugar beets, sugarcane, soya, rice, corn, forestry
Ecuador	Banana, sugarcane, oil palm, forestry
Paraguay	Soya, corn, wheat, livestock
Peru	Fruits, vegetables, sugarcane, oil palm
Uruguay	Soya, dairy, wheat, rice, livestock, forestry
Mexico	Corn value chain, sugarcane, fruits, flowers, coffee, barley, tequila
Costa Rica	Banana, pineapple, oil palm
Guatemala	Sugarcane, oil palm, forestry
Nicaragua	Livestock, rice, oil palm, sugarcane, citrus, tourism, forestry
Panama	Banana, coffee, rice, oil palm
Dominican Republic	Sugarcane, banana, fruits, vegetables
Guyana	Sugarcane, livestock, rice, pineapple, forestry
Trinidad & Tobago	Sugarcane, cacao, fruits

Source: 17 FAO country studies (see Annex 1), plus the summary paper (Gómez 2011). Table 2 is reformatted from Gómez (2011).

agriculture during the past decade. The level of these investments is high for most of the 17 countries, with only about three in the medium level (Costa Rica, Guatemala and Panama), while only one country is in the 'low-to-none' category (Trinidad and Tobago). Indeed despite the unevenness between and within countries, the recent vibrant investment atmosphere in land and agriculture is a region-wide phenomenon. Second, using the FAO definition of 'land grab' only two countries would qualify: Argentina and Brazil. Third, in all 17 countries studied, there is no single country case where food security was undermined by the surge of land and agricultural investments – at least not at a national scale, and not for the time being. Meanwhile, Table 2 shows that concentration of land and/or capital occurs, and it occurs in various food and non-food sectors, although there is a remarkable surge in some specific sectors, including soya, oil palm, sugarcane and corn, reforestation, and carbon offset projects, alongside land acquisitions related to mega conservation projects. Looking at a global picture in comparative glances, the dataset in Table 2 supports our argument against a too food-centered/food crisis-centred analysis of land grabs.

Overall, the definition used by FAO, i.e. large-scale land acquisitions involving foreign governments and undermining food security of a country, necessarily excluded from one's analysis many of the land grabs that occurred in the past and are occurring at present. But there are other possible explanations as to why land grabs do not happen, or cannot be detected easily in some countries in the region (in some instances even with broader lenses). For instance, Costa Rica had massive foreign investments in land pre-2000, in multiple sectors, including in foreign-funded conservation and carbon sequestration projects, tourism, beach homes, rice, sugar, cattle, flowers, as well as winter fruits and vegetables (see e.g. Edelman 1999). Costa Rica was one of the first countries to establish a market in carbon sequestration credits (long before REDD+). By 2000 it may have been 'saturated', i.e. no more land to acquire in a small country. Panama's economy is so heavily reliant on finance and canal-related services that agricultural investment there may not be as attractive as elsewhere (though it certainly exists). Trinidad and Tobago, similarly to Panama, has a mainly urban (and petroleum) economy. Unfortunately Honduras was not included in the 17 countries studied. Recent and ongoing agrarian conflicts in the Bajo Aguán (related to oil palm development, mostly Honduran capital) and Tela (tourist development, Honduran and foreign capital) have been significant.[1]

The challenge of defining land grabs

The three dimensions in the definition used by the FAO in the study are among the most controversial aspects of the current land rush. Defining land grab narrowly like this is bound to miss significant aspects of the scope and extent of contemporary land grabbing and possible trajectories of agrarian change. This is the reason why it is important to go beyond the food-centred/food crisis-centred analysis of land grabs. But defining land grab too broadly will miss what is distinct in the particular wave of contemporary global land grabbing. This is the reason why we exclude from our land grab definition small- and medium-scale everyday

[1]We gratefully acknowledge the suggestion made by Marc Edelman for the key points contained in this paragraph.

forms of *dispossession by differentiation* (see Araghi 2009), be it a Leninist or Chayanovian variant.

In order to avoid the problems cited above we offer the idea of three *key interlinked defining features of contemporary land grabbing*. First, a fundamental starting point is to clarify that land grabbing is essentially 'control grabbing': grabbing the power to control land and other associated resources such as water in order to derive benefit from such control of resources. It partly builds on Ribot and Peluso's (2003) 'theory of access'. Land grabbing in this context is often linked to a shift in the meaning or use of land and associated resources[2] as the new uses are largely determined by the accumulation imperatives of capital that now has the control over a key factor of production, land. 'Extraction' or 'alienation' of resources for external purposes (national or international) is often the character taken by land grabs (Wolford 2010). Control grabbing is inherently relational and political; it involves political power relations. Control grabbing manifests in a number of ways, including, 'land grabs' (capture of vast tracts of lands'), '(virtual) water grabs' (capture of water resources – see Woodhouse 2012, Kay and Franco (2012)), and 'green grabs' (resource grabs in the name of the environment – see Fairhead, Leach and Scoones 2012). This perspective addresses the problem of 'too land-centred perspective' in the current land grab thinking. Seen from the perspective of control grabbing, analytically and empirically land grab does not always require expulsion of peasants from their lands; it does not always result in dispossession.

Second, study of current land grabbing requires consideration of scale of land grabs. But scale of land grabs should not only be about the *scale of land acquisitions*, often within the dominant view that defines 'large-scale land acquisitions' as those that pass the 1000-ha benchmark (see e.g. Oxfam 2011, ILC 2011, World Bank 2010). For us, land grabbing entails large-scale transactions in two broadly distinct but interlinked dimensions: *scale of land acquisitions* and/or *scale of capital* involved. This framework necessarily considers various forms of acquiring control: purchase, lease, contract farming, forest conservation, and so on. In other words, taking the scale of capital as the unit of analysis necessarily includes land as central in the operation of capital, while a 'too land-centred' view (scale of land acquisitions only) on land grabs tends to miss or de-emphasize in its analysis the underlying broader logic and operation of capital. Our framework captures the diversity of biophysical requirements in capital accumulation dynamics: 300 ha of high-value vineyard, 500 ha of rare metal mining concession, 100,000 ha of land for industrial tree plantation, and 500,000 ha of grazing land for livestock may in fact have comparable scale in capital (and, for that matter, scale of labour involved) despite the huge discrepancies in physical land requirement among them.[3] Our framework brings capital back into our unit of

[2]It is, however, not always the case because land investors may simply grab control over an already existing enterprise that has a character that fit our profile of contemporary land deals.
[3]This also brings under a better light the owners of capital who are not directly engaged in land acquisitions, but are nevertheless involved in grabbing control and shifting land/resource use towards extractive nature, e.g. supermarket chains that favour capturing value chain (see e.g. Amanor 2012). It also throws light onto contested and grey issues such as the indirect but far-reaching implications of some distant policies, e.g. the European Union biofuels policy that has triggered speculative land investments in the oil palm sector worldwide, and the notion of 'indirect land use change' where crop use change in Europe (e.g. rapeseed in Germany converted to biodiesel) triggers land use change elsewhere to produce substitute

analysis, casting an interrogating gaze at what has emerged to be a flurry of 'land measurement-oriented accounting' of land grabs: Oxfam's 227 million ha (Oxfam 2011), ILC's 80 million ha (ILC 2011), World Bank's 45 million ha (World Bank 2010); Africa as hosting 70 percent of all land grabs, and indeed in the FAO conclusion that land grabs occur in only two countries in Latin America and the Caribbean.

Third, and finally, the first two features are more or less the same defining features of land grabs that happened worldwide, historically. What is distinct in the current land grabs is that these occur primarily because, and within the dynamics, of capital accumulation strategies largely *in response to* the convergence of multiple crises: food, energy/fuel, climate change, financial crisis (where finance capital started to look for new and safer investment opportunities) (McMichael 2012), as well as the emerging needs for resources by newer hubs of global capital, especially BRICS and some powerful middle income countries (MICs). The key mechanisms of land grabbing arise from this: food security, energy/fuel security, climate change mitigation strategies, and demands for natural resources by new centres of capital. One offshoot of this recent development is the emergence of 'flex crops': crops that have multiple uses (food, feed, fuel, industrial material) that can be easily and flexibly inter-changed: soya (feed, food, biodiesel), sugarcane (food, ethanol), oil palm (food, biodiesel, commercial/industrial uses), corn (food, feed, ethanol). It has resolved one difficult challenge in agriculture: diversified product portfolio to avoid devastating price shocks, but not easy to do and achieve because of the cost it entails. With the emergence of relevant markets (or speculation of such) and the development and availability of technology (e.g. flexible mills) that enables maximization of multiple and flexible uses of these crops, diversification has been achieved – within a single crop sector. When sugarcane prices are high, sell sugarcane; when ethanol prices are high, sell ethanol. When actual market for biodiesel is not there yet, sell palm oil for cooking oil, while waiting (or speculating) for a more lucrative biodiesel market to emerge (a feature not present in *Jatropha*). The emergence of flex crops is a logical outcome of the convergence of multiple crises. Hence, in a single crop sector we find multiple contexts of land grabs: food, energy/fuel, climate change mitigation strategies. It is these broader interlinked contexts that largely differentiate current land grabs from the ones that existed before.

In short, contemporary land grabbing is the capturing of control of relatively vast tracts of land and other natural resources through a variety of mechanisms and forms that involve large-scale capital that often shifts resource use orientation into extractive character, whether for international or domestic purposes, as capital's response to the convergence of food, energy and financial crises, climate change mitigation imperatives, and demands for resources from newer hubs of global capital. This work-in-progress definition rests on the three defining features discussed above that provide our analytical lenses, enabling us to navigate between a narrow parameter, and a perspective that is too broad that includes everyday forms of dispossession by differentiation. The framework offered here aims at de-focusing from a too food-centred/food crisis-centred narrative, a too

crops – in this case, oil palm (Franco *et al.* 2010). But we do not include these in our definition of land grabbing despite the fact that these two examples can arguably qualify as 'control grabbing'; it is simply relatively too 'indirect' for our purposes.

land-centred unit of analysis, too much weight given to transnational actors in land grabbing centred on China, South Korea and the Gulf States, on questions about physical acquisition of land through purchase and long term lease, and Africa being the main centre of gravity. Our analytical lens provides us with a better analytical handle in understanding land grab dynamics in Latin America and the Caribbean, a task we now turn to in the next section.

Key contexts and extent of land grabbing

Food production, energy/fuel production, other climate change mitigation strategies, and demands for resources from new hubs of global capital are key contextual processes of resource control grabbing in the region.

The expansion of soya, sugarcane and oil palm, all flex crops, in the context of recent changes in the global food-energy regime has led to significant expansion of commercial farms and plantations for these crops in Argentina, Bolivia, Brazil, Colombia, Ecuador, Paraguay, Peru, Uruguay and Guatemala. The phenomenal increase in demand for meat and other animal and dairy products (largely due to the dramatic demand from China), as well as for fruits and wines, have in turn led to the expansion of land for livestock and fruits and vineyards in Argentina, Bolivia, Chile, Uruguay and Nicaragua (Table 3). Table 4 shows the trend in Central America and South America in terms of the rising importance and expansion of flex crops (soya, sugarcane and oil palm) from 1961 to 2009 (see also Graziano da Silva *et al.* 2010, Hecht 2005).

The rise of flex crops has been significant (see Table 4). Dramatic increases in soya production are not only found in Brazil. In Argentina, the area planted to soya jumped from 3.77 million ha in 1980/81 to 48 million ha in 2007/08 (Murmis and

Table 3. Land grabbing by country, by sector.

Country	Flex crops and other food sectors	Non-Food
Argentina	Soya, wheat, livestock, sugarcane, fruit	Tobacco, conservation
Bolivia	Soya, livestock	Forestry
Brazil	Soya, sugarcane, poultry, livestock, fruit	Forestry
Chile	Fruit, dairy, wine, seeds, poultry	Conservation, Forestry
Colombia	Oil palm, sugar beets, sugarcane, soya, rice, corn	Forestry
Ecuador	Banana, sugarcane, oil palm	Forestry, minerals
Paraguay	Soya, corn, wheat, livestock	
Peru	Fruits, vegetables, sugarcane, oil palm	Minerals
Uruguay	Soya, dairy, wheat, rice, livestock	Forestry
Mexico	Corn value chain, sugarcane, fruits, coffee	Flowers, barley, tequila, conservation
Costa Rica	Banana, pineapple, oil palm	
Guatemala	Sugarcane, oil palm	Forestry
Nicaragua	Livestock, rice, oil palm, sugarcane, citrus	Tourism, forestry
Panama	Banana, coffee, rice, oil palm	
Dominican Republic	Sugarcane, banana, fruits, vegetables	
Guyana	Sugarcane, livestock, rice, pineapple	Forestry
Trinidad & Tobago	Sugarcane, cacao, fruits	

Source: Gómez (2011, 56).

Table 4. Flex crops in South America and Central America, Area harvested (ha), 1961–2009.

	Soya	Sugarcane	Oil Palm
South America			
1961	259,534	2,124,775	38,700
1970	1,443,590	2,485,528	57,081
1980	11,467,985	3,623,922	83,088
1990	17,725,284	5,290,929	210,906
2000	24,156,087	5,995,162	341,709
2009	42,792,479	9,878,744	448,313
Central America			
1961	9943	500,207	22,910
1970	111,844	761,258	21,986
1980	155,287	882,750	35,447
1990	309,996	875,047	57,197
2000	85,992	1,071,684	109,430
2009	83,444	1,231,025	239,204

Source: FAOSTAT (2012), constructed by the authors.

Murmis 2011, 28). In the region of Santa Cruz in Bolivia, the area planted to soya increased from 172,334 ha in 1990 to 1,003,690 ha in 2009 (Urioste 2011, 55). In this context and as mentioned earlier it is difficult, if not impossible, to make clear-cut differentiations about the actual use of these crops. The actual, potential, or speculated markets for any or all of the flexible uses of these crops are likely to have rendered investments in them safer by way of somehow achieving diversification within a single crop sector. This is a likely factor that partly explains the emerging popularity of these crops in several countries in the region. Yet, overall, other food sectors remain quite important, most especially livestock. In South America, there were 347 million cattle in 2009, up from 293 million in 1995; while in Central America there were 45.6 million in 2009, up from 41.5 million in 1995. Extrapolating these figures with a minimum of one ha per cattle, then we see the extent of land use this sector commands. They require far more land than any of the flex crops combined (see Wilkinson and Herrera 2010, and Novo *et al.* 2010 in the case of Brazil), although it is to be noted that we have witnessed in recent years significant technological advancement in the livestock and dairy sector where capital has started to move towards intensive, not extensive, production systems, as in the cases of Chile (Echenique 2011), Argentina and Brazil toward the direction of US-style feedlot cattle production (but are still mainly grass-based). In this transition to diversified livestock, cattle, in any case, are a kind of flex crop in that they can be fed with grass or any number of crops or crop and biofuel by-products.[4]

In addition, the share of non-food land grabs has been significant (see Table 3). The two most important sectors in terms of land use are fast-growing tree plantations (e.g. eucalyptus) and conservation. The expansion of fast-growing tree plantations (for pulp and other industrial and commercial demands within and outside the region) is seen in Bolivia, Brazil, Chile, Colombia, Ecuador, Uruguay, Guatemala, and Guyana. The Caribbean, South America and Central America are the three regions that registered the highest annual increase (in percentage terms) in tree plantations in the world, way beyond the world average (see Table 5). And while

[4]We thank Marc Edelman for pointing out the idea of 'cattle' as a kind of 'flex crop'.

Table 5. Regional plantation area and increase in (tree) plantation area during 2000 to 2010.

	Africa	Asia	Oceania	Europe	Caribbean	Central America	North America	South America	World
Plantation Area, 2010 (× 1000 ha)	15409	115783	4101	69318	547	584	37529	13821	264084
Annual Increase (× 1000 ha)	245	2855	78	401	15	16	809	376	4925
Annual Increase (%)	1.75	2.87	2.12	0.60	3.34	3.14	2.46	3.23	2.09

Source: UNEP (forthcoming, 2012).

large-scale conservation occurs mainly in two countries, Argentina and Chile, the scale of individual cases (e.g. United Colours of Benetton deal in Argentina with close to a million ha) and aggregate total are significant. Moreover, not included in this paper and in the 17 FAO studies are the data on and analysis of emerging trend of placing forests under the carbon-offset programs, i.e. REDD+. Such policies have important impacts in terms of land control and on the livelihoods of local communities (see e.g. Osborne 2011 in the case of Mexico; see also Corbera and Schroeder 2011). Large-scale conservation, fast-growing tree plantations, policies such as REDD+ among others, are being referred to collectively as 'green grabs' – land grabs in the name of the environment. Increasingly, this kind of land grabbing (and its impact on existing agrarian social relations) is getting integrated in the critical land grab academic literature, with Fairhead, Leach and Scoones (2012) as a landmark collection. Finally, rising regional and international demands for minerals and fossil fuels have led to large-scale mining concessions in Peru and Ecuador (Remy and de los Rios 2011, Martinez 2011, Bebbington *et al.* 2009).

That land grabbing and the accompanying land and capital (re)concentration occur in two broad mega-sectors, namely, the flex crop complex/food sectors and the broad non-food sector contradicts the dominant narrative that land grabs occur because of the food crisis of 2007–2008 and that such land deals are oriented to food production for export to food insecure countries. There is certainly some truth to this latter storyline, especially when we see some Chinese and Gulf State governments and/or companies negotiating with various governments in different parts of the world for possible land leases or contracts for food exports. But to focus on food in isolation to other factors is problematical. In general, what we witness in Latin America and the Caribbean are the changing character and demands from the newly emerged flex crop complex, to the continuing importance of livestock, to the sharp increase of demands for natural resources by newly emerged centres of capital, and responses to policies linked to climate change mitigation strategies (conservation projects, including REDD+).

Meanwhile, the extent of recent large investments in agriculture – not necessarily involving land and other resources, such as those focusing on technology and lending – is quite wide in terms of geographic scope. Empirical data from the FAO country studies show that recent large investments in agriculture are present in 13 countries. Only Trinidad and Tobago qualifies in the category of 'low-to-none', and Guatemala, Costa Rica and Panama in the 'medium category'; the rest are in the category 'high'. What Table 6 (first three columns from the left) shows is that indeed there is a renewed interest in agricultural investment across the region.

Columns 4, 5 and 6 of Table 6 show land grabbing sites – based on the aggregated empirical data from the 17 FAO country studies. It suggests that the phenomenon is wider than previously assumed: ten countries are currently experiencing a relatively wide extent of land grabbing and another three a medium extent. The ten countries where significant land grabbing is observable are: Argentina, Bolivia, Brazil, Chile, Colombia, Ecuador, Guatemala, Paraguay, Peru and Uruguay (i.e. all in South America, except for Guatemala which is in Central America), while the medium-level countries are Panama, Mexico and Nicaragua. Detailed empirical data for each country are not presented in this paper, but these are available in the 17 country studies, available in Spanish only.

Many of the recent land deals in the region are large-scale, such as those seen in Brazil, Argentina, Paraguay, Uruguay and Chile, transacted rather rapidly within a

Table 6. Presence of land grabbing in selected Latin American countries and the Caribbean*.

Presence of recent large investments in land			Presence of land grabbing (domestic & foreign capital)			Country with major land investors into other countries in the region
High	Medium	Low to None	High	Medium	Low To None	
Argentina Bolivia Brazil Chile Colombia Ecuador Paraguay Peru Uruguay Mexico Nicaragua Dominican Republic Guyana	Costa Rica Guatemala Panama	Trinidad & Tobago	Argentina Bolivia Brazil Chile Colombia Ecuador Guatemala Paraguay Peru Uruguay	Panama Mexico Nicaragua	Costa Rica Dominican Republic Guyana Trinidad & Tobago	Argentina Brazil Chile Colombia Panama Mexico Costa Rica

*Based on close reading of the 17 FAO country studies and the summary paper by Gómez (2011) – using as analytical lenses the three defining features of land grabbing discussed in the current paper.

Table 7. Bolivia – Advertisements of land for buying and selling, classified by size and year (in ha).

Year	Less than 4000 ha	4001 to 8000 ha	8001 to 12000 ha	12001 to 16000	16001 and more	Total
2004	191	49	17	9	26	292
2005	205	97	15	7	10	334
2006	164	32	7	4	2	209
2007	270	56	9	2	15	353
2008	290	83	16	1	9	399
2009	129	26	5	0	0	160
2010	149	45	10	4	4	213

Source: Urioste (2011, 62).

relatively short period of time. The case of Bolivia, concentrated in Santa Cruz, is insightful where the bulk of land *advertised* in the period of 2004–2010 for buying and selling had a size range between 4000 ha and 8000 ha (see Table 7). But in some settings, even more important than the *scale of land acquisitions* is the *scale of capital* involved. Most of the 17 studies have shown capital-intensive ventures in the usually land-extensive investments: e.g. capital-intensive sugarcane-ethanol flexible mills linked to extensive sugarcane farms in Brazil (Wilkinson *et al.* 2011), soya processing plants linked to land-extensive plantations, livestock processing infrastructures linked to land-extensive ranches, and so on. There are also highly capital-intensive ventures that do not require extensive lands, e.g. fruits and wine, cut flowers, vegetables, new emerging intensive livestock and dairy farms, and so on, where capital's control of the value chain becomes key, as in the case of Mexico (Robles 2011).

Furthermore, studies show that investors do not always look for marginal, empty lands, as the official narrative justifying land grabs claim (see e.g. World Bank 2010, Deininger 2011). Instead, they tend to go where there are existing productive lands with water supply and/or existing approximate road networks. Fruits and vineyards in Chile are concentrated in a few suitable geographic spots in the country (Echenique and Romero 2009, Kay 2002), as in Argentina (in the San Juan and Mendoza regions). Significant expansion of sugarcane plantations in São Paulo occurs partly at the expense of adjacent or nearby farms, including some land reform settlements (Wilkinson and Herrera 2010, Novo *et al.* 2010).[5] There is also an emergence of plantation corridors along national borders that usually have existing infrastructures: in Paraguay along the borders with Brazil and Argentina. This has in fact prompted national governments in the region to pass laws and policies prohibiting foreign investments within 50 kilometers of the borders, worried about the potential geo-political and national security implications of such foreign investments. It seems, however, that several of the non food-related land deals, i.e. fast-growing tree plantations, mining concessions, and large-scale conservation projects are carried out in places that more or less fit the profile of 'marginal lands' (see Oliveira 2011), although this cannot be said for REDD+ areas (see e.g. Osborne 2011 in the case of Mexico). But whether these marginal lands are automatically 'available' – which we contend is more of a political question than an administrative/technical one – is another issue, especially in the context where existing communities are found in these lands (see Cardenas 2012 on Colombia, Balleti 2012 on Brazil).

State and capital in intra-regional land grabbing

Attention has been focused on foreign private companies and foreign governments and their role in the global land rush. The role played by the central state is often (inadvertently) ignored or de-emphasized. In Latin America and the Caribbean, the role of the central state in either promoting entry of (foreign) land investments, or promoting national companies to invest abroad has been critical. Brazil is a good example. All States are engaged in systematic policy and administrative initiatives around the notion of 'available marginal lands',[6] and its role in facilitating land investments in these spaces include some, or all, of the following: (i) invention/justification, (ii) definition, reclassification, quantification, (iii) identification, (iv) acquisition/appropriation and (v) re-allocation/disposition of these lands to transform such scarce resources (most of which are within the political-legal control of central states) into productive factors of production for renewed large-scale land-based investments. Here, technical re-mapping and land use reclassification – both technical processes that are also very political – are important instruments employed by the state. A good example of state active involvement in ways explained above is the Colombian state and Afro-Colombian territory (Cardenas 2012). In some cases, coercion and violence have accompanied the state's effort at territorialisation, enforcement of its sovereignty and authority as well as its ardent support for private

[5]This is also partly based on Borras's fieldwork in São Paulo in 2008.
[6]The 'marginal land' discourse has interesting parallels in nineteenth-century Argentine discourses about unoccupied lands: the famous 'guerra del desierto' genocide against the native peoples there was not carried out in a desert; 'desierto' referred to empty lands, some of which were humid tropical forests. We thank Marc Edelman for pointing this out.

capital accumulation, as in the case of present Colombia (Grajales 2011) and in many countries in Latin America and the Caribbean during recent decades (Kay 2001). It is easy to surmise in the case of Latin America and the Caribbean that this practice of 'state simplification' (Scott 1998) is quite common, especially since many land deals involved opening new land frontiers. The cases of massive fast-growing tree plantations, livestock expansion and large-scale conservation involve this key role of the state, as well as in facilitating and/or brokering REDD+ contracts.

Stepping back, and looking at the bigger picture, there emerge three broadly distinct but interlinked areas of state actions that are relevant in understanding contemporary land grabs, namely: (i) Scott's 'state simplification process', (ii) the assertion of sovereignty and authority over territory, (iii) coercion through police and (para)military forces to enforce compliance, extend territorialisation, and broker for private capital accumulation. First, in order to administer and govern, states engage in a simplification process to render complex social processes legible to the state. The creation of cadastres, land records and titles are attempts at simplifying land-based social relations that are otherwise too complex for state administration (Scott 1998). This requires the state's official powers of recording land relations and (re)classifying lands (see Oliveira 2011 in the case of the Brazilian Cerrado and Amazon). This in turn brings us back to the notion of available marginal, empty lands. The trends in state discourse around land grabs are: if the land is not formally privatized, then it is state-owned; if an official census did not show significant formal settlements then these are empty lands; if the same official census did not show significant farm production activities, then these are unused lands. Currently, many of the land investments in Latin America and the Caribbean are in frontier regions, encroaching into indigenous peoples' territories, taking in grasslands such as the case of the Cerrado in Brazil (Oliveira 2011), or sparsely populated landscapes such as the Patagonia.

Second, beyond the economic benefits of land investment, land deals are also viewed as an essential building block for state-building processes where sovereignty and authority are extended to previously 'non-state spaces' (Scott 1998). Again, a good example in the region is the Afro-Colombian case (Ballve 2011) and the Amazon frontier in Brazil (Oliveira 2011, Balletti 2012). Third, coercion and violence, usually with the use of police and (para)military forces to enforce compliance to state simplification projects and the broader state-building process, have accompanied some of the land deals in the region, but most especially in Colombia (Grajales 2011) and Honduras.

This three-fold state role in land deals is carried out to a large extent on behalf of the dominant classes of capital (foreign or domestic), although it is always accompanied by the other task of the state to maintain a minimum level of political legitimacy. This makes capital accumulation and political legitimation inherently interlinked and contradictory, tension-filled, uneven and contested, across geographic spaces and over time (Fox 1993, Harvey 1998). This is the case across Latin America and the Caribbean.

The critical role of the central state in the current land rush makes the issue of land grabbing a murky one: taking what is yours is not considered 'grabbing'. It will also render efforts at global governance on land grabbing (such as the FAO's Voluntary Guidelines, or the World Bank's 'principles of responsible agriculture investments' – see Borras and Franco 2010) quite complicated and difficult to frame, implement and enforce. The centrality of the state and domestic elites (landlords and

Table 8. (Trans-)Latin American land investors in Latin America and the Caribbean (partial).

Country of origin of regional investors	Countries in which they are active
Argentina	Brazil, Uruguay, Paraguay
Brazil	Bolivia, Colombia, Paraguay, Uruguay, Chile
Chile	Argentina, Brazil, Uruguay, Colombia, Ecuador, Peru
Colombia	Bolivia, Peru
Panama	Brazil, Argentina, Paraguay
Mexico	Nicaragua, Guatemala
Costa Rica	Nicaragua, Guatemala

Source: 17 FAO studies, plus the summary paper (Gómez 2011).

capitalists) in the land grab process, and their embeddedness in a situation where there is a nationalist sentiment against land grabbing (that is often understood in terms of foreignization or denationalization of land – see Zoomers 2010) makes a class analytical lens even more relevant in terms of understanding dynamics and trajectories of agrarian change and 'state-capital' links (Bernstein 2010). Geopolitics and the nationalist dimension in the emerging politics of land grabbing in the region should always be examined within a complementary, if not fundamental, class analytical framework.

It is almost always the state-capital links that drive land grabs. This perspective offers a better view of the constellation of land grabbers, or classes of capital (in Bernstein's terms), that are in alliance with the state. The hallmark of land grabbing in Latin America and the Caribbean is the strong *intra-regional character* of the phenomenon – quite developed here compared to the South Africa-centred intra-regional process in Africa observed by Hall (2012), although probably the Southeast Asian condition scanned initially by Borras and Franco (2011) may be be somewhat closer to the Latin America and the Caribbean condition. This character differentiates the region from most other regions of the world, and, more generally, casts a different light on global land grabbing as we currently understand it.

In Latin America and the Caribbean today, there are seven countries that are host to land grabbing and are at the same time countries where land grabbers originate: *Land grabbed land grabbers*, to coin a crude but catchy phrase. This is in a scale not seen in any other region in the world, although, again, probably Southeast Asia may exhibit some aspects of this condition but to a comparatively limited extent (Borras and Franco 2011). The role of (Trans-)Latin American Companies (TLCs) is central to this set-up. The seven countries are: Argentina, Brazil, Chile, Colombia, Panama, Mexico and Costa Rica (see last column in Table 6). Table 8 offers a partial canvass of where the operations of (Trans-)Latina Companies are located.

The strong intra-regional character of land grabbing in Latin America and the Caribbean shows the key role played by regional investors. Following the discussion above, one of the most important changes in the global food-energy system and the phase of capitalist development today is the rise of powerful regional economic players seen in the emergence of the BRICS, combined with the rise in significance of several MICs in these regions. In Latin America and the Caribbean, this manifests in a situation where land investors are primarily, though not solely, (Trans-)Latin

Table 9. Area of land and plantations by the Chilean companies CELCO and MININCO.

Countries	Total area of land (ha)	Total area of forest plantations (ha)
Area of land and plantations by the Chilean company CELCO		
Chile	1,099,846	736,000
Argentina	257,722	129,000
Brazil	126,616	67,000
Uruguay	126,786	63,000
Total	1,610,970	995,000

Country	Total area of property (ha)	Area of forest plantations (ha)	Area for 'plantar' (ha)	Area for other uses (ha)
Area of land and plantations by the Chilean company MININCO				
Chile	716,590	498,000	33,326	185,264
Argentina	94,283	65,164	2724	26,395
Brazil	213,592	94,806	31,160	87,626
Total	1,024,465	657,970	67,210	299,289

Source: Gómez (2011, 33).

American Corporations (TLCs). The latter are of two types: a company with single origin in terms of nationality (Latin American), and an alliance of two or more nationals (Trans-Latin American). Either type is likely to have some tie-up with international finance, especially in the light of laws passed in Southern America aimed at curbing the foreignization of land that seems to have driven foreign capital to forge alliances with domestic capital (via local subsidiaries and other schemes) and to explore non-purchase land acquisition schemes to circumvent such legal limitations.[7] This is shown in the increasing share of mixed foreign-domestic capital acquisition of land in Argentina largely through renting ('pools'). It is an illustration of an earlier point we have raised: *control grabbing*. Regional investments can be in the form of natural or juridical persons. For example, Brazilians (with varying support and encouragement from Brazilian government and companies) buy up or lease lands in Paraguay to produce soya or to engage in livestock raising, creating recurring tensions between the locals and the Brazilians; or, many Brazilians have ended up owning a significant quantity of lands in Bolivia, a trend that started much earlier, but gained momentum in recent years, and the emerging production orientation has been linked to the changed global context.

In addition, intra-regional capital is not only active in the flex crops and livestock sectors, it is also active in non-food sectors. The Chilean company CELCO's industrial tree plantation has 26 percent of its operations outside Chile (in Argentina, Brazil and Uruguay, for a total of 259,000 ha), while the Chilean company MININCO has 38 percent of its industrial tree plantation operations outside Chile (in Argentina and Brazil, for a total of 114,000 ha) – see Table 9.

[7]It is not just a question of conforming to laws limiting foreign land ownership but also of navigating bureaucracies and managing perceptions that pave the way for the mixed foreign-domestic investment or capital alliance. Often national investors can provide skilled legal counsel and bureaucratic expertise that might be harder for a purely outside group to access. Obvious foreign ownership of prime agricultural land is politically delicate (hence all those laws trying to prohibit or regulate foreignization of land) at the on-the-ground level, so it is useful to have a national face on it to deflect resentment from neighbours, and nationalist politicians. We thank Marc Edelman for pointing out this important issue.

A necessary element in intra-regional land investment in the region is the active and central role played by national capital. Despite all the talks about the significance of foreign land investors and foreignization of land (Zoomers 2010), it is constant across Latin America and the Caribbean that national and domestic elites (landlords and capitalists) remain as key investors in land and agriculture. There has been a net increase of investments in agriculture during the past decade or so, coming from foreign (intra-regional and international) and national sources. And so, in places where we witnessed a surge in foreign investments in land, such as in the ten countries where land grabbing occurs, investments by national capital has also increased. This is most concrete in Argentina, Bolivia, Brazil and Paraguay. In many instances, the differentiation between foreigners and locals is blurred especially when it involves naturalized and/or dual citizens, as in the case of Brazilians or Japanese in Bolivia (Mackey 2011), the mixed foreigner-national category in Argentina (see Murmis and Murmis 2011), or probably some elements of 'diaspora' as in the cases of foreign investors from Europe and Japan located in Brazil, Argentina, Chile, and Paraguay. Whether and to what extent diaspora from outside the region has played an important role in 'foreign' land investments requires further empirical research.

The scale of direct investments from the Gulf States, China, South Korea, and India for land acquisitions is quite low, at least for the time being. There have been talks about land acquisitions for years now, in Argentina and Brazil primarily, but nothing significant has been clinched in this regard. It does not mean it will remain this way. Further reconfiguration in global capital may alter future opportunity structure in the region that may allow for a bigger role for investments from these countries. Meanwhile, conventional TNCs are currently entrenched in the region, and to a relatively significant extent engaged in land investments. They are generally North Atlantic-based. But these North Atlantic-based investors (European and American) are more common and dominant in Brazil, Argentina and Mexico, and to some extent Paraguay.

The net increase of investments from regional and international (including diaspora?) as discussed above is partly captured in the recent land acquisition transactions in Paraguay. This is seen in Table 10, showing that during the past five

Table 10. Lands purchased by foreign entrepreneurs (1000 ha and over) in the regions with the most intense foreign ownership. Years 2006–2010.

Department	Purchased (in ha.) 2006–2010	Total purchased (in ha) until 2010	Nationality of main property buyers (in order of importance)
Concepción	70,000	250,000	Brazilian, French, German
San Pedro	130,000	325,000	Brazilian, German, Portuguese
Caaguazú	180,000	370,000	Brazilian, German, Japanese, Spanish
Caazapá	50,000	180,000	German, Portuguese, Brazilian
Itapúa	100,000	400,000	Brazilian, German, Japanese, Spanish
Alto Paraná	150,000	700,000	Brazilian, Argentinean, Portuguese
Amambay	80,000	450,000	Brazilian, Portuguese
Canindeyú	170,000	650,000	Brazilian, German
Alto Paraguay	900,000	2,500,000	Brazilian, Uruguayan, North American
Total	1,830,000	5,825,000	

Source: Galeano (2011, 401).

years nearly two million ha of land were transacted involving foreign ownership (for a total of nearly six million ha in this ownership category by 2010): Brazilian, Argentinean, Uruguayan, German, French, Portuguese, Spanish, American and Canadian.

More generally, it is important to examine the constellation of international investors involved in the region today within the context of an emerging polycentric food-energy regime in contrast to previous food regimes anchored by empires on either side of the North Atlantic (Friedmann and McMichael 1989, McMichael 2009). Current trends suggest multiple centres of power, and a more diverse range of key international actors within the governance structure of the food-energy complex, both sectorally and geopolitically. Sectorally, what we witness is not only the conventional food-feed TNCs involved in agricultural input-output markets (Teubal 1995). TNCs involved in the region today include unconventional actors such as oil companies, car conglomerates, biofuel companies, and so on, reflecting the changed global food-energy regime (see Wilkinson and Herrera 2010 in the case of Brazil). Geopolitically, we are witnessing not only North Atlantic-based TNCs and empires, but a far more diverse range of actors, namely originating from the BRICS and several MICs both within and outside Latin America and the Caribbean, giving rise to combined international and regional capital, whether in competition or in alliance. It is not difficult to surmise that the implications of this changed global configuration are profound and far-reaching for formal and informal rules in the governance of production, distribution and consumption within the food-feed-fuel complex and other food sectors, plus non-food sectors examined (the 'food regime' by Friedmann and McMichael 1989). But it is a question that needs deeper empirical research and further theorizing.

The foreignization narrative and dynamics of land property and labour relations change

All 17 FAO studies have been framed within the perspective of and debates around 'foreignization' of land property (see also Zoomers 2010). It is relevant and important, especially because it strikes right into the heart of what is a controversial and politically sensitive dimension of global land grabbing. The political tension in Paraguay near the border with Brazil where native Paraguayans feel aggrieved of their dispossession amidst land takeover by Brazilians is one example of how potentially and actually explosive this issue is in the region. The tension-filled phenomenon of what is sometimes referred to as 'Brasiguayos' (Brazilian commercial farmers in Paraguay), however, is quite different from the less tension-filled Brazilian presence in Santa Cruz, Bolivia, as noted by Mackey (2011). The limited presence of Gulf States, Chinese and South Koreans recently controlling some pockets of land in the Cerrado in Brazil (in addition to earlier Japanese investors) is another example of the foreignization (or denationalization) creeping in the region.

But, as mentioned earlier, the extent of land acquisitions by investors from outside the region, especially those involving foreign governments, is not as significant as it is in Africa. In addition, it is also not always the case that there is clear-cut 'native-foreigner' animosity over 'foreign' ownership of land, as explained by Mackey (2011) in his study of Brazilians owning lands in Bolivia. Yet, overall, the foreignization of land remains a politically sensitive issue, driving nearly all South

American governments to formally prohibit or regulate such practice. But on its own the 'foreignization narrative' can be problematic as it is fraught with contradictions: recent foreignization of land is not acceptable, but foreignization of capital and investments which has been creeping in the region much earlier and which ultimately indirectly captures (land) resources is welcomed; exploitation by foreigners is not socially acceptable, but exploitation by national classes of capital is tolerable. To some extent, the narrative has inadvertently evolved into a nationalist *versus* class politics. It is easily blurred by capital's creative maneuver to forge alliances between international and national capital, resulting in the recent rise of mixed foreign-domestic land investments in the region.

There are at least four problematic dimensions in the 'foreignization of land' narrative. First, there are two extreme poles in this perspective, namely, the 'foreign government-as-land grabber' and the 'diaspora-as-foreign-land grabber'; both are indeed processes of foreignization. On the one hand, narrowly defining land grabs as those land investments with direct participation by foreign government will take one to account for a marginal portion of the global land rush phenomenon. On the other hand, automatically counting diaspora land purchases as land grabs may deflect our analytical focus away from the dynamics that we are interested in: the processes of agrarian restructuring due to recent changes in the global food-energy regime and the overall capitalist requirement for key resources for industrial/commercial purposes. It is certainly important to include these two poles in our analysis, but one should go beyond these. Second, a foreignization narrative tends to present a sharper focus on the transnational dimension of the key issues in land grabbing (e.g. foreigners chopping down natural forest to create industrial forest based on monocrop fast-growing trees). But what if land deals that cleared forests to open up massive soya plantations for export were carried out by mixed foreign-domestic, or by domestic, capital?

Third, a narrow current in the foreignization narrative focuses on a handful of 'foreign' investors – China, Gulf States, South Korea, and India – and has a less sharp understanding of the role played by the traditional North Atlantic empires, and by the regional capital. It also tends to be blind to (indirect) key policy drivers. The narrative then has difficulty capturing cases such as the European Union biofuel mandatory blending target that has sparked massive worldwide speculation for biofuel markets and thus the opening of new plantations for a variety of feedstocks (Franco *et al.* 2010) that has inspired the Colombian or Guatemalan government to aggressively promote oil palm expansion largely in anticipation of European markets for biodiesel (see, e.g. Salinas 2011 for Colombia, Carrera and Carrera 2011).

Finally, the foreignization narrative in the end is strongest in terms of objecting to 'foreign ownership of land' in a country that in turn partly drives recent national policy initiatives at curtailing such a phenomenon. But policy adjustments tend to happen without really addressing the logic of capital that underpins global land grabbing, as the latter continues amidst dominance of national and mixed foreign-national capital as in the cases of Brazil, Bolivia, Ecuador, Paraguay and Argentina (Teubal 2009). Does land grab necessarily and always requires foreignization of land property? Not really, as we have argued earlier using the notion of *control grabbing*. Conversely, does foreignization of land always imply land grabbing? – as one wonders about the small- and medium-scale individual land acquisitions by natural persons from Brazil in Paraguay or Bolivia – a process we categorize in this paper as part of the everyday forms of dispossession by differentiation which we exclude from our definition of land grabs.[8]

We argue that instead of overly focusing on foreignization of property in addressing changes in land property relations it is more useful to look into the *character* and *direction* of change in social relations of property. There are two key features of the dominant narrative in land grabs. One is the foreignization of spaces, as explained above, and the other is the assumption that land grabs lead to dispossession. For us the challenge is how to analyze these two interlinked processes.

In addition, land deals (involving foreign or national capital) have indeed resulted in the expulsion of people from their lands. We see this in many places in Colombia, Paraguay, and Argentina, among others, and where this happens it is usually carried out with violence. However, data from the 17 FAO studies suggest that the phenomenon of 'dispossession by displacement' (in Araghi's term) – or expulsion of people from their land – in Latin America and the Caribbean is at a scale that is relatively less in extent compared to processes of expulsion in other regions in the world – at least for the time being. There are a number of possible combinations of reasons for this, among which is that much of the expansion of new land investments occurs in extensive livestock ranches (where massive land grabbing, of a primitive accumulation type, occurred in these spaces much earlier – decades or even centuries ago). The latter then either shifts to more intensive production technology as in the case of many Chilean ventures (Echenique 2011), or moves to the frontier and engages in fresh forest clearing, or to a combination of both approaches, as in the case of Brazil (Novo *et al.* 2010).

Furthermore, land property relations change is better understood from a broader perspective, seen through a typology of directions of land-based wealth and power transfer, adapted from Borras and Franco (2012). The main idea is that land policies are not neutral instruments: when passed and implemented, they follow four broad ideal-type paths, namely, redistribution, distribution, nonredistribution, and reconcentration. The defining principle of the redistribution type is the transfer of land-based wealth and power from the monopoly control of either private landed classes or the state, to landless and near-landless working poor (poor peasants and rural labourers). It is a 'zero-sum' process. The conventional notion of redistributive land reform that is usually applied only to large private lands (*latifundia*) is the most commonly understood example of redistributive land policy. However, there are a variety of other policy measures that can change the relative shares of land held by social classes and groups, including land restitution, share tenancy, land tenure reform, land stewardship, indigenous land rights recognition, and labour reform, regardless of whether the policy is applied to private or public land.

Like redistribution, the distribution type is where the landless and near-landless working poor are beneficiaries of land-based wealth and power transfers. But in this

[8]What if the lands were previously in the hands of large owners? Does the transfer of ownership from one group of medium-large owners to another, with a change in land use, count as a grab? (We thank Jonathan Fox for raising this difficult point.) There are other possible related questions that are not easy to answer even with our working definition offered here. For us, there will always be a 'grey area' in the land grab definition, and one way to look at this is to consider a land deal character in relation to the three key components we offered in our working definition. From there, one can infer whether a land deal is closer to a land grab or to an everyday form of dispossession by differentiation involving large- to small-scale land owners.

type the original source of wealth and power is either the state or community, or a private entity fully compensated by the state. This is a 'positive sum' process; it does not confiscate resources from one social class to redistribute to another, and has been deployed in some cases precisely to avoid more radical redistributive policies (Fox 1993, 10). In other cases, this type of reform involves affirming and protecting pre-existing land access and occupancy by poor peasants whose tenure is insecure.

The non-(re)distribution type is one the defining character of which is the maintenance of a status quo, marked by land-based inequality and exclusion. The most typical land policy here is 'no land policy' which, in conditions of land-based inequality, perpetuates the existing distribution of land-based wealth and power. In other settings, a similar effect may be created when an existing land policy, even a redistributive land reform policy, is kept dormant 'from above' or becomes frozen or flounders in the course of implementation as it comes up against impediments within the state or in society or both. However, this kind of situation should not be confused with others involving active land policies that are categorically non-(re)distributive, to which we turn next.

The defining character of the (re)concentration type is that while land-based wealth and power transfers do occur, access to and control over land is further concentrated in the hands of dominant social classes and groups: landed classes, capitalists, corporate entities, and state or other dominant community groups such as village chiefs. This kind of change can occur in private or public lands. The organization of control over land resources can be through individual, corporate, state or community property rights. The transfer may or may not involve full land ownership. Different variations are possible, but the bottom line is the same: the beneficiaries of such transfers are dominant social classes and groups (or the state).

In the context of the typology above, the existing trend in Latin America and the Caribbean on the eve of the current land rush was away from (re)distributive land policies, and was towards non-redistribution (Kay 1998). When the contemporary land rush in the region started to gain momentum, the trend veering away from (re)distribution got consolidated, with non-redistribution standing tall (Borras, Kay and Lahiff 2007). During the same period, the reconcentration type has accelerated, a trend observed in nearly all 17 FAO studies. These studies point out varying forms and degrees of (re)concentration of land ownership and land-based wealth and power through resource control grabbing. Today, in many countries in the region, the Gini coefficient for land ownership remains very high. This is despite the long history of land reformism in the region (Kay 1998) that was followed by neoliberal agrarian restructuring and state agricultural policy reforms (Spoor 2002). The onslaught of land investments and land grabbing are likely to exacerbate this already problematic condition of land control. Moreover, one of the immediate effects of rising economic value of land is that it would make the already difficult (re)distributive land policies even more difficult. This is certainly the case for Brazil as pointed out by Sauer and Leite Pereira (2012). Private land owners' resistance will become even stronger. The state's calculation of its control over land resources and possible dividends is likely to block, not facilitate, pro-(re)distribution policy currents within central states. Contemporary Brazil is a good example of this.

Today, not many countries in the region talk about conventional redistributive land reform in any serious way, not even countries where some sort of land reform is still underway and being carried out with varying degrees of state support; namely, Brazil, Paraguay, Venezuela, Bolivia and Ecuador. Market-led agrarian reforms were attempted in Central America as part of the 1996 Peace Accords, but with dismal outcomes (see e.g. Gauster and Isakson 2007 for Guatemala, and de Bremond 2007 for El Salvador) as well as in Brazil (Servolo de Medeiros 2007). Auctioning state lands has been resorted to in Peru and has been heralded by mainstream economists as a good way to (re)allocate land resources for more efficient use (World Bank 2010). Formalization of titles to claimants, not necessarily along the ideological bias of conventional land reform, is favoured in some places (Eguren 2006). This is the case, for example, in Brazil, with the aspiration of *Terra Legal* to formalize some 300,000 homesteaders' land claims in the Amazon, a policy which is very much calculated in the context of rising investments in agriculture for the expansion of agribusiness into this agricultural land frontier. In the *Terra Legal* campaign the individual farm size ceiling has been increasingly adjusted over time to the current 1,500 ha which is relatively large even by Brazilian standards of family farms (Oliveira 2011).

Meanwhile, in settings where land deals did not result in the expulsion of people from their lands, these have facilitated the incorporation, adversely or otherwise, of smallholders and indigenous communities into the emerging commercial farms, plantation enclaves and value chains (Butler Flora and Bendini 2003). While peasants may retain some access to land they increasingly have to diversify their sources of income and seek a variety of off-farm employment opportunities, furthering the process of degrarianization (Bryceson 2000, Gómez 2002, Giarracca and Levy 2004, Brumer and Piñeiro 2005, Edelman 1999, 2008). In the case of Latin America and the Caribbean evidence suggests that this is likely to be the more prevalent condition and possibly the future broad trajectory (Kay 2000, de Grammont and Martínez Valle 2009).

Available evidence in Latin America and the Caribbean based on the 17 FAO country studies does not offer any conclusive insight regarding to what extent and how peasants and workers are being incorporated in the newly emerging production complex. There are earlier studies about existing commercial farms and plantations employing workers in not very good terms like casualization of labour, such as those we see in Chile and in the sugarcane sector of Brazil. There is some evidence that suggests possible adverse incorporation into the value chain, such as the Mexican food value chain controlled by US-based supermarkets (Rubio 2003, Teubal 2005, Robles 2011). There is nothing elaborated and conclusive on this issue from the 17 FAO studies. However, it is most likely that it is much more *differentiated terms of incorporation*, that is, 'adverse', 'favourable' and 'somewhere in between' (see also Du Toit 2004) largely depending on pre-existing agrarian structures and institutions (including histories of rural class formation, state-peasant relations, and rural trade union and agrarian movement building). The study done by Fernandes *et al.* (2010) about various experiences in small-scale, community-based biofuels projects linked and not linked to large-scale industrial processors in Brazil, where some are adversely incorporated, others not, is suggestive of such diverse conditions. But overall, this is one area that needs further empirical research.

Concluding discussion

Land grabbing – as resource control grabbing, explained in the beginning of the paper – has gained momentum in Latin America and the Caribbean during the past decade. There are four key contexts within through which it has been carried out: food security initiatives, energy/fuel security ventures, other climate change mitigation strategies, and recent demands for resources from newer hubs of global capital. The hallmark of land grabbing in the region is its intra-regional character: the key investors are (Trans-)Latin American Companies that in turn necessarily involve parallel or allied accumulation strategies by foreign and national capital, almost always in alliance within the central state. Initial evidence suggests that recent land investments have consolidated the earlier trend away from (re)distributive land policies in most countries in the region, and are likely to result in widespread reconcentration of land and capital.

Initial evidence of land grabbing in this region suggests that the phenomenon has taken different forms and character as compared to processes that occur in other regions of the world, especially Africa. It puts into question some of the most dominant assumptions in the emerging literature on land grabbing, suggesting these are too food-centered/too food crisis-centered, too land-centred, too centred on new global food regime players (China, South Korea, Gulf States and India), and too centred on Africa. In this region, land grabs occur not only in the food sector, but in non-food sectors as well. Land grabs in this region have certainly been boosted by the 2007–2008 food crisis, but the process gained momentum much earlier than the 2007–2008 global food price spike period. Investors from China, South Korea, Gulf States and India are definitely seeking some grounds in this region, but at the moment they are marginal in terms of direct land investments. Intra-regional and domestic capital is the key player in land grabbing in this region. Land grabbing occurs in ten countries in the region, wider than previously assumed; when brought into the broader picture of global land grabbing, data from this region is likely to change the relative share of Africa in global land grabbing.

While the empirical data gathered and assembled in the 17 FAO studies upon which we based our analysis in this paper are relatively fresh, robust and comprehensive, these studies and our current paper are initial studies. They have, however, facilitated our identification of important knowledge frontiers. It will be urgent and important to carry out scientific research on a number of important themes, including: the actual *scale of capital* (to complement accounting of the *scale of land acquisitions*) involved; the actual condition and trajectories on likely *differentiated terms of incorporation* of peasants, indigenous peoples and rural labourers in the emerging commercial farm and plantation enclaves; and *differentiated reactions* towards land deals from affected communities and the implications of these in reshaping and transforming land investments, as well as the *agrarian implications* of green grabs.

References

Alden Wily, L. 2012. Looking back to see forward: the legal niceties of land theft in land rushes. *Journal of Peasant Studies*, 39(3–4), 751–75.

Amanor, K.S. 2012. Global resource grabs, agribusiness concentration and the smallholder: two West African case studies. *Journal of Peasant Studies*, 39(3–4), 731–49.

Araghi, F. 2009. The invisible hand and the visible foot: peasants, dispossession and globalization. In: A.H. Akram-Lodhi and C. Kay, eds., *Peasants and globalization: political economy, rural transformation and the agrarian question*. London and New York: Routledge, pp. 111–47.

Balleti, Brenda. 2012. Ordenamento territorial: reordering the Amazon for development and conservation. *Journal of Peasant Studies* 39(2).

Ballve, T. 2011. Territory by dispossession: decentralization, statehood, and the narco land-grab in Colombia. Available from: http://www.iss.nl/ldpi [Accessed 25 May 2011].

Baumeister, Eduardo, 2011. Dinámica del mercado de la tierra en América Latina y el Caribe: El caso de Nicaragua. Santiago: FAO. Available from: http://www.rlc.fao.org/fileadmin/content/events/semtierras/acaparamiento.pdf [Accessed 30 December 2011].

Bebbington, A., *et al.* 2009. 'Contienda y Ambigüedad: Minería y Posibilidades de Desarrollo'. *Debate Agrario*, 44, 31–62.

Bernstein, H. 2010. *Class dynamics of agrarian change*. Halifax: Fernwood.

Borras, S. Jr. and J. C. Franco. 2010. 'From Threat to Opportunity? Problems with the Idea of a "Code of Conduct" for Land-Grabbing'. *Yale Human Rights and Development Law Journal*, vol. 13, no. 2, pp. 507–23.

Borras, S. Jr. and J.C. Franco 2011. *Political dynamics of land grabbing in Southeast Asia: understanding Europe's role*. Discussion Paper, January 2011. Amsterdam: Transnational Institute (TNI).

Borras, S. Jr. and J. C. Franco. 2012. Global land grabbing and trajectories of agrarian change: a preliminary analysis. *Journal of Agrarian Change*, 12(1), 34–59.

Borras, S, Jr., C. Kay and E. Lahiff, eds. 2007. *Market-led agrarian reform*. London: Routledge.

Brumer, A. and D. Piñeiro, eds. 2005. *Agricultura Latino-Americana: Novos Arranjos e Velhas Questões*. Porto Alegre: Editora da UFRGS.

Bryceson, D. 2000. Peasant theories and smallholder policies: past and present. In: D. Bryceson, C. Kay and J. Mooij, eds. *Disappearing peasantries? Rural labour in Africa, Asia and Latin America*. London: ITDG Publishing, pp. 1–36.

Butler Flora, C. and M. Bendini. 2003. Globalización en Cadenas de Valor Agroalimentarias. Relaciones entre el Mercado, el Estado y la Sociedad Civil. *In*: M. Bendini, S. Cavalcanti, M. Murmis and P. Tsakoumagkos, eds., *El Campo en la Sociología Actual: Una Perspectiva Latinoamericana*. Buenos Aires: Editorial La Colmena, pp. 341–65.

Cardenas, Roosbelinda. 2012. Green multiculturalism: articulations of ethnic and environmental politics in Colombia's black communities. *Journal of Peasant Studies*, 39(2),

Carrera, Jaime Arturo and Jaime Luis Carrera Campos, 2011. Dinámica del mercado de la tierra en América Latina y el Caribe: El caso de Guatemala. Santiago: FAO. Available from: http://www.rlc.fao.org/fileadmin/content/events/semtierras/acaparamiento.pdf [Accessed 30 December 2011].

Corbera, Esteve and Heike Schroeder. 2011. Governing and implementing REDD+. *Environmental Science and Policy*, 14(2), 89–99.

Cotula, Lorenzo. 2012. The international political economy of the global land rush: a critical appraisal of trends, scale, geography and drivers. *Journal of Peasant Studies*, 39(3).

De Bremond, Arianne. 2007. The politics of peace and resettlement through El Salvador's land transfer programme: caught between the state and the market. *Third World Quarterly*, 28(8), 1537–56.

de Grammont, H.C. and L. Martínez Valle, eds. 2009. *La Pluriactividad en el Campo Latinoamericano*. Quito: FLACSO.

Deininger, K. 2011. Forum on global land grabbing: challenges posed by the new wave of farmland investment. *Journal of Peasant Studies*, 38(2), 217–47.

De Schutter, O. 2011. Forum on global land grabbing: how not to think land grabbing: three critiques of large-scale investments in farmland. *Journal of Peasant Studies*, 38(2), 249–79.

Donoso, Santander Tristan, 2011a. Dinámica del mercado de la tierra en América Latina y el Caribe: El caso de Costa Rica. Santiago: FAO. Available from: http://www.rlc.fao.org/fileadmin/content/events/semtierras/acaparamiento.pdf [Accessed 30 December 2011].

Donoso, Santander Tristan, 2011b. Dinámica del mercado de la tierra en América Latina y el Caribe: El caso de Panamá. Santiago: FAO. Available from: http://www.rlc.fao.org/fileadmin/content/events/semtierras/acaparamiento.pdf [Accessed 30 December 2011].

Du Toit, A. 2004. *Forgotten by the highway: globalisation, adverse incorporation and chronic poverty in a commercial farming district of South Africa*. Cape Town: PLAAS Chronic Poverty and Development Policy Series No. 4.

Echenique, Jorge, 2011. Dinámica del mercado de la tierra en América Latina y el Caribe: El caso de Chile. Santiago: FAO. Available from: http://www.rlc.fao.org/fileadmin/content/events/semtierras/acaparamiento.pdf [Accessed 30 December 2011].

Echenique, J. and L. Romero. 2009. *Evolución de la Agricultura Familiar en Chile en el Período 1997–2007*. Santiago: Corporación Agraria para el Desarrollo.

Edelman, Marc. 1999. *Peasants against globalization: rural social movements in Costa Rica*. Stanford: Stanford University Press.

Edelman, Marc. 2008. Transnational organizing in agrarian Central America: histories, challenges, prospects. *Journal of Agrarian Change*, 8(2&3), 229–57.

Eguren, F., ed. 2006. *Reforma Agraria y Desarrollo Rural en la Región Andina*. Lima: Centro Peruano de Estudios Sociales (CEPES).

Fairhead, J., M. Leach and I. Scoones. 2012. Green Grabbing: a new appropriation of nature? *Journal of Peasant Studies*, 39(2), 237–261.

FAOSTAT. 2012. FAO Statistics: www.faostat.org. Downloaded and processed 15 February 2012.

Fernandes, B.M., A.W. Clifford and E.C. Gonçalves. 2010. Agrofuel policies in Brazil: paradigmatic and territorial disputes. *Journal of Peasant Studies*, 37(4), 793–819.

Fox, Jonathan. 1993. *The politics of food in Mexico: state power and social mobilization*. Ithaca: Cornell University Press.

Franco, J., L. Levidow, D. Fig, L. Goldfarb, M. Hönicke and M.L. Mendonça. 2010. Assumptions in the European Union biofuels policy: frictions with experiences in Germany, Brazil and Mozambique. *Journal of Peasant Studies*, 37(4), 661–98.

Friedmann, H. and P. McMichael. 1989. Agriculture and the state system: the rise and decline of national agricultures, 1870 to the present. *Sociologia Ruralis*, 29(2), 93–117.

Galeano, Luis, 2011. Dinámica del mercado de la tierra en América Latina y el Caribe: El caso de Paraguay. Santiago: FAO. Available from: http://www.rlc.fao.org/fileadmin/content/events/semtierras/acaparamiento.pdf [Accessed 30 December 2011].

Gauster, Susana and Ryan Isakson. 2007. 'Eliminating market distortions, perpetuating rural inequality: an evaluation of market-assisted land reform in Guatemala. *Third World Quarterly*, 28(8), 1519–36.

Giarracca, N. and B. Levy, eds. 2004. *Ruralidades Latinoamericanas: Identidades y Luchas Sociales*. Buenos Aires: CLACSO Libros.

Gómez, S. 2002. *La 'Nueva Ruralidad': ¿Qué tan Nueva?* Santiago: LOM Ediciones.

Gómez, Sergio, 2011. [Summary paper:] Dinámica del mercado de la tierra en América Latina y el Caribe. Santiago: FAO. Available from: http://www.rlc.fao.org/fileadmin/content/events/semtierras/acaparamiento.pdf [Accessed 30 December 2011].

GRAIN. 2008. *Seized: the 2008 land grab for food and financial security*. Barcelona: GRAIN.

Grajales, J. 2011. The rifle and the title: paramilitary violence and land control in Colombia. *Journal of Peasant Studies*, 38(4), 771–92.

Graziano da Silva, J., S. Gómez and R. Castañeda S. 2010. 'Boom' Agrícola e Persistência da Pobreza na América Latina, *Revista NERA*, 13(16), 7–21.

Hall, D., 2011. Land control, land grabs, and Southeast Asian crop booms. *Journal of Peasant Studies*, 38(4), 837–57.

Hall, R. 2012. The next Great Trek? South African commercial farmers move north. *Journal of Peasant Studies*, 39(3–4), 823–43.

Harvey, N. 1998. *The Chiapas rebellion: the struggle for land and democracy*. Durham: Duke University Press.

Harvey, D. 2003. *The new imperialism*. Oxford: Oxford University Press.

Hecht, Susana. 2005. Soybeans, development and conservation on the Amazon frontier. *Development and Change*, 36(2), 375–404.

Hofman, I. and P. Ho. 2012. China's 'developmental outsourcing': a critical examination of Chinese global 'land grabs' discourse. *Journal of Peasant Studies*, 39(1), 1–48.

ILC. 2011. *Land rights and the rush for land: a report*. Rome: ILC.

Kay, C. 1998. ¿El Fin de la Reforma Agraria en América Latina? El Legado de la Reforma Agraria y el Asunto no Resuelto de la Tierra. *Revista Mexicana de Sociología*, 60(4), 61–98.

Kay, C. 2000. Latin America's agrarian transformation: peasantization and proletarianization, *In*: D. Bryceson, C. Kay and J. Mooij, eds. *Disappearing peasantries? Rural labour in Africa, Asia and Latin America*. London: ITDG Publishing, pp. 123–38.

Kay, C. 2001. Reflections on rural violence in Latin America. *Third World Quarterly*, 22 (5), 741–75.

Kay, C. 2002. Chile's neoliberal agrarian transformation and the peasantry. *Journal of Agrarian Change*, 2(4), 464–501.

Kay, Sylvia and Jennifer Franco. 2012. The global water grab: a primer. Amsterdam: Transnational Institute.

Lavandier, Ingrid, 2011. Dinámica del mercado de la tierra en América Latina y el Caribe: El caso de Republica Dominicana. Santiago: FAO. Available from: http://www.rlc.fao.org/fileadmin/content/events/semtierras/acaparamiento.pdf [Accessed 30 December 2011].

Levien, M. 2011. Special economic zones and accumulation by dispossession in India. *Journal of Agrarian Change*, 11(4), 454–83.

Li, T. M. 2011. Forum on global land grabbing: centering labor in the land grab debate. *Journal of Peasant Studies*, 38(2). 281–98.

Mackey, L. 2011. Legitimating foreignization in Bolivia: Brazilian agriculture and the relations of conflict and consent in Santa Cruz, Bolivia. A paper presented at the Land Deal Politics Initiative (LDPI) conference on land grabs, IDS University of Sussex, April 2011. Available from: www.iss.nl/ldpi [Accessed 15 January 2012].

Martinez, Luciano Valle, 2011. Dinámica del mercado de la tierra en América Latina y el Caribe: El caso de Ecuador. Santiago: FAO. Available from: http://www.rlc.fao.org/fileadmin/content/events/semtierras/acaparamiento.pdf [Accessed 30 December 2011].

McMichael, P. 2009. A food regime genealogy, *Journal of Peasant Studies*, 36(1), 171–96.

McMichael, P. 2012. The land grab and corporate food regime restructuring. *Journal of Peasant Studies*, 39(3–4), 681–701.

Murmis, Miguel and María Rosa Murmis, 2011. Dinámica del mercado de la tierra en América Latina y el Caribe: El caso de Argentina. Santiago: FAO. Available from: http://www.rlc.fao.org/fileadmin/content/events/semtierras/acaparamiento.pdf [Accessed 30 December 2011].

Nalepa, R.A. and D.M. Bauer. 2012. Marginal lands: the role of remote sensing in constructing landscapes for agrofuel development. *Journal of Peasant Studies* 39(2), 403–22.

Novo, A., K. Jansen, M. Slingerland and K. Giller. 2010. Biofuel, dairy production and beef in Brazil: competing claims on land use in São Paulo State. *Journal of Peasant Studies*, 37(4), 769–92.

Oliveira, Gustavo. 2011. Land regularization in Brazil and the global land grab. A paper presented at the Land Deal Politics Initiative (LDPI) conference on land grabs, IDS University of Sussex, April 2011. Available from: www.iss.nl/ldpi [Accessed 15 January 2012].

Osborne, Tracey. 2011. Carbon forestry and agrarian change: access and land control in a Mexican rainforest. *Journal of Peasant Studies*, 38(4), 859–83.

Oxfam. 2011. *Land and power*. Oxford: Oxfam International.

Peluso, Nancy and Christian Lund. 2011. New frontiers of land control: introduction. *Journal of Peasant Studies*, 38(4), 667–81.

Piñeiro, Diego 2011. Dinámica del mercado de la tierra en América Latina y el Caribe: El caso de Uruguay. Santiago: FAO. Available from: http://www.rlc.fao.org/fileadmin/content/events/semtierras/acaparamiento.pdf [Accessed 30 December 2011].

Remy, Maria Isabel and Carlos de los Rios, 2011. Dinámica del mercado de la tierra en América Latina y el Caribe: El caso de Peru. Santiago: FAO. Available from: http://www.rlc.fao.org/fileadmin/content/events/semtierras/acaparamiento.pdf [Accessed 30 December 2011].

Ribot, J. and N. Peluso. 2003. A theory of access. *Rural Sociology*, 68(2), 153–81.

Robles, Hector Manuel, 2011. Dinámica del mercado de la tierra en América Latina y el Caribe: El caso de Mexico. Santiago: FAO. Available from: http://www.rlc.fao.org/fileadmin/content/events/semtierras/acaparamiento.pdf [Accessed 30 December 2011].

Rubio, B. 2003. *Explotados y Excluidos: Los Campesinos Latinoamericanos en la Fase Agroexportadora Neoliberal*. Mexico City: Plaza y Valdés Editores.

Salinas, Yamile Abdala, 2011. Dinámica del mercado de la tierra en América Latina y el Caribe: El caso de Colombia. Santiago: FAO. Available from: http://www.rlc.fao.org/fileadmin/content/events/semtierras/acaparamiento.pdf [Accessed 30 December 2011].

Sassen, S. 2010. A savage sorting of winners and losers: contemporary versions of primitive accumulation. *Globalizations*, 7(1–2), 23–50.

Sauer, Sérgio and Sergio Pereira Leite. 2012. Agrarian structure, foreign investment in land, and land prices in Brazil. *Journal of Peasant Studies*, 39(3–4), 873–98.

Scott, J. 1998. *Seeing like a state: how certain schemes to improve the human condition have failed.* New Haven: Yale University Press.
Servolo de Medeiros, Leonilde. 2007. Social movements and the experience of marker-led agrarian reform in Brazil. *Third World Quarterly*, 28(8), 1501–18.
Sommerville, M. 2011. The global land grab and marginalization in Canada: the case of One Earth Farms. A paper presented at the Land Deal Politics Initiative (LDPI) conference on land grabs, IDS University of Sussex, April 2011. Available from: www.iss.nl/ldpi [Accessed 15 January 2012].
Spoor, M. 2002. Policy regimes and performance of the agricultural sector in Latin America and the Caribbean during the last three decades. *Journal of Agrarian Change*, 2(3), 382–401.
Teubal, M. 1995. Internacionalización del capital y complejos agroindustriales: impactos en América Latina. *In*: M. Teubal, *Globalización y Expansión Agroindustrial: ¿Superación de la Pobreza en América Latina?* Buenos Aires: Ediciones Corregidor, pp. 45–79.
Teubal, M. 2009. La Lucha por la Tierra. *In*: N. Giarracca and M. Teubal, eds. *La Tierra es Nuestra, Tuya y de Aquel: Las Disputas por el Territorio en América Latina.* Buenos Aires: Antropofagia, pp. 205–29.
Teubal, M. 2005. Transformaciones Agrarias en la Argentina. Agricultura Industrial y Sistema Alimentario. *In*: N. Giarracca and M Teubal, eds. *El Campo en la Encrucijada. Estrategias y Resistencias Sociales, Ecos en la Ciudad.* Buenos Aires: Alianza Editorial, pp. 37–78.
UNEP. forthcoming 2012. Land. *In*: UNEP, *Global Environmental Outlook (GEO) 5.* UNEP.
Urioste, Miguel, 2011. Dinámica del mercado de la tierra en América Latina y el Caribe: El caso de Bolivia. Santiago: FAO. Available from: http://www.rlc.fao.org/fileadmin/content/events/semtierras/acaparamiento.pdf [Accessed 30 December 2011].
Visser, O. and M. Spoor. 2011. Land grabbing in post-Soviet Eurasia: the world's largest agricultural land reserves at stake. *Journal of Peasant Studies*, 38(1), 299–323.
White, Ben and Anirban Dasgupta. 2010. Agrofuels capitalism: a view from political economy. *Journal of Peasant Studies*, 37(4), 593–607.
Wilkinson, J. and S. Herrera. 2010. Biofuels in Brazil: debates and impacts. *Journal of Peasant Studies*, 37(4), 749–68.
Wilkinson, J., Bastian Reydon and Alberto Di Sabbato, 2011. Dinámica del mercado de la tierra en América Latina y el Caribe: El caso de Brasil. Santiago: FAO. Available from: http://www.rlc.fao.org/fileadmin/content/events/semtierras/acaparamiento.pdf [Accessed 30 December 2011].
Williams, Allan, 2011. Dinámica del mercado de la tierra en América Latina y el Caribe: El caso de Trinidad and Tobago. Santiago: FAO. Available from: http://www.rlc.fao.org/fileadmin/content/events/semtierras/acaparamiento.pdf [Accessed 30 December 2011].
Williams, Patrick, 2011. Dinámica del mercado de la tierra en América Latina y el Caribe: El caso de Guyana. Santiago: FAO. Available from: http://www.rlc.fao.org/fileadmin/content/events/semtierras/acaparamiento.pdf [Accessed 30 December 2011].
Wolford, Wendy. 2010. Contemporary land grabs in Latin America. Presentation at the FAO Committee on Food Security 36th Session, October 2010, Rome. Available from: www.tni.org [Accessed 15 January 2012].
Woodhouse, P. 2012. New investment, old challenges. Land deals and the water constraint in African agriculture. *Journal of Peasant Studies*, 39(3–4), 777–94.
Woodhouse, P. and A.S. Ganho. 2011. Is water the hidden agenda of agricultural land acquisition in sub-Saharan Africa? A paper presented at the Land Deal Politics Initiative (LDPI) conference on land grabs, IDS University of Sussex, April 2011. Available from: www.iss.nl/ldpi [Accessed 15 January 2012].
World Bank. 2010. *Rising global interest in farmland: can it yield sustainable and equitable benefits?.* Washington DC: World Bank.
Zoomers, A., 2010. Globalisation and the foreignisation of space: seven processes driving the current land grab. *The Journal of Peasant Studies*, 37(2). 429–47.

An earlier version of this paper, which was a study commissioned by FAO Latin America and Caribbean Regional Office, was presented at the regional workshop on land grabbing organized by FAO on 14–15 November 2011 in Santiago, Chile. We thank the participants for their comments, especially Jose Graziano da Silva and David Goodman. We also thank four peer reviewers for this journal for their tough, critical but constructive comments and suggestions. Any remaining errors are ours alone. We would like to thank Camilo Andrés Ardila Galvis and Fabio Andrés Díaz Pabón for translation assistance.

Saturnino M. Borras Jr. is Associate Professor at the International Institute of Social Studies (ISS), The Hague, Netherlands, Adjunct Professor at the College of Humanities and Development Studies of China Agricultural University in Beijing, and Fellow of the Amsterdam-based Transnational Institute (TNI) and of the California-based Food First.

Jennifer C. Franco is a Research Associate at the Transnational Institute (TNI) and Adjunct Professor at the College of Humanities and Development Studies of China Agricultural University in Beijing.

Sergio Gómez is a consultant at FAO Regional Office in Santiago, Chile. Director Magister Degree, Rural Territorial Development Program, at the Universidad Academia de Humanismo Cristiano, Chile. He coordinated the 17 FAO studies on land grabbing in Latin America and the Caribbean.

Cristóbal Kay is Emeritus Professor at ISS in The Hague, and Professorial Research Associate at the School of Oriental and African Studies (SOAS) in London. He is an Editor of the *Journal of Agrarian Change*.

Max Spoor is Professor of Development Studies at ISS, Visiting Professor at the Barcelona Institute of International Studies (IBEI) in Spain and Guest Professor at Nanjing Agricultural University in China.

Agrarian structure, foreign investment in land, and land prices in Brazil

Sérgio Sauer and Sergio Pereira Leite

The recent global rush for farmland in Latin America has produced a dramatic increase in the level of foreign investment in land in Brazil. The current trend accentuates the ongoing process of foreignization of agriculture associated with the production of grains, sugar, ethanol and other commodities, increasing land prices. In response, the Brazilian government reestablished a legal mechanism for 'controlling' land-based foreign investment which has proven neither efficient nor effective in solving land concentration. This paper examines this issue by analyzing the causes of the increase in investment as well as the consequences of this process with respect to land prices, critically situating land-based investments and the government's policy response in a broader discussion of the demands of agrarian social movements.

Introduction

The recent increase in the interest in and search for land on a global scale is largely driven by demand for food, agro-energy crops and raw materials. According to a World Bank study in 2010, global demand for land has increased notably since 2008 (World Bank 2010). This process has transformed 'territorial disputes' for land into a global phenomenon. Prior to 2008, transfers of cultivable or cultivated land represented approximately four million hectares (ha) per year. By contrast, in the period from October 2008 to August 2009 alone, more than 45 million ha of land were traded. According to some estimates, 75 percent of this land was in Africa and an additional 3.6 million ha were in Brazil and Argentina (World Bank 2010). This has propelled the phenomenon dubbed 'land grabbing', in reference to how these transfers extend beyond standard mechanisms of land purchase (Merlet 2010).

In the wake of the crisis of food prices in 2008, and given future outlooks on demand, it is not surprising to see the increasing interest by governments – most notably by China and several Arab countries – in the purchase of foreign land for the production of food for domestic consumption. These particular developments also highlight larger questions over the changing investment trends of a financial sector which had been historically averse to investments in land markets on account

This text expands upon and updates previous contributions of the authors to this theme, including: Leite and Wesz Jr. (2011), Sauer (2011), Sauer and Leite (2010), and Leite and Sauer (2011). The authors thank the reviewers for their timely comments.

of their relative capital illiquidity. Global interest in the relatively abundant land of Latin America (especially in Brazil, Argentina and Uruguay) as well as sub-Saharan Africa has led to an increase in the price of land. This is suggested in the Brazilian case by the above-mentioned World Bank study (2010) and has been regularly announced by the Brazilian press through anecdotal cases or local-level data. However, there are still no systematic studies that provide a nationwide or even a state-level panorama of these transactions and the resulting impact on land prices.

The main arguments of this paper are that: first, the recent global rush for land is not entirely new in Brazil nor in Latin America, but it does present new dynamics with respect to the concentration of control over the production of agricultural and non-agricultural commodities (Safransky and Wolford 2011, Leite and Wesz Jr. 2010, Hecht 2005, Murmis and Murmis 2010). Second, global scrambles for land are primarily driven by high global prices for agricultural and non-agricultural commodities, especially sugar and soybeans, but are also due to new opportunities that this presents for business (speculation) (Wilkinson *et al.* 2011). Third, rising demand for land deepens the historical imbalance between 'social' demand and supply, accentuates processes of concentration, and intensifies territorial disputes (Sauer 2011, Leite and Sauer 2011).[1]

In order to contribute to this discussion, the present paper approaches five aspects of this issue that we believe are essential to the debates and which address points that have not been dealt with in detail by existing literature. The first section discusses the historical demand for land in Brazil and calls attention to the fact that the current rush for land is not new, but rather is part of an existing pattern of land concentration that has persisted into the 21st century. In this context, foreign investment should be viewed as one among several factors that form a historical dispute for land and its related social, economic and political consequences. The second section examines the process of expansion by Brazilian agribusiness, focusing in particular on the territorial movements of sugarcane and soybean plantations in 'agricultural frontier' regions. This process, which is motivated by elevated global prices for commodities in the international market, promoted by national and foreign investments, produces new disputes over land and shapes existing frontier contests, especially in the Amazon region.

The role of foreign investment in land-based assets for the production of biomass and grains is the subject of the third section. Official data released by the National Institute for Agrarian Reform and Colonization (INCRA) highlights a process of foreign appropriation of land throughout the twentieth century. Despite the limited regulatory control of foreign investments over time, we observe a recent increase in this type of investment, and examine these dynamics in relation to the expansion of agrofuels and grains production. The fourth section of the essay explores the impact of rising investments and agribusiness expansion, associated to national and foreign investments, on the behavior of land markets through changes in land prices. Given the scale and regional diversity of land in Brazil, any analysis of land markets must

[1] It will not be dealt with fully in this article, but it is important to note that Brazil is not merely a 'victim', because the country does not 'fit the usual profile of a "fragile" or "weak" state' (as well put by Borras *et al.* 2011, 7), but also because it is a promoter of land grabbing in several Latin American countries, including Paraguay and Bolivia (Mackey 2011), and more recently, in African countries.

be careful to avoid generalizations. For this reason, our discussion of land prices focuses in particular on regions that are notable for the expansion of monocultures and commodities.

Finally, the paper analyzes measures that the Brazilian government has recently adopted with respect to land acquisition by foreign companies and individuals. We critically assess the political limits of such measures and examine these tools in relation to the demands of agrarian social movements that have been articulated in a campaign for a national land-ceiling.

Land dynamics: the historical demand for land in Brazil

The high level of concentration that characterizes the current structure of land ownership is not a new development in Brazil. This process was established in the colonial era and acquired its current form in the 1960s through the implementation of the Green Revolution and the modernization of large estates for agriculture and livestock production. In fact, land concentration, which is shaped by a lack of taxation on land, illegal private encroachment on public lands (vacant lands and secured lands), and the absence of official data on the actual situation of rural estates, can be seen as an important feature of the Brazilian countryside.[2]

Brazil's geographical area is equal to 850 million ha of land. Approximately half of this land area (436.60 million ha, or 51.35 percent) is registered as rural property in the National Database of Rural Registration (SNCR) of the National Institute for Colonization and Agrarian Reform (INCRA). Excluding the units of environmental conservation (102.06 million ha of national parks, extractivism reserves, national forests, etc.), indigenous peoples' areas (128.47 million ha), formally recognized public lands (4.20 million ha), urban perimeters and infrastructure perimeters (roads, hydroelectric plants, etc.), there is still an outstanding land area equal to 172.95 million ha (Sampaio *et al.* 2003).

This residual is confirmed by the official version of the 2nd National Plan for Agrarian Reform (PNRA): '50.86% of the total area of the Brazilian territory is yet to be registered' (Brasil 2005, 22). This immense area must be considered as public and vacant land ('*terras devolutas*'), that is, land without any legal possibility of private appropriation. This land has not yet been legalized, meaning that there is no formal registration of land ownership in any of the official data for approximately 20 percent of the Brazilian territory (Sampaio *et al.* 2003).[3]

If we consider the land situation of the Amazon region, these figures are even more alarming. From a total are of 509 million ha, 178 million, or 35 percent, of the

[2]The rural areas are the object of systematic measurement by two distinct public institutions, each using a different category of analysis. The National Institute of Colonization and Agrarian Reform (INCRA) accounts for the *rural estates* (located in the rural environment independently of their use or end activity). The Brazilian Institute of Geography and Statistics (IBGE) accounts for *agricultural and livestock establishments* (areas under one same administration, totally or partially dedicated to some exploitation activity linked to the agricultural and livestock sector) through the Agricultural and Livestock Census. Therefore, they both are self-declaratory (Sampaio *et al.* 2003).

[3]According to Wilkinson *et al.* (2010, 15), this level climbs to 24 percent if one considers only the lands of the Amazon, where many estates are not registered in any category, and "therefore, are technically considered as unallocated public lands". However, no official data is available on their real situation.

total of registered lands are privately occupied. However, of these 178 million ha that are declared

> ... as private property, 100 million hectares can be based on fraudulent documents. Another 42 million hectares of this area are categorized from the registry declarations as owned, and may or not be liable to a process of land regularization – depending, once more, on the circumstances of size, history and location. Thus, 30% of the area can be legally uncertain and/or become object of dispute (Wilkinson *et al.* 2010, 15).

One incongruous result of this situation is that 'cases of legal appropriation of vacant lands, collected lands, and irregularities in the registry lead to a situation in which landed estates represent an area that is greater than the total area of the state itself' (Brasil 2005, 22). In other words, 'land grabbing' or '*grilagem de terras,*' as it is known in Portuguese, is both a part of Brazilian history as well as a current reality.[4]

In this climate of statistical inconsistencies and illegal appropriations, one finds the classical concentration of land ownership in Brazil – a situation that is made possible by the total absence of rural taxation.[5] Data from the most recent Agricultural and Livestock Censuses (Brasil 2009), undertaken by IBGE (2006, 1995, 1985) confirm that land concentration, defined by estates with agricultural and livestock production greater than a thousand ha, exhibited relatively little change over the past twenty years (Brasil 2009). Gini indicators that measure the distribution of land continue to show the persistence of a high degree of land concentration in Brazil, with nearly constant indicators during this period: 0.857 in 1995/96, and 0.856, in 2006 (Hoffmann and Ney 2010).

Estates with less than ten ha represent over 47 percent of the total number of farm units but occupy only 2.7 percent, or 7.8 million ha, of the total area of rural establishments. At the other end of the land area spectrum, farms that are larger than one thousand ha correspond to only 0.91 percent of the total number of farms but concentrate more than 43 percent, or 146.6 million ha, of the total area (cf. Table 1). IBGE itself officially recognizes this high concentration of land when it concludes that 'the inequality in the distribution of land reveals both the past and the contemporary processes of natural resource appropriation in Brazil' (Brasil 2009, 107).

These data on land concentration also express underlying struggles for land by social groups for the purposes of life, production, consumption and citizenship – but whose demands for land continue to be denied. Attaining a greater level of access to land for these groups would require a more equitable distribution of rural estates as part of a process of social justice and land democratization (Leite *et al.* 2004, Leite

[4]According to Merlet (2010), the recent process of appropriation of large sections of land by foreign capital cannot be reduced to the mechanisms of "purchasing" land, since there is a movement of "appropriation and concentration of land – and natural resources – in large scale" (Merlet 2010). Other institutions, such as the *International Land Coalition* (ILC 2009; Taylor and Bending 2009), use the expression "commercial pressures on land", which may also restrict the comprehension of what we are currently observing (more details, see Borras *et al.* 2011).

[5]Brazil has the Rural Territorial Tax (ITR), but its levy is absolutely insignificant, as it represented, for instance, 0.01 percent of the total tax collection in 1996. According to Sabbato (2008), the reform of this tax in 1996 did not produce the desired effect of coupling this instrument of fiscal policy with the agrarian-reform program – quite to the opposite. An estimation presented by the author indicated that the tax evasion of the ITR in 1997 reached 90.3 percent (Sabbato 2008, 121).

Table 1. Number and area of agricultural and livestock estates by groups of total area: Brazil, 2006.

Groups of total area	Number of estates (units)	%	Area of the estates (ha)	%
Less than 10 hectares (ha)	2,477,071	47.86	7,798,607	2.36
10 to less than 100 ha	1,971,577	38.09	62,893,091	19.06
100 to less than 1000 ha	424,906	8.21	112,696,478	34.16
1000 ha and beyond	46,911	0.91	146,553,218	44.42
Total	5,175,489	100.00	329,941,393	100.00

Source: IBGE, Censo Agropecuário 2006 (Brasil 2009, 107).

and Ávila 2007). According to the IBGE data, there are 2.5 million small farms with land, but at an average of less than ten ha this is an insufficient area to generate income (Brasil 2009), to maintain a family and secure a minimum quality of life in rural areas (Brasil 2005, 18).

The 2nd PNRA estimated that in 2005 there was 'a total of 3.1 million families,' so-called 'landless people,' who are 'rural workers without access to land, including small-scale agricultural producers – proprietors, partners or leaseholders' (Brasil 2005, 17). This historical demand for land amounts to over 5.6 million families, with approximately 200 thousand landless families that are camping out in hundreds of land occupations in Brazil.

According to INCRA data, out of a total of 436.6 million ha of land that are registered in the SNCR, 120.4 million ha have been declared as under-productive or un-productive (Sampaio *et al.* 2003). Wilkinson *et al.* (2010, 14) affirm that 'there is a lot of land that is either idle or has an extremely low intensity of use' in Brazil. In other words, more than 14 percent of the land registered at the state office could become a supply of land to meet the demand from landless families. However, in an environment of juridical insecurity and illegal appropriation, this possibility has been suppressed through various means, particularly in the northern region of the country.

It is crucial to note here that the concept of 'under-productive' or 'un-productive' land[6] does not represent the concept of 'marginal land' used in the World Bank survey for 'empty, under-utilized and available' land (Borras and Franco 2010, 12), as used or suggested by the World Bank (2010, 56) to calculate total amount of available land to be used for supplying future demands for production. The majority of these lands are inhabited and used by indigenous people, pastoralists (Borras *et al.* 2011, 13), squatters, peasants and other traditional communities or social groups (Leite and Ávila 2007; Wilkinson *et al.* 2011), in addition to being appropriated (in many cases illegally) by large landlords as part of historical territorial disputes (Toulmin *et al.* 2011, 26). Simply put, these lands are not 'available for taking' (Borras *et al.* 2011, 14).

The recent expansion of Brazilian agribusiness toward the central and northern regions, including the Amazon, increases this appropriation by the private sector,

[6]Such concept of under-productive and/or un-productive is referred to legal standards for measuring the degree of use of land and the economic efficiency on cultivating it, but that does not mean that is vacant, empty or available (Borras and Franco 2010). It means the land is or is not accomplishing one criterion of its social function, resulting in one of the causes for conflicts and disputes over land (Delgado 2005, Sauer 2010b, Hoffmann and Ney 2010).

both domestic and foreign, and leads to an increase in land concentration (Leite and Wesz Jr. 2010). According to the World Bank (2010), the increase in land acquisition for agricultural production is related to only eight commodities: corn, soybean, sugarcane, palm oil, rice, rapeseed, sunflower and planted forest. The participation of Brazilian agribusiness occurs primarily in the three first products – corn, soybean and sugarcane – and is shaped by rising prices of agrofuels as well as government crop subsidies (World Bank 2010). Approximately 23 percent of the increase of worldwide agricultural production is attributed to the expansion of 'agricultural frontiers' such as those in Brazil, despite of the fact that the more pronounced increase (estimated at 70 percent) in production is the result of improvements in productivity (World Bank 2010).

According to the World Bank (2010), the reasons for the expansion in production and land transactions include: (a) demand for food, feed, cellulose and other industrial inputs due to increases in population and income; (b) demand for raw materials for agrofuels production that reflects the policies and demand by the main consumer countries; and (c) shifts in the production of commodities to regions with abundant land, cheaper prices and good prospects for productivity increases (World Bank 2010).

The expansion of agricultural production is an important determinant of the current land rush since the 2008 food price spike. However, this 'dominant, mainstream narrative' (Borras *et al.* 2011) does not fully explain it all, since there are 'broader international political economy processes involving lands beyond food production' (Borras *et al.* 2011, 6). The increasing demand for land is due not only to high commodities prices for food crops (Sauer 2011), but also includes 'land for conservation, carbon offset arrangements (e.g. REDD+), industrial tree plantations' (Borras *et al.* 2011, 14), extraction of minerals (Safransky and Wolford 2011), timber (Hecht 2005) and agrofuels production (Wilkinson *et al.* 2010).

Perhaps most importantly, investment in land expresses 'business opportunities' not only for the production of agricultural and non-agricultural commodities like we will see in the next topic related to the expansion of agribusiness, but also for land speculation (Delgado 2005). Speculative investments take advantage of lower land prices and develop in combination with the expansion of agricultural frontiers (Leite and Wesz Jr. 2011), government incentives (Sauer 2010a) as well as public infrastructure investments in areas such as roads, railways, waterways, ports, and energy (Safransky and Wolford 2011, Hecht 2005). Speculative and productive investments should therefore be viewed in unison as the financial appreciation of rural property (Rangel 1986) in the context of public investment creates 'a situation in which land itself became a medium for capturing economic rents' (Hecht 2005, 387), or, as we can see in the Argentine case, the agricultural production itself is object of the 'financialisation' process, like '*pool de siembra*' and '*fideicomizos*' initiatives (Murmis and Murmis 2010). It's curious to note that a major group of Argentine soybean producers invests now on land in Mato Grosso, Brazil – and reproduces the 'Argentine model' of management and production in the agricultural sector.

Recent expansion of Brazilian agribusiness

During the international crisis of the early 1980s, the rural sector in Brazil was under pressure to develop and expand the capacity to export processed agricultural products in order to generate revenue for foreign-debt obligations and produce a trade surplus. Agricultural exports during this period represented a positive contribution to the trade balance, and agriculture continued to play a vital role in the 'external offensive' (i.e. exports and outward investments) during subsequent years, especially after the devaluation of the currency (Real) between 1999 and 2004.

By the 1980s, shifts in grain production had established a powerful occupation of the Cerrados[7] biome that includes the center-west region comprising the states of Goiás, Mato Grosso do Sul and Mato Grosso, in addition to the state of Minas Gerais in the central region and the state of Bahia. In the past decade, a vigorous expansion of the 'agricultural frontier' also incorporated new areas located in the states of Maranhão, Piauí and Tocantins, as well as the Amazon region (Leite and Wesz Jr. 2011).[8]

In the following maps, we present the expansion of soybean and sugarcane production since these crops are widely known as representative of agribusiness and are two of the eight commodities identified with the increase in agricultural production according to the World Bank study (2010). The expansion of soybean production as well as that of sugarcane (albeit to a lesser degree) did not occur in isolation from other production systems (Leite and Wesz Jr. 2010). In many cases, these crops advanced into areas that were initially used for cattle raising, or, alternatively, woodlands and forests such as in certain regions of the state of Mato Grosso and the Amazon. These areas were later substituted by the production of rice in order to 'tame the land' and were subsequently converted to the production of soybeans (Heredia *et al.* 2010).

Figure 1 shows the dramatic process of territorial expansion for the production of soybeans between 1990 and 2008.[9] Initially concentrated in the southern region of the country, soybean production later extended forcefully into new areas of the Cerrado biome in the center-west as well as the central and northeastern regions. In the current decade, the expansion towards the Amazon represents 'a two-pronged strategy: agricultural *extensification* and *intensification*' (Borras *et al.* 2011, 15), especially in the Humaitá region of the state of Amazonas, and the southeastern area and western area surrounding Santarém in the state of Pará. This represents the 'heart' of the Brazilian Amazon (Leite and Wesz Jr. 2010).

This expansion forms part of an ongoing contest for land in these frontiers. Environmental organizations and native communities have engaged in protests in areas where the occupation of land for soybean production was based on deforestation (Hecht 2005) and/or a forced displacement of small farmers, squatters or indigenous people, as in the plateaus (chapadas) in the state of Piauí (Heredia *et al.* 2010). Agribusiness interests have pressed for changes in legislation in an

[7]The Cerrado is a vast tropical biome, which could be considered as savanna, but with the presence of gallery forests. It is an especially large eco-region of Brazil, located mainly in the central and north parts of the country, and characterized by rich biodiversity (an enormous range of plants and animals), but less than 3 percent of its area is protected by law.

[8]Nonetheless, we must highlight here the effort undertaken by the Brazilian Agricultural Research Corporation (EMBRAPA) in the exploitation of soils previously considered as inappropriate, and in the production of seeds adapted to the edaphoclimatic conditions that prevail in these regions (Heredia *et al.* 2010).

[9]The authors appreciate the kind contribution of Valdemar Wesz, Jr., in the creation of the cartograms. A part of the data of these paragraphs is based on the contribution of Heredia, Palmeira and Leite (2010).

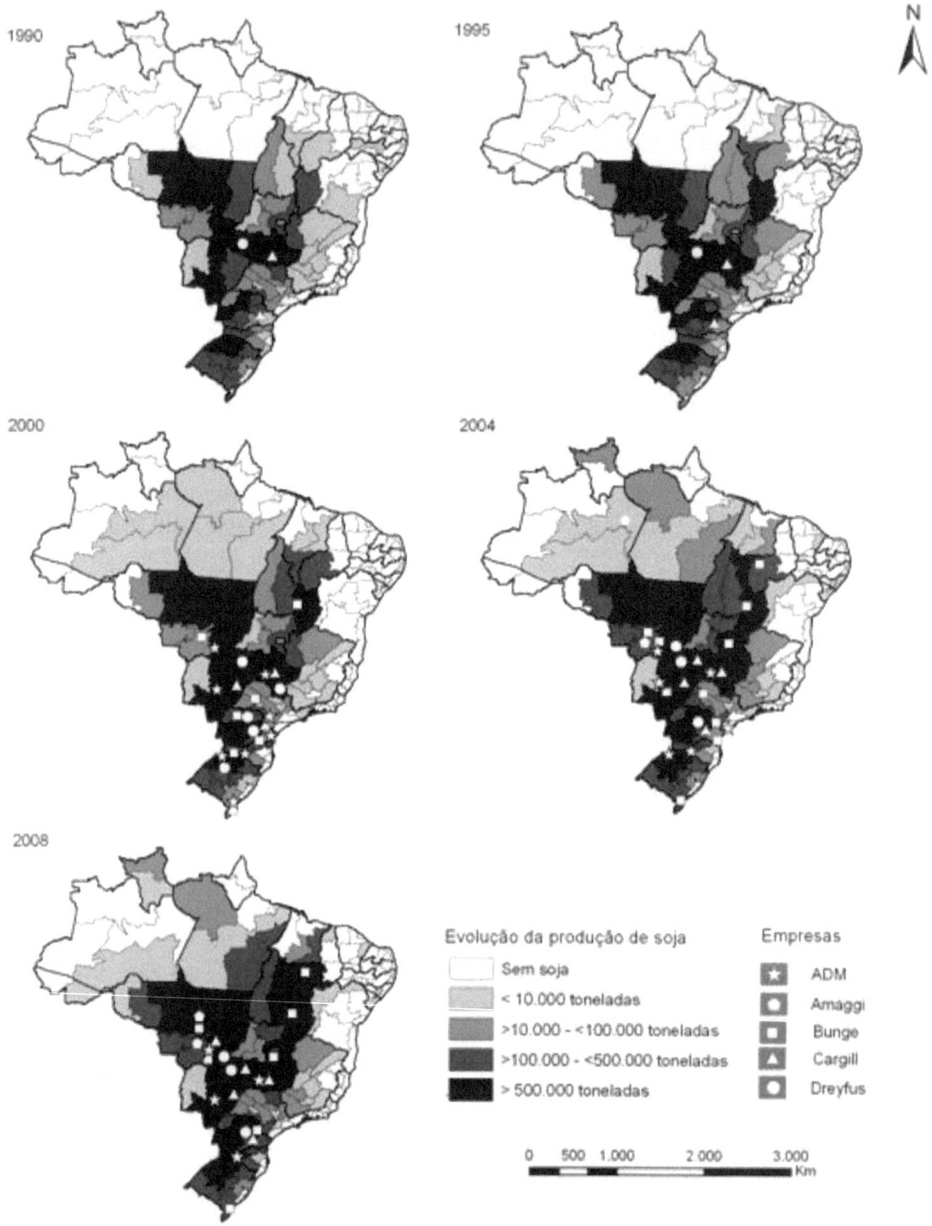

Figure 1. Expansion of soybean production and location of main agricultural industries. Color gradations indicate the presence of soybean production and its spread towards the north of the country. The darkest band, which resembles the letter 'Y', represents the areas with the highest concentration of production (Leite and Wesz Jr. 2010).
Source: IBGE (PAM) cited in Wesz Jr. (2011), and Heredia *et al.* (2010).

attempt to open new expansion fronts, most notably in the 'cerrados' and rain forest biomes (Leite and Wesz Jr. 2010). This consists of pressure to loosen the environmental legislation that establishes a Forest Code, Areas of Legal Reserve, Areas of Permanent Protection and a National System of Conservation Units, as well as territorial legislation that establishes the legal rights of native communities based on a collective/traditional form of occupation in 'Quilombo' (communities of afro-brazilian slave descendants), indigenous peoples' lands, common pasture, and extra-ctivist areas.

Figure 1 also shows the location of agro-industrial units that crush soybeans for the production of feed (bran), oil, feed and other products. The leading players during this period are represented by the so-called ABCD-group of multinational trading companies (ADM, Bunge, Cargill, and Dreyfus) (Wesz Jr. 2011). The share of international capital in the agro-industrial grain processing sector was equal to 16 percent in 1995 but increased dramatically to 57 percent in 2005 – highlighting the powerful tendencies of economic concentration and de-nationalization in this sector (Wesz Jr. 2011, Leite and Wesz Jr. 2010).

Figure 1 also presents the location of the Maggi group (or Amaggi) of companies headed by Blairo Maggi, the former governor of the state of Mato Grosso who is also a powerful domestic actor in the production of soybeans. A spatial shift in the position of these companies in the late 1990s expresses a change from the previous strategy of purchasing units from national groups in the mid-southern region to the current strategy of constructing new industrial plants in the 'cerrado' areas. This region is now a 'major center' that concentrates 72 percent of the production value of these units (Heredia *et al.* 2010, Wesz Jr. 2011). According to Wilkinson *et al.* (2010), this has also produced the emergence of new real-estate brokerage firms linked to the 'clearing', preparing and selling of new areas to both domestic and foreign private sector groups in the grain and sugarcane/ethanol industries (Wilkinson and Herrera, 2010).[10]

It is also important to briefly mention that the sugarcane and sugar-alcohol sector has engaged in a similar process of territorial expansion in areas that were previously producing food and non-food commodities, as well as areas that are still the object of conservation. The geographical expansion of sugarcane plantations is primarily concentrated in the state of São Paulo (Novo *et al.* 2010) as well as Minas Gerais, Mato Grosso do Sul, Goiás and Paraná (Pietrafesa *et al.* 2010).

This expansionary tendency is largely oriented to the production of ethanol which is an important component of the National Program of Agricultural Energy under the Ministry of Agriculture, Livestock and Supply (MAPA) and whose production 'has expanded sharply from 15 billion liters in 2003 to 25 billion liters in 2008–09' (Wilkinson and Herrera 2010, 751). This can be seen to rival – or complement, depending on competing perspectives – the Brazilian National Biodiesel Program (PNPB) that promotes the production of biodiesel.[11]

In addition to questions of food security, energy independence (Novo *et al.* 2010), and environmental legislation (which have led various movements and organizations to call for more sustainable environmental programs), the trend of agribusiness expansion poses additional key debates in Brazil. First of all, even though agro-industry is a central and dynamic element of economic growth for certain regions,

[10]The World Bank (2010) estimated that the total cultivated area with raw materials for agrofuels was of 36 million ha in 2008, that is, twice the 2004 area. Of this total, 8.3 million ha are located in the European Union (with rapeseed crops), 7.5 million in the United States (with corn) and 6.4 million ha in Latin America mainly areas of sugarcane crops in Brazil (Wilkinson and Herrera 2010).

[11]For further details and discussion on the Brazilian National Biodiesel Program (PNPB), see Wilkinson and Herrera (2010) and Novo *et al.* (2010).

this does not necessarily produce broader outcomes of regional development. Achieving a more inclusive process of development would entail governmental policies that ensure and enforce compliance with the provisions for social rights as an integral part of these strategies (Pietrafesa *et al.* 2010).

Part of this process of expansion has been driven by the appropriation of wealth by international capital (Wilkinson and Herrera 2010). Both in the purchase of agro-industrial processing firms, as well as in the purchase of vast rural estates, the transfer of national patrimony to the hands of foreign capital is an ongoing process that demands improved regulation by the Federal Government (Hermele 2011).

Historically, small-scale and family-based agriculture has had a relatively limited importance in the process of soybean expansion since this sector is responsible for only 16 percent of national production (Brasil 2009). The importance of soybeans for the production of biodiesel contradicts the initial goals of social inclusion that were envisioned in the PNPB through the incorporation of oilseeds production (castor oil plant, in particular) by family-based producers in the semi-arid northeastern regions (Wilkinson and Herrera 2010). Data from IBGE (Brasil 2009) clearly show that the regions with the highest degree of concentration of soybean production, as we discuss above, are dominated by commercial markets and agroindustrial firms rather than cooperatives of small and medium farmers concentrated in the South and North Regions of Brazil (Leite and Wesz 2010).

The majority of the soybeans produced in the Brazilian 'cerrados' (Mato Grosso, in particular) are directed to industrial processing destined for the international market, and we can infer that part of this production serves as raw material inputs for biodiesel since a significant portion of the established processing and production is located in the center-west 'cerrados', principally in Mato Grosso (Wilkinson and Herrera 2010).[12] This concentration of processing in the center-west also jeopardizes the attempt to make the semi-arid region (almost 80 percent of the area of Brazil's Northeast) a 'geographic priority' region of the PNPB, despite the recent efforts by the company Petrobras Biocombustível (Petrobras Biofuel) to concentrate its processing plants in this area (Flexor *et al.* 2010, Wilkinson and Herrera 2010).

The evolution in the expansion of sugarcane production can be seen in Figure 2. These maps exhibit the concentration of production in the southeast region, in particular, in São Paulo and Minas Gerais, and its expansion towards Mato Grosso do Sul, Goiás and Paraná (Pietrafesa *et al.* 2010). This occurs to the detriment of 'traditional' regions of production of this commodity, including the northern state of Rio de Janeiro, as well as the northeast region such as the coastal region in Pernambuco, northern Alagoas and southern Paraíba states (Leite and Wesz 2010).

Sugarcane data collected by IBGE show that the location of production is strongly linked to commercialization through processing plants and distilleries for the production of sugar and ethanol (Brasil 2009). Since sugarcane generally operates in more proximate distribution circuits (differently from the case of soybeans, sugarcane is not exported *in natura*), a territorial concentration is created in the regions of crop production, principally in the state of São Paulo (Brasil 2009).

[12]According to the information provided by Odacir Klein, President of the Brazilian Biodiesel Union (UBRABIO), in a public hearing in the Federal Senate in 2007, soybean was responsible for around 85 percent of the national biodiesel production, a datum that has not changed significantly in recent years.

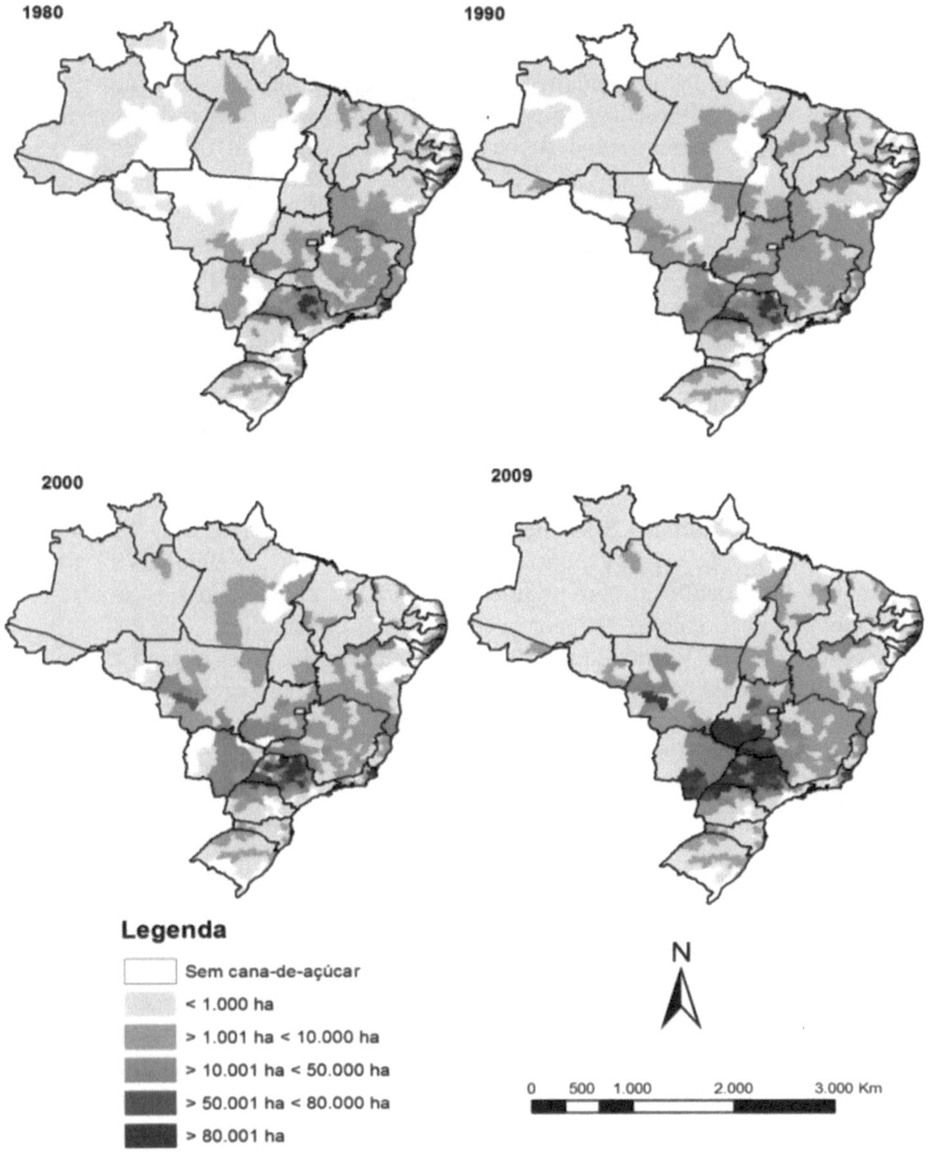

Figure 2. Territorial data on the production of sugarcane in Brazil, 1980 to 2009.
Source: IBGE (PAM).

The increasing land use for sugarcane plantations has displaced areas that were previously used for food production for the regional market, such as milk production or for raising cattle (Novo *et al.* 2010).[13] This phenomenon is particularly evident in the state of São Paulo and in the Triângulo-Mineiro region in Minas

[13]Like in South Africa (Hall 2011), there are several national and international investment funds investing especially in sugar production, which previously focused on hedge funds in financial markets, but now are buying sugar-cane mills and land in Brazil (Wilkinson and Herrera 2010, 751).

Gerais. In the case of the expansion to the center-west region (Goiás and Mato Grosso do Sul) as well as the region of Uberlândia in Minas Gerais, sugarcane plantations confront grain and livestock producers (poultry and swine) that are already established in these locations (Lima 2010). This situation produces territorial disputes between sugarcane and soybean strategies, which, in turn, has led some municipalities in these regions to seek limits on sugarcane plantations (Pietrafesa *et al.* 2010).

Sugarcane crops directed primarily to ethanol and agroindustrial production present a clear case of foreignization in terms of production and processing, perhaps less so in terms of large-scale land acquisition (Cotula 2012) but on agriculture and its products. International data and existing literature (Toulmin *et al.* 2011, World Bank 2010, Borras and Franco 2010) indicate that agrofuels demand has been a main driver in the global land rush (Cotula 2012, 13), but even more so in the process of concentration of agricultural production, as is shown in the case of sugarcane ethanol in Brazil. According to Wilkinson and Herrera (2010, 751), 'foreign investment has been traditionally low' in this sector in Brazil, but this changed in the period after 2003, and by 2010 the participation of foreign capital 'had already reached 20 percent'. This expansion process has been accelerated by foreign investments in agricultural commodities and directly in land, in the context of a wider increase in foreign investment in sectors such as mining.

According to the World Bank (2010) the current international land demanders are: (a) governments concerned with the ability of domestic production to achieve sufficient food for domestic consumption, especially after the food crisis of 2008; (b) financial entities that, in the current climate, find comparative advantages in the purchase of land; and (c) agro-industrial firms that, due to the high concentration level of trade and processing, seek to expand their businesses. Wilkinson *et al.* (2010, 32ff.) propose that the types of investors and motives for investment be divided into eight 'categories': (a) capital that represents core agribusiness activities; (b) agribusiness capital in related and/or converging sectors; (c) non-traditional capital in agribusiness in response to new synergies; (d) rural brokerage companies that emerged in response to the increase of land value and outlook of Brazilian agribusiness; (e) countries that are wealthy in terms of capital, but poor in terms of natural resources who seek to secure their food and energy supply; (f) investment funds attracted by the prospect of price appreciation for agricultural commodities; (g) investments in connection with incentives for environmental services; and (h) mining and oil-prospection companies. In general terms, a significant portion of the investment in land as part of the expansion for products such as soybeans and sugarcane has originated – directly or indirectly – from international capital or, at the very least, through ties with international capital. This reinforces the argument of Wilkinson *et al.* (2010) that international groups (companies, individuals or governments) have directed their resources to agriculture and livestock sectors not only for activities of processing raw materials, as it used to be in the past, but also in order to acquire rural property, such as in the initiatives known as 'Greenfield Projects' that we examine below.

The process of foreign appropriation of land in Brazil

Alvim (2009) shows a sharp upturn in the total level of foreign direct investment (FDI) towards Brazil, from USD 4.33 billion in 2002 to USD 8.98 billion in 2008 –

representing an increase of 107 percent. According to the newspaper *O Globo*, the Institute for Applied Economic Research (IPEA) finds that FDI in Brazil's primary sector leaped from USD 2.4 billion in 2000 to USD 13.1 billion in 2007, for a 445 percent increase which was led by mining (71 percent of the total investment received in 2008). There was also an increase in the foreign participation in agricultural and cattle-raising activities, such as the sugarcane and soybean crops, and the production of alcohol and biofuels, that occurred through purchases and mergers of Brazilian companies (Alvim 2009, Pretto 2009).[14]

Thus, following the trend of an increment of foreign investment in Brazil over recent years, there was an increase in external participation in agriculture and livestock activities,[15] such as, for instance, in the cultivation of sugarcane and soybean, and in the production of alcohol and biofuels (Alvim 2009, 53). A significant part of this investment was used for purchases and mergers of already-existing companies (Alvim 2009, 52), and 'the total FDI in the agribusiness sector was of 46.95 billion [US] dollars' between 2002 and 2008 (Alvim 2009, 47).

According to Alvim,

> ... the industrial production of biofuel was the activity that presented a clear growth-trend in the absorption of foreign investment in Brazil, and was mostly concentrated in the states of the Southeast Region. The FDI in alcohol and biofuels increased from USD 4 million in 2002 to USD 1.64 billion in 2008 (2009, 55).

Using distinct sources of information, including researches in the SNCR, but also along with consultancy companies in the field, among other sources, national newspapers have published data on this process of land purchase by foreigners, without counting on 'a due correspondence between the SNCR/INCRA databases' (Pretto 2009, 4). In mid-2000, the main national newspapers were publishing articles on the process of land appropriation by foreigners ('foreignization') in Brazil (Scolese 2008, Chade 2010).

From the SNCR data, the newspaper *Folha de São Paulo*, for instance, calculated the rhythm of this process of land appropriation by foreigners for a six-month interval. According this newspaper, 1533 rural estates were purchased by foreigners between November 2007 and May 2008 in Brazil, totaling an area of 226,920 ha. In the same period, there were also property sales, but the balance was of 1372 estates purchased by individuals of other nationalities, totaling an area of 205,320 ha (Scolese 2008).

In 2010, from the analyses of the INCRA Registry (SNCR), the newspaper *Folha* publicized data on the advance of foreign capital over Brazilian land. According to this article, 'companies and individuals of other countries are purchasing the equivalent to 22 soccer fields of land every hour in Brazil. In two and a half years,

[14]It is important noting that direct investments shall be seen embedded in projects carried on as paths for such investments. According to Safransky and Wolford (2011, 5), "extensive infrastructure corridors have been constructed across the continent" and "to link extractive frontiers to metropolitan areas and foreign markets" attracting investments and build "trends for land grabbing in the Americas".

[15]Until 2000, the share of the primary sector was only 2.3 percent of the total foreign resources invested in Brazil. However, this share reached almost 14 percent in 2007, while the expansion of foreign investment in the industrial sector was of 33.4 percent in the same period (Duarte 2008).

foreigners have purchased 1152 estates, totaling 515.1 thousand hectares' (Odilla 2010).

The investment forecast for the Brazilian primary sector, especially in the production of agro-energy forms, was already advertised in the mid-2000s. According to the data made available through the Internet,

> Market estimations point that, up to the mid-years of the decade, the sector shall receive something around US$ 25 billion of foreign investment. At least 40% of this total will be aimed at the sugar and alcohol sector. Along with ethanol and sugarcane, international investors are interested in taking part in reforestation projects, in rubber-tree plantations, and in the purchase of land in Bahia, Mato Grosso and Rondônia, in order to implement agricultural and cattle-raising projects (Guimarães 2006).

Despite the absence of a more systematic survey, it is possible to conclude that this foreign investment in Brazil's primary sector also results in a wide-scale purchase of land. According to a research in the INCRA database (SNCR), there were 34,632 entries of estates in the hands of foreigners in 2008, with a total area of 4,037,667 ha (Pretto 2009). These are quite impressive figures, considering that the study did not include the 'rush-for-farmland period' after the 2008 crisis (Pretto 2009). We also highlight that over 83 percent of this total refers to landed estates categorized as large farms (larger than 15 fiscal units).[16]

Table 2 was obtained through the study by Pretto (2009, 21), and illustrates the historical process of rural-estate purchase by foreigners. As can be seen in the table, the highest number of estates and the highest quantity of land registered by foreign individuals and companies took place in the recent period. Indeed, the 1980s, 1990s and 2000s show a concentration trend, corresponding respectively to 29.9 percent, 18.2 percent and 18.8 percent of the total number of estates registered by foreigners, and to 27.9 percent, 25.9 percent and 20.1 percent of the accounted area. From the 1990s we can note a slight fall in the total number of properties registered. This can be partly explained by the fact that foreign groups began to acquire land by resorting to forging alliance with 'national companies', as we shall see.

As a matter of fact, the process of purchasing Brazilian land by foreigners is not recent. An important landmark in this historical course was the Japanese-Brazilian cooperation in the implementation of distinct stages of the Japanese-Brazilian Cooperation Program for the Development of the Cerrados (so called Prodecer, or 'JICA' because it was financed by the Japan International Cooperation Agency), designed in the mid-1970s and executed in the 1970s and 1980s in several 'cerrado' regions, especially in the states of Minas Gerais, Goiás, Bahia and Mato Grosso (Bertrand *et al.* 1991, Salim 1986).[17]

An update of these figures for 2010, using the INCRA database (SNCR) too, indicates – as can be seen in Table 3 – 34,371 rural estates owned by foreigners,

[16]The fiscal unit is a measure expressed in ha for a family to dwell and work on, considering factors such as, for example, soil fertility, predominant agricultural exploitation, and income from this exploitation. Such aspects, among others, resulted in fiscal units from 5 or 10 ha, in the South or near large cities, to 100 or even 120 ha in the Amazon region. Estates larger than 15 fiscal units are, according to the Constitutional limit, considered 'large properties'.

[17]The same process of agricultural expansion was undertaken by Brazilian landlords who also moved to neighbor countries during the 1980s and 1990s. For example, according to Mackey (2011, 12), available estimations say that 'Brazilians have landholdings of 1.2 million ha, with 500,000 ha of agricultural land and 700,000 in cattle ranching in Santa Cruz', Bolivia.

Table 2. Land purchase (in number of estates and area) by foreigners in Brazil, 1900–2000.

BRAZIL – Relative share of registries and areas by decade				
Decade	% registries	Accumulated % of registries	% of area	Accumulated % of area
1900	0.012	0.012	0.0004	0.0004
1910	0.047	0.058	0.064	0.065
1920	0.143	0.202	0.170	0.235
1930	0.415	0.617	0.309	0.544
1940	1.207	1.824	0.971	1.514
1950	4.015	5.839	2.551	4.065
1960	8.773	14.612	6.194	10.259
1970	17.421	32.033	15.040	25.298
1980	29.888	61.921	27.940	53.238
1990	18.201	80.122	25.925	79.163
2000	18.783	98.904	20.152	99.315
No date	1.096		0.685	
Totals		100.000		100.000

Source: SNCR/INCRA – special research 2008 (Pretto 2009, 21).

Table 3. Number and area of rural estates owned by foreigners in Brazil, May 2010.

State	Number of Estates	%	Area (ha)	%
Rondônia	119	0.35	29,242.00	0.67
Acre	26	0.08	13,799.68	0.32
Amazonas	307	0.89	232,021.68	5.33
Roraima	66	0.19	27,729.49	0.64
Pará	1143	3.33	235,628.39	5.42
Amapá	15	0.04	6,228.00	0.14
Tocantins	181	0.53	109,517.18	2.52
Maranhão	184	0.54	70,135.35	1.61
Piauí	82	0.24	58,770.32	1.35
Ceará	401	1.17	34,734.45	0.80
Rio Gde Norte	128	0.37	20,806.69	0.48
Paraíba	248	0.72	6,828.47	0.16
Pernambuco	368	1.07	9,667.19	0.22
Alagoas	101	0.29	13,577.66	0.31
Sergipe	81	0.24	3,439.45	0.08
Bahia	2192	6.38	368,888.05	8.48
Minas Gerais	2639	7.68	491,548.57	11.30
Espírito Santo	304	0.88	19,770.66	0.45
Rio de Janeiro	2110	6.14	85,284.78	1.96
São Paulo	12,291	35.76	491,437.42	11.30
Paraná	5130	14.93	299,061.84	6.88
Santa Catarina	1290	3.75	54,605.77	1.26
Rio Gde Sul	1895	5.51	113,801.07	2.62
Mato Grosso Sul	781	2.27	473,325.65	10.88
Mato Grosso	1229	3.58	844,279.92	19.41
Goiás	843	2.45	230,629.91	5.30
Distrito Federal	217	0.63	4,314.36	0.10
Brazil Total	**34,371**	**100.00**	**4,349,074.00**	**100.00**

Source: INCRA, May 2010. Adapted by the authors.

encompassing a total 4,349,074 ha. By this table, it is possible to see that the largest numbers of estates are concentrated in the states of São Paulo (35.7 percent of the total), Paraná (14.9 percent), Minas Gerais (7.68 percent) and Bahia (6.38 percent). When we consider area, the figures change, and point to a leading role by the states of Mato Grosso (19.4 percent of the total area), Minas Gerais and São Paulo (with 11.3 percent each) and Mato Grosso do Sul (10.9 percent).

Our access to the database did not allow us to identify the dates of purchase/ access to these estates, which made difficult a more precise view of the trend of land purchase by foreigners in the recent years. Yet, considering the previously-mentioned World Bank study (2010), there is an intensification of this investment flow starting in 2008. Thus, with the data presented by Pretto (2009), in regard to the position of the registry in 2008, we can make an approximate comparison in order to verify the (absolute and relative) variance through almost two years (2008–2010).

We find an increase in the number of estates (an addition of 153 registers, representing a variance of 0.45 percent in relation to the data of 2008), but above all, a significant increase in the quantity of land incorporated by foreigners: approximately 312 thousand ha, representing an estimated variance of 8 percent in relation to the previous information, in a short period of time, as our data refer only to May 2010.

The strongest positive variation in the number of estates between 2008 and mid-2010 took place in the states of Minas Gerais, Roraima and Rio Grande do Norte. When we consider the area of the incorporated land, a quite different figure appears, indicating Amazonas (with a brutal increase of 120 percent in the area of registered land), Piauí (77.7 percent), Minas Gerais (57.4 percent) and Espírito Santo (43.9 percent) as the most expressive states. A part of this trend, one may expect, is linked to the expanding agricultural, livestock and agroindustrial activities in the sectors of grain and sugarcane, and also in the forest sector.

When it comes to arriving at conclusions, a measure of caution is needed in regard to the statistical data, as one may see in the figures on the origin of the capital invested in land assets (in the INCRA registry, around 35 percent of the data related to this item were either inexistent or invalid). In spite of that, it is possible to reach some considerations on the countries of origin of these land-investment flows, as shown by Table 4.

Table 4. Origin of the capital invested in land in Brazil, 1900–2010.

Country	Area (ha)	%	%*
Portugal	1,030,119.42	23.68	36.95
Japan	432,469.84	9.94	15.51
Italy	256,145.06	5.89	9.19
Lebanon	172,696.63	3.97	6.19
Spain	127,499.12	2.93	4.57
Germany	123,667.19	2.84	4.44
Netherlands	114,189.29	2.62	4.10
Subtotal	**2,787,713.56**		
Others	530,927.01	12.21	–
Inexistent Data	1,208,690.22	27.79	–
Invalid Data	352,598.26	8.11	–
Total	**4,349,002.04**	**100.00**	

Source: INCRA, May 2010. Data reworked by the authors.
*These percentages are calculated excluding invalid and inexistent data.

Table 4 shows all registers since 1900, including those with either inexistent or invalid information of origin. The table shows seven countries with the strongest participation in the total of registered estates: Portugal, Japan, Italy, Lebanon, Spain, Germany and the Netherlands. As the time-reference of the table includes registers since the beginning of the past century, it is evident that there is a significant presence of countries with a strong participation in the processes of colonization and immigration experienced by Brazil.

No other more detailed record exists that would allow us to make a cross examination of the origin of the capital, including the purchase dates, the size of the estates and their geographical locations. However, we can affirm that practically three-fourths of all real estate registers refer to the occupation or ownership by individuals, while the remaining (26 percent) refer to companies. As Wilkinson *et al.* (2010) point out, in this latter case, the identification of foreign capital is more complicated, because many investors make use of national companies in order to incorporate these areas.

Land market, fluctuations in the prices of estates and the dynamic of land

Another essential aspect of this discussion is the impact of this expansion of agribusiness and foreign investment on the land market and on price fluctuations. According to the newspaper *Valor Econômico,* the projects for sugar and alcohol executed between 2008 and 2010 led to an increase in the land value in new frontiers that are located mainly in the states of Tocantins, Goiás, Mato Grosso do Sul and Minas Gerais (Batista 2010). According to this article, a study by *NAI Commercial Properties*, a North-American multinational company specializing in real estate markets, shows that land value has strongly increased since 2009, with up to a 33 percent leap in the municipality of Pedro Afonso (state of Tocantins). Also according to the newspaper,

> Some regions of Goiás also recorded one of the highest value-increases, pushed by sugarcane. A market study by NAI in the region of the municipality of Edéia found that the hectare, which cost R$ 8.5 thousand in 2009, is now negotiated at R$ 10 thousand – a 17%-increase. It is in the municipality of Edéia that we find the project of the Tropical Plant, a partnership by British Petroleum and the Maeda Group – now sold to the Arion Capital fund – and by Santelisa Vale, which is now controlled by the French group Louis Dreyfus (Batista 2010, B12).

This increase in prices has impacts on other policies (along with the agricultural policy of incentives to the sector), but especially on the agrarian policies. Besides stirring up the territorial disputes, the land policies are harmed because, among other factors, it is more expensive to expropriate and indemnify the estates for the purposes of an agrarian reform, through the creation of settlements (Adams 2010, 2).

On the other hand, it is crucial to consider that a significant part of foreign investments are funded with public resources (Sauer 2010a), especially resources of the National Bank for Economic and Social Development (BNDES) (Wilkinson and Herrera 2010) and of the Constitutional Fund of the center-west (FCO). These loans and fiscal incentives are now allocated mostly in the region of expansion of sugarcane and ethanol (Midwest), and soybean (Midwest, Amazon, Bahia and Tocantins) (Pietrafesa, *et al.* 2010).

According to the magazine *O Focus* (2010), BNDES is the largest source of credit for sugar ethanol production since 'the sugarcane complex occupies a strategic position in Brazilian development strategies' (Wilkinson and Herrera 2010, 752). In 2007 and 2008, the bank 'funded 59 ethanol projects, and a good part of the funding was used in the purchase of equipment for co-generation of electric energy from bagasse, oftentimes in regions of sugarcane expansion, such as Mato Grosso do Sul and Goiás' (Focus 2010). In other words, public resources have kept the land market heated, also enabling foreign investments in the production of agricultural commodities.

It had long been known that land purchase represented a solid alternative for investment capital in Brazil (Brandão, 1988). This occurred either with a direct-appreciation intention[18] (in an investment type that, in spite of not reaching the highest peaks in terms of earnings, had no depreciation risk), or with the goal of using the land as a 'password' to access the strong subsidized rural-credit policy of the 1960s and 1970s (Delgado 2005). More recently, the increasing demands for agrofuels and agriculture commodities are 'thought to be a window of opportunity provided initially by natural resource competitiveness for assuming technological leadership in a globally dynamic sector' (Wilkinson and Herrera 2010, 752).

For a brief historical retrospective on the behavior of the prices of purchase and sale for rural estates, Wilkinson *et al.* (2010, 70) present a time-categorization that portrays a market cooling after Brazil's process of macroeconomic stabilization, with the adoption of Plano Real in 1994, after decades of war on inflation. Thus, in the mid-1990s, the market experienced an ebb in the prices of rural estates. However, as Wilkinson *et al.* (2010) astutely remark, already in the early 2000s, the prices had quickly recovered, in the wake of the good performance of the prices of the main commodities, spearheaded by soybean, until 2004. What followed next was a period of stability in the price of areas with crops, which was then greatly accelerated starting in 2007–2008 – according to the authors, backed by the investments made on the production of ethanol. With the financial crisis of 2008, there was a short-term negative impact on the behavior of prices, which nonetheless did not affect the more strategic regions for the production of commodities.

Thus, it is possible to state that, in general, the process of expanding the productive areas based on export-directed monoculture farms had an additional effect on the recent increase in the prices of rural estates of these regions (Leite and Wesz 2010). It is first necessary to note that, up to the late 1990s, the price of land areas with crops had leaped significantly in average terms for the Brazilian case, as we can see in Figure 3, which presents data on the total for the country and on a few selected states (São Paulo, Pará and Mato Grosso).

We can see that the average price per ha of crop lands, in the Brazilian case, leaped from R$ 1188.30 in June 1994 to R$ 7490.40 (an increase of more than 430 percent)[19] in June 2010. It is also possible to verify that the intense increase started in 2000, as a trend that was followed by the state of Mato Grosso (a strong grains

[18]Ignácio Rangel, an important Brazilian economist, minted the idea of a fourth type of land rent (complementary to the existing types of Marx's absolute land rent, and Ricardo's differential land rent I and II), derived from the strong financial appreciation of rural property in Brazil, pointing for the first time to the capacity of landed property to become something close to a financial asset, attracting to itself a type of capital that is fleeing from currency de-indexation, in particular during the economic crisis of the 1980s (Rangel 1986).

[19]Exchange rate in Brazil: 1 USD = 1.62 R$ (April 2011).

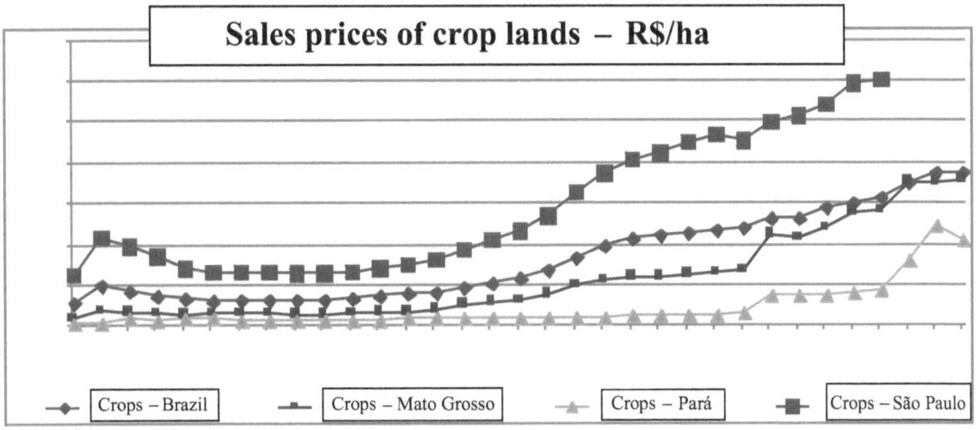

Figure 3. Variation in the sale prices of crop land (R$/ha) in Brazil and selected states, 1994/2010 – current prices.
Source: Fundação Getulio Vargas/FGVDados.

producer). At a distance, it was followed by Pará with a more recent increase in land value, due to the fact that its process of land occupation took place mostly through the 2000s. A noteworthy contrast is found in the strong increase in land value in the state of São Paulo, an area that concentrates the expansion of sugarcane and has the highest average price of land in comparison to the national average (R$ 12,020.00/ha compared to R$ 6283.00/ha in December 2008).

This trend is even more evident in some regions that have experienced an expansion in commodities production, such as the 'cerrados' regions in Minas Gerais, Goiás and Bahia. The strong value-increase in the price of land has also been stimulated by the massive wave of new investments in sugarcane plantations, especially in the two first states, aimed at the production of ethanol, in the wake of the emergence of new types of fuel based on natural resources.

It seems evident to us, therefore, that the expansion of agribusiness activities, along with the increase in foreign investments, resulted in a strong process of appreciation of land price and jeopardized the situation and demands of many social sectors (family farmers, peasants, indigenous people, etc.) that lived in these areas. Facing such reality, these segments were forced to move towards more distant areas, making the issue of access to land and to the territory significantly more complex.

If we associate the behavior of land prices in the process of legal flexibility for the acquisition of rural properties that occurred during the 1990s, we observe an intensive growth of values of land assets. This valorization is powered by speculative gains on financialization mechanisms for agricultural activities. Thus, the expansion of cultivated area on 'frontiers areas' is part of a larger strategy of globalization of agriculture, which includes, among other characteristics, the foreign ownership of lands (Hervieu and Purseigle 2009).

The demand for land: governmental restrictions and social struggles

In response to study conclusions that 'the demand for land has been enormous', and 'the "land rush" is unlikely to slow', the World Bank has made a series of recommendations that position this demand as 'a business opportunity' (World Bank 2010, xiv, 14). The main concern is then how to achieve sustainability and promote investments in a 'responsible' way (World Bank 2010, x). This can also be said to be valid for analysts of the Brazilian case.

Nassar (2010) corroborates the conclusions of the World Bank study and highlights this increasingly strong demand as an important 'window of opportunities' for Brazil. Drawing on examples from the telecommunications and finance sectors, Nassar defends regulatory mechanisms that could secure foreign investment while at the same time contributing to the country.[20] According to him,

> In the same way, measures can be taken in order to make the land investor produce income and work, not concentrate the possession of the land, fulfill environmental duties above the average of the sector, and promote technology transfer to less prepared producers. With the correct incentives, foreign investors can become an example for the agricultural sector (Nassar 2010).

However, this 'rush for farmland' has frightened even representatives of agribusiness, as was the case of Glauber Silveira, President of the National Association of Soybean Producers (APROSOJA). When Silveira took office as APROSOJA president in May 2010, he estimated that one million ha of Brazilian land (cultivated with soybean) was in foreign hands. In his words, 'even if it is advantageous for producers with land leasehold, the foreign surge is worrisome, as it pushes Brazilian competitors away from business and allows territorial occupation of Brazil' (Tavares 2010).

With this perspective, the editors of the newspaper *O Estado de São Paulo*, in its article entitled 'China buys land in Brazil', affirmed:

> Former minister Antônio Delfim Netto is right when he recommends caution with selling of lands to Chinese State-controlled companies, or to companies with a State participation. Foreign investments are welcome as a rule and may make important contributions to the country's growth. Foreign groups can make good deals while strengthening Brazilian economy with additional resources, and, occasionally, with technological inputs. But 'businesses' undergo a change of meaning when the investments are subordinated to the strategic reasons of a foreign State. In the cases of natural resources and agriculture-land, a proper evaluation of such strategy becomes a security-issue (*O Estado de São Paulo* 2010, A3).

In spite of a certain consensus that it is necessary to be cautious with the voraciousness of foreign investments, there is no agreement on concrete measures to be adopted. Even given the general concern over the force of this trend, analysts represent a range of positions, from a constant defense of free circulation of capital (including land purchase), through propositions of institutions and rules in order to restrict profit transfers abroad (Nassar 2010), all the way to demands for stronger control by the State (Alfonsin 2010).

[20]Analyzing the distinct proposals of 'codes of conduct' for land purchase, as proposed by FAO, FIDA and the World Bank, Borras and Franco point out a series of dangers, concluding that they are based on the 'belief, bereft of all criticism, in the essential benevolence of the measures with a formal and juridical nature, such as clearer contracts, clearer and safer property rights (understood, in general, as private and individual rights), transparent contracts, the existence of a free, previously thought out and well-informed consent, and the establishment of partnerships between the State and civil society' (2010, 14).

The struggle to reduce the degree of concentration of land has been led, among others, by the National Forum for Agrarian Reform (FNRA).[21] The entity promoted a national land-ceiling campaign in 2010. The campaign included a plebiscite (popular consultation) and a signed petition (supporting an act on the issue), with the goal of defining a maximum limit of 35 fiscal units for each and every estate, and not only for foreigners.

According to the campaign materials, setting a maximum limit would have two central points that were not meant to threaten private property, but to allow a diminishment of the land concentration: (a) defining a clear mechanism of limited use of common good (land); (b) establishing justice in the rural environment, as all must have the right to land, and other rights, which become secured starting from access to land (FNRA 2010).

According to these materials, 'a limitation of the size of rural estates is justifiable also as a stimulus to an increase in food production, in the preservation of natural resources, in rural employment and in settling people in the countryside, and also to hinder further offenses to the Brazilian territorial sovereignty' (FNRA 2010).

When it comes to access to land, it is still crucial to recognize that Brazil experiences a situation of illegality and juridical instability, leading to what Delgado (2005) has dubbed as 'the laxity of agrarian politics', including its effects on land purchase by foreigners,[22] which was aggravated by the National Congress' approval of Constitutional Amendment 6, in 1995.[23]

Related to this Congressional decision, the Brazilian Federal Executive asked the Office of Solicitor General (AGU) to produce an Informed Opinion to establish clear legal procedures regarding land deals by foreigners. The Informed Opinion GQ181 was issued in 1998 and demobilized 'any form of effective control on land purchase by foreign companies in Brazil' (Pretto 2009, 7).

This loss of any legal control by the federal government is explicitly mentioned in the new Informed Opinion of the AGU issued in 2010. According to the document, the result of the previous Informed Opinion of 1998 was that the State 'lost the objective conditions to exert an effective control on the purchase and leasehold of land by Brazilian companies whose stock control and management control are in the hands of foreigners who do not reside in the national territory' (Adams 2010, 2, item 6).

Facing an increasing demand for land, and the finding that the INCRA had no concrete mechanisms to exert an adequate control on land purchasing, the working group formed to evaluate the situation concluded that it was necessary 'to review the informed opinions, in order to provide the State with better conditions to

[21]The FNRA is a national wide net gathering over 40 different agrarian movements, rural organizations (including rural trade unions and church's groups) and non-governmental organizations, representing together with Via Campesina the main demands and struggles for land reform in Brazil.

[22]Act 8629 of 1993, the so called 'Agrarian Law', for instance, in its article 23, § 1, determines restrictions to the leasehold of land to foreigners, even not having a practical effect, especially due to the lack of supervision and control of land purchase.

[23]This Congressional decision eliminated Article 171 from the Brazilian Constitution that had made a distinction between national and foreign companies. By analogy, the Informed Opinion issued in 1998 by the AGU determined that, once this distinction had been discontinued, 'it would eliminate the barrier that blocked the establishment, by force of law, of restrictions to the activities of companies', thus reinforcing the movement of economic liberalization of that period, regarding also to the land market (Wilkinson *et al.* 2010).

supervise the land purchases by Brazilian companies controlled by foreigners' (Adams 2010, 3).

The AGU issued Informed Opinion LA-01, on August 19, 2010, reestablishing the possibilities of limiting, or better yet, of regulating processes of land appropriation by foreigners in Brazil. This document restates the provisions of Act 5709, of 1971, affirming that this law must be accepted under the scope of the Constitution of 1988. This act was created to regulate 'the purchase of rural property either by a foreign individual who lives in the country or a foreign company with a license to operate in Brazil', setting the maximum limit of purchase of 50 fiscal units (art. 3), and establishing that the sum total of the estates of a foreign individual cannot be more than one-fourth of the area of the municipality (art. 12) (Adams 2010, 4).[24]

Alfonsin (2010), stating legal validity of the AGU's Informed Opinion, affirms that the Federal Constitution itself underlines 'two fundamental rights in connection to the national territory and entrusted by its people: the first of them, quite obvious, is the sovereignty over its land (art. 1, single paragraph of the Constitution); and the second is the necessary reflex of such sovereignty on national security' (2010, 4). From this constitutional provision, Alfonsin affirms that this new Informed Opinion 'had no function other than acknowledging the efficaciousness of the constitutional norms to defend and protect our territory and our people' (2010, 20), therefore the adoption of control measures by the State is of fundamental importance.

Without questioning the juridical and legal importance of such Informed Opinion, we argue that it does not arrive at the solution to the problem. First, there are problems in the contents of Act 5709/71, such as, for instance, the limit of 50 units (three times more than the Constitutional limit of large estate) or the restriction to one fourth of a municipality's area. There are huge municipalities in Brazil, especially in the north and midwest regions, which are main targets of the search for land and of the expansion of agribusiness (see Heredia *et al.* 2010).

Second, land issues and popular demands widely transcend the problem of land deals well known as 'farmland grab' or land grabbing of the farmland rush and of the ensuing appropriation of land by foreigners. It is vital not to forget the historical levels of concentration of landed property in Brazil, newly corroborated by the Agriculture and Livestock Census of 2006, as we highlighted earlier. Such concentration of land will not be reverted with the adoption of mechanisms to control land purchase by foreigners, as the great majority of large areas is in the hands of a few Brazilian individuals.

On the other hand, the State has failed to implement a comprehensive agrarian reform policy. It must be acknowledged here that there have been expropriations of land that did not fulfill its social function, and landless families have been settled in settling projects, but it is far from meeting the historical demands. Lately, the settlement process has slowed down, and increased its sluggishness to acknowledge the territories of Quilombo communities and to demarcate Indigenous lands, territorial rights that shall be considered as parts of any agrarian reform process.

[24]Act 5709 (in its art. 7) also rules about land ownership in Brazilian borders, which is an important theme in discussion these days in the National Congress. There are lots of pressures for reviewing the law, diminishing from 150 to only 15 kilometers the zone inside the borders where land shall not be privatized.

The territorial rights – already secured in international treaties as well as in the Brazilian Constitution – must be effectively secured in governmental actions, because 'land has a totally different significance from what the mass-culture of the West understands. It is not only about housing, which can be substituted by the individual with no bigger traumas, but about the link that keeps the group united and allows its continuity in time through successive generations' (Sarmento 2008, 7).

This understanding should be extended to include other countryside sectors that fight for the right of access to land. For many of them, land is not only a means of production, but indeed it is a place for living and constructing an identity (being a rural worker, for instance) (Sauer 2010b). In this sense, it is crucial to deepen the debate on the 'right of the family-based, peasant producers' as a key human right for social reproduction and quality of life in the countryside.

Final remarks

Through history, the Brazilian countryside has been a site of territorial disputes as well as a source of resistance against expropriation and exploitation through popular demands for land access. These contests ascribe a special value to territory in Brazil but have also materialized alliances between capital and land that are reinforced and expanded as part of the increasing demand and inflow of investments towards land in Brazil.

In spite of being viewed by some as an opportunity for business, land cannot be analyzed in terms of its reduction to the status of a mere asset or means of production. Global demand for land confronts historical demands for agrarian reform in Brazil, both in the sense of a direct competition for a finite good as well the reality of increasing prices and increasing costs for public policies that would seek to achieve a more equitable access to land. The reduction of land to a mere means of production is also a tangible product of the 'productive-land' concept that was introduced in the 1988 Constitution as a mechanism to block agrarian reform. By restricting the social function of land to an economic dimension (the sole criterion for expropriation being rational use in terms of production) this enshrined and increased the distance from the concept of land as a means and place of production as well as a source of identity, self-recognition, historical occupation, and life.

The increasing flow of foreign investment in Brazilian land has been met by concerns and protests, from sometimes bitterly-opposed social groups. Increasing investments in land assets threaten food security and sovereignty. These investments concentrate agriculture and livestock production even further in a few commodities, and this concentration – both of products and of their ownership – favor monopolies and the increase in the control of the production of food and agro-energy by a few. However, representatives of Brazil's so-called 'agribusiness', as well as opinion articles in the mainstream press, have also voiced alarm over trends of 'foreignization'. Even as these sectors defend a 'pro-market' perspective, there is clearly a voice of concern with respect to the intensity and quantity of land purchases by foreigners. In many ways this presents a far more complex scenario that envisioned by the World Bank recommendations that aimed at exploring the windows of opportunities of these new areas through 'responsible investments'.

The legal action of the government (the above-mentioned Informed Opinion of the AGU of August 19, 2010) reestablished regulatory capacities in this field and also created the impression in some international circuits that the State could effectively intervene in such a strategic sector. However, the solution to this problem, as we have seen in the preceding discussion, is not addressed by this instrument. Without overlooking the juridical-legal importance of this action, such an initiative should be viewed as an opening so that discussion of these issues may expand in depth and breadth across the country. It is crucial to remember that the historical levels of concentration of land in Brazil (newly corroborated by the last Agriculture and Livestock Census of 2006) will not be reverted with the adoption of control mechanisms to land purchase by foreigners, as the vast majority of large estates are in the hands of a few Brazilian individuals. This situation increases the urgency for policies of redistribution and territorial registration related to agrarian reform and recognition of areas as belonging to indigenous and traditional populations.

References

Adams, L.I.L. 2010. *Parecer LA-01 Aquisição de terras por estrangeiros.* Brasília, Subchefia para Assuntos Jurídicos da Casa Civil da Presidência da República, August (unpublished).

Alfonsin, J.T. 2010. *Apontamento sobre parecer da AGU relacionado com a aquisição de terras rurais do Brasil por empresas estrangeiras e empresas brasileiras com maioria de capital estrangeiro.* Porto Alegre, (unpublished).

Alvim, A.M. 2009. *Investimentos estrangeiros diretos e suas relações com os processos, causas e efeitos da concentração e estrangeirização das terras no Brasil.* Project of technical cooperation on 'Apoio às políticas e à participação social no desenvolvimento rural' (PCT IICA/NEAD), Brasília, NEAD (unpublished).

Batista, F. 2010. 'Cana mantém aquecido mercado de terras'. *O Valor Econômico*, São Paulo, 2 August, p. B12.

Bertrand, J.P., H. Théry and P. Waniez. 1991. Les Japonais et la mise en valeur des Cerrados au Brésil: pour quelle maîtrise de l'espace et de l'approvisionnement alimentaire? *Économie Rurale*, 202–203, 58–64.

Borras, S.M. and J. Franco. 2010. La política del acaparamiento mundial de tierras: replanteando las cuestiones de tierras, redefiniendo la resistencia. *ICAS working paper Series*, No. 001. Transnational Institute, LDPI, ICAS.

Borras, S.M., J. Franco, C. Kay and M. Spoor. 2011. *Land grabbing in Latin America and the Caribbean viewed from broader international perspectives.* Santiago: FAO Regional Office.

Brandão, A.S. 1988. Mercado da terra e estrutura fundiária. Brandão, A.S. (org.). *Os principais problemas da agricultura brasileira: análise e sugestões.* Rio de Janeiro: PNPE/IPEA.

Brasil. Instituto Brasileiro de Geografia e Estatística 2009. *Censo agropecuário 2006.* Rio de Janeiro, Available from: http://www.ibge.gov.br/home/estatistica/economia/agropecuaria/censoagro/default.shtm [Accessed October 2011].

Brasil. Ministério do Desenvolvimento Agrário 2005. *II Plano Nacional de Reforma Agrária: paz, produção e qualidade de vida no meio rural.* Brasília: INCRA.

Chade, J. 2010. Estrangeiros avançam na aquisição de terras. *O Estado de São Paulo*, 13 September.

Cotula, L. 2012. The international political economy of the global land rush: A critical appraisal of trends, scale, geography and drivers. *Journal of Peasant Studies*, 39(3–4), 649–80

Delgado, G. 2005. A questão agrária no Brasil, 1950–2003. *In*: L. Jaccoud, ed. *Questão social e políticas sociais no Brasil contemporâneo.* Brasília: IPEA, pp. 51–90.

Duarte, P. 2008. O apetite dos estrangeiros por terra e petróleo. *O Globo*, 22 June, p. 36.

Flexor, G., B. Rocha, K. Kato and M.S. Lima. 2010. *A economia política do PNPB: uma problemática de ação pública.* Rio de Janeiro: IM e CPDA/UFRRJ (unpublished).

Focus – Visão Brasil. 2010. Funbio and Instituto Arapyaú, Rio de Janeiro. Available at www.visaobrasil.org [Accessed 18 May 2010].

FNRA (Fórum Nacional pela Reforma Agrária e Justiça no Campo). 2010. *Campanha nacional pelo limite de propriedade da terra.* Brasília. Available from: www.limitedaterra.org.br/forum.php?cod = forum [Accessed 13 March 2011].

Guimarães, J.S. 2006. *Multinacionais irão investir US$ 25 bilhões no agronegócio brasileiro.* São Paulo, 14 June. [Available from: www.agrovigna.com.br [Accessed 20 September 2010].

Hall, R. 2011. The next Great Trek? South African commercial farmers move north. Paper presented at the International Conference on Global Land Grabbing, Brighton, University of Sussex, 6–8 April. Available from: www.future-agricultures.org/index [Accessed 10 October 2011].

Hecht, S.B. 2005. Soybeans, development and conservation on the Amazon frontier. *Development and Change*, 36(2) 375–404.

Heredia, B. M. Palmeira and S.P. Leite. 2010. Sociedade e economia do 'agronegócio' no Brasil. *Revista Brasileira de Ciências Sociais*, 25(74), 159–96.

Hermele, K. 2011. Sustainable grofuels, land use change and certification schemes. *In*: J.P. Pietrafesa and S.D. Silva, eds. *Transformações no Cerrado: Progresso, consumo e natureza.* Goiânia: PUC, pp.189–215.

Hervieu, B. and F. Purseigle. 2009. Pour une sociologie des mondes agricoles dans la globalisation. *Études Rurales*, 183(1), 177–200.

Hoffmann, R. and M. Ney. 2010. *Estrutura fundiária e propriedade agrícola no Brasil.* Brasília: NEAD/MDA.

ILC (Internatiotal Land Coalition) 2009. *Commercial pressures on land worldwide: issues and conceptual framework.* Roma: ILC.

Leite, S.P. and R.V. Ávila. 2007. *Um futuro para o campo: reforma agrária e desenvolvimento social.* Rio de Janeiro: Vieira & Lent.

Leite, S.P. and S. Sauer. 2011. Expansión de agronegocios, mercado de tierras y extranjerización de la propiedad rural em Brasil. *Mundo Siglo XXI.* No. 26(VII), 3–63.

Leite, S.P., Heredia, B., Medeiros, L., Palmeira, M. and Cintrão, R. 2004. *Impactos dos assentamentos: um estudo sobre o meio rural brasileiro.* São Paulo: Ed. UNESP.

Leite, S.P. and V. Wesz Jr. 2010. *Modèle de développement et dynamiques foncières au Brésil: analyse de l'expansion de l'agribusiness du soja et ses effets sur le millieu rural.* Montpellier: CIRAD (unpublished).

Lima, D.L. 2010. *Estrutura e expansão da agroindústria canavieira no Sudoeste goiano: impactos no uso do solo e na estrutura fundiária a partir de 1990.* Campinas: IE/UNICAMP (unpublished).

Mackey, L. 2011. Legitimating foreignization in Bolivia: Brazilian agriculture and the relations of conflict and consent in Santa Cruz. Paper presented at the International Conference on Global Land Grabbing. Brighton, University of Sussex, 6–8 April. Available from: http://www.future-agricultures.org/index [Accessed 10 November 2011].

Merlet, M. 2010. *Les appropriations de terres à grande échelle: analyse du phénomène et propositions d'orientations.* Paris: Comité Foncier et Développement.

Murmis, M. and M.R. Murmis. 2010. *Dinámica en el mercado de tierras en América Latina: el caso de Argentina.* Santiago del Chile: FAO (unpublished).

Nassar, A.M. 2010. Terras agrícolas na alça de mira. *O Estado de São Paulo.* 20 October, p. A2.

Novo, A., K. Jansen, M. Slingerland and K. Giller. 2010. Biofuel, dairy production and beef in Brazil: competing claims on land use in São Paulo state. *The Journal of Peasant Studies*, 37(4), 769–92.

Odilla, F. 2010. Estrangeiros compram 22 campos de futebol por hora. *A Folha de São Paulo*, 2 November.

O Estado de São Paulo. 2010. Editorial: China compra terras no Brasil. São Paulo, 3 August, p. A3.

Pietrafesa, J.P., S. Sauer and A.E.A. Santos. 2010. Políticas e recursos públicos na expansão dos agrocombustíveis em Goiás: ocupação de novos espaços em áreas de Cerrado. *VIII Congresso Latino Americano de Sociologia Rural.* Porto de Galinhas.

Pretto, J.M. 2009. *Imóveis rurais sob propriedade de estrangeiros no Brasil* (project of technical cooperation PCT IICA/NEAD). Brasília: NEAD (unpublished).

Rangel, I. 1986. A questão da terra. *Revista de Economia Política*, 6(4), pp. 71ff, out./dez.

Sabbato, A. 2008. *A intervenção no mercado de terras rurais no Brasil: um estudo sobre o crédito fundiário e o Imposto Territorial Rural no período 1997–2002.* Rio de Janeiro: CPDA/UFRRJ (unpublished).

Safransky, S. and W. Wolford. 2011. Contemporary land grabs and their alternatives in the Americas. Paper presented at the International Conference on Global Land Grabbing. Brighton, University of Sussex, 6–8 April. Available from: www.future-agricultures.org/index [Accessed 10 November 2011].

Salim, C.A. 1986. As políticas econômica e tecnológica para o desenvolvimento agrário das áreas de cerrados no Brasil: avaliação e perspectivas. *Cadernos de Difusão de Tecnologia*, 3(2), 297–342.

Sampaio, P.A. *et al.* 2003. *Proposta do Plano Nacional da Reforma Agrária*. Brasília: MDA/INCRA (unpublished).

Sarmento, D. 2008. *Territórios Quilombolas e Constituição: A ADI 3.239 e a Constitucionalidade do Decreto 4.887/03*. Rio de Janeiro: Ministério Público Federal, 03 of March (unpublished).

Sauer, S. 2010a. Dinheiro público para o agronegócio. *Le Monde Diplomatique Brasil*, 3(33), 8–9.

Sauer, S. 2010b. *Terra e modernidade: a reinvenção do campo brasileiro*. São Paulo: Expressão Popular.

Sauer, S. 2011. Demanda mundial por terras: 'land grabbing' ou oportunidade de negócios no Brasil? *Revista de Estudos e Pesquisas sobre as Américas*. Brasília, CEPPAC/UnB. Available from: www.repacm.org [Accessed 20 January 2012].

Sauer, S. and S.P. Leite. 2010. A estrangeirização da propriedade fundiária no Brasil. *Carta Maior*, December. Available from: www.cartamaior.com.br.

Scolese, E. 2008. Aumenta compra de terras brasileiras por estrangeiros. *Folha da São Paulo*, 7 July, p. A10.

Tavares, V. 2010. Um milhão de hectares já são de estrangeiros, diz Glauber Silveira. *Olhar direto*. Brasília. Available from: www.olhardireto.com.br/noticias [Accessed 10 November 2011].

Taylor, M. and T. Bending. 2009. *Increasing commercial pressure on land: building a coordinated response*. Roma: ILC.

Toulmin, C., P. Bindraban, J. Borras, E. Mwangi and S. Sauer. 2011. *Land tenure and international investments in agriculture*. Report prepared for the High Level Panel of Experts (HLPE) on Food Security and Nutrition (CFS), Committee on World Food Security. Rome: HLPE Report 2, July.

Wesz Jr., V.J. 2011. *Dinâmicas e estratégias das agroindústrias esmagadoras de soja no Brasil*. Rio de Janeiro: E-papers.

Wilkinson, J. and S. Herrera. 2010. Biofuels in Brazil: debates and impacts. *The Journal of Peasant Studies*, 37(4), 749–68.

Wilkinson, J., B. Reydon and A. Sabbato. 2010. *Dinâmica do mercado de terras na América Latina: o caso do Brasil*. Santiago: FAO, December (unpublished).

World Bank 2010. *Rising global interest in farmland: Can it yield sustainable and equitable benefits?* Washington, DC: World Bank.

Sérgio Sauer has a PhD in Sociology and is a professor at the University of Brasilia (FUP/UnB) in the Master Program for Environment and Rural Development (MADER/FUP), National Rapporteur to the Human Right to Land, Territory and Food of DhESCA Brazil Network. His latest books published on agrarian issues are: *Land and territory in the Amazon: demands, challenges and perspectives* (2011) and *Land and modernity: the reinvention of Brazilian countryside* (2010).

Sergio Pereira Leite is associate professor in the Graduation Program in Social Sciences in Development, Agriculture and Society (CPDA), Federal Rural University of Rio de Janeiro (UFRRJ), Brazil. Post-doctor in Social Sciences at the Ecole des Hautes Etudes en Sciences Sociales (EHESS), and holding a PhD in Economics at the University of Campinas (UNICAMP), he is a member of the French Society of Agriculture.

Oligarchs, megafarms and land reserves: understanding land grabbing in Russia

Oane Visser, Natalia Mamonova and Max Spoor

This paper seeks to unravel the political economy of large-scale land acquisitions in post-Soviet Russia. Russia falls neither in the normal category of 'investor' countries, nor in the category of 'target' countries. Russia has large 'land reserves', since in the 1990s much fertile land was abandoned. We analyse how particular Russia is with regards to the common argument in favour of land acquisitions, namely that land is available, unused or even unpopulated. With rapid economic growth, capital of Russian oligarchs in search of new frontiers, and the 2002 land code allowing land sales, land began to attract investment. Land grabbing expands at a rapid pace and in some cases, it results in dispossession and little or no compensation. This paper describes different land acquisitions strategies and argues that the share-based land rights distribution during the 1990s did not provide security of land tenure to rural dwellers. Emerging rural social movements try to form countervailing powers but with limited success. Rich land owners easily escape the implementation of new laws on controlling underutilized land, while there is a danger that they enable eviction with legal measures of rural dwellers. In this sense Russia appears to be a 'normal' case in the land grab debate.

1. Introduction[1]

The topic of large-scale land acquisition or 'land grabbing' has recently received growing attention among academics and policy makers (von Braun and Meinzen-Dick 2009, World Bank 2010a, HLPE 2011). Specifically, land grabbing has been studied in Africa (Cotula *et al.* 2009, Hall 2012), South-East Asia (Julia and White 2012), Latin America (Zoomers 2010, Borras *et al.* 2012) and China (Ding 2007, Hofman and Ho 2012). However, until now, the vast land masses of the former Soviet Union constitute a blank spot in the global investigation of land grabbing, as noted by Visser and Spoor (2011a) in a first overview of land grabbing in this region. The current paper builds on this with a more elaborate investigation of land relations and mechanisms of land acquisition, focusing on the case of Russia. Furthermore this paper is based on recent fieldwork, whereas Visser and Spoor (2011a) was

[1]We thank Ruth Hall and three anonymous reviewers for their valuable comments. The assistance in language editing by Martha Jane Robbins is also acknowledged here. Part of the research for this article was made possible by a Land Deal Politics Initiative (LDPI) research grant.

predominantly based on media and web research. The current literature on land grabbing predominantly tries to map the magnitude and speed of land grabbing, the main types of countries or private actors involved worldwide, and the general drivers of the global land grab. Following new research lines advocated by Borras *et al.* (2011, 211) this paper pursues a closer look at the actors involved, their motivations and interests, and especially the mechanisms through which large-scale land deals are enacted. The link with the global land grab debate will be made in particular through analyzing the 'availability of land' and what this means in the Russian context.

Russia is the largest agricultural producer in post-Soviet Eurasia, and worldwide it is one of the countries with the largest land reserves for agricultural production, as well as the country with the potential for the largest increase in grain production worldwide (World Bank 2010a). Russia has one of the highest levels of arable land in the world, namely 0.9 hectares (ha) per capita (World Bank 2010a, b). This ratio is higher only in Canada and Australia, but these countries have less potential to increase production, as most agricultural land is being used, with high land productivity. Moreover, it is to be expected that climate change will reduce the acreage of arable land, whereas in Russia the acreage of cultivable land might expand northward, for instance. According to the World Bank, water availability will increase (5–15 percent) in the north of Russia and Kazakhstan (FAO 2011), which may offset the higher occurrence of droughts in the south of Russia (HLPE 2011, Dronin and Kirilenko 2011). Whereas most countries in the world face declining availability of agricultural and arable land per capita, in Russia the availability of land per capita has actually *increased* over the last two decades, because of stagnation of the population.[2] Moreover, a substantial amount of Russia's agricultural land was taken out of production in the 1990s. According to the president of Russia Dmitry Medvedev, the amount of unused or inefficiently used agricultural land may total 30 million ha (Wegren 2011, 152). However, reports differ on the amount of land taken out of production (IAMO 2011, Ioffe and Nefedova 2004). This abundance of available 'unused' but often very fertile land has raised the interest of investors, not only of domestic investors (the Russian oligarchs amongst others), but also of foreign companies and governments. However, whether all of the land classified as such really is 'unused' or 'abandoned' needs further scrutiny.

The phrase 'global land grab' has become a catch-all to describe and analyze the current explosion of large-scale (trans)national commercial land transactions. Some see land grabs as a major threat to the livelihoods of the rural poor and oppose such commercial land deals. Others see them as a necessity to modernize agriculture and as an economic opportunity for the rural poor, although they are wary of corruption and negative consequences, and hence, call for improving land market governance (Borras *et al.* 2011). Often land grabbing has been defined as large-scale land acquisitions by foreigners (GRAIN 2008), but currently a substantial share of land grabbing is conducted by domestic investors, or by coalitions of foreign and domestic actors. Following Borras *et al.* (2012), we define land grabbing as 'the large-scale acquisition of land or land-related rights and resources by a corporate,

[2]Russia's population peaked in the early 1990s (at the time of the end of the Soviet Union) with about 148 million people in the country. Today, it is approximately 143 million. The United States Census Bureau estimates that Russia's population will decline from the current size to a mere 111 million by 2050, a loss of more than 30 million people and a decrease of more than 20 percent (Rosenberg 2012).

non-profit or public buyer for the purposes of resource extraction geared towards external consumers (whether external simply means off-site or foreign)'. Extraction and alienation are essential to this definition rather than the type of capital invested, the intended market or the act of commodification or privatisation of land per se.

The widespread farm restructuring and share-based land reform in Russia in the first half of the 1990s 'opened up' a huge and fertile area of land for investors in agricultural land. However, for more than a decade this sparked little interest among investors. The picture changed when Russian oligarchs and Western investors started to search for new frontiers of investment with domestic economic growth and the influx of Russian offshore capital as well as due to the global financial crisis (Jouko and Granberg 2011).[3] Moreover, rapidly increasing local demand for livestock products (related to demand for grains) in Russia, and subsequent legal and policy changes stimulating agricultural investment, caused high demand for agricultural land. Large-scale land investments became popular among Russian and foreign companies. Foreigners are not officially allowed to acquire agricultural land in Russia. However, they do so by means of their Russian subsidiaries, which are considered as Russian domestic companies by Russian law. 'We are seeing a land grab bigger than anywhere else in the world, and it has attracted a mighty cast of characters', stated Kingsmill Bond, a chief strategist at the Moscow Brokerage firm Troika Dialog (Kandell 2009).

This current rush of investors for land (and the earlier experience with industry) raises serious concerns about the transparency, inclusiveness and fairness of land acquisitions. There is a danger of widespread dispossession of rural inhabitants who recently acquired entitlements to property in the form of land shares in the 'cosmetically' privatised, large-scale successors of the state and collective farms, which are developing into megafarms. Will land grabbing cause rural dwellers to lose their land rights, soon after they received them in the 1990s? In the absence of studies on land grabbing in the former Soviet Union, we know relatively little of the various effects.

This paper focuses on the investors in land and their strategies. However, effects of these land acquisitions for the rural population regarding land ownership (and potential dispossession) will also be addressed. We focus on the main players in the Russian land market and the strategies and mechanisms they use for land acquisition, such as deliberate bankruptcy of collective farms in order to buy their lands, the massive purchase of land shares acquisition after land law (2002), and using the inequality in power relations. Finally, the role of the Russian state vis-à-vis large-scale land acquisitions is discussed.

The paper is structured as follows. In the second section, the specificities of research on land grabbing in Russia are described, with special attention to difficulties with data collection. In the third section, it will be shown that the land rights ('shares') the population received during the restructuring of state and collective farms were rather insecure, which facilitated land grabbing later on. The fourth section investigates the magnitude of the large-scale land acquisitions in Russia, and the actors and structures behind them, such as Russian oligarchs, megafarms and foreign investors.

Section five looks at the motivations of investors and the land acquisition strategies that are used to avoid different restrictions on land sales in legal (using

[3]The financial crisis led to temporary problems with access to finance in Russia; however, more fundamentally it stimulated, globally as well as in Russia, a major shift of capital towards the agri-food sector, and the primary sector more broadly.

loopholes in Russian legislation) and illegal ways. The sixth section looks at the issue of abandoned land, land reserves and dying villages, distinguishing between discourse and reality. Section seven analyses land accumulation strategies and responses of rural dwellers attempting to counter or mitigate unfair land acquisition. In section eight the role of the Russian state in the facilitation of land grabbing is analysed, which is often crucial. Finally, the concluding section looks at land grabbing in Russia in the years to come, arguing that it is expected to continue at great pace, and that proposed government interventions to constrain land grabbing will likely not prevent illegal land acquisitions, and might even have adverse effects.

2. Doing research on land grabbing in Russia: data resources

To study land grabbing empirically in Russia is complex. The information about domestic and foreign land transactions provided by Russian statistical institutions such as Rosstat and the cadastral agency Rosreestr is incomplete and provides only general information about land ownership and usage. Furthermore, the rapid increase in international transactions in agricultural land renders available information quickly outdated (Visser and Spoor 2011b). Moreover, according to Uzun *et al.* (2009), statistics on landholdings and land deals are collected at the agricultural enterprise level and are generally not available for agroholdings (Uzun *et al.* 2009), which are the crucial actors in land acquisitions. An agroholding includes a number of agricultural organizations whose controlling blocks of shares are owned by the holding company. It acts as an umbrella for the subordinated (not necessarily only agricultural) units and controls their policies and management. Agroholdings often combine agricultural production enterprises with other upstream and downstream firms in the agri-food chain, such as producers of concentrated fodder, elevators, processing units and wholesalers. As a rule, investors (both domestic and foreign) invest in Russian land via agroholdings.

There are no databases in Russia that include full information about Russian agroholdings, their capital or the acreage they control. As these agroholdings are currently the main actors in the land grabbing process, this shortcoming seriously complicates research on land grabbing in Russia. However, even with more statistical data at this level, the reliability of such data would be questionable due to the lack of transparency of Russian agroholdings and the concealed nature of much of the domestic and foreign investment that is taking place.

Therefore, in this paper we study land grabbing based on the following methods. First, through continuing our web research: gathering many dispersed pieces of information from the Russian media, reviews in web journals, and examining web pages of agroholdings and foreign investment funds operating in Russia. Second, we held in-depth interviews with leaders of some foreign and domestic agroholdings operating in Russia, government officials, rural inhabitants and representatives of emerging rural movements (Visser 2010, Visser and Mamonova 2011).[4] Within the

[4]Most of the interviews were conducted by Mamonova and Visser in the autumn of 2010 and winter and spring of 2011 in the Moscow region. Most of the agroholdings which were included in the research have (or formerly had) foreign investment, and operate in multiple Russian regions. The paper also benefited from interviews with foreign investors conducted by Steggerda and Visser in the autumn of 2011 and winter of 2012. Michelle Steggerda is an honours student in the department of Anthropology and Development Studies, Radboud University Nijmegen. Finally, the paper uses data from interviews with domestic and foreign

scope of this research paper we can only pay limited attention to the regional differences in land acquisitions within Russia and aim to focus on some of the general tendencies. For studies that deal extensively with regional differences we refer to the work by Pallot and Nefedova (2007) that covers a wide range of regions, as well as the work of Visser (2006, 2008) and Lindner (2008), which provides a close up study of farm enterprises in three contrasting regions.

3. Post-Soviet land reform: the political economy of insecurity

3.1. Post-Soviet land reform after 1991

The Russian land reform was aimed at changing ownership of farmland from state into private property, while the dominant large state and collective farms were to be restructured towards 'individualized' production. No land restitution policy was followed as it was not possible to determine who the previous land owners were before the forced collectivization of farms and the nationalization of land in the 1930s. This was only done in the Baltics and in some countries of Central Europe. Moreover, the collectivization had often been carried out together with imprisonment, deportation and killing of numerous 'rich' peasant farmers (the so-called *kulaks*). The various farm enlargement campaigns in the later Soviet period were accompanied by multiple relocations of rural settlements, further adding to the alienation of the rural population from their former or ancestral land. Therefore, among both policy makers and the rural population, there was no serious interest for restitution of land to the pre-communist owners. Instead, land was to be distributed equally among the current rural inhabitants (World Bank 1992), although, as we will see below, this was not done in a physical sense, but through share-based entitlements to land (Spoor and Visser 2001).

In the early 1990s only one percent of the agricultural land was privately used (these were the subsidiary household plots of the employees on the state and collective farms). The other 99 percent of land belonged to the state (Poshkus 2009, 68). This situation changed dramatically with the post-1991 land reform. Land was given to the former *kolkhozes* and *sovkhozes* (collective and state farms) for further distribution among their members in the form of paper shares. During that time investors in agriculture were not able to buy land, but they were not completely passive. Some of them waited until the end of this moratorium (2002), in the meantime making contacts with the management of attractive farms and lobbying with local authorities.

Land reform in Russia was carried out in two stages. First, in 1990 the law 'On land reform' focused on taking 10 percent of *kolkhoz* and *sovkhoz* land and transferring it to local authorities for distribution among peasants who were willing to establish a private family farm. However, at the end of 1992 there were only 50,000 private family farms that owned less than 1 percent of agricultural land. According to Poshkus (2009, 68) problems arose due to the unwillingness of *kolkhoz* leadership to hand over the land to local authorities and the lack of understanding of land reform among peasants. Second, at the end of 1991 the Presidential Decree

investors in agriculture conducted earlier by Visser in the Moscow, Rostov, Pskov and Saint Petersburg regions. Among the rural movements, Krestyansky Front (Peasant Front) is the largest. This movement fights for the protection of peasants' rights, in the face of land dispossession by large-scale investors.

No. 323 ('On urgent measures for the implementation of land reform in the RSFSR[5]') was issued by the first Russian President Boris Yeltsin. This involved the transformation of state and collective farms into new juridical forms (mostly closed joint-stock companies or CJSC) with the employees becoming 'shareholders'. Subsequently, each employee received non-land asset shares and land shares (on average three to fifteen ha) for free. However, the shares did not include real, individual ownership. The land shares that employees received were paper certificates, that substantiated their rights to unspecified land plots on the territory of their former state or collective (Poshkus 2009, 68, Spoor and Visser 2001). Gustav Wetterling (Agro-Invest Group – a subsidiary of Swedish Black Earth Farming Ltd) calls this process 'semi-privatisation', meaning that peasants didn't become real owners of the land, as they only received the right to be owners (Interview, 23 September 2010, Moscow).

Ownership of such land could occur in two cases. First, the peasant transferred his/her land share to the charter capital of the restructured farm enterprise. In this case, land became property of the farm. Second, the peasant created a separate private farm, by registering the land as his/her ownership. However, private land registration required significant time and financial resources due to the highly bureaucratic system in the country. According to Poshkus (2009, 69) the main weakness of the reform was 'the problem with the definition of land rights and their usage'. This provided a loophole facilitating future land grabbing in Russia and is increasingly generating land conflicts in the country. However, more clearly defined property rights alone would not have been a solution, as land reform took place in a situation in which markets were still in the making, political interference remained strong, and distribution of power was highly unequal (Spoor and Visser 2004).

During President Vladimir Putin's first term the Federal Law N136-FZ 'Land Code of Russian Federation' (2001) was adopted that modified property rights for land with the exception of agricultural land, and in 2002, the Federal Law N 101-FZ 'On Agricultural Land Transactions' came into force. This law specified procedures for selling agricultural land and served as a template for regional legislation (Wegren 2009b). Tamara Semenova (vice-president of the peasant movement Krestyansky Front) suggests that this law 'was lobbied by a group of oligarchs who had already bought or planned to buy land. This law was worked out to legalize their purchases' (Interview, 29 September 2010, Moscow). During Putin's second term (2004–2008), Russian and foreign investors became interested in agricultural land and commenced buying land shares.

3.2. High costs of land registration

Land reform in Russia during the 1990s did not lead to clearly defined land rights and tenure security. According to Lerman and Shagaida (2007), most (84 percent) of the peasant farmers in the more agriculturally developed regions in Russia regarded first, complex procedures of land transactions, and second, high registration costs as major problems in the development of their farming operations. In order to register

[5]The Russian Soviet Federative Socialist Republic (RSFSR) is the republic of the former Soviet Union, which after the collapse of the Soviet Union in December 1991 was renamed the Russian Federation.

their land, peasants had (and have) to pass three organizational levels: the aforementioned district committees, cadastral chambers and registration chambers.

After this long procedure the land shares are transferable: they can be leased or sold, transferring the rights to the underlying acreage to another operator or owner. However, peasants have to spend considerable costs to pass through the entire procedure. 'Often people in remote areas do not even have money in order to formalize their land shares', as Tamara Semenova of Krestyansky Front stated (Interview, 29 September 2010, Moscow). First Deputy Premier Viktor Zubkov revealed in February 2009 that of the 12 million land shareholders, only 400,000 have been able to register their land plots. With the emergence of outside investors (see next section), the weaknesses of land reform in securing land rights of the rural population have become more and more apparent.[6]

4. The magnitude of land grabbing and the types of actors involved

4.1. Domestic investors

In the 1990s, most Russian companies considered the agricultural sector as unprofitable and preferred to invest in other branches, while interest from foreign investors was also limited (with a few exceptions; see below). The new Russian government sharply curtailed investment in the agriculture sector (Barnes 2006) and the large farm enterprises (LFEs), successors of the *kolkhozes* and *sovkhozes,* faced severe financial difficulties.

The privatisation of oil, gas, and mineral resources and industry sectors in the early 1990s led to a rush to acquire former state enterprises. A fierce struggle for the most prized assets evolved between the Russian oligarchs and sometimes between worker collectives and oligarchs. Later, foreign investors and the Russian state (especially in the oil sector) became involved in this struggle over acquisition of property (Barnes 2006). However, as indicated above, land reform had been 'largely cosmetic', involving little more than 'changing the name plate of the farm' (Brooks *et al.* 1996). Comparing agrarian property transformation in the 1990s with privatisation in other sectors, Barnes wrote that:

> ...the struggles in the sector could also seem underwhelming, lacking even the guilty fascination provoked by the decadence of conflicts in industry. [...] Private farmers did not drive around Moscow in Mercedes cars or pull rolls of hundred-dollar bills from their pockets in casinos. Newspapers did not brim with stories of upstart speculators battling corrupt managers and organized crime bosses for control over agricultural production. Instead, the sector often seemed interesting only for its pity value (2006, 141).

Indeed, Russian agriculture was in dire straits (with roughly 80 percent of the farm enterprises incurring losses by the mid-1990s). The market-oriented agrarian reform took place in a period of sharp economic decline of the economy at large and drastic cuts in government spending on agriculture in particular (Wegren 2009a, 2011). In the 1990s land prices were very low and often land even had a negative value because of costs in the form of land tax and informal social obligations attached to land, such as to provide feed for private livestock of the villagers or provide social infrastructure (Visser 2006). During this period the sparse outside

[6]This large gap between the number of unregistered and registered land shares/plots indicates that there is still widespread tenure insecurity.

'investors' (from outside the agribusiness sector) who took over farm enterprises did so often with the goal of asset stripping. Equipment and buildings were sold off, while the land was mostly left unused (Kalugina and Fadeeva 2009).

Due to the dire straits of the large farm enterprises, part of their lands ended up in the hands of creditors in order to pay off farm debts, in particular to suppliers of fuel such as the energy giants Gazprom and Lukoil (Barnes 2006, 161). The first substantial agricultural land bank, operated by Gazprom, started when farm enterprises were unable to cover their fuel debts and had to pay back with their land. The total property of Gazprom is more than 500,000 ha. Gazprom is currently engaged in a process of selling off its agricultural land again, as the company was not successful in agribusiness due to the absence of skills and knowledge in agriculture (Uzun *et al.* 2009).

'Attendant with strong economic growth that ensued from 1999 through 2007, land began to be perceived as a valuable commodity from which to build wealth. Agricultural land became the new frontier for those with money' (Wegren 2009b, 3). New changes in Russian land laws, focusing on attracting direct and indirect investments to the agricultural sector by guaranteeing private property rights on the land, gave impetus to the creation of large-scale agroholdings or megafarms at the expense of rural dwellers. In Russia most agricultural land continues to be controlled by LFEs, and since the early 2000s, a growing percentage of these are controlled by large agroholdings and corporations, previously state- and now privately owned.

According to the All-Russian Agricultural Census of 2006, private family farmers owned only 13 percent of agricultural land in Russia in 2005 and household plots accounted for less than 5 percent (Rosstat 2006, 36). All other land is controlled by LFEs and agroholdings, both national and international, although a large share of this land is still formally in the hands of (former) employees cum shareholders.[7] Nearly 60 percent of agricultural land is now privately owned, but most of this land (51 percent) is represented by land shares – pieces of paper corresponding to virtual plots of specified size in an unspecified location (Lerman and Shagaida 2007). These land shares are transferable: they can be leased or sold, transferring the rights to the underlying acreage to another operator or owner. Only 6 percent of agricultural land is in the form of physically demarcated plots, (that means land shares which were actually registered by the possessor), most of it owned by individuals (household plots and peasant farms) (Lerman and Shagaida 2007).

The number of agroholdings (as well as the size of their landholdings) has rapidly increased since the early 2000s. According to the Russian Ministry of Agriculture in 2003, more than 90 agroholdings were active in 25 regions. By 2006, 319 private agroholdings were already registered (Uzun *et al.* 2009, 159). By the mid-2000s in various fertile Black Earth regions, such as Belgorod, Lipetsk, Voronezh and Tambov, there was practically no 'free' land available that was not yet controlled by an agroholding (Didenko 2009). By mid-2008, according to the Institute for Agricultural Market Studies, 196 large agroholdings controlled 11.5 million ha (BEFL 2010, 9). Of these agroholdings, 32 had landholdings of over 100,000 ha.

It is hard to say how many land transactions are conducted in Russia exactly, let alone those of specifically agricultural land. But definitely the number of land transactions is growing. In the mid-2000s about five percent of agricultural land was

[7]Also, individual entrepreneurs account for about 1.5 percent of agricultural land. A small part of the land used by large-scale farm enterprises is still owned by the state.

transacted annually (Shagaida 2005), a figure which has risen further in the last few years to at least eight percent (see Table 1) with a strong rise in the number of land deals (25 percent in 2009) between citizens and legal entities. These data of course do not include shady, unregistered deals.[8]

4.2. *Foreign land acquisitions: a brief overview*

Foreign land acquisitions are clearly on the rise in Russia since the mid-2000s (Visser and Spoor 2011a), although already in the mid-1990s some Western companies attempted to enter Russian agribusiness. However, it is hard quantify foreign land acquisitions. Foreign companies are not allowed to directly buy land. The Federal

Table 1. Land transactions (sale and lease) in Russia (2006–2009).

	2006	2007	2008	2009
Transactions with state and municipalities				
Number of transactions	3,921,393	3,875,650	3,985,846	3,887,691
Change in amount		−1%	3%	−2%
Amount of land (ha) in transactions	105,214,779	114,622,029	115,351,558	114,018,863
Approx ha per deal	27	30	29	29
Transactions between citizens and companies				
Number of transactions	623,747	643,443	794,792	996,462
Change in amount		3%	24%	25%
Amount of land (ha) in transactions	1,231,830	1,221,829	2,415,179	18,761,609
Approx ha per deal	2	2	3	19
Total transactions				
Number of transactions:	4,545,140	4,519,093	4,780,638	4,884,153
Change in amount		−1%	6%	2%
Amount of land (ha) in transaction	106,446,609	115,843,859	117,766,738	132,780,473
Total amount of land				
Total amount of land in Russia	1,709,800,000	1,709,800,000	1,709,800,000	1,709,800,000
Total amount of agricultural land in Russia			402,300,000	400,000,000
Land transacted as % of the total amount of land in Russia			7%	8%

Source: Authors' compilation based on Rosreestr (2010).
Notes: Land transactions include buying, selling or leasing of all categories of land: agricultural land, settlement, industrial land, special protected land, forest, and reserved land. Therefore, we calculated land transacted as percentage of the total amount of land in Russia, which somewhat understates the share of land transacted. However, if we would consider the land in transaction as only agricultural land, we would arrive at nearly 30 percent of agricultural land being bought, sold or leased per year in Russia, which likely overstates the percentage transacted.

[8]In 2010 the upward trend was temporary affected by the financial crisis. According to some estimates, the registered transactions in 2010 were almost a third less than during the same period before the crisis of 2008, due to problems of investors with financing land deals in context of the global financial crisis, which hit Russia hard.

Law N136-FZ 'Land Code' (2001) prohibits ownership of agricultural land by foreign citizens and companies, as well as Russian companies with foreigners owning more than 50 percent of the shares. To avoid the application of this law, foreign investors establish Russian subsidiaries that are legally not regarded as foreign companies (a strategy which is actively promoted by the Russian government, as will be shown in a later section). According to our most recent estimates, roughly 50 foreign companies control up to an estimated 3.5 million ha of Russian agricultural land. However, accumulation of land is a highly sensitive issue for both the leasing and the hosting countries, and therefore it is likely that the process of land grabbing in post-Soviet Eurasia has advanced further than official statements and media reports suggest (Billette 2009, Visser and Spoor 2011a). In terms of the total size of the agricultural land obtained, Russia is not a top target of foreign investment, but investment by outside investors is indeed substantial and the size of average land deals, including actually implemented ones, is very large. Having access to large financial resources, foreign investors acquire the most fertile land, use modern technology to achieve high productivity, and take leading positions in regional and sometimes even national markets.

The origins of the companies that are interested in Russian land acquisitions are quite diverse. Roughly we can distinguish the Western companies investing primarily in the European part of Russia, and the Asian (most notably Chinese companies) in Siberia and the Far East[9] of Russia (Visser and Spoor 2011b). Investors from the Middle East (in particular from the Gulf States) have only been searching for land in some of the former Soviet republics very recently and have not yet concluded major deals (Visser and Spoor 2011b). Until now Gulf States investors have focused more on Ukraine and Kazakhstan rather than Russia. When discussing the origin of investors it is important to realize that the identity of investors might be different, and more complex, than what it seems at first glance. Various recent studies in Africa have shown that what appears to be foreign investment is in fact investment by representatives of the diaspora from the target country. Furthermore, Hall (2012) has shown that the widespread land acquisitions by South African investors across the African continent are often indirectly based on Chinese investments in South African companies that are funding these acquisitions. Part of what seems like foreign investment appears to consist of investment by Russian oligarchs (Visser and Spoor 2011).

Finally, a note about the aim/orientation of foreign farmland acquisitions. Globally an important share of the large-scale farmland acquisitions is not (solely) for food or fodder but also for biofuel production, for instance in Brazil and large parts of Africa.[10] In Russia, the share of land acquisitions aimed at biofuel production is rather insignificant. Various factors seem to play a role here. Russia, which is itself a prime oil and gas exporter with low internal fuel prices, does not actively stimulate biofuel production as is done, for instance, in Brazil and the USA. This makes the Russian case special, although the mechanisms of land grabbing are often comparable. Foreign investors who want to produce biofuels for export

[9]The Russian Far East concerns the easternmost parts of Russia, between Lake Baikal in Eastern Siberia and the Pacific Ocean. In Russia, the region is usually referred to as just the 'Far East'.

[10]Disentangling the food, fodder and biofuel functions of agricultural production is sometimes difficult, as it consists in part of 'flex crops' which can be used for multiple purposes.

(mainly to the nearby EU) prefer Ukraine, which has better port infrastructure and a government that is more interested in stimulating biofuel production and processing, being very oil-dependent on neighbouring Russia.[11]

5. Motivations for large-scale land deals: finance and the political economy of the land rush in Russia

The recent land rush in Russia has clear characteristics of a new frontier for investors. In the creation of this new frontier, global factors play a role (such as the financial crisis and the search for an inflation hedge) but domestic factors seem to be more decisive. Also domestic land acquisitions predominate over foreign ones.

After the slump of the 1990s, the Russian economy has shown impressive growth rates throughout most of the 2000s and demand for high quality food products, such as livestock products, is continuously rising. An important factor on the supply side is the large amount of capital in the hands of the Russian oligarchs, which after the appropriation of the energy and industry sectors are looking for new frontiers of development. Whereas in the 1990s they transferred their capital to tax havens in the West, with the economic recovery of the 2000s these oligarchs increasingly started investing their offshore capital again in Russia. By 2002, a senior economist of the Moscow Brokerage firm Aton Capital Group concluded, 'Russians are starting to trust Russia' and therefore 'money is coming back' (Starobin and Belton 2002).

The land rush can be seen as part of the larger 'hunt for the Next Big Thing', the new accumulation drive that started in the 2000s with the privatisation of assets which were left in state ownership in the 1990s, such as railways, electricity and financial services (Starobin and Belton 2002). The Russian state plays an important role in enabling the boom in large-scale land acquisitions, in legal terms by adopting the 2002 land law allowing free sale of land and, even more important, by financial stimulation through mechanisms that lower taxes and increase subsidised loans.

The fact that the current boom in agriculture takes the form of large-scale land acquisitions with the predominance of huge agroholdings is often portrayed as an inevitable process or the most efficient mode of farming for Russia, by investors and authorities alike. While it is true that the Russian landscape and the crops cultivated allow for a high degree of mechanisation and economies of scale, there are also various diseconomies of scale associated with these large farm enterprises (Nikulin 2005, Visser 2006, 2008), such as those related to monitoring costs. Another justification for large-scale land acquisitions and mega-farming given by the major actors in Russia is that the rural population is not willing to take up independent farming, that small or medium private family farms are not a feasible form of production, or do not have the finances to expand and modernise production. 'Today only agro-industrial holdings can be profitable in farming, because it requires huge financial resources', stated Zorigto Sakhanov, chairman of Agro-Invest Group, the subsidiary of Swedish Black Earth Farming (Bush 2008).

Indeed, the growth of private farms has been below expectations. However, it is important to stress that many of the farmers face obstacles which are not simple

[11]In addition, a probably unexpected motivation for biofuel production is the following: in Ukraine biofuel production is used as a means to generate value from the Northern area around Chernobyl, which due to the radioactive contamination can no longer be used for food and/or fodder production.

natural problems related to their size, but a direct problem of a political economy (and government policy) targeted at large-scale farming (Visser 2008). The lack of investment in extension services, as well as the large-scale nature of input and output channels, form obstacles for private farms (Visser 2008). Further, some regional authorities have set limitations on the minimum size of land deals, such as in Krasnodar Krai where there is a threshold of 300 ha for land deals, hindering the emergence and expansion of small and medium sized farms.

A very important factor that skews the agricultural boom towards large-scale farming is the financial system. There is an urgent lack of accessible finance/credit. Obtaining loans is virtually only possible through take-over and capitalisation by rich investors who have built up their capital in another sector, or via state-subsidised credit. Most commercial banks are more vehicles for the investment projects of oligarchs than accessible sources of credit. Interest rates are very high and agricultural land is mostly not accepted as collateral. Furthermore, banks are generally not much oriented toward agriculture.

From the early 2000s, the state has increased finance for the agricultural sector. A network of state-owned and operated banks for agribusiness was set up, but in some regions these bank branches were only established when agriculture had already virtually disappeared. State-subsidised loans have been targeted predominantly towards large farm enterprises, and within this group, towards the largest and most successful ones. For instance, Uzun (2005) states that '1.4 percent of the largest corporate farms received 22.5 percent of all subsidies'. Also, the more indirect forms of subsidisation seem to stimulate mostly the largest LFEs.[12] Various requirements of the loans, such as the need for matching of resources and often brief repayment terms, skew the loans towards the largest farm enterprises. Due to the above-mentioned problems, it is difficult to develop agriculture without investors that already gained capital in lucrative sectors outside agribusiness. Whereas in the 1990s, the few investors in agriculture and land were mostly coming from that sector (food processors, food wholesalers, or providers of inputs for farming), now they often do not have any existing link to the sector.

5.1. Motivations for large-scale land acquisitions

The precise motivations of outside investors/oligarchs to acquire land are not easy to discover since the whole process is highly non-transparent, not least for the villagers, who often know little more than that 'a rich investor from Moscow' obtained their land (D'Hamecourt 2010, 13). However, it is clear, as we will elaborate below, that the current land rush cannot be explained sufficiently by economic incentives or rising food demand only, and we will proceed to disentangle a few other factors. With regard to the global food prices and land grab debate, several authors have argued that the idea of the growing global population and rising food demand as an explanation for rising food prices and subsequently the drive for land acquisitions is a mystification of reality (Jouko and Granberg 2011). This is clearly the case for Russia. Except for the more obvious economic reasons for investment, such as increased demand for livestock products, and aspects of investment portfolio management

[12]However, it should be noted that by the mid-2000s the national budget for the first time since the mid-1990s set aside a substantial amount of subsidies for household plots and private farms (Wegren 2007).

(integration, differentiation and risk-spreading), there are several other motivations for investment. The following will be discussed: high subsidies, speculation, state pressure and state-business agreements, tax evasion and money laundering.

5.1.1. High subsidies

Acquiring land and investing in agriculture has been made attractive by the policy of the Russian government. Since the early 2000s the state has stimulated agriculture through a range of instruments such as a debt restructuring programme, the establishment of a state-financed agricultural bank, subsidised crop insurance programmes, simplified and lowered taxes on agriculture, and subsidised loans for capital investment (Wegren 2007, 517). Whereas in the 1990s subsidies for the agricultural sector were sharply curtailed, under Putin, from 2006 onwards, agriculture became one of the four priorities of the 'national project'. The government set ambitious goals for domestic food security, targeting first of all the livestock sector, which experienced the most dramatic decline during the 1990s. As a result, cheap, state-subsidised credit is available for investments in livestock and especially dairy production. Asked about the motivations of the investors he had worked with, a Dutch interim farm manager and consultant, Han van Riel, who has worked in several agroholdings in Russia, answered:

> ... it could be, normally you get 10 million from the state, with a guarantee by Putin. The interest rate is zero percent'. [...] Yes, there are enormous funds. Look, normally the interest rate is 15 to 16 percent. But then it is zero percent (Han van Riel, Interview, 16 November 2011, The Netherlands).

Western observers often hasten to characterise state intervention in Russia as a legacy of, or return to, the Soviet past, but it is more productive to see it as an example of the wider global tendency of various states to stimulate a boom in agriculture. Whereas in some countries the subsidies focus on biofuel/diesel and energy security (such as the biofuel (corn) subsidies in the US, or biodiesel policies in Germany), in Russia, as a major fossil energy exporter and a weak food (livestock) producer, the subsidies focus on livestock production.

5.1.2. Speculation and land conversion

Land brokers and representatives of agroholdings searching for investment widely mention the expected value appreciation of agricultural land in Russia. Also, part of the land is obtained in order to sell it off later with a large premium as land for construction sites (in a later section several such cases will be discussed):

> I have, [...] tried to cultivate cabbage, on 20,000 ha, for a real estate agent. He wanted it in ownership, at least, the right of usage. But he had to cultivate it, otherwise he would not get it [...]. Just what I thought. At least 50 to 80 percent of his territory is now expansion area for construction for Moscow for the next 25 years. He knew that already by then. Thus, yes, for him it was just to have it, due to its location in Moscow [region].... (Han van Riel, Interview, 16 November 2011, The Netherlands).

5.1.3. Pressure by the government

Furthermore, many political and other, often hidden, motives seem to play a role, such as creating loyalty and political support among regional governors and a tacit agreement with the Kremlin, which seems to have promised not to investigate the dubious practices of the oligarchs in return for their investment in the countryside (Boldyrev 2001, 21). Interviews with consultants in agriculture suggested that regional authorities made agreements with the largest oligarchs that the latter should invest in the cumbersome agricultural sector in return for earlier or forthcoming privileges such as tax breaks, cheap credit or other forms of state support (Visser 2008). Obviously it is difficult to find clear evidence for such influence, but anecdotal evidence suggests that this indeed played a role. The quoted Dutch interim farm manager confirmed this pressure: 'I think one of the motives is the choice you have: either you pay tax arrears or you start producing food' (Han van Riel, Interview, 16 November 2011, The Netherlands).

For instance, one investor, the director of a fur factory described by Kalugina and Fadeeva, started to invest in a near-bankrupt farm enterprise after multiple requests by the district authorities, with promises of state support and privileges for its development. The investor described the interaction: 'The district head insisted that I who come there [to invest]. He said "nobody except from you can revive this enterprise". I tried several times to run away from it' (quoted in Kalugina and Fadeeva, 2009, 165).

A German investor even stated that 'the land was almost forced on us' (Winter 2012).

5.1.4. Tax evasion and money laundering

Furthermore, tax evasion, and probably money laundering, seems to play a role (Boldyrev 2001). Agriculture is a complex sector, with large fluctuations in production and profitability year to year, which allowed for extensive creative accounting already in the Soviet period (Visser and Kalb 2010). Furthermore, the tax on agricultural production is low to virtually zero.

6. Abandoned land, dying villages and investment: discourse and reality

Whereas the previous section looked into the motivations behind the land rush, this section will turn to a major topic in the legitimisation of these deals: the concept of 'abandoned', 'unused' or 'available' land.

6.1. The discourse of abandoned land

The idea of widely available abandoned land waiting to be (re)cultivated is very widespread among media and investors describing the Russian countryside:

> You should take the car. Drive to the North, [further] across Belarus, and then towards the South, Smolensk, and in the direction of Moscow. Or even go further. [...] Then you will meet millions of hectares which are just abandoned. Nobody is living there anymore (Interview with Dutch consultant/farmer Han van Riel, 16 November 2011, The Netherlands).

and:

> Back then this tradition-rich region was on its knees. The giant state agricultural collectives of the Soviet era were all bankrupt; their ancient equipment was in disrepair, and their land stood largely fallow. Young people fled to urban areas, leaving behind

> aging parents, decrepit farm houses and weed-choked gardens. Today, however, the region has burst to life. The capitalist upheaval that years ago began sweeping through Russia's big cities and oil and gas fields has finally made its way to the country's rural heartland. Newly imported tractors and harvesters with the latest ground-positioning satellite systems navigate vast privatized estates, producing bumper crops of wheat, barley, rapeseed, sugar beets and corn. (Kandell 2009, 1)

The director of the largest German investor in Russian agriculture (Ekoniva) states:

We change the attitude to land. In every farm we start restoring abandoned, neglected, often uncultivated lands and some points which were marked as fields on the map even appeared to be forest-belts in reality (Zaslavskaya 2010).

6.2. *Is a win-win situation possible?*

If anywhere, it would appear that a 'win-win' situation proposed by investors and for example studies such as World Bank (2010a) on land deals would be possible in Russia. A few features of agriculture in Russia suggest, at first glance, that this is indeed the case. First, in general there is no food shortage or hunger. Russia is even a main exporter of grain (although importer of livestock products).

Second, much land is not cultivated. Indeed, rates of land that turned from agriculture into non-agriculture were very high in Russia in the 1990s, with this process already starting in the 1980s during the last decade of the Soviet Union (Ioffe and Nefedova 2004). Prishchepov *et al.* (forthcoming) show, based on detailed Landsat satellite data, that 52 percent of agricultural land in the North-Western Smolensk region was 'abandoned' between 1989 and 2000. Ioffe, Nefedova and Zaslavsky (2006) estimate that about 30 percent of Russia's rural settlements 'have either died out or are about to do so'. Ioffe and Nefedova (2004) suggest that as a rule farm enterprises are not profitable in areas where population density drops below 10 people per square kilometre. Low population density, an aging rural population, and out-migration of working-age people all confirm the image of abandoned land. Ioffe, Nefedova and Zaslavsky (2004, 934) even speak about 'black holes' in the countryside. In some villages most of the titles to land were in the name of people who had disappeared or owners who had died. In one district, Shagaida (2005) found that nearly half of the land shares were owned by people who had died.[13]

Third, the population often appears willing to rent out or sell the land. A farm consultant for large-scale farms in Russia states:

> Indeed many do sell their land. Some [of the rural population] do nothing with their land, which is just laying fallow. They don't get any revenue from it. And as I just said, if they hand it over to the large enterprise – let me put it that way – then they get some return from it. They earn something from it (Simon de Schutter, Interview, 13 January 2012, Holten, the Netherlands).

There is no widespread desire to become an independent farmer. In fact, in the 1990s, when some new independent family farmers emerged, a large share of the population was quite hostile to these efforts as they feared for the break-up of the LFEs and the loss of their low-paid, but until then 'secure', jobs and the whole social infrastructure, pensions, etc., connected to it (Visser 2008). The majority of the

[13]In addition to an indication of aging in the countryside, it is also an indication of the difficulties of registering land for rural dwellers.

working-age people (and a lot of people above pension age trying to supplement their low pensions) prefer to have a job at a LFE. Consequently, many of them are indeed happy if an investor approaches them to buy or rent the shares with the promise to lift up production and re-create employment. Quite a large part of the population is above working age and would physically be unable to work the land.

6.3. A critique of the discourse of abandoned land

Despite the arguments above, one should not jump to the conclusion that therefore land acquisitions in Russia cannot be but beneficial for rural dwellers next to the investor.

First, empty or uncultivated land is predominantly in the north (with poorer soils, and worse climatic circumstances) where hardly any investors are interested in obtaining land, with the exception of land near the large cities (Uzun 2011). In the well-endowed south, there is fallow land in the peripheral regions but here investors are also mostly not interested in obtaining land. A Dutch farmer/investor, who on a trip to Russia in search for agricultural land, met with a mayor showing him a potential investment site, stated in an interview:

> ...we said, ok, that is not worth investing in anymore. If we want to start with a big unit, than we like to build it near a big road. There we have good infrastructure... [...] we didn't want to build it in the village. We wanted to put it in the middle of the fields. That works most efficient (Farmer De Boer, Interview, 13 January 2012, The Netherlands).[14]

Second, when land is not cultivated it does not mean that land is unused. Often villagers do use it in some ways and have (informal) entitlements to land for functions other than agricultural cultivation, involving such functions as grazing, hunting, bee keeping, and mushroom or berry collection. It is important to note that in the Soviet era, a symbiosis existed between the collective farm and the household plots of the employees and other villagers (see Visser 2006). Households were mostly allowed to let their private livestock graze or bees forage at the fields of the collective farm. Further, the collective farm, as a rule, ploughed the household plots and provided free inputs such as manure or fodder for private livestock. Most of these entitlements were mutually agreed between farm management and employees, although employees and villagers also had one-sided claims on collective property, which were considered stealing by the farm management. The land reform and reorganisation of the farm enterprises in the 1990s did not take into account these existing property relations.

Research on property in other countries has shown that informal property rights are often overlooked during property reform, and it is not easy to do justice to the complex web of pre-existing property rights when formalizing them (von Benda Beckmann 2003, von Benda Beckmann *et al.* 2006). While many of the farm directors continued part of these longstanding, informal entitlements after the land

[14]The mayor did not agree with starting a new investment site further away from the village, and as a result the investor continued to search elsewhere. The investor could understand the position of the local authorities and mentioned their care for the rural inhabitants with some appreciation. The word 'mayor' is used here following the terminology of the interviewee.

reform of the 1990s (Spoor and Visser 2004, Visser 2009), with the emergence of outside investors farm property is often fenced off, support to households is curtailed and guards are installed (Visser 2008, Nikulin 2010). Nikulin (2010) shows in his case study of a large investor in Perm region that while villagers quite easily sold their land shares as they did not cultivate the land, they often persistently opposed the new production technologies (industrial ploughing and chemical fertilization) brought along by the investor, which disturbed the livelihood they had built on the 'abandoned' land. The investors stated:

> Now we... will increase the ploughing area up to 7 thousand, [...] Meanwhile, many people here have formed a certain life style, the life style of living in the forest. They go to the woods, they keep bees. When we arrive here with an active production cycle, to some extent we disrupt their social rhythm. Now they cannot already wander in the fields and ride as they please, because the fields are ours, and they are ploughed by us, [...]. Instead of their wild strawberries there is, let's say, our pea. We begin ploughing the young forest with which the fields are overgrown, but there are mushrooms there (Nikulin 2010, 26).

Third, even if land is uncultivated usually there is an owner. Although part of the population is prepared to sell or rent out their land, not all of them want to do so. Some would be able to work the land physically. Some of them sell the land because the institutional environment, described earlier, is such that they have difficulty getting finance or registering their land, and also sometimes there are regional limitations (minimal limits to the size of farms).

Fourth, the rural dwellers are often in unequal power relations *vis-à-vis* the investors and rarely get a good price for their land. Here the difficulties mentioned earlier, of registering land for rural dwellers, play an important role. Costs of registration hamper them in their opportunities to sell the land freely. Often, the costs of registration are prohibitive and investors buying the land take up these costs. Even when agroholdings 'freely' buy up the shares of the employees, as is often the case with LFEs, the management and employees often sell under pressure from the investor. In a survey among 200 farm employees who sold off their land shares to an agroholding, 156 answered that they did so under pressure of the holding group (Gerasin *et al.* 2003, 176).

7. Land accumulation strategies

A major consequence of the share-based privatisation of land was that large farm enterprises (LFEs) retained control over former state-owned agricultural land because most land shareowners rented their share allotments back to the large farm in return for payment (incidentally, villagers rented out their land to private family farms but in many cases there was no other actor to rent the land to than the local farm enterprise). As a result, investors in search of land normally encounter large farms with integrated landholdings, but with actual ownership being very dispersed in the form of land shares. A relatively small amount of agricultural land concerns land shares concentrated in the hands of one, or a few, owners. Normally these are farm managers or outsiders who managed to obtain shares from the rural population. Finally, as mentioned earlier, a minority of agricultural land is still held by the state (often regional authorities or municipalities).

There are several strategies to acquire land in Russia. Land can be leased, or it can be bought in various direct and indirect ways. Buying up (or leasing) land from a large number of land shareholders takes quite some time. It was shown earlier that the registration procedure leads to high transaction costs. Sometimes investors try to avoid the time-consuming and bureaucratic procedures of acquiring land, in order to grab land at a very low cost.[15] Frequently, the strategies of land acquisition within the framework of the law are combined with or substituted by means that contradict the national laws or obey the letter of the law but not the spirit, using loopholes of Russian legislation to legalise unfair land acquisition.[16] Thus, one could discern a continuum of land acquisition strategies, running from acquisitions within the framework of the law, strategies at the fringe of the law (including misinformation, deceit and pressure), and finally outright fraud and dispossession. Due to limitations of space, the discussion below will cover strategies at the two opposites of the spectrum and largely leave aside the in-between category.

7.1. Land accumulation strategies within the framework of the legal system

The first way of obtaining land is leasing or buying land from (dispersed) land shareholders. Lease was already possible before the 2002 land law on land transactions. Buying and selling of agricultural land by individuals is minor compared to leasing land shares from shareowners (or the state). According to the research of Lerman and Shagaida (2007) in corporate and peasant farms, the share of leased land is on average 60 percent of the total area of agricultural land used. In corporate farms, three-quarters of the leased land is in the form of land shares and only one-quarter is leased as land plots.

Currently, much of the land acquired by agroholdings is rented, but there seems to be a growing tendency to buy land, which is of course partly due to the law of 2002. Based on our web research and interviews, we can conclude, not surprisingly, that foreign investors prefer buying to leasing. In our interview with Gustav Wetterling of Agro-Invest Group, he said:

> We have significant costs related to getting rid of all the weeds, all the trees… So that's why we want ownership. So we know that we have it for a long period, so we will get this money paid back. There is a significant investment in the beginning… (Interview 23 September 2010, Moscow).

Many agroholdings do not disclose information about land in use and ownership, but the official web pages of the main ones which do so suggest that the above-mentioned statement by the representative of Black Earth Farming is indicative of a wider trend.

This company buys land rights directly from every individual shareholder, making an agreement with rural dwellers to register the land unit in their name and on their behalf and then buying this land according to a sale-purchase agreement. When Black Earth Farming began land acquisitions in 2004, it paid 90 Euros per ha (Kandell 2009). According to the director, the shareholders were happy to sell their rights as they were not able to register the lands themselves due to bureaucratic

[15]It should be noted that these strategies are mainly used by domestic agroholdings or ones that are offshore investment funds of Russian oligarchs.

[16]Also some deals which are within the framework of the law, such as a deal involving a farm director selling his land shares to an outsider, may have been preceded by a fraud in obtaining these shares from the rural population but with the legal period for filing a complaint on the early fraud expired.

obstacles and the costs of registration, which can be too high for poor villagers; hence, their land did not provide them with any income. Gustav Wetterling stated that 'during the last two years we got one offer per week to buy a land plot. There is no shortage of land for sale in Russia right now' (Interview 23 September 2010, Moscow).

The same mechanism was used by Agro-Invest Brinky BV. The CEO of this Dutch company, Willeke van den Brink, stated that it was difficult to acquire the shares of former collective employees. 'It was problematic not only to agree on the (conditions of the) sale, but even to find all these people: someone died, someone had left, et cetera' (Luchev 2009). Acquiring shares for a controlling stake in the hands of Agro-Invest Brinky took close to one year. Dealing with large numbers of shareholders is a common downside for investors building up an agroholding in this way. The UK company Heartland Farms Ltd, owner of 18,500 ha in 2009, concluded contracts with around 1500 land owners (BTA Analitika 2008, 15).

A way to obtain large amounts of land in one go without numerous negotiations with individual land shareholders is, first, to buy a whole farm enterprise (in which land shares were already accumulated by the farm management). Buying land in this way is in some cases not done completely legally (see next sub-section). The villager-shareholders do not always want to invest their shares in the companies. Investors used tricks, and some started preparing the documents of the enterprise and placing their managers in the farm staff several years before they announced their wish to acquire the farm.[17]

Second, a minor, but increasing, percentage of land has been accumulated by companies and is offered for purchase. For instance, Gazprom, which ended up with half a million ha of land, is selling off much of this land. Also some agroholdings which grew rapidly sell off some of their land due to financial problems or to cluster their land in a few regions. In these two variants, a (foreign) investor faces fewer transaction costs in obtaining land, but of course it has to deal with all the other organizational issues of starting up a new farm enterprise, which can be quite a laborious process in Russia.

With the emergence of increasing numbers of domestic agroholdings in the past few years, there are foreign and domestic investors that have taken equity (whether a minority or majority share) in these entities. Several large domestic agroholdings in Russia and Ukraine are even quoted at stock exchanges in Western Europe and others recently have announced plans to do so. Also some Western agribusinesses operating in Ukraine and Russia are quoted on stock exchanges, such as the French farming group AgroGeneration. Although this process of share emissions only started after the mid-2000s, currently Russia and Ukraine already have more agroholdings at the stock exchange than do the large agricultural powers of Latin America, Brazil and Argentina (World Bank 2010a). In this sense, Russia, which was still a socialist country dominated by state and collective farms two decades ago, is now at the 'forefront' of financialisation and large-scale global commodification of agriculture and land. An expert with Renaissance Capital stated that Russian agriculture will increasingly be dominated by agrocompanies with capitalisation of over half a billion dollars (Vasilyeva 2008, 2).

[17]Based on interview with Tamara Semenova, vice-president of the peasant movement Krestyansky Front, discussing the case of the farm enterprise Matveyevskoye. Interview was conducted 29 September 2010 in Moscow, Russia.

In February 2011, US-based PepsiCo, the second largest beverage and food company in the world, obtained a majority share in OJSC Wimm-Bill-Dann, the largest dairy agroholding in Russia. Wimm-Bill-Dann owns 37 food factories, many dairies around Moscow and elsewhere in Russia and even in other countries of the former Soviet Union, as well as port infrastructure. It also controls 40 large farm enterprises, with 20,000 ha in ownership and 250,000 ha in leasehold. The legal obstacle concerning foreign land ownership has been solved by transferring the land bank formally to the owners-founders, who had the controlling share before PepsiCo came in. One of the largest agroholdings in Russia, Cherkizovo Group OJSC, which holds numerous pig farms and fodder producing farms, is now predominantly owned by US investors, namely JP Morgan Chase Bank and MB Capital Partners. Also in the Far East of Russia, agroholdings are taken over by foreign investors (Visser and Spoor 2011b).

7.2. Land accumulation outside the legal system

There is increasing evidence of illegal land deals in Russia, in a land market which is generally far from transparent. According to official statistics, the number of land disputes in the first half of 2008 compared to the first half of 2007 increased by 63 percent for the whole system of arbitration courts of the Russian Federation.[18] An analysis of crimes associated with illegal registration of land transactions also indicates growing abuses associated with the misappropriation and illegal registration of land transactions (Moskalkova 2011). The 'leader' in terms of criminal investigations is the Moscow region, due to its advantageous location close to the capital and due to the high land prices, which reach 1000–1500 USD per square metre as a consequence of demand for land to build *dachas*[19] (fieldwork of Spoor and Visser 29/10/2011). In 2011, the Governor of Moscow region said that 'almost of all of them [the investors who buy large plots of land] have problems with documents about the land rights' (Moskalkova 2011). Tamara Semenova (Krestyansky Front) stated:

> [S]ince the moment that, in 2002, the state law about the turnover of agricultural lands came into effect and land became a commodity, illegal court practices have become popular in the Moscow region. Before 2002 the land rights were given to peasants, after 2002 – the other way around' (Interview, 29 September 2010, Moscow).

In almost every district of Moscow region, there are land conflicts between peasants and the new owners of the land, according to the Krestyansky Front. Moreover, the illegal land deals do not only happen in the Moscow region, but also in the well-endowed regions of south Russia (for instance Krasnodar) and in various other regions (Visser and Mamonova 2011), where land acquisitions focus less on conversion into construction sites than in Moscow region. However, due to the fact that these regions are covered less by the mass media, the situation there is less well known.

According to Tamara Semenova, 'invaders' usually pretend to be investors, offering to invest in the development of agricultural enterprises by buying shares. If the peasants refuse to sell their shares to these investors, then various illegal schemes

[18]Report of Arbitration Court of Bashkortostan Republic 2008.

[19]*Dacha* is a Russian word for seasonal or year-round second homes often located in the exurbs of the country. It is estimated that about 50 percent of Russian families living in large cities have dachas.

are used to achieve the goal. Manipulations with land units through forgery of farm enterprise documents on land ownership is the most common illegal method of weaning the land from the peasants, because technically and time-wise it is most simple to realize. For example, employees of former state farm Zaoksky in the Serpukhov district of Moscow region have lost their land as a result of the enterprise bankruptcy. As a result of bankruptcy proceedings, more than six thousand ha of land were purchased for only 10.8 million roubles (approximately 270,000 euros) by Center-Capital CJSC. As a result, over two thousand people lost their legal right to the land shares.

Another criminal case has happened with CJSC Matveyevskoye in the Odintsovo district. This case can illustrate an important mechanism through which land grabbers acquire land. First there was the falsification of a meeting of shareholders, at which they ostensibly gave their shares to the capital of the enterprise. This already took place in the 1990s. There was also a case between the two main shareholders of the farm. Supposedly, they were going to share their land shares. This court case was specially fabricated to get control over the land shares. During this court case, Matveyevskoye acted as a third party. Matveyevskoye came with a counterclaim to recognize the rights to the entire land mass for Matveyevskoye. Finally, the court recognized the property rights of Matveyevskoye on the land shares of a common area of 712.4 ha of land of agricultural value (land value of more than 600 million roubles – approximately 15 million euros). Shareholders came to know about this court case, which was carried out in another region, only three years later when the statute of limitations had passed. In that period the former CEO of Matveyevskoye N. Dubrovsky and the next CEO V. Bobynin were assassinated, murder cases that remained unresolved.

For six years, the shareholders of Matveyevskoye have been seeking the recovery of their land shares. However, the prosecutors and investigative bodies, one after another, refused to initiate criminal proceedings. Currently, Matveyevskoye is owned by a company registered in Britain, Millhouse Capital UK Ltd, which manages the capital of the famous oligarch Roman Abramovich and the company Inteko (headed by Elena Baturina – wife of Moscow ex-Mayor Yuri Luzhkov). There seem to be no real plans to develop agriculture as had been promised. The lands are used to build VIP country houses and business centres. On 1 November 2008, appealing to the article 159 Part 4 of the Criminal Code ('fraud, committed by a group of persons on a large scale') the criminal case No. 152983 was initiated against the 'raiders' of Matveyevskoye. However, a few months later the case was dismissed at the request of the Moscow region Prosecutor's Office. Currently the victims are appealing to higher authorities for help.

7.3. Countering illegal land acquisitions

In a few cases, mobilization by dispossessed rural dwellers who also pursued court cases resulted in some effect, but only after very long legal procedures (often involving multiple court cases at different levels). Former employees of a poultry farm in Krasnogorsk, Moscow region, were in a similar situation to Matveyevskoye above. The farm was acquired after the bankruptcy of the Agricultural Land Corporation Znak, co-owned by oligarch Nikolai Tsvetkov. Since 2001, people have been trying to restore their rights, but had not succeeded. However, in the middle of March 2010, the equity holders were paid 190 thousand roubles – approximately

4,750 euros – (per share of 0.39 ha), to make them withdraw their complaints. The following successful case gives more insight into the enormous efforts and kind of strategies that are required by rural dwellers to achieve some form of justice.

'It is impossible to fight land grabbing with classical legal instruments alone', says Tamara Semenova of Krestyansky Front (Interview, 29 September 2010, Moscow). According to her, the courts in Russia, especially those that deal with land issues, are bought by raiders. The rural social movement, which was founded by dispossessed shareholders of a Moscow farm, is the most active among the recently emerging rural movements in Russia (Visser 2010, Visser and Mamonova 2011). It started by organizing public events (pickets) in Moscow city in order to draw the attention of federal authorities to this issue. It did not work particularly well. The state did not respond to their repeated appeals to address the large-scale dispossession of land held by rural dwellers. Recently the movement has developed another strategy. The movement decided to seek a solution on a case-by-case basis, trying to mobilize those rural dwellers directly involved and targeting the companies and authorities related to each case. Tamara herself was a leader of the movement of shareholders (600 people) of CJSC Agrocomplex Gorky-2 in reaction to an illegal action of acquiring land shares from shareholders without any compensation. For two years, the group of deprived shareholders has almost never left the Rublyovo-Uspenskoe highway where the management office of Agrocomplex Gorki-2 is situated.

> We had been standing for two years three days a week from 8 a.m. to 5 p.m. with placards along Rublevka.[20] Well, because it was Rublyovo-Uspenskoe highway, many officials and government members use that road every day and they had been asking our raiders: 'When will you pay, when will these protesters leave the road?' It was the sore point that we found. Courts didn't work for us, of course. In courts we lost every case. Totally absurd decisions were made by courts, even sometimes hard to believe: is that a circus or a court? (Interview, 29 September 2010, Moscow).

In total the fight against the Agrocomplex Gorki-2 took five-and-a-half years. According to the Krestyansky Front leadership they experienced threats and violence and were offered bribes in attempts to stop their efforts (Tamara Semenova, Interview, 29 September 2010, Moscow). But the activists continued their activities to get compensation. Finally in 2008 the shareholders got financial compensation for their lands.

The previous two cases led to compensation being paid, but restitution of the dispossessed land was not achieved. In Moscow, we have found only one court case of illegal land acquisition that has led to land being returned to the former, rightful owners. One of the main landlords of the Moscow region, JSC Vash Finansoviy Popechitel, has more than 63,000 ha in Ruza district in the Moscow region. The Chairman of the Board of Directors Vasily Boyko is currently under home confinement. He is charged under the articles of 'large-scale crime' and 'legalization of monetary funds or other property acquired as a result of crime'. Vasily Boyko used a standard way to assemble the lands: buying up the collective farm land units. He acquired nine of eleven farms in Ruza district with a plan to develop a real estate project called 'Ruza Switzerland'. The resort should occupy 40 ha in Ruza district. The project included a country hotel with a golf club, equestrian parks and spas,

[20]The name 'Rublevka' is derived from the name of Rublyovo-Uspenskoe highway. It is an unofficial name of a prestigious residential area west of Moscow, Russia.

sports facilities and yacht clubs, a ski resort, a mini-airport for sports and business aviation, helicopter pads, a safari park and elite cottage villages.

However, due to the activities of resentful peasants and the end of the support from local authorities, the project had to be postponed. Nevertheless, it is questionable whether peasant mobilization in the form of a court case and demonstrations alone could have achieved this result. Some authors say that a number of the deprived peasants had private connections with bodies of internal affairs and that due to these connections their request was met (Kozyrev and Abakumova 2008). Others argue that Boyko lost the support of the local authorities and that is why he lost the court case (Solomatina 2007). Meanwhile the CEO of Vash Finansovy Popechitel, Maria Loboda, expressed her point of view that the arrest of Boyko is linked to the struggle for control over the land in the Ruza district. 'The purpose of this attack is to take away our business by competitors, and law agencies are used as a tool to achieve this goal,' she said in an interview to Rambler Media Group (Rambler 2007). According to Boyko's lawyers, the criminal case was initiated by organizations that are interested in obtaining contracts for the construction of the Central Ring Road and have close contacts with the government of the Moscow region. In this last version, which might be compatible with the earlier explanations, the restitution of land is not a sign of the impartial functioning of the rule of law, but rather the result of a power struggle within the arena of the business and political elites that offered a rare window of opportunity for the rural dwellers, an example of 'divided elites', one of the political opportunity structures for social movements as described by Tarrow (1998). Therefore, it is important to take a look at the role of the state.

8. The state: policies regarding foreign land investment and land grabbing

8.1. Attracting foreign investment

Agrarian policy of Russia in the Putin/Medvedev era (the 2000s and beyond) has become characterized by what has been called 'economic nationalism' (Wegren 2009a, b); with more focus on food security, increased customs duties (formal, or informal through quality controls), and increased use of food policy as a political tool in international relations.[21] Nevertheless, the authorities seem to be convinced that modernization and production increases could be enhanced with foreign investment. As a consequence, over the past few years the Russian government is actively encouraging this.

By Russian law, foreign persons and companies are not allowed to own land. However, the legal loophole to avoid this, by establishing a majority or fully owned daughter company in Russia, is not at all frowned upon by state officials. In fact, this loophole is actively promoted. During a seminar on investing in Russian and Ukrainian agriculture, a Dutch farmer, operating in Russia, complained about the fact that a foreigner cannot own land. The Russian vice-minister of Agriculture responded that it is no problem, as a foreigner 'you just open a subsidiary in Russia, and buy as much land as you want'.[22] On the regional and local level the stance of

[21]Examples of food policy as a political weapon in Russian international relations have been the wine ban for Georgia and the 'milk war' with Belarus (Wegren 2009a, b, 2010).
[22]This is based on observations by Visser during a seminar on investment opportunities in the Agribusiness of Russia and Ukraine, held in Wassenaar, The Netherlands, November 2010.

the authorities differs, depending to a large extent on the governor in charge (Zimin 2010), but most authorities encourage both domestic and foreign investment in agriculture.[23]

8.2. Policies to address illegal land grabbing

There are some attempts by the Russian federal state to prevent illegal land grabbing, at least in discourse. High-ranking Russian officials have recently condemned illegal land grabbing and announced measures to prevent it. Former Minister of Agriculture Aleksei Gordeev criticized urban 'raiders' and warned of 'wars' over rural land. As land 'raiding' became more frequent Gordeev advocated greater government regulation of land relations in order to protect the property rights of land shareholders (Wegren 2009b). In 2011, president Medvedev condemned land 'raiders' in a speech and proposed laws to limit conversion of agricultural land into construction land as a main tool to deal with the issue. Overall, there is very little indication that the regulations by the state are guided by a concern for the dispossession of land held by the rural population. Farmer movements fighting in defence of the rural dwellers' land rights have not been allowed a single meeting with federal government officials. In discussions on land governance between officials and academics from FSU countries within the framework of meetings on the voluntary guidelines for land grabbing by FAO in Moscow (November 2010), of all the officials, Russian officials were least inclined to introduce policies to allow free or transparent access to land registration for the population.[24]

Instead, current policies (and proposed new regulations) are largely oriented at keeping intact (and furthering) the concentration of land and the dominance of agroholdings, with more power for government land governance bodies (see also Wegren 2009a, b). In 2010 the Ministry of Economic Development and Trade began working on amendments to the Land Code, although there is no indication as to when these might be considered by the Duma. Draft legislation suggested by the Ministry of Agriculture will: (1) give right of first refusal to large farms to lease reregistered land; (2) give preference to municipal and regional governments to convert unclaimed, unwanted, and abandoned land to state property; (3) increase fines and/or land taxes for land that is used inappropriately or is not used for its intended purpose; (4) create a unified system of state monitoring of agricultural land.

At the moment, two main ways of controlling the land market are implemented. First, conversion of agricultural land into non-agricultural is becoming more regulated. As Anton Mitrofanov, director of the real estate agency Housing Strategy, stated, 'it was relatively easy, in negotiation with the municipality and with limited expenses, to convert agricultural land into another category, for instance for the construction of dacha (summer house) complexes' (BFM 2011, 3). Beginning in 2011 it became more complicated as Federal Law (2010) N 435-FZ 'On Amendments to Certain Legislative Acts of the Russian Federation with regard to improvement of agricultural lands' entered into force, which contained additional requirements to a

[23]For instance, Visser had various conversations with Dutch businessmen who told him that they were offered agricultural land to start a farm by various regional authorities, on visits for business in other sectors.

[24]Observations by Visser at the FAO Voluntary Guidelines regional consultation for the CIS in Moscow.

transaction if the land is converted from agricultural to construction purposes, making such a conversion more difficult. In the Moscow region, where illegal land deals are particularly widespread, to prevent widespread conversion of land for agricultural purposes to construction purposes the Federal Law No. 172-FZ 'On the transfer of land or land plots from one category to another' (2004) has been further elaborated. In the near future there will be annual quotas on the sale of land to prevent the sale of agricultural enterprises (to agroholdings). However, there is a danger that such measures will only add to further bureaucracy and corruption, without substantially reducing illegal conversions.

Second, regulations combating ineffective use of agricultural land have been introduced, and are more often enforced. Authorities can penalize owners for 'ineffective use' of agricultural land by penalties and even dispossession by the state of agricultural land that is not used for at least three years.[25] In 2009, several regions started to enforce the first part of this law. In Leningrad region, the authorities checked 170,000 ha, of which 10,500 ha were not properly used, thus total penalties for these land plots amounted to over half a million roubles (40 roubles was about one euro) (BFM 2011, 2). Unfortunately, this enforcement can also be used to the detriment of land right holders who do not have the funds to 'properly use' their allocated land, and could therefore be dispossessed by law.

An important question therefore arises: namely, to what extent will these new laws and state policies of closer monitoring of land use (and penalizing or even dispossessing land in case of misuse) lead to constraints on illegal land acquisitions? Or will they facilitate dispossession and further land grabbing? First of all, the attempt to address the downside of land grabbing through handing more power to the state encounters the problem that the local authorities are strongly in favour of large-scale land acquisitions and are often part and parcel of the problem. It seems that land acquisitions are not possible without 'friendship' with, or at least ignorance by, the government. According to Tamara Semenova (Krestyansky Front):

> [W]hen the raiders came and started buying land, they came not just off the street. It was by prior arrangement with the regional and district authorities. Who would let the strangers do this business in the Moscow region!? Well, we suppose they got the possibility to do this business for '*otkati*'[26] (Interview 29 September 2010, Moscow).

If authorities were willing to address the problem, what are their chances of succeeding? As mentioned earlier, authorities have conducted legal procedures to withdraw land that is not used for over three years from its owners. However, in practice it is not easy for authorities to succeed. In interviews with authorities in Moscow region, it was stated that it is very difficult for the state to win such cases, and the interviewees did not know of any example in the region where land was actually taken by the state. Sources from elsewhere in Russia confirm these statements (BFM 2011, 2). Due to breaches of the legal procedures, the state loses these disputes (BFM 2011) in court cases with rich investors employing highly skilled lawyers.

Moreover, owners of agricultural land with speculative aims go to great lengths to circumvent above-mentioned laws. Some landowners are even building (partly)

[25]According to article 284 of the Federal Law N 136-FZ 'Land Code; see: www.souzsad.ru/index.php/gl/39-sgt/60-2011-01-26-19-51-51 of Russian Federation' (2001).
[26]*Otkati* (in English, kickbacks) – a kind of bribe (money or gift) to authorities.

virtual agroholdings to accumulate low-priced agricultural land for other purposes.[27] Vasily Boyko of JSC Vash Finansoviy Popechitel has created the 'agricultural' holding Ruzskoe Moloko (Milk from Ruza), which sells packaged milk. However, the pastures are empty as livestock is sorely lacking. It appears that the company buys milk from farmers in neighbouring areas, which it (re)packages to sell under its own brand. Another example is oligarch Nikolai Tsvetkov's agricultural company Znak, which pretends to be a poultry farming agroholding. The bulk of the eggs the company 'produces' it actually buys from a poultry farm in another region and subsequently resells under its own brand.

Thus the law enabling confiscation of ineffectively used land so far does not pose a major obstacle for rich land owners, who can hire expensive lawyers and devise creative business strategies to prevent dispossession. At the same time, the law may form a threat for the rural population with their small land shares' land plots. Small-scale landowners who are temporarily unable to cultivate their land may easily lose their land as they are not able to hire lawyers to dispute confiscation of their land by the state. The situation of a school director in a village in Moscow region gives an indication of the uncertainty among the rural population due to this law. The director was close to his pension, but had no time to work the land currently. He was afraid that the state would take this land, and therefore he considered hiring someone to plough the land, to prove that he was cultivating the land.[28] Proposals in the Russian parliament (which to date have not been accepted) to have the state confiscate all land shares not yet registered (Wegren 2009a, 2011) would primarily lead to dispossession of rural dwellers, instead of limiting dubious land acquisitions by outsiders.

9. Conclusions

This research paper has analysed the post-Soviet land reform, the emergence of domestic and foreign investors, their types and the forms of land acquisitions, and the role of the state in the emerging land grabs. Russia (and the former Soviet Union in general) is largely overlooked in the land grab debate, although the process of land grabbing is occurring there at a great pace. Russia is a special case in the global land grab debate. It has vast amounts of land reserves and therefore is attractive to domestic and foreign investors. Russian companies, in particular in the energy and mineral resources sector, are very active in land acquisitions. Interestingly enough, these land deals are not for biofuels production as is the case in other parts of the world, since Russia is blessed with ample energy resources and low domestic energy prices as a consequence. Land has a speculative value and is meant for food production, urbanization (construction land) and building dachas ('dacha-ization').

Russia does not fit in the mainstream division of 'land grabbing countries' (oil-rich, and highly populated) versus 'target countries' (poor and land abundant). Russia is both an oil-rich country and at the same time one of the most land-abundant countries. Also, the case of Russia confirms recent critiques of the idea that land grabbing is predominantly carried out by foreigners. In addition, Russia

[27]For more on the Soviet and post-Soviet mechanisms leading to 'virtual production' see Visser and Kalb (2010) and Kalb and Visser (forthcoming), and see Lindner (2008, 133–139) for a discussion of virtual accounting within agriculture.

[28]Interview with school director, Moscow region, 14 February 2011.

does not conformably fit into the two main drivers of land grabs distinguished so far: food security (by governments) or commercial motives (by companies). In fact, in Russia the government plays an important role in stimulating large-scale investments (despite some occasional rhetoric to the contrary), but not only with the (primary) aim of food security (at least not in the crop sector), but also with the aim of using food export as a political tool in international relations (cf. Visser and Spoor 2011a, Wegren 2009a, b).

Although high levels of corruption and government intervention are a major drawback of Russia in terms of risk for investors, there are also a range of features of Russia that are likely to further raise the interest of domestic and foreign investors in large-scale land acquisitions. The huge size of landholdings and farm enterprises, and related infrastructure, is attractive for investors. Increasingly, land is available not only as large-scale land plots jointly owned in the form of hundreds of individual land shares, but also as ownership concentrated in the hands of farm directors, agroholdings or land brokers, although often not in accordance with Russian legislation. Also, foreign investors can take equity in large agroholdings registered on the stock exchange (Visser and Spoor 2011b). Despite the very recent emergence of Russia as an agricultural producer and major grain exporter, together with Ukraine, it already has more agricultural enterprises listed at stock exchanges than the established agricultural powerhouses Brazil and Argentina together. An increasingly high degree of 'financialization' of the agribusiness sector in Russian is observed, which furthermore facilitates large-scale land acquisitions or land grabs by global investors.

The important question is whether this investment will be for the better or worse. On the positive side, the agricultural sector might receive an additional boost when investors introduce new technology to Russian farm enterprises. Foreign agricultural companies might set up new business standards in the agriculture sector in Russia. Gustav Wetterling of Agro-Invest Group said that he 'doesn't think that it would be possible without foreign investments, without foreign companies that operate on Russian agricultural market, to achieve these results in agribusiness' (Interview, 23 September 2010, Moscow). Although a foreign company like this one, controlling 330,000 ha, has an important impact at a district or regional level, on a countrywide scale the predominance of foreign investors is restricted to some sub-sectors, such as poultry.

On the negative side, with growing numbers of domestic and foreign investors interested in agriculture, land dispossession and conflicts over land have risen as many (predominantly domestic) investors turned to illegal ways of acquiring land. While in Russia the 1990s represented the decade of struggle over property in industry and energy, in the first decade of the new millennium (2000–2010) attention for agriculture gradually increased, and the second decade (2010–2020) increasingly looks like the decade of struggle over land. Indeed, as Barnes (2006) pointed out in the quotation at the Section 4 of this paper, in the 1990s the property changes in agriculture still seemed underwhelming. Few investors from outside agribusiness (mainly domestic processors and among the foreign investors, former food importers to the SU) acquired land at that time. However, our research suggests that in quite a number of the illegal land grabs, farm directors, sometimes in cooperation with outsiders, already in the late 1990s and early 2000s paved the way for the land grabbing that became possible after the 2002 land code. In various cases, already in the 1990s meetings of shareholders and other documents were falsified to transfer land shares of the shareholders illegally to the charter capital of the farm enterprise (*de facto* to the farm director).[29] Farm managers who forged documents in the 1990s cashed in on these early moves in the 2000s when they could sell the land for huge profits, while rural dwellers were left landless without compensation. In summary, there is a wide range of strategies that are being used by investors and oligarchs to get land, from within the existing Russian legislative framework, to outright use of pressure, violence and fraud.

It is likely that land acquisitions will proceed at a great pace in Russia. There is little indication that the latest laws and the increased enforcement of laws limiting land conversion and 'ineffective use' of land will seriously contribute to constraining semi-legal and illegal land acquisitions. In fact, they might even contribute to the reverse – namely, increased dispossession of land of rural dwellers, as in the discussion on legalization by Peluso and Lund (2011, 674–675). They argue that rule of law often legalizes and legitimates the dispossession of the powerless. Governments and large corporations frequently 'operate with virtual impunity, while weaker actor may see rights whittled away' in the name of formalization (Peluso and Lund 2011, 675). In the Russian case there is an urgent need for improvement of the impartiality of courts, and the improvement of governance by authorities (in particular the transparency of deals between authorities and large-scale agroholdings). Without these measures and without stronger rural social movements defending the rights of the rural population, regulations and laws most likely will not reduce the risk of further land grabbing and dispossession.

The post-Soviet land reforms actually set the stage for the land grabbing that is taking place now, at least *de facto*, in the way they were implemented. Land basically remained under control of the large farms during the reform period of the 1990s, despite the rhetoric of equal distribution of land shares. Also, part of the farm management together with the help of outsiders already changed legal documents to allow for quick accumulation of land shares, once the land code was introduced. After the approval of the land code in 2002, and with the rising profitability of agriculture, the large farm enterprises (and the agroholdings) are also acquiring *de jure* control over land, as they increasingly take over land through legal and illegal practices. However, *de facto* control of land by the LFEs should predominantly be understood as effective control *vis-à-vis* the rural population, whereas in relation to the state, their control remains insecure and depends on cultivating good ties with authorities. With recently introduced and proposed legislation giving the state more power over land, insecurity of land for rural dwellers, but also for LFEs, tends to increase. Only the largest agroholdings have such a financial power that the power-balance between them and the state remains in favour of the agroholdings.

While it is shown that Russia is particular because it has huge land reserves, and therefore there would be a case to be made that 'win-win' land deals are possible, this availability of (unused) land thesis is undermined by the fact that much of the land reserve is available in areas that are not attractive for investors, and that where there is fertile land in better endowed regions, the very unequal power relations between investors and rural dwellers, and the weak legal framework (and enforcement of the law), leads often to unfair deals, dispossession and no or low compensation. In that sense Russia should be considered as a 'normal' case in the global land grab debate.

[29]Whereas in Moscow region farm directors often did so with the perspective of selling it on to outsiders, in most other regions in the 1990s rising land prices were not (yet) anticipated and control over the LFE seemed to be the main motivation.

References

Barnes, A. 2006. *Owning Russia. The struggle over factories, farms and power*. Ithaca, London: Cornell University Press.

BEFL. 2010. Russian market in agricultural land. PowerPoint presentation by Vladislav Novoselov, Director of BEFL, presented at the conference for agricultural producers and suppliers of agricultural inputs and services, Radisson SAS Slavyanskaya, 26 January, Moscow.

BFM. 2011. Direktor zavoda v Posmoskov́e obvinen v makhinatsiyakh s zemlei [Factory director in Moscow region guilty of illegal land acquisition]. Available from: www.bfm.ru/realty/2011/07/01/direktor-zavoda-v-podmoskove-obvinen-v-mahinacijak-s-zemlej.html [Accessed 2 July 2011].

Billette, A. 2009. Après une offensive discrète au Kazakhstan, la Chine lorgne les terres russes inexploitées [After a quiet aggressive social resistance in Kazakhstan, China is looking for Russian lands]. *Le Monde*, 20 April. Available from: www.farmlandgrab.org/2917 [Accessed 1 April 2010].

Boldyrev, F. 2001. Roga i kopyta. Pochemu oligarkhi zainteresovalis' sel'skim khozyaistvom [Horns and hooves. Why oligarchs became interested in agriculture]. *Kompaniya*, 188(42), 19–22.

Borras S. Jr., R. Hall, I. Scoones, B. White and W. Wolford. 2011. Towards a better understanding of global land grabbing: an editorial introduction. *The Journal of Peasant Studies*, 38(2), 209–216.

Borras, J., R. Hall, I. Scoones, B. White and W. Wolford. 2012. The new enclosures: critical perspectives on land grabbing. *Journal of Peasant Studies*, 39(3).

Borras, S. Jr., P. McMichael and I. Scoones. 2011. *The politics of biofuels, land and agrarian change*. London: Routledge.

Brooks, K., E. Krylatykh, Z. Lerman, A. Petrikov and V. Uzun. 1996. *Agricultural reform in Russia. A view from the farm level*. World Bank Discussion Paper no. 327, Washington, DC: The World Bank.

BTA Analitika. 2008. Ezhenedelniy obzor/Rynok zemli v Kazakhstane: sostoyanie, tendetsii razvitiy [Weekly review/Land market in Kazakhstan: status and trends]. Available from: http://www.bta.kz/files/wr_1_december.pdf [Accessed 1 July 2010].

Bush, J. 2008. Farming makes a comeback in Russia. *Business Week*, 9 October. Available from: www.farmlandgrab.org/2537 [Accessed 20 September 2010].

Cotula, L., S. Vermeulen, R. Leonard and J. Keeley. 2009. *Land grab or development opportunity? Agricultural investment and international land deals in Africa*. London/Rome: IIED/FAO/IFAD.

D'Hamecourt, P. 2010. *Toen Napoleon Maly verliet* [When Napoleon left Maly]. Schoorl: Uitgeverij Conserve.

Didenko, L. 2009. Pyat' gektarov – I korova, [Five hectares and a cow]. *Kommersant Guide*, 16(1), 16–17.

Ding, C. 2007. Policy and praxis of land acquisition in China. *Land Use Policy*, 24(1), 1–13.

Dronin, N. and A. Kirilenko. 2011. Climate change and food stress in Russia. *Regional Environmental Change*, 11(1), 167–78.

FAO. 2011. Food security and the transition region. Working paper prepared under the FAO/EBRD Cooperation for presentation at EBRD's Grant Planning Meeting. 29 November, London.

Gerasin, S., O. Rodionova and E. Schulze. 2003. On management of agro-industrial complexes, In: A. Balmann and A. Lissitsa, eds. *Large farm management*. Bergen/Dumme: AgriMedia, IAMO Studies on the Agricultural and Food Sector in Central and Eastern Europe, pp. 159–75.

GRAIN. 2008. Seized: The 2008 land grab for food and financial security. *Briefing*. Available from: http://www.grain.org/briefings/?id=212. [Accessed 20 September 2010].

Hall, R. 2012. *Journal of Peasant Studies*, 39(3).

HLPE. 2011. Land tenure and international investments in agriculture. *Committee on World Food Security High Level Panel of Experts on Food Security and Nutrition*. REPORT, July. Rome. Available from: http://www.fao.org/fileadmin/user_upload/hlpe/hlpe_documents/HLPE-Land-tenure-and-international-investments-in-agriculture-2011.pdf. [Accessed on 27 February 2012].

Hofman, I. and P. Ho. 2012. *Journal of Peasant Studies*, 39(1).

IAMO. 2011. *The agricultural potential of European Russia*. F. Schierhorn, K. Hahlbrock and D. Müller. Available from: www.iamo.de [Accessed 25 January 2012].

Ioffe, G. and T. Nefedova. 2004. Marginal farmland in European Russia. *Eurasian Geography and Economics*, 45(1), 45–59.

Ioffe, G., T. Nefedova and I. Zaslavsky. 2004. From spatial continuity to fragmentation: the case of Russian farming. *Annals of the Association of American Geographers*, 94(4), 913–43.

Ioffe, G., T. Nefedova and I. Zaslavsky. 2006. *The end of peasantry? The disintegration of rural Russia*. Pittsburgh, PA: University of Pittsburgh Press.

Jouko N. and L. Granberg. 2011. The global food crisis and changes in the Russian agricultural regulation. Paper presented at the International Radboud University/ISS/MSSES conference: Land accumulation in the former Soviet Union and beyond: Global investors and local communities. 27 October, Moscow, Russia.

Julia and B. White. 2012. *Journal of Peasant Studies*, 39(3).

Kalb, D. and O. Visser. In press. Oligarchy and state capture: Notes on the Soviet style mechanisms of contemporary finance capitalism. *In:* J.G. Carrier, ed. *A handbook of economic anthropology, second edition*. Cheltenham: Edward Elgar publishers.

Kalugina, Z. and O. Fadeeva. 2009. *Rossiiskaya derevnya v labirinte reform. Sotsiologicheskie zarisovki* [Russian village in the labyrinth of land reforms]. Novosibirsk: Institut Ekonomiki i Organizatsii Promyshlennogo Proizvodstva Sibirskogo otdeleniya RAN [Institute of Economics and Industrial Production Management of Sibirian department of Russian Academy of Science].

Kandell, J. 2009. Betting on the Russian farm. *Institutional investor*. Available from: http://farmlandgrab.org/13810 [Accessed 1 April 2011].

Kozyrev, M. and M. Abakumova. 2008. Zona riskovannogo zemlevladeniya [Zone of risky land tenure]. *Journal Forbes*. Available from: http://www.lobbying.ru/content/sections/articleid_3770_linkid_58.html [Accessed 11 October 2010].

Lerman, Z. and N. Shagaida. 2007. Land policies and agricultural land markets in Russia. *Land Use Policy*, 24, 14–23.

Lindner, P. 2008. *Der Kolchoz-Archipel im Privatisierungsprozess: Wege und Umwege der russischen Landwirtschaft in die globale Marktgesellschaft* [The collective farms in the archipelago- privatization process: paths and detours of the Russian agriculture in the global market society]. Bielefeld: Deutsche Forschungsgemeinschaft.

Luchev, N. 2009. Interview of the CEO of Agro-Invest Brinky Willeke van den Brink, *Journal Agroinvestor* 11. Available from: http://www.agro-investor.ru/issue/75/5251/ [Accessed 20 December 2010].

Moskalkova, T. 2011. Zemelnaya i sovdovodcheskaya mafiya [Land and garden mafia]. Available from: www.souzsad.ru/index.php/gl/39-sgt/60-2011-01-26-19-51-51 [Accessed 10 October 2010].

Nikulin, A. 2005. Noveichaya gigantomaniya [New gigantomania]. *Politichesskii Zhurnal*, 63 (12). Available from: http://www.politjournal.ru [Accessed 25 January 2012].

Nikulin, A. 2010. *Oligarkhoz kak preemnik postkolkhoza. Ekonomischeskaya Sotsiologiya*, 11 (1), 17–33. Available from www.ecsoc.msses.ru [Accessed 5 February 2012].

Pallot, J. and T. Nefedova. 2007. *Russia's unknown agriculture: household production in post-socialist rural Russia*. Oxford: Oxford University Press.

Peluso, N. L. and C. Lund. 2011. New frontiers of land control: introduction. *Journal of Peasant Studies*, 38(4), 667–81.

Poshkus B. 2009. *Stanovlenie rinochnih otnosheniy v selskom khoziaystve* [Creation of market relationships in Russian agriculture]. Moscow: VIAPI Press.

Prishchepov, A., V. Radeloff, M. Dubinin and C. Alcantara. Forthcoming. The effect of satellite image dates on land-cover change detection and the mapping of agricultural land abandonment in Eastern Europe. *Remote Sensing of Environment*.

Rambler Media Group. 2007. V Moskve arestovan zastroyschik kurortnoy zoni Ruzskaya Shveytsariya [The developer of the resort Ruza Switzerland was arrested in Moscow]. Lenta.Ru/News. Posted 26 February. Available from: http://lenta.ru/news/2007/02/26/Boiko/ [Accessed 10 October 2010].

Report of Arbitration Court of Bashkortostan Republic. 2008. *Obobschenie sudebnoy praktiki po rassmotreniyu sporov sviazannih s primeneniem sudebnogo zakonodatelstva za 2007 god I za 1 polugodie 2008 goda* [Generalization of court practice in the consideration of disputes involving the use of land laws in 2007 and for the 1st half of 2008]. Available from: http://ufa.arbitr.ru/pract/2961.html [Accessed 20 November 2010].

Rosenberg, M. 2012. Population decline in Russia. Available from: http://geography.about.com/od/obtainpopulationdata/a/russiapop.htm [Accessed 27 January 2012].

Rosreestr (Federal Service for State Registration, Cadastre and Cartography). 2010. State (national) report on the status and use of land in the Russian Federation in 2009. Available from: http://slon.ru/documents/doklad_ob_ispolzovanii_zemel_2008.pdf [Accessed 21 November 2011].

Rosstat (Federal State Statistics Service). 2006. *Vserossiyskaya selskokhoziaystvennaya perepis 2006. Part 3. Zemelnie ressursi i ih ispolzovanie*. [The outcome of the All-Russian Agricultural Census of 2006, Book 3: Land resources and their use]. Available from: http://www.gks.ru/news/perepis2006/T3/tabl_t3.pdf [Accessed 10 October 2010; 20 November 2011].

Shagaida, N. 2005. Living with constraints: Russia's agricultural land market. *BASIS Brief*, 36(10), 1–4.

Solomatina O. 2007. Zolotaya sotka [Golden hectares]. Available from: http://moszem.ru/text/1180045737.html [Accessed 1 November 2011].

Spoor, M. and O. Visser. 2001. The state of agrarian reform in the FSU. *Europe-Asia Studies* (former Soviet Studies), 5(6), 885–901.

Spoor, M. and O. Visser. 2004. Restructuring postponed? Large Russian farm enterprises 'coping with the market'. *Journal of Peasant Studies*, 32(2), 515–51.

Starobin, P. and C. Belton. 2002. Russia's new wealth. *Business Week* 5 August. Available from: www.businessweek.com/print/magazine/content/02_31/b3794013.htm?chan=mz [Accessed 1 March 2010].

Tarrow, S. 1998. *Power in movement: social movements and contentious politics*. Cambridge: Cambridge University Press.

The Federal Law N 136-FZ 'Land Code of Russian Federation', article 284 (2001). Available from: www.souzsad.ru/index.php/gl/39-sgt/60-2011-01-26-19-51-50 [Accessed 27 January 2011].

Uzun, V. 2005. Large and small business in Russian agriculture: adaptation to market. *Comparative Economic Studies*, 47(1), 85–100.

Uzun, V. 2011. Klassificatsiya selhozproizvoditeley Rossii i agrarnaya strukltura regionov [Classification of agricultural producers in Russia and agrarian structure of its regions]. Paper presented at the Guidar Forum. 18 March, Moscow.

Uzun, V., E. Gataulina, V. Saraikin, V. Bashmachnikov, O. Pavlushkina and O. Rodionova. 2009. *Tendentsii razvitiya i mekhanizmy vzaimodeistviya kruphnogo i malogo biznesa v agropromyshlennom komplekse* [Trends in the development and mechanisms of interaction large and small businesses in the agricultural sector]. Moscow: VIAPI Press.

Vasilyeva, N. 2008. Russian farming: from basket case to breadbasket? *Associated Press* 19 September 2008. Available from: www.farmlandgrab.org/2514 [Accessed 20 September 2010].

Visser, O. 2006. Property, labour relations and social obligations in Russia's privatised farm enterprises. *In:* K. von Benda-Beckman, F. von Benda-Beckman and M.G. Wiber, eds. *The changing properties of property*. London, New York: Berghahn, pp. 126–46.

Visser, O. 2008. *Crucial connections: the persistence of large farm enterprises in Russia*. PhD Thesis, Department of Anthropology and Development Studies, Radboud University Nijmegen.

Visser, O. 2009. Household plots and the symbiosis with the large farm enterprises. *In:* M. Spoor, ed. *Land, peasants and rural poverty in transition: the political economy of rural livelihoods in transition economies*. Studies in Rural Livelihoods Series. London, New York: Routledge.

Visser, O. 2010. Rights over land and obstacles to family farming in Russia: a comparison with Argentina. *Laboratorium*, 1(2), 420–4.

Visser, O. and D. Kalb. 2010. Financialised capitalism Soviet style? Variants of state capture and crisis. *Archives Européennes de Sociology*, L(2), 171–94.

Visser, O. and N. Mamonova. 2011. Emerging rural movements in Russia. Genuine movements or virtual vehicles with concealed aims? Paper presented at the master class: Social Movements in the South: how to study and write about them. Wageningen School of Social Sciences, 27 May, Wageningen University.

Visser, O. and M. Spoor. 2011a. Land grabbing in post-Soviet Eurasia: the worlds largest land reserves at stake. *The Journal of Peasant Studies*, 38(2), 299–324.

Visser, O. and M. Spoor. 2011b. Land governance in post-Soviet Russia: conflict and cooperation between local authorities and agroholdings. Paper presented at the ISS/ Cornell workshop Comparative Land Institution, January 20–21, Institute of Social Studies (ISS), The Hague.

von Benda-Beckmann, F. 2003. Mysteries of capital or mystification of legal property? *Focaal-Journal of Global and Historical Anthropology*, 41, 187–91.

von Benda-Beckmann, F., K. von Benda-Beckmann and M.G. Wiber, eds. 2006. *Changing properties of property*. London, New York: Berghahn.

von Braun, J. and R. Meinzen-Dick. 2009. *'Land grabbing' by foreign investors in developing countries: risks and opportunities*. Policy Briefs 13. Washington, DC: International Food Policy Research Institute (IFPRI).

Wegren, S. 2007. The State and agricultural reform in post-communist Russia. *Journal of Peasant Studies*, 34(3), 498–526.

Wegren, S. 2009a. Russian agriculture in 2009: continuity or change? *Eurasian Geography and Economics*, 50(4), 464–79.

Wegren, S. 2009b. Russia's incomplete land reform. *Russian Analytical Digest*, 64(09), 2–5.

Wegren, S. 2010. Russian Agriculture in the Late Putin Period and Beyond, in: S. Wegren and Dale R. Herspring (eds.) *After Putin's Russia: Past Imperfect, Future Uncertain*. Landham, MD: Rowman and Littlefield, pp. 199–222.

Wegren, S. 2011. Food security and Russia's 2010 drought. *Eurasian Geography and Economics*, 52(1), 140–56.

Winter, S. 2012. Pigs and protection money: German farmers seek their fortunes in Russia. *Der Spiegel*. Posted 12 January. Available from: www.farmlandgrab.org/post/view/19872 [Accessed 12 January 2012].

World Bank. 1992. *Food and agricultural policy reforms in the former USSR: an agenda for transition*. Studies in Economies in Transformation, no. 1. Washington, D.C.: The World Bank.

World Bank. 2010a. *Rising global interest in farmland. Can it yield sustainable and equitable benefits?* Washington, D.C.: The World Bank.

World Bank. 2010b. Online database, section: Arable land (hectares per person) 2005–2009 Available from: http://data.worldbank.org/indicator/AG.LND.ARBL.HA.PC/countries, [Accessed 12 October 2010].

Zaslavskaya, O. 2010. Ekoniva-agro revives cattle breeding in Liski district. *Ekonomika i Zhizn' Chernozemie newspaper*, 7(30), 10–23.

Zimin, B. 2010. Promoting investment in Russia's regions. *Eurasian Geography and Economics*, 51(5), 653–68.

Zoomers, A. 2010. Globalization and the foreignization of space: seven processes driving the current land grab. *Journal of Peasant Studies*, 37(2), 429–47.

Oane Visser is Assistant Professor in Anthropology and Development Studies at Radboud University Nijmegen, the Netherlands. His research focuses on land grabbing, poverty and social movements in Eastern *Europe and the former Soviet countries. He has published in journals including Europe-Asia Studies, Focaal – Journal of Global and Historical Anthropology and the European Journal of Sociology*. Recently he published, with Don Kalb, on the financial crisis, financialised capitalism in the West and the unexpected similarities with the Soviet planned economy. He published with Max Spoor the first exploratory study of land grabbing in post-Soviet Eurasia (*Journal of Peasant Studies*, 2011).

Natalia V. Mamonova is a PhD candidate at the International Institute of Social Studies (ISS) of Erasmus University Rotterdam and at Radboud University Nijmegen, the Netherlands. Her doctoral research focuses on land grabbing in the post-Soviet countryside, and on land conflicts and rural social movements in Russia and Ukraine. She worked as a research associate at Radboud University Nijmegen in 2010–2011 under the supervision of Dr. Oane Visser. The projects she was involved in are 'Land grabbing in Russia' (2010, LDPI grant) and 'Large-scale land acquisition in the FSU' (2011, LANDac grant).

Max Spoor is Professor of Development Studies at the International Institute of Social Studies, The Hague/Erasmus University Rotterdam. His research focuses on rural and environmental issues, poverty, and inequality in Asian transition economies, Central Eastern Europe and former Soviet Central Asia. His most recent books are *The political economy of rural livelihoods in transition economies* (Routledge, 2009) and, with Murat Arsel, *Water, environmental security and sustainable rural development: conflict and cooperation in Central Eurasia* (Routledge, 2010). Together with Oane Visser he published the first exploratory study of land grabbing in post-Soviet Eurasia (*Journal of Peasant Studies*, 2011).

The land question: special economic zones and the political economy of dispossession in India

Michael Levien

Special Economic Zones (SEZs) have become the epicenters of 'land wars' across India, with farmers resisting the state's forcible transfer of their land to capitalists. Based on 18 months of research focused on an SEZ in Rajasthan, this paper illuminates the role of 'accumulation by dispossession' (ABD) in Indian capitalism today and its consequences for rural India. It argues that the existing theories of land grabs do not adequately explain why dispossession becomes necessary to accumulation at particular times and places, and seeks to reconstruct Harvey's theory of ABD to adequately account for it. It then shows the specific kind of rentier- and IT-driven accumulation that dispossession is making possible in SEZs and the non–labor-absorbing, real-estate–driven agrarian transformation this generates in the surrounding countryside. Land speculation amplifies class and caste inequalities in novel ways, marginalizes women and creates an involutionary dynamic of agrarian change that is ultimately impoverishing for the rural poor. Given the minimal benefits for rural India in this model of development, farmer resistance to land dispossession is likely to continue and pose the most serious obstacle to capitalist growth in India. The agrarian questions of labor and capital are, consequently, now rejoined in 'the land question.'

Introduction

Special Economic Zones (SEZs) have, over the past five years, become synonymous in India with grabbing land from farmers. In March 2007, 14 people were killed and many more raped and injured by police and party-thugs in Nandigram, West Bengal, for refusing to give their land for a petrochemical SEZ promoted by an Indonesian company. The uproar that followed shook the state and central governments (contributing to the eventual downfall of the former) and led to a cancellation of the project, a temporary moratorium on SEZs and a reduction in their maximum allowed size. Nandigram was the tip of the iceberg, as farmers across the country were resisting the government's use of *eminent domain* to acquire and transfer their

The author would like to thank for their generous comments, criticisms and advice: Amita Baviskar, Henry Bernstein, Daniel Buch, Michael Burawoy, Eli Friedman, Suchi Pande, Ian Scoones, Michael Watts, participants in the Land Deal Politics Initiative conference at the Institute of Development Studies in April 2011, the editors of JPS and the two anonymous reviewers.

land to private companies for the development of these hyper-liberalized enclaves. These 'land wars' have led to the cancellation, delay and downsizing of projects across the country, including two massive SEZs for Reliance Industries outside of Mumbai and Gurgaon, the South Korean POSCO steel SEZ in Orissa (supposed to be India's largest ever foreign direct investment)[1] and all the SEZs approved in the state of Goa. These conflicts and the stoppage or stalling of high-profile investments have created great concern within the state and capitalist class that farmers will be the largest obstacle to India's emergence as a 'world class' economic power.

That an export policy has exposed the land question as perhaps the biggest obstacle to capitalist development in India today may seem strange. After all, land has been acquired for industrial estates and townships in India since the first Five-Year Plan, and subsequently for early Export Processing Zones and various export promotion parks. Why an SEZ question has become an explosive land question has in part to do with how one novel feature of India's SEZs – the privatization of their development – opened up a lucrative investment opportunity in a particular phase of Indian capitalism. The SEZ Act of 2005 provided a framework for building hyper-liberalized economic enclaves – with minimal taxes, tariffs and regulations – on the Chinese model, with the avowed purpose of promoting exports, attracting FDI, developing infrastructure, and generating employment. But whereas China's SEZs were state-developed,[2] in India, the private sector would be enticed with offers of cheap land acquired from farmers to develop the zones themselves and create 'first-class' industrial and commercial infrastructure for exporting companies.[3] The SEZ Act only required that 35 percent (later raised to 50 percent) of the area acquired for SEZs be used for productive purposes, giving developers freedom with the remainder. Given that the highest value land use is by far housing, most developers would use that area for constructing high-end housing colonies and accompanying 'social infrastructure', like shopping complexes, private schools and golf courses, for what would essentially be privately developed cities on farmland in the peri-urban periphery. This was the real draw of the SEZ for most developers and central to the business model.

In the name of privatized infrastructure development, the SEZ Act thus created an opportunity for windfall real estate gains precisely as the India economy started

---

[1]At the time of writing, the outcome of the POSCO standoff remains uncertain. The steel plant, which was to be set up as a captive SEZ of the South Korean steel company, has been stalled for over five years in the face of stiff resistance by local agriculturalists and fishermen living on government 'forest land' that is the proposed location for the project. Police forces have been amassed outside of the villages for over a month, threatening to evict them by force. However, so far the resisting villagers have been able to hold their ground by erecting barriers, forming human chains, and drawing enough media attention to make violent removal politically difficult. Despite clear violation of the Forest Rights Act (2005), which mandates approval for such forest conversion by local assemblies (*gram sabhas*), the Central Forests and Environment Ministry finally gave its approval for the project in May 2011.

[2]While China's original SEZs were state-developed, privately financed 'economic development zones' and 'industrial parks' have become major engines for rural land seizures and urban evictions (Walker 2008, 471).

[3]Under the SEZ Act (2005), state governments can still establish public sector SEZs and many are developing their own SEZs through parastatal industrial development corporations. However, these publicly developed SEZs are typically not that different – from a land point of view – from the old industrial areas aside from higher boundary walls. What is really new about SEZs, and what has generated the big land rush, is the opportunity for their private development.

accelerating toward nine percent growth rates and, crucially, as a liberalized real estate market began a dramatic ascent. Indian real estate companies and large diversified corporate houses jumped at the opportunity of developing unprecedentedly large pieces of land that, in many cases, would be acquired for them at cheap rates by state governments keen to attract industry. Within the first 16 months after the SEZ rules were established in 2006, 464 SEZs were approved and the figure now stands at almost 600. While it is a vast underestimate, the government claims that the currently proposed SEZs will require over a half million acres of land (Government of India n.d.).[4]

While much ink has been spilled on the 'land wars' being generated by SEZs, there have been comparatively few in-depth studies of actually existing ones that would enable a careful empirical analysis of their productive and distributional consequences.[5] Through an 18-month study – using ethnography, interviews, surveys and government documents – focused on one of the few operational 'greenfield' SEZs[6] in India, the Mahindra World City outside of Jaipur, Rajasthan, and the surrounding villages whose lands were acquired for the project, this paper tries to illuminate the new political economy of dispossession embodied by SEZs and the peculiar kinds of agrarian changes they are unleashing across India. In doing so, I hope to show what a reconstructed theory of accumulation by dispossession can contribute to the emerging literature on 'land grabs'.

Theorizing dispossession

There is some danger in the emerging scholarship and policy discourse on 'the global land grab' that the novel interest in them is mistaken for novelty of the phenomenon itself. The dispossession of land from peasants, of course, has a long history and an almost equally long history of thinking about it. While most acknowledge this, there seems to be a noticeable tendency to take the current phenomenon of transnational agricultural deals as the basis for more general theorization about land dispossession. Thus Zoomers (2010) argues that the current upsurge of cross-border agricultural land grabs is part of a larger contemporary phenomenon of the 'foreignization of space', of which she cites India's SEZs as another example. Aside from the empirical considerations that transnational agricultural land deals have a long history (e.g. Grandin 2010), which demands more elaboration on what is new

---

[4]However, the Ministry of Commerce and Industry's (MoCI) statistics are very misleading. First of all, they reflect the land officially notified as an SEZ, not the total land acquired for the company. Many companies are building large 'Domestic Tariff Areas' attached to the SEZs for companies wishing to produce for the domestic market. This acreage would not be included in the MoC totals. More importantly, the Ministry's totals only reflect the size of the SEZ at the time of notification. It is common practice for SEZ developers to seek official notificiation status as soon as they have the minimum amount of land necessary, and then add on to it later. These 'notified' areas can be one-tenth the total size of the project. A more realistic estimate of total acres would easily be in the several millions of acres.

[5]Two of the major exceptions are Seethalakshmi's (2009) comprehensive overview of SEZs in Andhra Pradesh and Jamie Cross' (2009, 2010) ethnography of a factory in the Vishakhapatnam (also Andhra Pradesh) SEZ, an older EPZ converted to an SEZ. However, the focus of the latter is on the labor rather than land question.

[6]By 'greenfield' I mean an SEZ built from scratch after the SEZ policy came into effect in 2000, rather than one of the old Export Processing Zones converted to an SEZ after the policy came into effect.

about the contemporary ones (White and Dasgupta 2011), and that SEZs in India, rather than representing 'a strategy of global capital' (Banerjee-Guha 2008, 51), are primarily developed and financed by domestic capital,[7] it is not clear why a theory of land grabs should be based on the origin of the capital orchestrating them; this should rather be a factor in explaining variations in their character, frequency and consequences at different times and places. In creating a theory of contemporary dispossessions from land or other resources, the key question is not the origin of the capital, but the reasons why capital in general requires – or more precisely attempts and achieves – forceful expropriation at any given place and time to sustain accumulation. We might add that the term 'land grab' – whatever its political uses – provides little more than a self-evident descriptive label for this phenomenon.

I believe that a sounder basis for conceptually understanding contemporary processes of dispossession can be found in Harvey's reconstruction of Marx's 'primitive accumulation' as 'accumulation by dispossession'. Suitably reconstructed to iron out the ambiguities and lacunae in Harvey's rendition, ABD provides the beginning of a more powerful analytic concept regarding the role of dispossession under advanced capitalism, the variations of which in space and time can be empirically studied and form the basis of a dynamic and important research program. While ABD may not capture all varieties of contemporary land grabs – such as those motivated by logics other than profit (Hall *et al.* 2011, 13)[8] – it can certainly capture the great majority and allows for the term 'accumulation' to be

[7]Almost all of the private SEZ developers are Indian companies. While a high-level Ministry of Commerce and Industry official admitted to me that, 'honestly, we don't know where the finance is coming from', he said that the level of FDI had been disappointing (Interview, 17 Jan 2011), and the relatively incomplete MoCI data given to the author shows very little – only 11 of 42 'notified' SEZs for which there is data list any FDI at all in either their development or in their productive units. Another top official in charge of regulating SEZs in nine states told me that there had been 'very little' FDI and that most of the SEZs were being financed by internal accruals and domestic bank loans – which is the case with the Mahindra World City SEZ analyzed here. Of the 100 approved SEZs in his region, he added that perhaps only three or four had FDI in their development, with maybe 20 percent of units in the IT sector having some FDI, with very little in other sectors (Interview, 25 Jan 2011). It may also be added that the push for the SEZ Act came entirely from Indian capitalists and Ministry of Commerce bureaucrats and the stamp of the World Bank and other IFIs is conspicuously absent. All of this makes Moore and Schrank's (2004) characterization of EPZs as 'imperial overstretch' inapplicable to the Indian case, and Banerjee-Guha's statement that SEZs are 'part of the recent *strategy* of *global* capital to overcome the chronic problem of over-accumulation' misleading (2008, 51, italics added).

[8]Hall *et al.* (2011, 13) provide examples of land grabs for conservation areas or 'ethno-territories' (also see Li 2007). If there is no accumulation, of course, ABD would not be the correct term. However, conservation often does have a direct or indirect link to accumulation – for example, where conservation is part of a strategy of tourist-led development – an accumulation strategy based on what Lefebvre calls 'the consumption of space' (Lefebvre 1991, 349, 352–354; 2009, 188) – or is part of an effort to off-set accumulation elsewhere through carbon credits, compensatory forestation, or 'debt for nature swaps' (Hall *et al.* 2011, 78–79). Detailing the many linkages between conservation and capitalist development, Geisler suggests that 'conservation might even be seen as green-tinted structural adjustment' (2003, 72). However, the main point is that a good concept cannot and should not try to cover every conceivable phenomenon that on the surface seem related, but should be restricted to those phenomena for which it can provide a compelling explanatory logic at a level of abstraction that provides analytic leverage on important questions. Hall *et al.*'s preference for the term 'exclusion' has the merits of applicability to a wide range of phenomenon involving the social relations around land – beyond what would be characterized strictly as dispossession – but perhaps at the price of a diminution in explanatory power.

specified in any given context. However, there has been a great deal of confusion in the emerging literature on what exactly defines ABD – a confusion that can in part be traced to two central ambiguities in the way scholars have traditionally understood Marx's theory of 'primitive accumulation'. Since I believe these are of considerable significance to current debates, they call for some elaboration.

The most important ambiguity – which continues to haunt discussions of ABD – is whether primitive accumulation is defined above all by its *function* for capitalism or by the *means* specific to it. On the one hand, Marx's theory of primitive accumulation referred to the process of creating the *pre-conditions* for capitalism by effecting the 'two transformations, whereby the social means of subsistence and production are turned into capital, and the immediate producers are turned into wage-laborers' (Marx 1976, 875). On the other hand, Marx also distinguished primitive accumulation by the use of blunt, extra-economic force in contrast to 'the silent compulsion of economic relations' that characterized fully developed capitalism (Marx 1976, 899). In the former, 'It is a notorious fact that conquest, enslavement, robbery, murder, in short, force, played the greatest part' (1976, 874). Sheer violence and state-driven expropriation were necessary to generate the two major prerequisites of capitalist social relations: a 'prior' accumulation of capital on one side and a group of wage laborers who had nothing to sell but their labor power on the other. Here we might note that within this first ambiguity, a second one emerges – which also plagues contemporary discussions of ABD – as to how much emphasis is to be placed on each of the two transformations: the forces seeking to turn the land into capital (e.g. the gentry turning the commons into sheep walks), and how much on its result – the creation of a class of wage laborers 'freed' from their means of production. If more emphasis were to be placed on the first, would not primitive accumulation be a continuous process as more land and natural resources were needed to feed a burgeoning capitalism that already had a sufficient reserve of wage-laborers? However, because Marx's analysis of primitive accumulation remained a retrospective, historical account of how the pre-conditions of capitalism came into being that was not grafted to his theory of developed capitalism (Balibar 1979, 279),[9] this tension was not adequately resolved. And as violent means and the 'two transformations' converged in Marx's classic example of the British enclosures (if not some of his other examples), the first tension between the two potentially

[9]There are, to be sure, suggestions in *Capital* that primitive accumulation was not simply a one-off event. Thus, Marx states that, 'The history of this expropriation assumes different aspects in different countries, and runs through its various phases in different orders of succession, and at different historical epochs' (1976, 876). He acknowledges, for instance, that primitive accumulation is far from completed in the colonies (1976, 931). Nevertheless, the overwhelming textual evidence from both the *Capital* and the *Grundrisse* is that Marx was trying to argue that primitive accumulation comprises the historical origin of capitalism and, to the extent that it continues, it does so *serially* as different countries transition to capitalism. Thus, Marx calls it, 'an accumulation which is not the result of the capitalist mode of production but its point of departure.' (1976, 873). In the more Hegelian idiom of the *Grundrisse*, he states, 'The conditions and presuppositions of the *becoming*, of the *rising*, of capital presuppose precisely that it is not yet in being but merely in *becoming*; they therefore disappear as real capital arises, capital which itself, on the basis of its own reality, posits the conditions for its realization' (1973, 459). Thus primitive accumulation creates the presuppositions of capital; once established, capitalism re-creates its own conditions of existence.

distinct ways of defining primitive accumulation – by its means or ultimate function – remained latent.

However, subsequent scholarship questioned the centrality of 'enclosures' to transitions to capitalist agriculture in other parts of the world (Byres 1991), and even to the 'classic' transition to capitalism in England (Wood 2002). In much of the 'agrarian question', primitive accumulation came to mean any process that separated peasants from their means of production, which more often proceeded through a gradual process of class differentiation (Lenin 1967, Kautsky 1988, 17, Adnan 1985, 57). Economic processes like debt, often working through 'inter-locking markets', came to be seen as equally effective levers of primitive accumulation as the extra-economic enclosure of land (Bhaduri 1983). In sum, for many, primitive accumulation came to be defined by its results – first and foremost proletarianization – rather than its extra-economic means.

It is not the point of this very cursory discussion to assess this strand in the conceptual development of 'primitive accumulation'. The point, rather, is to show how this ambiguity in the concept has colored contemporary receptions of ABD in ways that, I believe, obscure the novel theoretical advancement it offers. While some scholars continue to see primitive accumulation and ABD as synonymous (Arrighi *et al.* 2010, 411), I believe that in fact Harvey's concept of ABD marks a definitive break with primitive accumulation traditionally conceived by unmooring it from the historicism of modes of production and thereby freeing it for application to a panoply of contemporary forms of dispossession of private and social wealth – for SEZs, slum clearances, large-scale agricultural plantations, dams, real estate development, infrastructure projects and all manners of privatizations of natural resources and public wealth – that may have little to do with agriculture and that emanate from, rather than create the pre-conditions for, advanced capitalism. ABD has more to do with the multiple forces seeking to turn land and other resources into capital (Marx's first transformation) than about what may or may not be its result: adding to the pool of wage-laborers. As Sassen bluntly puts it, in many cases the land is 'more valuable to the global market than the people on it' (2010, 23). The significance of Harvey's reconstruction of primitive accumulation as ABD lies, above all, in its attempt to explain the contemporary upsurge in political struggles centered on the dispossession of land and various other resources rather than the exploitation of labor (Levien 2007, Burawoy 2010, 309).

However, there has been a notable persistence among scholars in seeing ABD not in the multiple dispossessions being generated by the demands for land and resources of contemporary capitalism but in a typical 'transition' process of proletarianizing the peasantry. This has generated some baffling conclusions. For example, some have argued that China is a case of 'accumulation without dispossession' because its industrialization was preceded by thorough land reform (Arrighi 2007, 361, Arrighi *et al.* 2010). While it is true that China's industrialization was not *preceded* by a full-scale enclosure of the peasantry (Hart 2002), as several accounts have shown (Hsing 2006, 2010, Walker 2006, 2008), its *process* of capitalist development has made the Chinese state quite likely the largest motor of ABD in the history of the world (followed by India), with some estimates of the dispossessed for factories, urbanization, mining and all sorts of infrastructural projects reaching into the hundreds of millions (Walker 2008, 472). Inversely, Arrighi *et al.* (2010, 434) argue that excessive dispossession in South Africa has created severe developmental pathologies – the solution to which might lie in the mining sector, one of the oldest and most notorious sources of land dispossession, meaning that the solution to the problems created by ABD is more ABD! While their overall argument that excessive dispossession has negative long-term development consequences may be correct, their confusing application of the concept of ABD reflects a persistence in seeing ABD as 'more of the same' – a wholesale agrarian transition whose significance rests in the generation of wage laborers – when what makes ABD a significant advancement on primitive accumulation is precisely its ability to capture diverse contemporary dispossessions (like mining or housing or factories) that often take sector-specific and geographically dispersed forms, and whose significance for capital, at least in the proximate sense, lies more in the expropriated asset than in the dispossessed owner. It is not that a name change is necessary for this reconceptualization – other scholars before and since Harvey have advanced conceptions of 'ongoing primitive accumulation' that mark a similar break (Perelman 2000, De Angelis 2001, Hart 2002, Luxemburg 2003, RETORT 2005, De Angelis 2007, Sanyal 2007) – but it does help in giving the concept a contemporary valence that makes clear we are no longer dealing with what Marx called the 'pre-history' of capital (Marx 1977, 875) or simply the serial repetition of transitions to capitalism in the Global South.

However, if it is Harvey's great merit to harness the phenomenon of dispossession to a sophisticated theory of *advanced* capitalism, it is also not clear that the result is completely satisfactory. Harvey notes (2003, 145) that aside from the enclosures, Marx includes a wide range of phenomena in his discussion of primitive accumulation (colonialism, the slave-trade, modern taxation, usury and the national debt, for example) and he himself wants to include a broad array of processes under the category of ABD: land grabs, privatizations of collective social assets, biopiracy and the various predatory machinations of finance capital. According to Harvey, these mechanisms have increased in importance relative to 'expanded reproduction' under neoliberalism.

But what do all these processes share that make them examples of the same phenomenon? There is not a clear and succinct definition of accumulation by dispossession in either *The New Imperialism* (2003) or *A Brief History of Neoliberalism* (2005). Harvey instead provides lists of processes or categories of processes that count as accumulation by dispossession.[10] What separates these from expanded reproduction is not fully explained. Since ABD is clearly no longer creating the preconditions for capitalism, we might expect Harvey to define ABD by the deployment of extra-economic coercion in the process of accumulation, which is how Glassman (2006) interprets both primitive accumulation and ABD. Such a definition would capture at least many – though perhaps not all[11] – of the examples that Harvey provides, which entail the use of non-market power (often by the state) to bring new assets within the circuits of capital accumulation, or to profit from the de- and

[10]For example, in *A Brief History*, Harvey offers the following four general forms: privatization and commodification of previously non-commodified assets, financialization, the management and manipulation of crises, and state redistributions of wealth from the poor to the rich (2005, 160–165).

[11]It is not clear, for example, whether the operations of finance capital would fit in this definition. As Brenner (2006, 101) states in critiquing what he sees as Harvey's overly expansive definition of ABD, 'the huge redistributions of income and wealth away from workers that are indeed sometimes entailed by the operations of financial markets are, for the most part, no less straightforward results of the capitalist game than is exploitation through the purchase of labour-power'.

re-valuation of already commodified assets. Harvey (2003, 154) comes close to this when he states that ABD in the Global South often occurs 'with the strong backing of state powers'. However, in a subsequent defense against critics, Harvey (2006a, 159) explicitly rejects the notion that extra-economic coercion is the defining feature of accumulation by dispossession. Instead, he claims, 'It is most importantly exercised through the credit system and financial power. I construe this as primarily economic rather than extra-economic'. Here, the concept definitively falls apart.

Without the means-specific definition of ABD, Harvey must replace the earlier functional interpretation of primitive accumulation with a new one: ABD is no longer whatever creates the pre-conditions for capitalism but whatever re-creates the conditions for its expansion. The closest he comes to providing a definition of ABD is in discussing its role in offsetting crises of over-accumulation: 'What accumulation by dispossession does is release a set of assets (including labor power) at very low (and in some instances zero) cost. Over-accumulated capital can seize hold of such assets and immediately turn them to profitable use' (Harvey 2003, 149). But if ABD is defined as whatever provides an outlet for over-accumulated capital, it is no longer clear what separates it from other 'spatio-temporal fixes' and the ordinary operation of capitalist expansion (Brenner 2006). The concept's boundary with 'expanded reproduction' is destroyed and its analytical specificity lost.

I believe that the only coherent way of defining accumulation by dispossession is the use of extra-economic coercion to expropriate means of production, subsistence or common social wealth for capital accumulation. It is not simply an economic process of over-accumulated capital seizing hold of under-commodified assets (Harvey 2003, 2006a), but fundamentally a political process in which states – or other coercion wielding entities – use extra-economic force to help capitalists overcome barriers to accumulation. It is a form of class struggle that appears 'as a crystal-clear relation of expropriation' lacking 'the fetishistic character assumed by capital's normalization, or the ordinary run of things' (De Angelis 2007, 139).[12] While analyzing circuits of capital may explain why there is more or less pressure toward dispossession, it does not tell us why capital would need to dispossess land rather than purchase it through the ordinary operation of real estate markets, or whether it will be successful, which is ultimately decided by the balance of class forces (Brenner 1976, De Angelis 2007). Rather than assuming its economic role into the definition, this means-specific understanding invites comparative research into its conjunctural economic role and political outcome in different times and places.

[12]De Angelis provides an original and generative conception of ongoing primitive accumulation that is coherently differentiated from 'expanded reproduction' and that more forcefully than Harvey emphasizes class struggle as the animating force of, and limit to, primitive accumulation. However, I hesitate to adopt his theoretical framework more fully because it is structured around the idea of enclosures whose referent is always the commons, defended by 'commoners.' The problem is that ABD entails the dispossession of multiple property forms, including private property, and may even entail the dispossession of more 'traditional' capitalist forms by more 'advanced' ones (Sassen 2010, 24). In India, ABD entails the widespread expropriation of not only various commons, but also privately held peasant landholdings. Many – though not all – of the movements resisting it demand no more than the protection of this small-scale private property. The romantic conflation of resistance to primitive accumulation with 'defense of the commons' – shared by others in the 'new enclosure' school (Linebough 2008) – obscures the fact that resistance to dispossession can be animated by diverse property claims or, in De Angelis' terms, 'value practices.' Thus, I believe that it makes more sense to look at the enclosure of commons as a sub-category of ABD, which has the merit of being able to capture a broader range of dispossessions of multiple property forms.

Moreover, it is by highlighting the widespread role of extra-economic expropriation under advanced capitalism that ABD provides significant analytic leverage on multiple questions of great contemporary relevance. The centrality under neoliberal capitalism of direct and transparent extra-economic intervention into the appropriation of value (in the form of land) – typically thought to be the defining characteristic of how feudalism extracts surplus labor (Marx 1981, 926–927, Anderson 1974, Wood 1981, Teschke 2003) – has far-reaching consequences for our understanding of contemporary state structures, politics and ideology, only some of which will be developed here.[13]

This concept of ABD would apply to the vast majority of what are now called 'land grabs'. After all, it only makes sense to talk about a 'grab' when land is expropriated using means other than voluntary market purchase. When sellers are unwilling, or where possession or use is not accompanied by recognized legal ownership (such as with government land or commons), land can typically only be alienated to capital with the backing of the state, though it might also be done by other agents capable of exercising coercion – mafias, hired thugs, paramilitaries or rural elites with 'their little independent methods' (Marx 1977, 885). In India, accumulation by dispossession is quite rationalized – probably more so than in China[14] – and above all involves the state restructuring itself under neoliberalism as a land broker for capitalists in a situation where rising demand for land confronts a rigid supply controlled by a large number of small peasants. Capitalists increasingly look to the state to expropriate land through *eminent domain* because farmers are often not interested in selling their land and/or do not have clear titles to it. Accumulation by dispossession is, in this instance, a process whereby state force is mobilized by the capitalist class to overcome the barriers to accumulation presented by land markets that do not provide an 'open field' for the circulation of capital (Harvey 2006b, 371).[15] As alluded to earlier, by seeking a transparently political solution involving state force to overcome the difficulty of acquiring land on the market, the state and capital have unleashed a secondary barrier to accumulation in

[13]The implications for our understanding of the role of the state and politics in dispossession are further developed in an article currently under review for the forthcoming special issue of *Development and Change* on 'land grabs and the state.'

[14]For excellent accounts of the state's land brokering role in the different legal and political context of China, see Hsing (2006, 2010) and Walker (2006, 2008). Some of the main differences between land dispossession in India and China would appear to stem from the greater significance of state-owned property and unclear property rights in the latter. This appears to give land grabbing in China a more extra-legal character, involving intra-bureaucratic competition and also greater reliance on non-state–means of coercion (such as the mafia and 'relocation companies'), which Walker identifies as part of China's 'gangster capitalism.' In India, while these elements are not absent (so-called land mafias do operate as more decentralized agents of coercive land alienation) and there is certainly often corruption involved in any large-scale land development project, land dispossession typically occurs quite legally through the routinized, bureaucratic application of the Land Acquisition Act (LAA) that has been in place since 1894. This is not to say that laws are not broken – for example, states often violate what should be protected forest rights or have are guilty of procedural violations of the LAA – but generally expropriations for increasingly for-profit and speculative purposes have been accommodated quite legally by stretching the interpretation of 'public purpose' and through the bureaucratic conversion of different categories of state land.

[15]ABD can be used not only to overcome the problem of private property that people are not interested in selling, but also in obtaining land or other resources for which markets do not exist (state or common land for example) and/or where property or use rights are ill-defined, unprotected, or more complex than those recognized by the state, which is the case in many parts of the world.

the form of powerful anti-dispossession 'countermovements' (Polanyi 2001) that are calling the outcome of ABD into question.

While much of the preceding has unfolded on a high-level of abstraction, the remainder of the paper will try to provide more 'concrete specification' to our understanding of ABD (Hart 2006) through the empirical example of the Mahindra World City SEZ in Rajasthan and the case of Indian SEZs more generally. I try to develop with more precision the specific type of rentier accumulation that dispossession makes possible in privately developed SEZs and put forward the concept of the 'rate of accumulation by dispossession' to capture its quantitative dimension. Then, bringing together Harvey and the tradition of agrarian political economy, I examine the developmental consequences of ABD, not in the macro-sense of assessing the productivity and international competitiveness of the labor force it helps to create (Arrighi *et al.* 2010), but by studying the welfare of the people dispossessed (Li 2011). I examine the peculiar type of agrarian transformation via land commodification that this type of ABD is unleashing in Rajasthani villages and its consequences for different social classes. With the labor of dispossessed peasants marginalized more than exploited by the an Information Technology (IT)- focused SEZ, the main effect on the rural hinterland has been dramatic commodification of land. This leads me to extend Harvey by showing how accumulation by dispossession does not always simply pit capital versus peasants – or 'commoners' (De Angelis 2007, Linebaugh 2008) for that matter – but can create a whole chain of rentiership that incorporates well-placed fractions of the peasantry. I show how existing class and caste inequalities in the agrarian social structure provide the basis for unequal gains from speculative land markets, which, in the absence of productive linkages from the SEZ, are invested overwhelmingly and with uneven success in forms of rentiership and exchange, creating an *involutionary* (Geertz 1963, Burawoy 1996, 2001, Burawoy *et al.* 2000) dynamic of agrarian change. The resulting pattern of class differentiation looks very different from the agrarian transitions via the development of agricultural capitalism that has historically dominated the attention of scholars of the 'agrarian question' (e.g. Kautsky 1988, Lenin 1967, Barrington Moore 1966, Brenner 1976, Patnaik 1990, Byres 1996, Akram-Lodhi and Kay 2010). Analyzing the peculiar non-agricultural agrarian changes being generated by SEZs and their developmental implications will then allow us to contextualize and evaluate the widespread peasant resistance to them and the contradictions they are posing for capital. I conclude that in India it is not so much that the 'agrarian question of capital' has been de-linked from that of labor and solved (Bernstein 2009) so much as it has been transformed from a question over the surpluses from agriculture to one over access to rural land. As farmers resist the expropriation of their means of production in part out of bleak assessments of their non-farm opportunities in the current growth trajectory of India, the 'agrarian question of labor' has direct and serious implications for that of capital.

Dispossession

While land has always been the most coveted asset in rural India, that asset is now increasingly desired by capital for industry, residential and commercial real estate and privatized infrastructure development. Such non-agricultural demand for land has skyrocketed post-liberalization and especially after 2005 as the Indian economy surpassed eight percent growth rates and a liberalized real estate sector experienced a spectacular boom (Searle 2010). However, this skyrocketing demand for land has exposed a latent supply problem. The problem (for capital) is that the majority of available land is in the hands of India's small-holding peasantry and significant enough sections of that peasantry remain, for various reasons, uninterested in selling it. Land, as Polanyi observed, is a 'fictitious' commodity not only because it is a non-produceable asset with location-specific qualities but also because it is valued in multiple ways (for example, as a habitation and long-term source of security) that are not easily reducible to exchange value (Polanyi 2001, 187). Even where farmers would, in principle, like to exit agriculture, they are often reluctant to surrender land where non-farm economic opportunities appear unpromising (this reluctance, we will see, might be well-founded). The possession of alienable private property is thus no guarantee that people will *value and treat* their land as a purely financial asset, which makes its supply particularly 'inelastic'.[16] While a capitalist looking to purchase hundreds of acres for a large factory or SEZ will surely find some sellers, they will also surely find some hold-outs that will prevent contiguity. This is compounded by the fact that even when there are willing sellers, a high degree of land fragmentation means that a buyer might have to negotiate with hundreds of farmers.[17] Finally, land titles in rural India are sufficiently unclear as to create high rates of litigation in rural land purchases (World Bank 2007). Such cases can take decades to resolve in India's backlogged courts and indefinitely delay projects.

It is for these reasons that any capitalist looking to establish a large greenfield project in rural India would much prefer to have the government acquire land for them.[18] If a project is large enough (say over a few hundred acres), it is very unlikely that the project will get off the ground without the land being acquired through *eminent domain* and/or state transfer of government land. The official in charge of

[16]The barriers to capitalist development posed by these specific qualities of land as a means of production, particularly when it is held as small-holding private property, has long been recognized within agrarian political-economy; however, it was seen as being a constraint on the development of capitalist agriculture that was not applicable to industry (Kautsky 1988, 147, Adnan 1985, 58). In today's context of increasing demand for large land development projects, these peculiar features of small-holding peasant property in land have become above all a significant barrier to industrial and commercial capitalism. On the distinction between alienability and valuation in the process of commodification, see Castree (2003).

[17]To illustrate, in the village of Rajpura where I concentrated my research, the 1,259 hectares of privately owned land was, before the SEZ came, divided into 2,950 parcels (*khasras*) owned by 391 different people (Rajpura land records, 2004). This means that the average holding was 3.2 hectares, and that one would have to negotiate with 313 families for every 1,000 hectares you wanted to purchase. Rajasthan is the least densely populated state in India with larger than average landholdings, so the difficulties would be much greater in states like West Bengal that have higher population densities and very small, fragmented holdings.

[18]Of course, World Bank-sponsored land titling initiatives are another way of grappling with the problem of making land available for capital. In its report, *India: Land Policies for Growth and Poverty Reduction*, the Bank estimates that 28 percent of land parcels in peri-urban areas are subject to legal disputes and recommends the updating and modernizing of land titling systems, which would, 'make available a huge amount of high-value peri-urban land for productive investment and development' (World Bank 2007, 2). However, it is clear that reducing the incidence of litigation by titling programs and land records modernization would in itself not solve the problem of capitalists looking to create large projects (involving hundreds or thousands of acres) on peri-urban or rural land, which requires persuading hundreds of farmers to sell. Thus, in the context of recent debates over a new land acquisition bill in India, the Confederation of Indian Industries (CII) released a press report that, while recommending the digitization of land records, also insisted that, 'Government cannot absolve its responsibility in land acquisition' (CII, 6 June 2011).

SEZs for Andhra Pradesh's Industrial Development Corporation frankly explained why land acquisition is necessary:

> If they have to talk to farmers, it will be a problem for industrial people to procure the land. If you go and ask the farmer, can it happen? No! They will come and cut your head [off]! If someone comes and asks for 500 acres for industry, of course they will deny it.
>
> ML: Even if they offer a good price?
>
> Even if they want to pay the [market] price to him, it is a problem. (Interview, 20 March 2010).

The analysis that a state role in land acquisition is necessary if very large industrial and infrastructure projects are to be built – the necessity of the projects themselves is, of course, a separate question – was repeated in less colorful terms by all the industrial development officials I interviewed in seven different states. Companies making large enough investments to have leverage thus insist on a state commitment to providing land in their Memorandums of Understanding (MoUs) with state governments. The ability of state governments to furnish this land has become the most important factor in the inter-state competition for investment.[19]

State acquisition of land for capital is in itself not new. Since independence, state governments – often through state-owned industrial development corporations (IDCs) or urban development authorities – have been acquiring land for industry under the colonial-era Land Acquisition Act (1894), which provides for the use of *eminent domain* for 'public purposes'.[20] However, post-liberalization, acquisition of land for the private sector has by all accounts increased dramatically and shifted qualitatively. Quantitatively, industrial development corporation officials from seven major states unanimously told me that they have had to acquire much more land post liberalization, and particularly in the past five years, to keep up with the rising demand. My own analysis of 40 years of land acquisition data from the Rajasthan Industrial Development and Investment Corporation (RIICO) shows that land acquisition doubled over the 1990s (when neoliberal reforms were introduced) from the previous decade and spiked considerably again during the boom of 2005–2008.[21] Moreover, RIICO has plans to acquire almost as much land in the next five years as they have in their entire 40-year history (Interview, 15 Dec 2010). The most comprehensive national analysis of development-induced displacement in India (for all projects, not just for private companies) has also shown a substantial increase post-liberalization (Fernandes 2008).

Not only has land acquisition for capital increased dramatically in India's neoliberal era, but its character has also strayed further from anything identifiable as a 'public purpose'. Much of the land acquired for industry outside of mining has historically been in state-run industrial areas developed by state industrial development corporations.

[19]Officials at the Indian Chamber of Commerce (ASSOCHAM), industry consultants, high-level state bureaucrats, and industrial development corporation officials all told me in interviews that land had become the most important factor in attracting large investments to one's state. Gujarat is often held up as a model state that is able to provide land while maintaining 'law and order' (see Sud 2009), while West Bengal is the negative model with its highly politicized agitations over land acquisition. This was dramatically illustrated when the Tata Company was forced to relocate its 'Nano' car factory to Gujarat after facing stiff farmer opposition to the West Bengal government's attempt to acquire 1,000 acres of land for the project.

[20]In India's federal system, land is a state subject under the Constitution.

[21]I assembled this data from 40 years of RIICO newsletters and annual reports.

The land in these industrial areas is owned by the government, which sells plots on a lease basis to private and public companies for purely industrial purposes. Very large projects like steel mills, cement plants and other large factories – many of them in the public or joint sectors – had large amounts of land acquired for them outside of these zones, though these were largely heavy industries thought to be crucial to the country's development. This type of state-led industrialization and the accompanying infrastructural projects had significant legitimacy under the Nehruvian era of nation-building (Khagram 2004; Roy 2007), and thus it was at least more plausible to the general public that land acquisition served a 'public purpose,' even if many of those brutally displaced without adequate compensation never agreed.

Post-liberalization, however, and particularly in the past seven years, industrial development corporations have been increasingly acquiring land outside of industrial areas for all kinds of private initiatives, including those with commercial, residential and even recreational components. While land has historically been acquired for Industrial Townships to provide housing and amenities for employees of large industries set-up far from existing urban centers, most prominently around India's public-sector steel mills (Parry 1999, 132, Roy 2007, 139–140, Parry and Struempell 2008, 54–55), land is now being acquired for IT office buildings, elite residential colonies, private colleges and shopping malls on the periphery of existing cities. Many infrastructure projects, increasingly being built on a public–private partnership (PPP) basis, incentivize private developers with land set aside for real estate development (IDFC 2008). The activities of industrial development corporations have thus begun to meld into those of urban development authorities who have themselves moved from acquiring land for publicly developed housing to blanketly amassing land banks on the peri-urban fringe to be auctioned off to real estate developers. SEZs are the culmination of this drift as IDCs and urban development authorities have both been acquiring unprecedentedly large chunks of land (sometimes in excess of 10,000 acres) for private SEZ developers who can then re-sell developed plots for industrial and residential development. Land acquired for a 'public purpose' can thus be used to build luxury housing, golf courses, hotels and shopping malls. SEZs thus complete the transition to a land broker state in which the chief economic function of state governments is to acquire land for unrestricted private capital accumulation (see also Sud 2009, Goldman 2011, Levien 2011).[22]

[22]Sud and Goldman both provide good accounts in different contexts of the changing role of the state with respect to land under neoliberalism. Sud's (2009) study of land acquisition for a cement plant in Gujarat highlights the ways in which the state's aggressive role in acquiring land for business under neoliberalism departs from the previous era of land regulation and agrarian reform, and cannot be accommodated by theories of 'state withdrawal' or 'governance'. Goldman (2011) advances the concept of 'speculative urbanism' to describe the broader reorientation of governments and parastatals around speculative real estate. These analyses are largely complementary and not opposing. However, I would place more emphasis on how this restructuring of the state towards what we might call a new 'regime of dispossession' is pushed by the impulse towards forcible land acquisition that is created when escalating demand under neoliberalism confronts the obstacles to accumulation posed by small peasant property and the 'fictitious' quality of land as a commodity (though Sud's case, it should be pointed out, is one of diverted forest land). With respect to Goldman's argument, the land brokering functions I am describing could be seen as part of the overall re-structuring of urban governance around real estate that he describes as 'speculative urbanism.' However, real estate is only one – though perhaps an increasingly defining – aspect of ABD in India today (as Sud's case also illustrates) and an analysis of the political economy of land dispossession must extend beyond municipalities and urban development authorities to industrial parastatals, which are often operating in areas that even after dispossession may not be 'urban.' In other words, the state's role as a land broker for

This role is well illustrated in my case of the Mahindra World City, a multi-purpose 'integrated industrial city' 25 km from Jaipur, which is supposed to include the largest IT-SEZ in India. To facilitate this project by the real estate subsidiary of the $7 billion Mahindra and Mahindra Company in partnership with RIICO, the Rajasthan government (through the Jaipur Development Authority) acquired 3,000 acres of land in 9 villages, of which 2,000 acres was privately owned farmland and 1,000 acres was common grazing land (officially owned by the government). As government land, the latter was transferred to the company with no compensation to the surrounding villages that were highly dependent on the livestock economy it supported. The private land, under a special Rajasthan state policy, was compensated by awarding farmers small developed plots of land next to the SEZ totaling one quarter the size of their original land, thus giving them a small stake in the inevitable real estate appreciation. This did not create a 'negotiated' consensus around a 'fair deal,' as was portrayed in the Indian press (NDTV 2008); aside from some political elites, farmers were anyway not consulted. But by individualizing people's relationship to the project and offering the prospects of easy real estate gains, the policy did divide the villages and break any potential for united opposition. As a consequence, the Mahindra World City became one of the first large greenfield SEZs to become operational while the majority remained stalled by farmer opposition.

Both Mahindra and RIICO officials told me that without a government role in land acquisition, the project would have never gotten off the ground. As a Mahindra official explained, '[Purchasing land] is always difficult .... You have to negotiate with too many people and you always find a maverick. Say you want to buy 100 acres, if you find someone who is smart and educated, he says, "I'm not going to sell my land." So government has to take it.' There is also, he explained, the possibility of litigation and disputes. 'You have to do it through government. They are the necessary evil' (Interview, 14 July 2010). What this reveals is the difficulty confronted by capitalists when land is not treated as a purely financial asset and does not provide an 'open field' for the circulation of capital (Harvey 2006b, 371). Ideologies of private property and 'free markets' (that would typically see government intervention as an 'evil') are pragmatically sacrificed to the calculation that state force in expropriating peasant property is necessary for large projects in India.

For projects requiring large amounts of contiguous land, state-sponsored dispossession is thus a necessary pre-requisite for accumulation with the result that state governments keen to attract investments have turned into coercive land brokers for capital. Many of my respondents – from chief secretaries of states to industrial development officials to chamber of commerce representatives – insisted that the ability of states to furnish this land has become *the* most important factor in the inter-state competition for investment (see Levien 2011). Indian states are thus competing with each other to be the best expropriators of the peasantry. It is this dynamic of institutionalized state expropriation that differentiates accumulation by dispossession from the ordinary operation of real estate markets. I now turn to the character of the accumulation that dispossession makes possible in an SEZ.

---

capital is surely a part of 'speculative urbanism,' but it is also much broader in both geographic and sectoral terms.

Land and accumulation in an SEZ

While the land in the Mahindra World City ultimately still rests with the government, Mahindra has a 99-year lease and can sell 99-year renewable leases to private companies for production/processing as well as for housing and commercial use. As a Mahindra executive succinctly explained their business, 'You can say our product is developed land parcels' (Interview, 29 October 2010). Mahindra has already sold 'developed land parcels' to over 42 companies in its IT, handicrafts and light-engineering zones. Most notable among the tenants are Infosys and Deutsche Bank, who have built large campuses for their back-office operations. Once enough industry is up and running, Mahindra will create luxury residential townships – in what they call the 'Lifestyle Zone' – on about 40 percent of the area. Mahindra executives were clear to me that they will make most of their money on the housing after their large initial investment in industrial infrastructure (Interviews, 24 October 2009 and 29 October 2010). An SEZ developer is thus a capitalist rentier who commodifies land for other capitalists and urban real estate development.

But it must be remembered that the SEZ developer is a state-appointed rentier and receives its land via dispossession rather than the market, which makes its land artificially cheap. It is normal practice that when states acquire land for industry, they transfer it at its nominal agricultural value as determined by the government rather than the price of commercial, industrial, or residential land to which it will soon be converted. The agricultural value is itself always an underestimate as it is based on average listed sales prices over several previous years, and in India, these are *always* less than the real purchase price, which is under-reported to avoid taxes. No consideration is given to the land's value under its proposed use. Government grazing land – often characterized as wasteland though it is hugely important to rural livelihoods – is sold even more cheaply, with no avenue for public consultation. The price at which Mahindra was able to acquire the land was consequently much less than would have been possible on the market. Privately developed SEZs are thus a form of arbitrage through which capitalists receive artificially cheap land acquired by the state and then re-sell it at many times that value. I call this ratio between the cost of dispossessed land handed over to capital by the state and its ultimate appreciation after development costs the 'rate of accumulation by dispossession.'

Using documents obtained through the Right to Information Act and interviews with Mahindra officials, I have attempted to calculate this rate for the MWC. If we group together the state and private land, the average price paid by Mahindra to the state government (including administrative fees for acquisition) was $22,679 per acre. Mahindra officials told me that their development costs were $66,000 per acre (to build 'world-class infrastructure') and they are currently selling industrial land for $55 per sq. meter or $224,000/acre. This makes for a profit of $135,000 per acre for industrial land parcels; it will be many times greater when they start developing more expensive residential space.[23] If we take as a benchmark for the latter the

[23]The price at which the government sold the land to the company is taken from correspondence between the Rajasthan State Industrial Development and Investment Corporation and the Jaipur Development Authority (2006) and between Mahindra World City and the Government of Rajasthan (2007), which the author obtained through Right to Information petitions filed by civil society organizations. This price per acre is an average of the price of the private acquired farmland (Rs. 245 crore for 2000 acres) and state (grazing) land (Rs. 61.17 crore for 1,000 acres). 1 crore = 10,000,000. Thus, the total cost for the 3,000 acres was Rs. 306.17 crore or Rs.

$137 per sq m (or $554,420 per acre) rate for residential plots in the nearby Jaipur Greens housing development just adjacent to the SEZ (which itself should rise over time as the SEZ develops), then the profit will ultimately be over $465,000 per acre in this section of the project. We can say, then, that the rate of accumulation by dispossession (the ratio of the final sale price of the land to its cost of acquisition and development multiplied by 100) will be 253 percent on the industrial land and 625 percent on the residential land, or a simple average rate of 439 percent.

More important than the exact figures, however, is the conceptual point that accumulation by dispossession for SEZs – though it could also be for other industrial and infrastructure projects – occurs through the creation of capitalist rentiers who profit from the transfer of under-valued assets dispossessed from farmers. Dispossession is instrumental to this accumulation in two senses: (1) As a necessary pre-requisite, releasing uncommodified or under-commodified land[24] that is difficult or impossible to obtain on the market and without which no accumulation could occur; and (2) As a mechanism for releasing that land more cheaply than if it were purchased on the market.

Labor and accumulation in an SEZ

Of course, the SEZ developer has to sell the land to other capitalists who are producing goods or services for export. These companies accumulate on the basis of labor rather than land, though of course they provide the ultimate demand for the (industrial) land as their 'spatial basis of operations' (Marx 1981, 916). By expropriating agrarian land with the help of the state, 'improving' it with modern infrastructure (through a system of contractors), and making it available by the square foot in a fully capitalist land market inside the zone (where land is exchanged on a purely financial basis), the SEZ developer is able to command a portion of the producing companies' profits as rent, which is capitalized as the land price. Given that the main focus of the MWC (like two-thirds of the approved SEZs) is IT and IT-enabled services (ITES), that profit comes from the labor of largely middle-class urban youth with bachelors, management and accounting degrees (see also Upadhya 2007, Radhakrishnan 2011). Since the price of such labor is becoming expensive in large cities like Bangalore, Mumbai and Delhi, IT/ITES companies are increasingly locating in Tier-2 and -3 cities like Jaipur, where they can find youth sufficiently well educated (and, if necessary, fluent in English) to do back-office work like accounting or programming for large multi-national clients (NASSCOM 2010).[25] These youth

1,020,567 ($22,679) per acre. It should be noted that the government rates for state land (used as commons) is cheaper than private land, making the rate of accumulation by dispossession steeper for commons than private land. Mahindra's development costs and the rate at which it is selling land were obtained through email correspondence (29 July 2010) and interviews (14 July 2010) with Mahindra officials. Note: from here onwards, I have converted all rupees amounts into dollars assuming an exchange rate of Rs. 45 per dollar.

[24]I use commodification here to mean not simply an asset that is legally available for being bought and sold on the market (as private land with clear titles has been in Rajasthan since the abolition of feudal intermediaries after independence), but as an asset that is treated as such by its owners. This orientation towards land, as we will see, shifts dramatically in these villages with the arrival of the SEZ.

[25]This was also explained to me by officials at Mahindra, 14 June 2010, Infosys, 2 April 2010, and the National Association of Software and Service Companies (NASSCOM), 18 October 2010.

are chauffeured in from Jaipur in SUVs to work night shifts corresponding with US and EU markets. The result is the rather surreal scene of middle-class youth from Jaipur settling the accounts of Deutsche Bank's global investment banking transactions in the middle of the night on the former grazing land of several Rajasthani villages.

Where do the former users of that land fit into this picture? Given their low levels of education and total absence of English fluency, mostly as security guards, gardeners, janitors and drivers. These jobs are relatively few in number compared to the number of people dispossessed; my survey – a random sample of 12 percent of households in 4 villages ($N = 94$) – indicates that only 18 percent of families who were dispossessed (and 14 percent of families overall) received some employment in the SEZ. As many of these families – and in particular women – pointed out, this was one job (typically for a male son) when the whole family was rendered 'unemployed' by dispossession. Without exception, this local employment consists of temporary, insecure, and low-paying (approximately $78 per month) work sub-contracted out by the companies to a system of local contractors (*tekedars*) (see also Cross 2010). Besides these service jobs, a small number of people from the surrounding villages are doing daily construction work for $3 per day through dozens of fly-by-night labor contractors, though, as is often the case, the vast majority of workers are more easily disciplined migrants shipped in from other states or far-flung districts of Rajasthan (see Breman 1996). Needless to say, there are no unions or safety precautions. While the modern eye might see the transition from being a farmer to a security guard for Infosys as an improvement, such work yields less income than two milk-yielding water buffalos, which most families had to sell after the enclosure of the grazing land. Women feel particularly marginalized by this process and repeatedly told me that they had been rendered 'unemployed' (*berozghar*) by the loss of land and livestock. Overall, 75 percent of those families who had their land acquired – including 78 percent of those who received employment after being dispossessed – reported receiving more 'loss' than 'benefit' from the SEZ. While from the economist's point of an IT campus is surely 'the highest and best use' of rural land compared to single-crop farming and livestock grazing, from the farmers' point of view, their land and livestock have more value than the employment available to them in this new economy.

So, while Marxists have traditionally see the creation of a class of propertyless wage laborers as the key function of primitive accumulation, the farmers dispossessed by the MWC have seen their labor marginalized more than exploited. The surplus value fueling the rent of the SEZ developer comes from moderately educated middle-class youth, while many of those dispossessed without the means of acquiring other land or livelihood assets have joined the ranks of the underemployed rural proletariat – the women 'stuck' at home or working on the government's National Rural Employment Guarantee Scheme (NREGS) and the men 'passing time' at the village *chai* stalls. In the context of India's non–labor-intensive, IT-led growth (Bhaduri 2009, Bardhan, 2010), the emphasis in Marx's two-fold account of primitive accumulation is clear: the land is above all being turned to capital while turning the peasant into wage-labor is a matter of indifference (see also Sanyal 2007). Without significant absorption of rural labor, the real transformation that the SEZ has wrought on the surrounding villages has been via the dispossession and commodification of land.

Agrarian change 1: dispossession and commodification

To understand the peculiar agrarian transformation the SEZ has generated through land dispossession and commodification, it is now necessary to introduce a more thorough understanding of the social structure and political economy of the villages affected, and particularly the village of Rajpura, where I focused my ethnographic research (the survey included three other surrounding villages). The villages are located in Sanganer Tehsil of Jaipur District, 25 km outside of Jaipur and close to a newly expanded national highway. While technically falling within the 'urbanizable' city limits of Jaipur since 1997, the economy of these villages was overwhelmingly agrarian until the arrival of the SEZ in 2005 and, to a large extent, still is outside of the SEZ boundary. The villages were largely dependent on rain-fed agriculture (centered around coarse grains) and livestock rearing, which provided cash-income through selling milk, supplemented by most small-holders with casual wage labor in and around Jaipur. With little irrigation – 27 percent of the land was irrigated according to the 2001 Census (Registrar General 2001) – and insecure rains, capitalist relations of production in agriculture were muted with 91 percent of households self-cultivating, and only 18 percent hiring wage labor for some part of the season (survey by author). Instead, most non-cultivating landowners leased their land to sharecroppers at the prevailing rent of one- to two-thirds of the crop, depending on who advanced the cost of production (the larger tenants being in a better position to do so). Both the lower caste – Scheduled Castes (SCs) and Tribes (STs) in Indian parlance – semi-proletariat and the dominant cultivating Jat caste[26] leased in land at similar rates (approximately 30 percent), the former to augment their meager holdings for greater food security, the latter to expand the surplus-generating production from their already significant holdings. Tractor ownership was limited to 30 percent of families and concentrated along with most assets at the top of the caste structure – 69 percent among the large land-holding Jats and 9 percent among the SC/STs.

With minimally capitalist agriculture, most of the semi-proletariat and landless found irregular work in construction or other informal industries in Jaipur, the block-printing mills of Sanganer, or a government industrial estate 15 km away. While only seven percent of households were landless, 44 percent were dependent on such casual wage labor, including 85 percent of the SC/STs (see Table 1). Traditional caste-based occupations absorbed some of the OBCs and SCs as village potters, blacksmiths, carpenters, sweepers, barbers, shoe-sellers and musicians. In addition to these were small provision shops, traditionally run by the village's Jain *bania* (merchant) caste, though increasingly over the years by Jats and Brahmins diversifying from agriculture. As Table 1 shows, the ownership of informal sector businesses was fairly widespread at 37 percent of households, but concentrated at the top of the caste hierarchy. Twenty-four percent of families had formal sector employment (private or government), though this was also concentrated among the general castes (who are mostly 'Haryana' Brahmins) and Jats (50 and 31 percent of families, respectively) with very little among SC/STs (nine percent) in spite of reservations in government jobs. This is, of course, related to large inequalities in

[26]After political agitation, Jats are now considered OBCs in Rajasthan. However, as dominant landowners in the area, I have separated them out from the other OBCs for analytical purposes.

Table 1. Distribution of Socio-Economic Attributes of Peasant Households by Caste in Villages Rajpura, Jatpura, Shivpura and Neempura before the SEZ.

Caste Category	Number and proportion of families in sample	Average size of family landholding (Ha.)	% who are landless	% who own a shop or business	% who have formal sector employment	% who have a business or formal employment outside of village	% doing informal wage labor	% with family member educated to 10th class or higher
General	24 (26%)	4.6	0	54	50	58	8	83
Jat	16 (17%)	7.3	0	44	31	44	13	69
OBC	21 (22%)	3	10	43	14	19	48	43
SC/ST	33 (35%)	2.2	15	15	9	6	85	24
TOTAL	**94 (100%)**	**4**	**7**	**37**	**24**	**29**	**45**	**50**

Note: These figures are based on a household level survey ($N = 94$) and thus show the percentage of households not individuals with these characteristics (i.e. column six shows the percentage of families that have at least one member with formal employment). Landholdings are the undivided family holdings, not the stake of individual siblings. Education levels reflect the attainment of the most educated adult person (over 18 in 2005) in the family. General castes include principally Brahmins and Banias (merchants). 'Otherwise Backward Castes'(OBCs) include Kumar, Gujjar, Mali, Lakhera, Soni, Nai, Raona Rajput, Kumavat, and administratively Muslims. Jats, while officially OBC, have been listed separately for analytical purposes. Schedule Castes (SC) include Raegar, Balai, Khatik, Sangat, and Beirwa. Scheduled Tribes (ST) include Meenas and Dhankas. An earlier version of this table appeared in Levien (2011).

education, as 84 percent of general castes and only 24 percent of SC/STs had someone in their family who had finished the 10th grade. As Table 1 also shows, the upper castes were also much more likely to have a business or formal employment outside of the village (58 percent of general castes and 44 percent of Jats compared to 19 percent of OBCs and six percent of SC/STs), providing social networks that, as we will see, became even more useful with the arrival of the SEZ.

Crucially to our story, the distribution of land in these villages was inherited from Rajasthan's insufficiently reformed system of feudal land tenure (*jagirdari*). Until 1949, Rajasthan consisted of 22 princely kingdoms in which land was granted to Rajput lords (*thakurs*) who extracted rent and corvée labor from peasant tenants. Land to the tiller policies post-independence abolished the role of feudal intermediaries and distributed land in surplus of land ceilings to the tenants (Rudolph and Rudolph 1984). However, the *thakur* in Rajpura – as in other parts of Rajasthan (Chakravarti 1975, 95, Rosin 1987, 94) – retained a large amount of land (over 375 hectares) by transferring it into the names of various family members and trusted retainers and continued to be by far the largest landholder in the village until the MWC. Because land was redistributed according to the amount of land being tilled under the *jagirdar* system (Singh 1964), the largest beneficiaries were the main cultivating Jat caste and the Brahmins, with the lower castes receiving much less (Chakravarti 1975, 82–110). Given the central importance of land to people's livelihoods and social standing, and the lack of any non-agricultural demand, this post-independence distribution of land changed only marginally through subdivision and periodic distress sales over the 50 years until the SEZ came along. As Table 1 shows, at the time of the SEZ's arrival, the average size holding for Jats was 7.3 hectares of land per family, followed by 4.6 hectares for the general castes, 3 hectares for the OBCs and only 2.2 hectares for SC/STs.[27] While SC/STs constituted 35 percent of the population (Registrar General 2001), they owned just 15 percent of the land.[28]

The SEZ transformed this distribution of, and relationship to, land in two important ways. The first of course was through forcible dispossession, which stripped the productive assets of the village while creating the basis for accumulation in the SEZ. People lost their farmland, their source of grain and fodder, and all the improvements that had been built up through generations. Those living on their land (the majority live in the village settlements which were not acquired) lost their habitation as well. The enclosure of the village grazing land – which Deutsche Bank now sits upon – compounded the loss of private land in making fodder and fuel unavailable, with particularly disastrous effects for the poor. Among those who had land acquired, 86 percent had to sell livestock for want of fodder and, overall, enclosure decimated 66 percent of livestock herds, including 56 percent of buffalo

[27]These landholdings reflect the size of the undivided family property at the time of the survey. This means that in many cases it is the holding of the living grandfather, while there might be two generations of adult heirs with their own households whose individual stakes will be correspondingly much lower. If we go by the individual stakes of the households in the survey to property that is not yet divided, then my survey indicates that the average size landholdings for each caste would be: 3 hectares for Jats, 1.6 hectares for general castes, 1.2 for OBCs and 1 for SC/STs.

[28]This is based on the census data and land records for the three villages (Rajpura, Shivpura, and Neempura) for which complete data was available to the author. In the village Rajpura alone, SC/STs were 41.4 percent of the population and owned 20 percent of the land.

and, crucially for the poor, 76 percent of goats (Survey by author).[29] Moreover, because wells were acquired along with the land, many were left having to purchase drinking and household water from tankers. In sum, people lost their means of production and had their daily needs commercialized, a situation compounded by drastic inflation in food prices. While in absolute terms the loss of land was proportionate to land ownership, larger farmers were already more diversified into off-farm activities and, as we will see, were more able to profit from the market-based compensation model they were offered. The poor and mostly lower caste semi-proletariat lost the only asset separating them from the insecure penury of landlessness and proved much less successful in navigating commercial land markets.

This leads us to the analytically distinct effect of commodification. As mentioned earlier, farmers were compensated for their acquired land with smaller pieces of developed (commercial and residential) land which was, in the words of one official, designed to 'buy peace' by giving farmers a stake in the real estate appreciation or enhanced market demand (if they use it to start a business) that would inevitably accompany multi-national companies setting up in a rural village. This neoliberal compensation model was in some ways an application to land acquisition of the De Sotoean (2000) proposition that the best solution to poverty is to give the poor secure property rights that they can leverage in markets to harness their micro-entrepreneurial abilities.

In fact, the SEZ provided more ideal conditions for success than De Soto would assume, as the value of these (not yet constructed) assets appreciated massively in the real estate boom set off by the SEZ. Even before the SEZ was officially announced, real estate companies and outside parties tipped off by government officials (who may have been silent partners in some of these shell companies) flooded into the area to buy up cheap land that would appreciate many times over after the SEZ came into being. Once the announcement came, even more buyers from Jaipur, Delhi and Mumbai snatched up land. Large national real estate developers – including a joint venture with Dubai's Emaar – bought up farms to build luxury housing colonies adjacent to the SEZ. A market for the rights to the not-yet-constructed compensation plots also quickly developed. A whole stratum of local land brokers (*dalals*) emerged in the villages to facilitate these transactions for two percent commission in coordination with Jaipur-based brokers. As Figure 1 shows, land transactions in Rajpura increased out of all proportion to anything that had happened before – reaching a peak of 91 in 2006, over nine times the pre-SEZ level – as land prices surged from roughly \$16,000 per hectare to over \$280,000 per hectare in some areas by 2008. This land commodification – by which I mean the increased treatment of land as a commodity – hit Rajpura like a sudden exogenous shock.

This shock refracted through the local social structure in complicated ways, and with some amount of randomness that is inherent to real estate speculation. But the ability of local farmers to profit from this speculative boom – either through selling their compensation plots or land they still had outside of the SEZ – was largely based on existing inequalities in economic, cultural and social capital. Land markets are essentially a social game that people have differential abilities to play. As Table 2 shows, size of land holdings, education and social networks are all highly associated

[29]To provide some historical perspective (though for obvious reasons the events are not comparable), the 1899–1900 famine that struck Rajasthan decimated 40 percent of the livestock in Jaipur district (Kachhawaha 1992, 200).

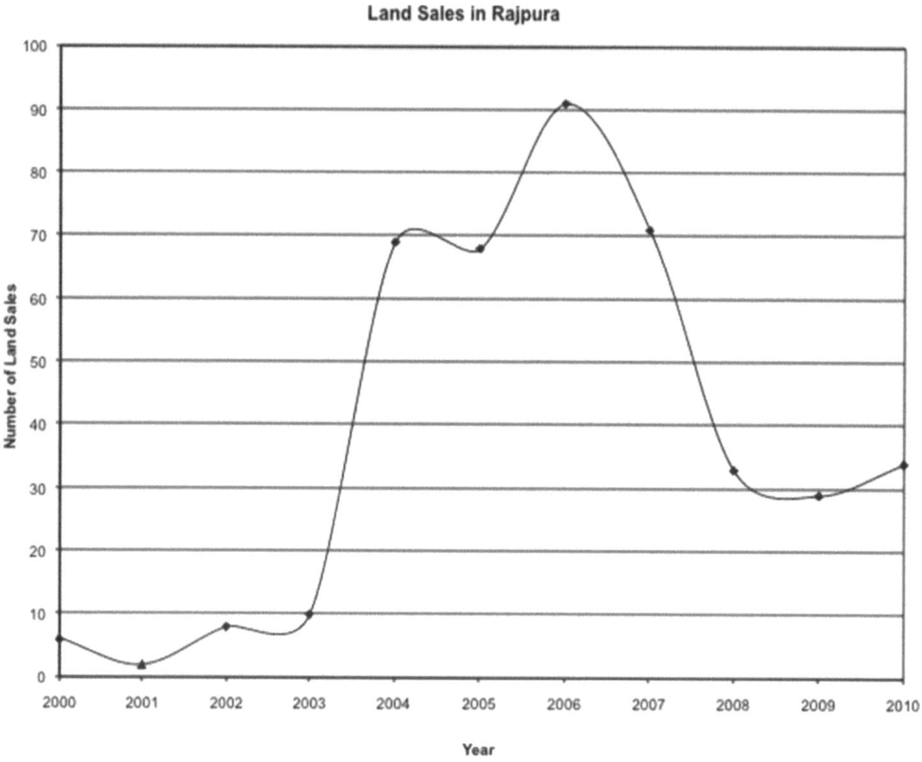

Figure 1. Land sales in Rajpura.
Note: This figure previously appeared in Levien (2011). The numbers are taken from the official village land records kept with the village patwari (2005–2010) in Sanganer and at the district collectorate in Jaipur (before 2005).

with success in navigating land markets, measured by whether people already sold their plot at the time of the survey in February 2011 (thus getting a lower price than those still holding on to them), the ultimate selling price of those who did sell, their propensity to become land brokers, their degree of relative proletarianization by the process (whether they are now doing more wage labor), and whether their food insecurity increased or decreased after the process. All of these initial inequalities are highly associated with each other and congeal around caste (as shown in Table 1).

Those with little land and few other sources of income had to sell the rights to their compensation plots quickly and at a consequently lower price. Thus, at the time of the survey, 70 percent of those with less than two hectares of land had already sold their plots, compared to only 23 percent of those possessing more than four acres. When the larger-holders did sell their plots, they received a median sales price that was over $44,000 per hectare more than the small-holders. A much larger percentage of the larger holders also became land brokers, appropriating commission from the land sales of other villagers; 25 percent of large-holding families contained brokers compared to eight percent of small-holding families. As a result of these uneven gains from land markets – combined with an uneven ability to invest, as we will see later – 39 percent of the small-holders with less than two hectares were pushed further towards proletarianization compared to 15 percent of the large holders. Moreover, 61 percent of the small-holding families reported having ‘less food’ after

Table 2. Differences in Capability of Peasant Households to Cope with Consequences of Land Dispossession for SEZ by Forms of Capital Possessed.

Form of Capital	% who sold compensation plot	Median price received for plot (US$)	% who became land brokers	% doing more wage labor after land acquisition	% with less food after land acquisition
Economic					
Landholding (ha)					
More than 4	23	$106,666	25	15	38
2.1 to 4	46	$62,222	17	23	46
Less than 2	70	$62,222	8	39	61
Cultural					
Household education					
10th class or higher	40	$62,222	17	12	40
Less than 10th	63	$62,222	11	46	63
Social					
Formal job or business outside of village					
Yes	24	$88,888	29	6	25
No	64	$62,222	8	39	66
TOTAL	**51**	**$62,222**	**14**	**29**	**51**

Note: The figures in all columns except that of brokers are based on those farmers in the sample who had land acquired for the SEZ ($N = 49$). The number of brokers is based on the sample as a whole ($N = 94$).

land acquisition compared to the still significant 38 percent of large-holding families. The incompletely reformed feudal land structure thus appears as a crucial determinant of the uneven outcomes of people's forcible entry into land markets.

Cultural capital, measured imperfectly in this case by formal education, but in theory also including non-formal market savvy that enable people to better understand what an SEZ heralds for local land markets, was also important. Many illiterate farmers report being cheated and many did not anticipate that land prices would keep multiplying, while those with more education better optimized the timing of selling their land or plots. Thus 40 percent of those households that had a member educated till the 10th grade or higher had sold their plot at the time of the survey, compared to 63 percent of families without such educational attainment. When they did sell, however, there was no difference in the median sales price between those with more or less education (there were, however, more extremely low outliers among the less educated, with a few selling their land at half the median price). The incidence of becoming a land broker was higher among the more educated – 17 percent compared to 11 percent of the less educated – though the relatively small difference suggests that other factors may be more important. Along with the previous observations, this suggests that education up to the 10th grade may not be the most relevant threshold (which may be that of literacy) and/or that formal education is much less relevant than informal business savvy. However, education did have a substantial association with people's ultimate post-SEZ welfare: while 46 percent of the less educated families became relatively proletarianized by land acquisition, only 12 percent of families with education above the 10th class experienced a similar fate. This suggests that formal education is probably significant because it is associated with higher initial diversification and people's subsequent ability to profitably invest their cash earnings. However, it is sobering that 40 percent of those families educated to the 10th grade still wound up with less food, which was better than the 63 percent of less educated families who found themselves in that position, but bad nonetheless.

Besides the distribution of land and education, the possession of social networks spanning the village and city – what we might call social capital – also seems to have had a significant bearing on the outcomes of forcible land acquisition. By 'social capital' I do not mean normative bonds adhering to groups that enable cooperative action for collective benefit (e.g. Putnam 2001, Krishna 2002), but on the contrary an unequally distributed form of power based upon access to networks that convey real or potential advantages (Bourdieu 1986). Both the quantity of those networks and their quality (the capital of those within one's network) are unequally distributed among individuals within any social unit and this is certainly true of a class- and caste-divided Rajasthani village. Some only go to Jaipur to perform wage-labor, while others have more economically advantageous connections based on formal employment or business. While caste and kinship are central to social networks and economic cooperation (Harriss-White 2003), what I have found to be particularly crucial to profitably navigating land markets are connections built through non-menial work and business to those outside of the village, and particularly the city. While these connections may or may not be within caste, Table 1 shows that the upper castes have much more of them.

As Table 2 illustrates, differences in social capital – operationalized as whether people had prior business or non-menial employment outside of the village – were more highly correlated than education with the speed at which people sold plots, the price they received, their propensity to become brokers, and showed a roughly equivalent correlation with the outcomes of proletarianization and food security. Only 24 percent of those with formal employment or a business in Jaipur had sold their plots, compared to 64 percent of those did not have such an urban foothold. The former received a median sales price that was over $26,000 per hectare more than that received by the latter. Only six percent of those with these economic connections beyond the village were relatively proletarianized, compared to 39 percent of those who lacked them. Twenty-five percent of the former wound up with less food, compared to 66 percent of the latter.

Of course, business ownership and urban employment can also be considered forms of economic and cultural capital, and it seems quite possible that they capture the relevant cultural capital (an informal business or market savvy) better than formal education. However, the survey data combined with my ethnographic observations strongly suggest that the social networks associated with these urban economic footholds were, if nothing else, decisive in the ability of people to become land brokers. Land brokering requires little or no capital since it is largely based on commission (though more lucrative deals can be had by advancing money). It does require some aptitude for dealing with land documents and registration procedures, though formal education, as we have seen, is not necessary. As many of them told me, what a prospective land broker needs above all are contacts (*sampark*) beyond the village, particularly to the city. This, incidentally, was reinforced by the repeated attempts of people to bring *me* into land deals: despite my protestations that I had no savings and knew nothing about business, they insisted that that did not matter – all I needed to do was to bring in an 'excellent' outside party.

The survey indicates that 29 percent of those with formal employment or their own business outside of the village became land brokers, compared to seven percent of those who did not. Almost all of the several dozen *dalals* I have interviewed had previous business or non-menial work experience in Jaipur through which they told me they established contacts with Jaipur-based brokers. Many of them were former milk-middlemen, continuing their old role as intermediaries in a new economic context. Others were village *chai* stall proprietors, an obvious point of contact and source of information for any outsider looking to buy land. These networked individuals used their ability to establish connections between their village and the city to build trust between both buying and selling parties in return for handsome commissions. Many villagers ultimately felt cheated by these brokers who they see as the main local beneficiaries of the SEZ. Ironically, this manipulation by brokers of their social capital in the Bourdieusian sense of unequal social networks has led to a dramatic decline in social capital in the Putnamite sense of trust and norms, effectively undermining collective action.

The distribution of all these forms of capital congeal to a large degree in caste hierarchies (as Table 1 shows) and the inequalities they have produced in the context of real estate-driven agrarian change are perhaps most visible in the differential abilities of the members of different castes to cope with land acquisition, as shown in Table 3. We should start our analysis, however, with the family of the former Rajput *thakur* – the outlier not included in the survey – whose abundant land, highly sophisticated business acumen and many social and political connections have enabled them to gain the most from the SEZ. Instead of expropriating millet from sharecroppers on their 375 hectares of land as they have been doing since independence, they can now turn their remaining land into real estate and do what

Table 3. Differences in Capability of Peasant Households of Different Castes to Cope with Consequences of Land Dispossession.

Caste	% who sold compensation plot	Median price received for plot (US$)	% who became land brokers	% doing more wage labor after land acquisition	% with less food after land acquisition
General	19	$71,111	21	6	31
Jat	25	$66,666	25	25	25
OBC	58	$62,222	10	25	58
SC/ST	82	$57,777	6	53	75
TOTAL	**51**	**$62,222**	**14**	**29**	**51**

they will with the hundreds of appreciating compensation plots they have received for the land acquired from them. The *thakur's* grandson, a highly educated former investment banker for a multi-national company in Mumbai, is overseeing his family's interests as the functional village *sarpanch* (elected head) after his mother was elected on a reserved ticket. He has not sold any of their plots, is building his own housing development near the SEZ and is acting as a large land broker for outside buyers (Interviews, 9 July 2010, 16 Feb 2011). Thus, the former feudal class is proving quite adept at moving from pre-capitalist to capitalist ground-rent in the neoliberal era of real estate–driven growth.

After the Rajputs, as Table 3 shows, the mostly Brahmin general castes – with significant land holdings and high overall economic diversification, more abundant networks beyond the village, and the highest education levels – are mostly (81 percent) still waiting for their compensation plots to appreciate, got the highest prices when they did sell (with a median price of over $13,000 per hectare more than SC/STs), were well represented among the land brokers (second only to Jats), and were pushed in fewer numbers towards wage labor and food insecurity (though a substantial 31 percent still reported having less food than before acquisition). The Jats, with the largest landholdings and with economic diversification and educational levels second only to the Brahmins, were also shrewdly hanging on to their compensations plots (75 percent), received almost $9,000 per hectare more than SC/STs when they did sell (though because of the small number of Jats in the sample who did sell, we should treat this number with caution), and were over-represented (25 percent compared to six percent of SC/STs) among the land brokers 'eating commission' (in the words of local farmers) off the plot and land sales of others. With their larger landholdings more often spilling beyond the SEZ border, many were also shrewdly selling small fractions of their remaining land for huge returns. As we will see, they – along with the Brahmins – also did well in reinvesting these returns. My ethnographic and survey data, however, make clear that there is a stratum of relatively poor and uneducated Jats, and the survey shows that 25 percent were relatively proletarianized and left with less food, which is hardly insignificant. While this increased food insecurity among even the dominant castes is a further indictment of forcible land acquisition, for some it may reflect a transition period between losing their land and realizing the fruits of land appreciation (in other words, being momentarily rich in assets but poor in food and income).

Among the lower 'scheduled castes' (SC) and artisanal castes falling into the 'otherwise backward' (OBC) category things are more bleak. As Table 1 shows, these castes contain the vast majority of the marginal peasantry and semi-proletariat, with average holdings of 2.2 to 3 hectares. Educational levels are lower and fewer have non-farm business or employment outside of manual wage labor. Many also carry very significant debt. Because of both economic compulsion and lack of market savvy, 58 percent of OBCs and 82 percent of SC/STs sold their compensation plots early (many also sold their land outright before the land was acquired at very low prices). They received substantially lower prices than the upper castes ($14,000 per hectare less than the general castes for SC/STs and almost $9,000 less for OBCs) and were hardly represented among the land brokers (particularly the SCs) and thus more often paying than 'eating' commission. Over half of SC/STs and a quarter of OBCs were pushed further towards proletarianization by the land acquisition and a staggering 75 percent and 58 percent, respectively, were left with less food. This is perhaps the most damning indictment of forcible land acquisition for the SEZ and of the application to compensation of De Sotoean principles of poverty alleviation.

To complicate this analysis somewhat, I should add that my ethnographic fieldwork clearly shows that the process has exacerbated inequalities not only between castes but also within castes and families. The unprecedented sums involved in land speculation create the basis for inequalities of a magnitude that was not possible in an economy dominated by single-cropped agriculture and wage labor. It is now common to see huge concrete mansions next to *kaccha* (earthen) houses in the same caste *muhallah* (settlement). The two houses are in some cases owned by brothers, one of whom became a broker while the other still does wage labor. Inequalities in economic, cultural and social capital are highly associated with caste, but such inequalities (particularly in cultural and social capital) are also not absent within castes and even families. Factors such as where one's land is situated, and whether it is inside or outside the SEZ, also introduce an element of randomness into social trajectories. Some Brahmins have been ruined while a few Dalits have become *crorepatis* (rupee millionaires). Most significantly, and cutting across castes, women are almost universally excluded from negotiations over land sales, even in cases where the title is in their name. While their level of input into land-related decisions varies by household, the replacement of an agricultural economy over which women exercised some control with a real estate economy controlled exclusively by men – and involving much greater sums – has, by many accounts, weakened their position and exacerbated domestic abuse.[30]

What should in any case be apparent is that the formulation of accumulation by dispossession as a phenomenon that pits capital versus the peasantry (or

[30]The 'causal' link is, of course, difficult to pin down and by all accounts domestic violence was endemic before the SEZ. However, more than a few women I spoke with blamed the SEZ for an increase in the related phenomena of alcoholism and domestic abuse. On one level, alcoholism increased because those who cashed in on their land or plots had much more money to buy it, and domestic abuse increased because men were more often drunk. On another level, domestic abuse was seen to increase because the men were now unoccupied, making them idle. Once the land money ran out, they would resort to violence to expropriate women's wage earnings from NREGA to finance their drinking (this was common practice for a few men in the *muhallah* I lived in). Finally, because women were relegated to the margins of land transactions, men had more power over how it was spent. As one OBC woman said about domestic abuse, 'It was less before… Now, if they make a hundred it's only with a fight that they give (the women) fifty' (Fieldnotes, 6 April 2010). Whatever the combination of mechanisms, it seems apparent that the position of (especially poor) women in the village has deteriorated.

'commoners') is overly simplistic. Accumulation of this sort generates a whole feeding chain of rentiers that extends from multinational capital to the savvy farmer-cum-commission agent. While it is clearly the case that the SEZ developers, producing units, real estate companies, middle-class investors and city-based brokers (with their greater amount of capital, business savvy and in-house economists and elite connections) capture the lion's share of such land-based rents, it would be a mistake to ignore the way in which fractions of an inegalitarian rural society unevenly tap into this process. The rampant real estate speculation that swept through the villages because of the SEZ – and in which dispossessed farmers were given a small stake by the neoliberal compensation model – produced not a singular 'neo-rentier' class (Adduci 2009) but a chain of rentiership in which speculative opportunities were unevenly exploited by those (men) with different endowments of economic, social, and cultural capital. As we will see, these inequalities also shaped the ability of farmers to turn cash from land sales into capital.

Agrarian transformation 2: involution

We have so far covered only one aspect of the great transformation generated by land dispossession and commodification in the villages surrounding the SEZ: the quantitative distribution of money from land sales. Equally interesting from the point of view of the agrarian question is the qualitative ways in which that money circulates through and transforms the rural economy and social structure. A predominant concern of studies of agrarian change has been the conditions for, and consequences of, the development of capitalism in agriculture. However, in the case of the SEZ under question (and it would not be alone), we are confronted instead with the transformation of a minimally capitalist agrarian economy by the exogenous force of speculative land commodification. Capital is seizing hold of land rather than agriculture. As we have seen, this has led to the destruction of productive agricultural assets and a redenomination of agrarian land as real estate. But this has also injected unprecedented amounts of cash into the local economy. To what extent and in what forms has this money generated from land commodification been turned into capital? Put simply, how do farmers invest or otherwise allocate their money from land sales?

What is striking is the largely non-productive character of most of the investments made by farmers, as the agricultural economy is replaced by informal livelihood strategies based on rentiership, merchant capital and usury. Following Burawoy's (1996, 2001) adaptation of Geertz's (1963) concept of involution[31] to understand the economic collapse of post-reform Russia, we might call this an involutionary process of economic change in which the sphere of exchange cannibalizes production and 'antediluvian' forms of capital are not transformed but reproduced on a greater scale (see also Harriss 1991). As Table 4 shows, after paying off debts, paying for marriages, and building a concrete house (which generates only temporary employment), the most

[31]Geertz (1963) originally used the term involution to describe how Javanese peasants adapted to the twin pressures of the Dutch agro-export industry and increased population on existing land by intensifying rather than transforming the existing pattern of agriculture. Geertz saw this as a 'stagnant' over-elaboration of detail. The empirical basis of Geertz's argument was subsequently called into question by White (1982). However, Burawoy (2001, 270) gives the concept a more general and dynamic meaning as a situation in which economic change is not accompanied by transformation, and in which production degenerates while the sphere of exchange expands.

Table 4. Avenues of investment/allocation of cash gains from the sale of land or compensation plots, and brokerage commissions.

Caste Category	Paid off debt (%)	Built house (%)	Married children (%)	Bought land (%)	Became absentee landlord (%)	Bought urban plot (%)	Started shop/ small business (%)	Drilled tubewell (%)	Bought tractor (%)	Bought tanker (%)	Gave loans (%)
General	56	78	78	67	33	22	44	44	33	22	22
Jats	80	60	40	80	40	30	50	70	40	30	10
OBC	85	61	46	54	46	30	23	0	15	8	0
SC/ST	92	68	28	48	44	32	12	12	2	12	12
TOTAL	**82**	**67**	**42**	**58**	**42**	**30**	**26**	**25**	**25**	**16**	**11**

Note: Percentage of households who received any type of income from land transactions – land sales, compensation plot sales, or brokerage commission – ($N = 57$) who invested in any of the above. An earlier version of this table appeared in Levien (2011).

common investment of farmers who made money from land transactions (compensation plot sales, voluntary land sales and commissions) was to purchase equivalent or more land in more remote areas where land is cheaper. Overall, 58 percent of farmers bought land, though Jats (80 percent) and general castes (67 percent) were much more successful at it than OBCs (54 percent) and SC/STs (48 percent). Since most cannot or do not want to shift their houses to far off villages, the majority of these have become absentee landlords, expropriating one-third of a monsoon millet crop from local tenants. For the smaller holders, this has actually been quite an inconvenience and net loss as they receive less grain than they previously had when they were cultivating less land with family labor and now have to incur substantial transportation costs. Moreover, the available cheap land in remoter parts of Rajasthan is typically more marginal and less productive. But some of the bigger farmers (particularly Jats) have become quite large absentee landlords, receiving not insignificant rent-in-kind from sharecroppers. However, for many, their share of the millet crop is only part of the calculation; they have bought land where they think the forces of urbanization will eventually drive up land prices. Expropriating grain as landlords (pre-capitalist ground rent) is thus a short-term strategy until they can reap the gains of real estate speculation (capitalist ground rent). With a similar logic, 30 percent have also bought as yet vacant plots in some of the many residential colonies that began sprouting up around Jaipur or outlaying towns during the real estate boom. Others have made additions to their houses to rent rooms to construction laborers and very recently the first batch of students at for-profit colleges that have come up on land around the SEZ (the latter being more significant than the former for the rental market). Significantly, there is almost no investment in agricultural production. While 25 percent have used the money to drill tube-wells on remaining land (almost all of them Jats and Brahmins), most have done so in order to sell scarce groundwater in tankers to Mahindra for construction and to their fellow villagers for consumption (this is what most of the tractors are used for). Water is thus being diverted from production to exchange, and ultimately diminished. Overall, investment in land is largely of a rentier nature (in capitalist or pre-capitalist ground-rent) and is not accompanied by an investment in its productivity. We may note the irony that the entry of the most advanced sectors of the Indian economy into rural Rajasthan has led to an *expansion* of rent-in-kind sharecropping as previous self-cultivators become large or petty landlords.

After land, the next most common investment (26 percent of those receiving land money) is in small shops (*dukans*) selling daily provisions or various consumer items. The market in Rajpura has been transformed from a handful of shops run by the traditional *banias* and artisanal castes to a bustling profusion of sometimes redundant enterprises run by non-traditional merchants. These shops include general provision stores, cell-phone and photo shops, chai and juice stalls, small vegetable carts, tire-puncture stalls, motor-cycle parts and repair shops and two small eateries (*dhabas*). Most supply the newly commercialized needs of the dispossessed farmers or the growing consumer aspirations of those with land money in the bank (or dresser). Some are quite profitable (earning over $200/month), others yield only the equivalent of wage labor but with the benefit of avoiding hard manual work, and a significant number go out of business. The overall result is a growing stratum of shopkeepers, though one that is internally highly stratified and whose lower echelons hardly qualify as 'petty bourgeois.' For the upper castes, these enterprises often yield substantial mercantile profits, while for the lower-caste semi-proletariat they are often 'strategies of minimal survival' (Burawoy 2001, 47). If we consider the proprietors of enterprises that yield the equivalent of minimum wages as disguised 'wage laborers' (in terms of assessing their welfare, not their structural position) then those relatively proletarianized by the land dispossession would be much greater than what Table 2 suggests. This 'informal' or 'need economy' of micro-entrepreneurship is hardly liberating for the majority of rural poor and, on a larger scale, it is far-fetched to imagine this kind of economic activity absorbing the mass of dispossessed rural agriculturalists whose numbers increase by the day (Harriss 1991, Breman 2009).

Finally, those with remaining liquidity lend it out to other villagers at the prevailing rate of 2–3 percent per month or 24–36 percent annually. Eleven percent of respondents and 21 percent of the general castes reported becoming money lenders (though because of the stigma attached to this activity, there was likely substantial under-reporting). Many of the small shops also provide merchandise on credit, which adds to their margins. Thus the infusion of cash has actually expanded the pool of village moneylenders who usuriously fill the credit gap created by land loss and the general drying up of rural credit following the liberalization of India's banking sector (Shah *et al.* 2007). The village's biggest land broker –the *sarpanch* at the time the SEZ came – also gives loans at higher rates to other brokers for land deals. Land commodification has thus fed usurer's capital, a form of pre-capitalist exploitation which, as Marx observed, adaptively 'clings on to' and 'impoverishes' new modes of production without itself being transformative (Marx 1981, 731, Adnan 1985, 56).

Consequently, the entry of the much-heralded IT industry and accompanying real estate finance into rural villages via the SEZ has had the effect of expanding what Marx would call 'antediluvian' forms of capital (1981, 728), resulting in economic *involution* rather than productive transformation. By generating speculative rather than productive links with the rural economy, the SEZ has brought about 'an expansion of the sphere of exchange at the cost production' (Burawoy 2001, 270) as the proceeds of land sales (forced and voluntary) have found relatively few productive outlets, and instead fed landlordism, further land speculation, petty mercantile activities and usury. Accumulation by dispossession for the SEZ has dissolved the productive forms of the old economy while preserving and expanding its rentier, mercantile and usurious elements (while adding capitalist to pre-capitalist ground rent). This is not to say that this expansion and elaboration of pre-existing economic forms is 'rigid' or 'stagnant' as in Geertz's formulation; it is a wildly dynamic process as local people constantly and creatively devise and revise strategies to adapt to the extraordinarily rapid commodification of land and the marketization of their means of reproduction. But it is a dynamic of economic change that is fragile (in its dependency on speculative land markets), wildly unequal (in its distributional effects) and, for the poor, evidently incapable of sustaining the levels of well-being and consumption provided even by single-cropped agriculture.

The result is a peculiar agrarian transformation indeed, one that appears distinct from agrarian transitions via agricultural capitalism. Capital is seizing hold of land but not production as agriculture is sidelined for urban rents and knowledge-intensive production; rural land is commodified while peasant labor power is marginalized; the class and caste structure undergoes dramatic differentiation on the basis of people's ability to survive in speculative land markets though without polarization into 'traditionally' recognizable agrarian classes; and an enclave of high-tech capitalism and urban real estate feeds a process of involutionary growth in the surrounding rural hinterland as different classes of dispossessed and marginalized peasants look with unequal success to forms of rentiership and informal exchange for survival.

Conclusion

In constructing a theory of land grabs in contemporary capitalism, the key question is not the origin of the capital, but why land dispossession becomes necessary or at least useful and possible for capital accumulation in general at a particular time and place. Accumulation by dispossession, in contrast to the normal functioning of land markets (globalized or not), is above all a political process whereby states (or other bearers of coercion) exercise *extra-economic* power to transfer to capitalists means of production, subsistence or common social wealth that are difficult or impossible to obtain on the market. Reconstructing Harvey, I have offered a definition of accumulation by dispossession centered on the use of extra-economic force in process of accumulation, which has the merit of being flexible enough to allow for research into variations in both of its terms (forms of accumulation and means of dispossession) in different times and places. Factors such as different state structures and legal frameworks, the presence or absence of foreign finance capital, the type of industry involved, pre-existing agrarian social structures and the balance of class forces will shape the character, degree and consequences of such dispossession in different contexts. This diversity is well illustrated by the articles in this issue of *JPS*.

In India, SEZs are the culmination of a long transformation of the state (or states to be more precise) toward becoming the chief land broker for capital. As increased demand for land – driven both by higher growth rates in general and real estate markets in particular – has confronted an inelastic supply in rural land markets, capitalists increasingly turn to the state to use non-market means for making land available for capital accumulation. This is the crux of accumulation by dispossession in India today. Unlike the older developmental state that expropriated large amounts of rural land for public infrastructure and heavy industries, land brokering in the neoliberal era – culminating with SEZs – proceeds under an expansive definition of 'public purpose' that is indistinguishable from private capital accumulation. Elite housing colonies, IT parks, malls and amusement parks have joined the hydroelectric dam and steel mill as causes for expropriating the peasantry.

What SEZs represent is thus the ascendance of real estate (the developers) and high-tech services (the producing units) over heavy industry and irrigation projects as levers of dispossession in India. The SEZ developer is a state-appointed capitalist landlord who receives windfall returns by commodifying artificially cheap land expropriated from farmers. I have put forward the concept of the rate of accumulation by dispossession to define and measure such accumulation based on the expropriation of land rather than the exploitation of labor. In return for turning farmland into 'developed land parcels', SEZ developer-landlords command a portion of the profits generated by the producing firms inside the SEZ. In the Mahindra World City, as in the two-thirds of SEZs that are devoted to IT and IT services, these profits largely derive from the labor of middle-class back-office workers rather than the urban or rural proletariat. The production that occurs in the SEZ marginalizes more than it exploits the labor of the dispossessed peasantry. Accumulation by dispossession is, in this case, about capitalizing the land while not exploiting the peasant.

With rural labor power a matter of indifference, the main impact of the SEZ is through the dispossession and commodification of rural land. Instead of capital seizing hold of agriculture – the traditional problematic of agrarian political economy (Kautsky 1988, Akram-Lodhi and Kay 2010) – capital is seizing hold of the land, creating a new pattern of agrarian transformation driven by real estate speculation. The dispossessed see their assets disappear and are very unequally equipped to 'play the game' of land markets even in the rare case examined here in which they are given a stake in it. The legacy of feudal class and caste structures creates the basis for unequal upward mobility via land prices. Inequalities in economic, cultural and social capital congeal in caste inequalities, which perhaps express most clearly the differential ability of people to profit from land markets and turn cash from land sales into capital. However, the lack of perfect correspondence between different forms of capital and an inescapable amount of randomness in real estate speculation has generated a complex recomposition of the local class structure, introducing greater inequality not just between but within castes, classes and families (including between men and women). Rather than simple class polarization or the generation of a singular 'neo-rentier class' (Adduci 2009) there is a multi-scaled chain of rentiership that individuals try to tap into, with very uneven results. The more successful minority become landlords, brokers, shopkeepers and moneylenders of different magnitudes, while the less successful become redundant members of the underemployed rural proletariat. In the middle are those dispossessed marginal farmers who try to survive in this involutionary dynamic off of a vegetable cart or chai stall, the small returns of grain from a distant piece of land, or place their hopes in an appreciating plot in a gritty peri-urban development. While not entirely wrong in the broadest sense, binaries such as capital versus peasants or enclosers versus commoners do not adequately capture this complexity.

In the terms of classical economics, the process of transforming rain-fed agricultural and grazing land into an IT-SEZ is undoubtedly efficient. For the majority of rural inhabitants, however, it is hardly developmental. The two-thirds of India's population that live in rural areas see few opportunities in this new economy of the urban and educated middle classes (Upadhya 2007). The chasm between rural India and Infosys is so great that such development creates few employment opportunities for the agrarian population. Without significant employment, the sole spillover effect of such an IT/real estate–driven SEZ is land speculation, the money from which, in the absence of any productive linkages, feeds an expansion and elaboration of old forms of rentiership and exploitation while creating some new forms as well. The result is what we might call a real estate–driven agrarian *involution* in which exchange displaces production and a dispossessed peasantry looks, with very unequal success, to opportunities for rent, interest and mercantile profits in the 'informal economy' for survival. That over 50 percent of families – and 75 percent of SC/ST families – report having 'less food' after having their land acquired under one of the most 'inclusive' compensations models on offer in India, is a grim and tragic verdict on the development consequences of accumulation by dispossession in the neoliberal era.

So while the Indian peasantry is being chastised in some quarters for its 'emotional attachments' to the land, the above suggests that they are in fact not so ignorant. While in Rajpura the enticement of quick real estate gains was enough to dissolve organized resistance to the SEZ (to the subsequent regret of many), in many more parts of India, farmers are militantly refusing to relinquish their land to capitalists, putting a halt to many SEZs and other industrial projects. While it may be true that there are many Indian farmers who would in theory like to exit agriculture (Bardhan 2011), the important point is that many of them can see – quite rightly – that the types of development proposed for their farmland will create no place for them, or at least not a good enough place to warrant surrendering their land.[32] The 'agrarian question of labor' is thus still intimately connected to that of capital; but their intersection has shifted from the surpluses from agriculture to the control over land. Without attractive exit options from agriculture, for many farmers keeping their few *bighas* of land and buffalo seems like a much better option than handing it to capital. They will consequently continue to pose the most serious contradiction for capitalism in India today.

References

Adduci, M. 2009. Neoliberal wave rocks Chilika Lake, India: conflict over intensive aquaculture from a class perspective. *Journal of Agrarian Change*, 9(4), 484–511.

Adnan, S. 1985. Classical and contemporary approaches to agrarian capitalism. *Economic and Political Weekly*, 20(30), PE53–PE64.

Akram-Lodhi, A.H. and C. Kay. 2010. Surveying the agrarian question (part 1): unearthing foundations. *The Journal of Peasant Studies*, 37(1), 177–202.

Anderson, P. 1974. *Lineages of the absolutist state*. London: New Left Books.

Arrighi, G. 2007. *Adam Smith in Beijing: lineages of the twenty-first century*. London: Verso.

Arrighi, G., N. Aschoff and B. Scully. 2010. Accumulation by dispossession and its limits: the Southern Africa paradigm revisited. *Studies in Comparative International Development*, 45, 410–438.

Balibar, E. 1979. Elements for a theory of transition. *In*: L. Althusser and E. Balibar, eds. *Reading capital*. London: Verso, pp. 273–308.

Banerjee-Guha, S. 2008. Space relations of capital and new economic enclaves: SEZs in India. *Economic and Political Weekly*, 43(47), 51–59.

Bardhan, P. 2010. *Awakening giants, feet of clay: assessing the economic rise of China and India*. Princeton: Princeton University Press.

Bardhan, P. 2011. Industrialization and the land acquisition conundrum. *Development Outreach*, April, 54–57.

Bernstein, H. 2008. Agrarian questions from transition to globalization. *In*: A.H. Akram-Lodhi and C. Kay, eds. *Peasants and globalization: political economy, rural transformation, and the agrarian question*. London: Routledge, pp. 239–261.

Bhaduri, A. 1983. *The economic structure of backward agriculture*. London: Academic Press.

Bhaduri, A. 2009. *The face you were afraid to see: essays on the Indian economy*. New Delhi: Penguin.

Bourdieu, P. 1986. The forms of capital. *In*: J.E. Richardson, ed. *Handbook of theory of research for the sociology of education*. New York: Greenword Press, pp. 46–58.

Breman, J. 1996. *Footloose labour: working in India's informal economy*. Cambridge: Cambridge University Press.

Breman, J. 2009. Myth of the global safety net. *New Left Review*, 59, 29–36.

Brenner, R. 1976. Agrarian class structure and economic development in pre-industrial Europe. *Past and Present*, 70, 30–75.

[32]While arguing that land acquisition is necessary to transfer agricultural land to higher value uses, Bardhan cites surveys showing that 'the overwhelming majority of farmers children want to leave agriculture' (2011, 56). However, the important question is not what people would like to do in theory (if asked on a survey, most people would say that they would rather be a millionaire than be in their current occupation), but what choices they would make *among the options available to them*. In a situation where the trajectory of economic growth is not creating sufficiently remunerative opportunities for people from farming backgrounds, then maintaining possession of land to ensure household reproduction while making forays into irregular and low-paying labor markets is highly preferable to being a landless wage-laborer. I found the desire to keep land strong among most people in Rajpura, including youth, in spite of the fact that they were trying to find footholds in the urban economy. This study demonstrates the easy slide into greater poverty and food insecurity that can arise when land is removed from diversified livelihood strategies, even when it is compensated at higher than normal rates.

Brenner, R. 2006. What is, and what is not, imperialism? *Historical Materialism*, 14(4), 79–105.
Burawoy, M. 1996. The state and economic involution: Russia through a China lens. *World Development*, 24(6), 1105–1117.
Burawoy, M., P. Krotov and T. Lytkina. 2000. Involution and destitution in capitalist Russia. *Ethnography*, 1(1), 43–65.
Burawoy, M. 2001. Transition without transformation: Russia's involutionary road to capitalism. *East European Politics and Societies*, 15(2), 269–290.
Burawoy, M. 2010. From Polanyi to Pollyanna: the false optimism of global labor studies. *Global Labour Journal*, 1(2), 301–313.
Byres, T. 1991. The agrarian question and differing forms of capitalist agrarian transition: an essay with reference to Asia. *In*: J. Breman and S. Mundle, eds. *Rural Transformation in Asia*. Delhi: Oxford University Press, pp. 3–76.
Byres, T. 1996. *Capitalism from above and capitalism from below: an essay in comparative political economy*. London: Macmillan.
Castree, N. 2003. Commodifying what nature? *Progress in Human Geography*, 27(3), 273–297.
Chakravarti, A. 1975. *Contradiction and change: emerging patterns of authority in a Rajasthan village*. Delhi: Oxford University Press.
Confederation of Indian Industries (CII). 2011. CII recommends government's proactive role in land acquisition. Available from: http://www.ciionline.org/pressreleases [Accessed 6 June 2011].
Cross, J. 2009. From dreams to discontent: young men and the politics of work at a Special Economic Zone in Andhra Pradesh. *Contributions to Indian Sociology*, 43(3), 351–379.
Cross, J. 2010. Neoliberalism as unexceptional: economic zones and the everyday precariousness of working life in South India. *Critique of Anthropology*, 30(4), 355–373.
De Angelis, M. 2001. Marx and primitive accumulation: the continuous character of capital's enclosures. *The Commoner*, 2, 1–22.
De Angelis, M. 2007. *The beginning of history: value struggles and global capital*. London: Pluto Press.
De Soto, H. 2000. *The mystery of capitalism: why capitalism triumphs in the West and fails everywhere else*. New York: Basic Books.
Fernandes, W. 2008. Sixty years of development-induced displacement in India. *In*: H.M. Mathur, ed. *India Social Development Report 2008: development and displacement*. New Delhi: Oxford University Press, pp. 89–102.
Geertz, C. 1963. *Agricultural involution: the process of ecological change in Indonesia*. Berkeley: University of California Press.
Geisler, C. 2003. A new kind of trouble: evictions in Eden. *International Social Science Journal*, 55(175), 69–78.
Glassman, J. 2006. Primitive accumulation, accumulation by dispossession, accumulation by extra-economic means. *Progress in Human Geography*, 30(5), 608–625.
Goldman, M. 2011. Speculative urbanism and the making of the next world city. *International Journal of Urban and Regional Research*, 35(3), 555–581.
Government of India, Ministry of Commerce and Industries, n.d. *Special Economic Zones in India*. Available from: http://sezindia.nic.in/ [Accessed on 1 February 2011].
Grandin, G. 2010. *Fordlandia: the rise and fall of Henry Ford's forgotten jungle city*. London: Icon Books.
Hall, D., P. Hirsch and T. Li. 2011. *Powers of exclusion: land dilemmas in Southeast Asia*. Singapore: NUS Press.
Harriss, J. 1991. Agriculture/non-agricultural linkages and the diversification of rural economic activity: a south Indian case study. *In*: J. Breman and S. Mundle, eds. *Rural transformation in Asia*. Delhi: Oxford University Press, pp. 429–457.
Harriss-White, B. 2003. *India working: essays on society and economy*. Cambridge: Cambridge University Press.
Hart, G. 2002. *Disabling globalization: places of power in post-apartheid South Africa*. Berkeley: University of California Press.
Hart, G. 2006. Denaturalizing dispossession: critical ethnography in the age of resurgent imperialism. *Antipode*, 38(5), 977–1004.
Harvey, D. 2003. *The new imperialism*. Oxford: Oxford University Press.
Harvey, D. 2005. *A brief history of neoliberalism*. Oxford: Oxford University Press.

Harvey, D. 2006a. Comment on commentaries. *Historical Materialism*, 14(4), 157–166.
Harvey, D. 2006b. *The limits to capital*. London: Verso.
Hsing, Y. 2006. Brokering power and property in China's townships. *Pacific Review*, 19(1), 103–214.
Hsing, Y. 2010. *The great urban transformation: the politics of land and property in China*. Oxford: Oxford University Press.
Infrastructure Development Finance Corporation. 2008. *India Infrastructure Rreport 2008: business models of the future*. New Delhi: Oxford University Press.
Kachhawaha, O.P. 1992. *History of famines in Rajasthan, 1900 A.D.–1990 A.D.* Jodhpur: Research Publishers.
Kautsky, K. 1988. *The agrarian question*. London: Zwan Publications.
Khagram, S. 2004. *Dams and development: transnational struggles for water and power*. Cornell University Press.
Krishna, A. 2002. *Active social capital: tracing the roots of development and democracy*. New York: Columbia University Press.
Lefebvre, H. 1991. *The production of space*. Oxford: Blackwell Publishers.
Lefebvre, H. 2009. *State, space, world: selected essays*. Minneapolis: University of Minnesota Press.
Lenin, V. 1967. *The development of capitalism in Russia*. Moscow: Progress Publishers.
Levien, M. 2007. India's double movement: Polanyi and the National Alliance of People's Movements. *Berkeley Journal of Sociology*, 51, 119–149.
Levien, M. 2011. Special economic zones and accumulation by dispossession in India. *Journal of Agrarian Change*, 11(4), 454–483.
Li, T. 2007. *The will to improve: governmentality, development, and the practice of politics*. Durham: Duke University Press.
Li, T. 2011. Centering labor in the land grab debate. *Journal of Peasant Studies*, 38(2), 281–298.
Linebaugh, P. 2008. *The magna carta manifesto: liberties and commons for all*. Berkeley: University of California Press.
Luxemburg, R. 2003. *The accumulation of capital*. London: Routledge.
Mahindra World City (MWC). 2007. Letter to the principal secretary of industries, government of Rajasthan, 4 June.
Marx, K. 1973. *Grundrisse*. London: Penguin.
Marx, K. 1977. *Capital*. Vol. I. New York: Vintage.
Marx, K. 1981. *Capital*. Vol. III. New York: Vintage.
Moore, B. 1966. *Social origins of dictatorship and democracy: lord and peasant in the making of the modern world*. Boston: Beacon Press.
Moore, P. and A. Schrank. 2004. Overstretch and underdevelopment: export processing zones, US hegemony, and the limits of soft power in the third world. Paper presented at the Annual Meeting of the International Studies Association, Montreal, Quebec, Canada, 17 March.
National Association of Software and Service Companies (NASSCOM). 2010. *The IT-BPO sector in India: strategic review 2010*. New Delhi: NASSCOM.
NDTV. 2008. Mahindra's SEZ style in Jaipur, 26 Aug. Available from: http://www.youtube.com/watch?v=7gRu_IMjZok [Accessed on 20 September 2009].
Parry, J. 1999. Two cheers for reservations: the Satnamis and the steel plant. *In*: R. Guha and J. Parry, eds. *Institutions and inequalities: essays in honor of Andre Beteille*. New Delhi: Oxford University Press, pp. 128–169.
Parry, J. and C. Struempell. 2008. On the desecration of Nehru's 'temples': Bhilai and Rourkela compared. *Economic and Political Weekly*, 43(19), 47–57.
Patnaik, U., ed. 1990. *Agrarian relations and accumulation: the 'mode of production' debate in India*. Mumbai: Oxford University Press.
Perelman, M. 2000. *The invention of capitalism: classical political economy and the secret history of primitive accumulation*. Durham: Duke University Press.
Polanyi, K. 2001. *The great transformation: the political-economic origins of our time*. Boston: Beacon Press.
Putnam, R. 2001. *Bowling alone: the collapse and revival of American community*. New York: Simon and Schuster.
Radhakrishnan, S. 2011. *Appropriately Indian: gender and culture in a new transnational class*. Durham: Duke.

Rajasthan State Industrial Development and Investment Corporation (RIICO). 2006. Letter from the financial advisor of RIICO to the Secretary of the Jaipur Development Authority, 3 October.
Registrar General. 2001. *Census of India 2001*. New Delhi: Registrar General.
RETORT, 2005. *Afflicted powers: capital and spectacle in a new age of war*. London: Verso.
Rosin, R.T. 1987. *Land reform and agrarian change: study of a Marwar village from Raj to Swaraj*. Jaipur: Rawat Press.
Roy, S. 2007. *Beyond belief: India and the politics of post-colonial nationalism*. Durham: Duke University Press.
Rudolph, L. and S. Rudolph. 1984. *Essays on Rajputana*. New Delhi: Concept Publishing.
Sanyal, K. 2007. *Rethinking capitalist development: primitive accumulation, govermentality, and post-colonial capitalism*. New Delhi: Routledge.
Sassen, S. 2010. A savage sorting of winners and losers: contemporary versions of primitive accumulation. *Globalizations*, 7(1), 23–50.
Searle, R. 2010. Making space for capital: the production of global landscapes in contemporary India. PhD Dissertation, University of Pennsylvania.
Seethalakshmi, S. 2009. *Special economic zones in Andhra Pradesh: policy claims and people's experiences*. Hyderabad: ActionAid
Shah, M., R. Rao and P.S. Vijay Shankar. 2007. Rural credit in 20th century India: an overview of history and perspectives. *Economic and Political Weekly*, 42(15). 1351–1364.
Singh, D. 1964. *Land reform in Rajasthan: a study of evasion, implementation, and socio-economic effects*. New Delhi: Planning Commission, Government of India.
Sud, N. 2009. The Indian state in a liberalizing landscape. *Development and Change*, 40(4), 645–665.
Teschke, B. 2003. *The myth of 1648: class, geopolitics and the making of modern international relations*. London: Verso.
Upadhya, C. 2007. Employment, exclusion, and 'merit' in the Indian IT industry. *Economic and Political Weekly*, 42(20), 1863–1868.
Walker, K. 2006. 'Gangster capitalism' and peasant protest in China: the last twenty years. *Journal of Peasant Studies*, 33(1), 1–33.
Walker, K. 2008. From covert to overt: everyday peasant politics in China and the implications for transnational agrarian movements. *Journal of Agrarian Change*, 8(2–3), 462–488.
White, B. 1982. Population, involution and employment in rural Java. *In*: J. Harriss, ed. *Rural development: theories of peasant economy and agrarian change*. London: Hutchison and Co., pp. 300–320.
White, B. and A. Dasgupta. 2010. Agrofuels capitalism: a view from political-economy. *Journal of Peasant Studies*, 37(4), 593–607.
Wood, E.M. 1981. The separation of the economic and political under capitalism. *New Left Review*, 1(127), 66–95.
Wood, E.M. 2002. *The origin of capitalism: a longer view*. London: Verso.
World Bank, 2007. *India: land policies for growth and poverty reduction*. New Delhi: Oxford University Press.
Zoomers, A. 2010. Globalisation and the foreignisation of space: seven processes driving the current global land grab. *The Journal of Peasant Studies*, 37(2), 429–447.

Michael Levien is a PhD candidate in the Department of Sociology at the University of California-Berkeley. His dissertation is based on an ethnography of a Special Economic Zone in rural Rajasthan, and examines the political economy of land dispossession under neoliberalism in India.

Land expropriation and displacement in Bangladesh

Shelley Feldman and Charles Geisler

This paper examines land grabbing in Bangladesh and views such seizures through the lens of displacement and land encroachment. Two different but potentially interacting displacement processes are examined. The first, the *char* riverine and coastal sediment regions that are in a constant state of formation and erosion, are contested sites ripe for power plays that uproot small producers on their rich alluvial soils. The second examines new patterns of land capture by elites who engage gangs, corrupted public servants and the military to coerce small producers into relinquishing titles to their ever more valuable lands in and near urban areas. These historically specific and contingent land grabs draw attention to *in situ* displacement, where people may remain in place or experience a prolonged multi-stage process of removal. This contrasts with *ex situ* displacement, a decisive expulsion of people from their homes, communities and livelihoods. In both the *char* and peri-urban case, we signal new forms of collective action in response to involuntary alienation of land resources in a rapidly and violently transforming political economy. We conclude with a caution against naturalizing displacement, casting it as an 'inevitable' consequence of changing weather conditions in the former and population dynamics in the latter.

Introduction

Many recent investigations of land grabbing draw attention to foreign investment and resource deals in the Global South.[1] Their aim is to document how these transactions alter social and cultural landscapes as investors scramble to claim allotments in 'the global farm' (World Bank 2010). A high-stakes farms race has ensued, with support from the advocates of a new 'green revolution', opposition from nationalists and land reformers and ambivalence by many others who are mindful of the complexities at hand (Robertson and Pinstrup-Andersen 2010). Contrary to the widely shared discourse on such land deals that places current food and fuel demand in the foreground, there is an array of more banal, less legible dealings that are internal to particular national contexts. A focus solely on high-profile, off-shore 'land grabbing' runs the risk of consigning internally generated land deals and everyday displacement practices of vast numbers of people to semi-oblivion. Alternatively, drawing attention to locally perpetrated land deals and

---

We thank the reviewers for their helpful comments, the participants at the Global Land Grabs conference at the University of Sussex for their thoughtful engagement with the issues, and Wendy Wolford for her encouragement.

[1]See, for example, Borras *et al.* (2011), De Schutter (2011) and Peluso and Lund (2011).

expropriations, particularly as they are expressed in current practices of land capture and class predation, can shed important light on their headline-hungry global equivalents.[2] These practices, we suggest, can be explored through the lens of human displacement and therein help to reveal the complex and contested ways in which dispossession and alienation shape the current social relations of landed property.

The literature on human displacement is replete with accounts of everyday land deals that arise in response to public policy choices as well as to private behaviors. By definition, human displacement entails the uprooting of people, not only as refugees and asylum seekers but also, and more often, within their homelands. Acknowledging this phenomenon, the UN High Commission for Refugees (UNHCR) distinguishes between transboundary political victims or refugees, and internally displaced persons (IDPs).[3] For at least two decades, scholars, journalists and human rights activists have expanded the latter category to encompass development-induced displacement, most often in reference to mega-projects and infrastructure growth (Cohen and Deng 1998, Brookings Institution 2005, Vandergeest *et al.* 2007). Processes of internal displacement at times are accompanied by conspicuous land losses that remain understudied and downplayed by development institutions and may be discounted as the collateral damage of modernity (Porter 2009).

Moreover, there is a tendency to overlook the intense though gradual processes of land and livelihood loss because, in large measure, the conditions that lead to internal displacement are generally understood as a by-product of development projects, civil wars, or natural disasters (Cernea and McDowell 2000, Cernea 2003). This form of internal dislocation includes the displacement of people-in-place as they lose their resource base yet remain precariously in their communities. At some point – perhaps triggered by a debt coming due, a crop failure, a market calamity or radical revision in land prices – their exodus follows. Yet, most view such departures not as evictions but rather as out-migration, moves made by rational but insecure inhabitants voting with their feet. Simply put, the foreclosing forces underlying the process and often coercing people to leave their homes go unnoticed as the casualties of land grabbing, despite the toll they take on selective groups in society.

[2]These global media will be counterpointed by more localized media reports from Bangladesh to offer a window on the various expressions of land predation in the country. Such reports offer a portal on the everyday land violences that attend to life at the two sites of interest – the *chars* and peri-urban spaces.

[3]The 1951 Convention on Human Rights 'offers a general definition of the refugee as including any person who is *outside* their country of origin and unable or unwilling to return there or to avail themselves of its protection, on account of a well-founded fear of persecution for reasons of race, religion, nationality, membership of a particular group, or political opinion' (Goodwin-Gill 2008). The Convention does not address internally displaced persons. In 1998 Guiding Principles on Internal Displacement were ratified by the U.N. and define IPDs as 'persons or groups of persons who have been forced or obliged to flee or to leave their homes or places of habitual residence, in particular as a result of or in order to avoid the effects of armed conflict, situations of generalized violence, violations of human rights or natural or human-made disasters, and who have not crossed an internationally recognized State border' Global Database: Guiding Principles on Internal Displacement 2007). The *involuntary* character of one's departure and the decision to remain in one's own country are the two main elements determining who is an internally displaced person. According to some, however, IDPs remain hard to define and quantify and thus receive only ad-hoc attention (Stavropoulou 1998).

To illustrate this perspective, we turn attention to Bangladesh, a nation only slightly larger than England but with a population almost three times the latter's population (155 versus 60 million people). Bangladesh warrants attention as a displacement epicenter for several reasons. Not only does it have the highest density among non-insular countries in the world (Mongabay 2008),[4] but the World Bank (2010) estimates that 61 percent of its land is arable and three-fifths of Bangladeshis are employed in agriculture.[5] Yet, only 3.7 percent of the overall land base is permanent cropland, suggesting an enormous discrepancy between what is continuously and provisionally cultivated. Even as its population surges, the arable lands of Bangladesh have fallen from roughly nine million hectares in the 1990s to eight million hectares at present (World Bank 2010). This reflects the conversion of arable land to other uses and entails both *ex situ* and *in situ*[6] displacement for large masses of people. Figure 1 highlights the dramatic shifts in arable land over the period 1968 to 2008.[7] It is striking to note that by 2025, human settlement is predicted to consume as much as 50 percent of the arable land base that remains (SAMATA 2006).[8]

In what follows, we showcase the displacement effects of land expropriation in two political ecological zones, the first in coastal *char* areas and the second in peri-urban areas surrounded by agricultural holdings. The land transformations in both locations are dramatic, as are the human displacements that accompany them. Yet, as noted earlier, these processes do not get the attention they deserve; too often the political ecology that attends to them, including land predation by elites, the state, and the military, are ignored. The expansive *char* areas are characterized by fluctuations in riverine and coastal sedimentation, something to which Bangladeshi farmers appear to have adapted over centuries. To be sure, settlement in this zone is increasingly precarious because of global climate change, rising sea level changes, and inland storm surges from cyclones and erratic weather events (Sarwar 2005). But the story does not end here. Predatory interests also uproot and displace small producers, intensifying the conditions and relations of *char* insecurity. Such powerful interests often depend on a mixture of government incapacity and complicity to realize their goals (Islam 2009, Adnan 2011). Eventually, a large number of peasant producers are dispossessed by more powerful interests that seek to benefit by usurping their small, precarious holdings.

In the second instance, land predation also occurs, but in and around urban areas. It is again mobilized by elites, parlaying their privileged access to government and military networks, and entails bribery, intimidation, and coercion in a burgeoning

[4]These figures exclude some small islands including Bermuda, Maldives and Malta, or Singapore, Hong Kong and Bahrain.

[5]Agricultural land is the share of land area that is arable, under permanent crops and under permanent pastures. Arable land includes land defined by the FAO as land under temporary crops, meadows or pasture, land under market or kitchen gardens and land temporarily fallow. See also http://www.tradingeconomics.com/bangladesh/agricultural-land-percent-of-land-area-wb-data.html for slightly different figures [Accessed 25 April 2011].

[6]*In situ* displacement or close approximations to it have appeared in other research (e.g. Feldman *et al.* 2003, Doutriaux *et al.* 2008).

[7]Trading Economics, Agricultural land (% of land area) in Bangladesh. Available from: http://www.tradingeconomics.com/bangladesh/agricultural-land-percent-of-land-area-wb-data.html [Accessed 25 April 2011].

[8]While land grabs play a prominent role in this process, erosion, salinization, mangrove loss and the inland effects of storm surges contribute as well, as they have historically.

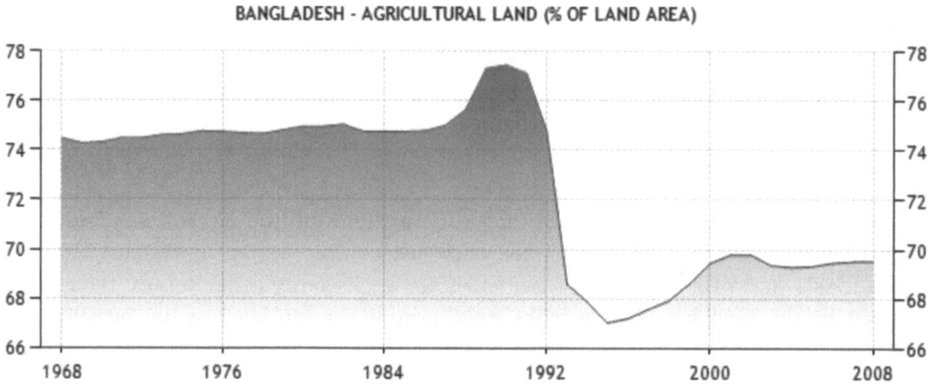

Figure 1. Bangladesh – Agricultural Land (& of Land Area).
Source: http://www.tradingeconomics.com/bangladesh/agricultural-land-percent-of-land-area-wb-data.html.

urban real estate market. Many farmers who operate in the urban shadows of Chittagong and Dhaka experience *in situ* displacement as their rights to work, to belong in stable and secure communities, and to reproduce their social worlds are lost. Eviction is sometimes an ambiguous process, as the value of farmland appreciates and triggers land grabs through various means over which small farmers have little control. One must ask, as we do, how do displacement and dispossession, as long-term features of land relations in Bangladesh, provide useful information to the current analyses of land grabbing? In our conclusion, we revisit the twin effects of *in situ* and *ex situ* displacement for what they reveal about new venues of accumulation and rent seeking.

Displacement: an analytic lens

In 2003, we offered a definition of *in situ* displacement following Arendt's (1973) insights on the widespread 'loss of the Rights of Man' across the twentieth century (Feldman *et al.* 2003). We posited that most accounts of displacement, including those that are internally generated, overemphasize physical dislocation and construe dispossession as involuntary removal from one's land and locus of survival. We suggested that such thinking overdetermined displacement as movement from particular places, thereby ignoring other relations and sites of production, consumption, renewal and resilience. We also suggested that, as Arendt perceived it, displacement could simultaneously occur at multiple scales (home, community, nation), thereby revealing the complex processes that constitute exclusionary practices from particular locales, as well as from rights, securities and identities. To restate this, we realized that displacement need not be characterized *only* or even primarily by flight, involuntary migration or the abandonment of one's worldly possessions, at least in the first instance. Such an understanding helps to reveal land grabs as both physical appropriations and, perhaps most crucially, as diminishments in the capacity to socially reproduce everyday lives and livelihoods. Simply put, we suggested that one could be displaced without a 'change of address' or a catastrophic dispossession that comes with land seizure, foreclosure, or violent removal. While remaining at home, in other words, one can experience rapid as well as the slow-motion loss of entitlements, social exclusion and alienation from the rights and identities that had been successfully fought for in social contracts between states and individuals.

More recently, our thinking on *in situ* displacement expanded to encompass human in/security. In a project devoted to understanding how insecurity accumulates in banal but important ways, we equated *in situ* displacement with failures in social reproduction, by which we meant how people internalize harmful shifts in economic, social and political well-being as well as altered subjectivities vis-à-vis powerful people and institutions (Feldman *et al.* 2011). In short, *in situ* displacement can take many forms: as workers face redundancy, superfluity and disposability in their employment, as citizens become permanently disenfranchised or deprived of rights in states of exception, and as inhabitants of ecological systems lose life-support services, ecosystem health and resiliency. For us, *in situ* displacement references the critical impairment of social reproduction (Martin 2010, Feldman *et al.* 2011).

Using a displacement analytic, then, land dispossession assumes multiple forms and entails the wielding of considerable power, both private and public. The better known cases are *ex-situ* displacement examples, whereby people are forced to move from places where they are secure, often by powerful developers, investors, and speculators, and with little warning, consultation, or transparency. This process of forced resettlement, usually orchestrated by state institutions, is what Woodhouse (2003) acknowledges as a default mode of development. Cernea and McDowell (2000) concur and insist that such resettlement requires compensation. Examples of *ex situ* displacement range from familiar assertions of eminent domain for urban renewal, infrastructure corridors, enterprise zones, government complexes, buffer zones, roads, and military facilities, to the denial of land rights to indigenous or nomadic communities or the usurpation of land and other critical resources (e.g. Li 2011, De Schutter 2011, also in this volume papers by Amanor, Woodhouse, Levien, and Filer). By comparison, *in situ* displacement does not depend on resettlement 'out of place', but rather finds that people may remain in place, and even in the same homestead, but with diminished and highly uncertain forms of livelihood, that jeopardize their reproducibility. This less obvious form of predation, in other words, may occur gradually and remain relatively invisible within longer processes of dispossession.

Criminal and opportunistic elites often are the perpetrators of these forms of displacement as they capture key parts of the legal system, manipulate land records, call upon the government to remove squatters and slums with police or army assistance and otherwise encroach on lands ripe for development or re-development.[9] These lands may be public or private. In some ways, as our case studies will show, Bangladesh has excelled in legislation that discriminates against particular landholder groups whether on public (*khas*) or private land. Celebrated cases of the former are the Enemy Property or Vested Property Act of 1965, used to justify the expropriation of Hindu-held lands after Independence (Yogeshsaxsena 2010) and the government's settler program relocating non-tribal people to the Chittagong Hill Tracts (Plant 1994, Blaikie and Sadeque 2000). But the country's disturbing record of violence (Wang *et al.* 2009, Fagan 2010) and corruption (bribery, misuse of public office, preferential discharge of government functions in the judiciary, executive, law enforcement, and tax policy) complicate and darken both displacement processes.

[9]Illegal measures, sometimes blatant and sometimes subtle, are not limited to Bangladesh (see, for example, Borras and Franco 2010).

Bangladesh 'down under'

Formerly East Pakistan, Bangladesh became an independent nation in 1971. It is a low-lying delta country that is constantly faced with the adverse effects of monsoons, ocean tides and the mood swings of powerful rivers descending from the world's highest mountains (Kamal *et al.* 1999). As many have noted, much of the country is in harm's way as oceans warm and its levels rise due to global climate change (Nishat 2008, Nicholls 1999, Germanwatch 2004, Sarwar 2005). It is estimated that the 660,000 square-mile delta of the Ganges–Brahmaputra–Jamuna Rivers is among the more vulnerable zones on earth to globally induced sea level fluctuations. Predictions are that the sea level will rise three meters by 2050 and that storm surges will carry saline waters far inland (IWM 2009). This increasingly turbulent environment will rob a fifth of the country of fertile lands and add to a fresh water crisis. Salinity will affect surface and ground waters, with fields up to 40 km from the new coastline rendered useless for growing crops (Buerk 2004, Nishat 2008).[10] In 2009, Bangladesh's Prime Minister said before the UN General Assembly that by 2050, one in every seven people in Bangladesh could be displaced. In fact, estimates suggest that 15 million people in Bangladesh could be displaced from farms and villages by 2050 because 11 percent of its land mass is expected to be lost due to climate change (Population and Sustainability Network 2011).

But not all displacement in Bangladesh from 'natural' causes is new, nor is it limited to extreme weather events and river bank/coastal erosion, however extraordinary these prove to be (Hutton and Haque 2004).[11] Human vulnerability to displacement is more complicated and involves deliberate land grabbing tainted by inter-class predation as well (Wisner *et al.* 1994, Lein 2008). Social forces also are at work that capitalizes on the misfortunes of those whose tenure in the delta is made tenuous by global meteorological forces and provide an enabling context for quasi-legal patterns of corruption to control valuable resources. These forces thrive on sustained political instability in Bangladesh, even after 1991, when democracy replaced15 years of military and authoritarian rule. Since 1991, two political parties, the Awami League and the Bangladesh Nationalist Party, have dominated bureaucratic institutions and remain relatively unaccountable to the citizenry. The result is a form of crony capitalism that depends on and reinforces a class politics built on collusion among politicians, a bureaucratic administration, law enforcement agencies, and the judiciary (Titumir 2007). Such collusion and control are operative in the empirical focus we explore below.

But first, how is land, a prerequisite for survival in Bangladesh, distributed and how secure is it? Inequitable ownership is the norm in Bangladesh, despite land

---

[10]Bangladesh's Institute of Water Modeling (IWM 2009), with Danish financial assistance, found the country needs at least US$4.17 billion (Tk 287,634 million) for polder construction to save the lives of coastal people from combined sea-level change and storm surge. 'There will be water-logging on 18 percent of land in the country's costal region by 2050 while on 25 percent by 2100 even if the embankments are raised at least by 4–6 metres', the study says (*Daily Star* 2009). The IWM report also notes that according to the Intergovernmental Panel on Climate Change (IPCC), sea-level change will wipe out more cultivated land in Bangladesh than anywhere in the world. By 2050, rice production is expected to drop by 10 percent and wheat production by 30 percent in response to water logging.

[11]Floods, droughts and cyclones are familiar forces on Bangladesh's delta. An estimated one million people are displaced by riverbank erosion each year (Zaman 1991).

reforms in the 1950s, 1960s and following Independence in 1971.[12] On paper, these laws address tenancy reforms, ceilings on landholdings and distribution schemes of public (*khas*) land to the landless (Uddin and Haque 2009). The Constitution, as originally written, and as amended in 2004, states that all citizens shall have the right to acquire, hold and dispose off property. Yet, over half of the rural population, almost 75 percent of the country's total population, is landless or holds less than half an acre, and the amount of land available for cultivation per household has been steadily declining (World Bank 2010, GOB 2010). The urban condition is similarly unstable and unequal. Dhaka City is among the ten most populous cities in the world, and the percentage of Bangladesh's urban population living in slum conditions is the second highest in South Asia (USAID 2010). Housing and industrial build-outs have grown without planning or regulation, claiming agricultural land and wetlands and filling in water bodies (World Bank 2007). This dearth of planning has allowed much land in residential areas to be contaminated with heavy metals and solid wastes from industry (UNEP 2005). Urban land prices are high, and formal land transactions are manipulated by the elite. The poorest households most often squat on public land with little security, the poor and middle-class are subject to relentless overcrowding, and the general distribution of land reflects national socioeconomic disparities. Seventy percent of the urban population lives on only 20 percent of the residential land in urban centers (USAID 2010, Shafi and Payne 2007, CPD 2003).

On paper, the government maintains a *khas* land allocation program for landless households with 99-year rights to plots at no cost. This distribution entails a lengthy process, however, and applicants for land often pay multiple bribes. Corruption is endemic, as powerful persons seize possession of *khas* lands and intimidate those holding them to vacate (Shafi and Payne 2007, Uddin and Haque 2009). As we elaborate below, violence is common in what appear to be 'market' transactions (IMF 2005, GOB 2008), both between and within classes. Still other forms of land grabbing occur when encroachers create false documents and secure 'legal' documents to confirm their ownership. Such encroachers often are influential persons who secure government compliance in forging documents (Uddin and Haque 2009) and in enabling illegal evictions (Shafi and Payne 2007, World Bank 2006, UN-Habitat 2007).[13] In a word, the potential for displacement is rife – both immediate and in stages, given the lawlessness that pervades everyday land dispossession in Bangladesh. To reduce the country's displacement juggernauts to the turbulence of nature, the drowning effects of population tides and normal market misfortunes is almost deliberately disingenuous.

[12]According to the USAID (2010, 4), 'Bangladesh has a long history of inequitable access to land. In rural areas, one percent of landowners own more than 7.5 acres. Ten percent of landowners own between 2.5 and 7.5 acres. The remaining 89% of landowners own less than 2.5 acres. Thirty-nine percent have less than .5 acres. This inequity exists despite a series of land reforms in the 1950s and 1960s that included tenancy reforms, imposed ceilings on landholdings, and provided for the distribution of public land to the landless'.

[13]Contributing to such landlord behaviors are the cumbersome and expensive registration requirements that impede land transactions. An individual or entity (either domestic or foreign) purchasing a real estate plot in a peri-urban area near Dhaka City can expect to pay 10 percent of the value of the property for registration and face long delays in completing the transaction.

Char areas

Some 60 percent of Bangladesh's land mass is less than six meters above mean sea level and, when combined with floods from its many rivers, between 30 and 70 percent of the county is flooded in most years (Agrawala *et al.* 2003). Literally thousands of square kilometers of riverbank face erosion annually, especially during the cyclone season. These riverine zones are janus-faced in that accretion accompanies erosion in a cycle as old as human settlement. The islands and sandbars resulting from these processes provide fertile alluvium for cultivation, despite punishing circumstances and unceasing vulnerability. In large measure, *chars* are inhabited by the country's poorest people, dispossessed of their lands elsewhere in the country (Wisner *et al.* 2004, Akter 2009). Land tenure in the *chars* is radically unstable. On one hand, they are extremely fertile and have access to water. Thus they are of interest to both precarious peasants and to commercial interests intent on harvesting mangrove forests, inserting commercial crops behind protective polders, or making fortunes in agro-fisheries and shrimp farming. On the other hand, the government legally owns *char* land. This being so, land grabbers must either keep the government at bay or enlist its considerable assistance. Rarely, as we shall see momentarily, has the government met its obligation of allocating *char* land to the landless.

Intermittent government intervention and laxity in this zone have a history worth recalling. In the colonial period, the British administered riverine areas of Bengal as early as 1825, when the Bengal Alluvion and Diluvion Regulation was implemented. Claims and disputes over emergent *chars* were decided upon by considering the local use of *payasty* (alluvion) and *sikosty* (diluvion) (Hutton and Haque 2004, 42–3). Land gained by accretion when waters receded after flooding belonged to the person whose estate it adjoined (Ali 1980, 296). Yet re-submergence would interrupt this tenure even if it re-emerged (Hutton and Haque 2004), making *char* tenure nearly as fluid as the waters surrounding them. In the face of such ambiguity, powerful property owners prevailed and the marginal existence of most small cultivators was taken as normal.

This situation changed little under unified Pakistan (1947–1971). The State Acquisition and Tenancy Act of 1950 provided rights to riparian landowners to accretions to their land. Following Independence, in 1972, amendments to this legislation were added (Sections 86 and 87 of the 1950 Act) to allow any landowner the right to utilize submerged diluvial land for 20 years. But, in the same year, under Presidential Order No. 135 of 1972, the rights of diluvial and alluvion were abolished entirely and adjudication[14] fell to the state and *char* land became *khas* land. According to the Order, 'All *char* lands will thus be at the disposal of the government and *will be available for settlement by the poorer classes of agriculturists* in accordance with government policy and the provisions of the law' (Mukherji 1976, 459–60, emphasis added). The result has been to render the *char* an open-access resource, vast and valuable, and in large measure a stateless place. The tenure belongs to those powerful enough to possess and retain it. Acquisition and retention are subject to the

[14]The reappearance of lands on account of alluvion, when such lands were earlier lost by diluvion, is a special case. Here, title and interest of the original tenant or his successor's interest are protected if such lands reappear in the *same place* within 30 years of their loss (Ali 2003).

eminent domain not of the public sector, but rather of the private sector, with government support or acquiescence.

To be fair, the central government has attempted to stabilize the *char* by engaging a mix of planning and engineering strategies. The result was the controversial Flood Action Plan (FAP) of the late 1980s. This enormous 26-component World Bank initiative built on studies of regional flooding in the region by the UNDP, a French engineering consortium, USAID and JICA (Japan International Cooperation Agency), under the assumption that embankments (levies and polders) would form the basis for controlled flooding. This assumption was forcefully challenged by a host of researchers at the time, and mounting costs and feasibility problems ultimately hampered the project's completion (Adnan 1991, Pearce 1991, Custers 1993). Also well known, the government has worked with numerous nongovernmental organizations to reduce risk, poverty and insecurity among the poor in the region (Leckie 2010, Chars Livelihood Programme 2004).

But, crucially, government lapses in the sizeable coastal zone have propelled the displacement of large masses of people that garner little publicity as land grabbing (total displacements from natural calamities in Bangladesh is predicted to increase from 49 million people today to 78 million in 2020, according to Akter (2009)). Two points must be underscored. At least where coastal displacement occurs, the toll may be conservative because of the multi-stage nature of the exodus (Adnan 2011). Additionally, the on-going human displacement can be too easily attributed to 'unruly nature'. Global climate change and other incremental calamities, important as they are, can become convenient masks for more routine, banal displacement effects resulting from the violent political ecology that characterizes the lives of many inhabitants of these areas. For poor peasants, sharecroppers and wage laborers, social reproduction itself is being transformed by those who profiteer from the land, water and other natural resources in the area, and against which the poor are unable to protect themselves. The region is thus changing as much from these transformations in the social relations of land ownership as from global environmental changes that characterized *char* lives and livelihoods a generation ago (Lynch and Alcorn 1994, Shafi and Payne 2007, World Bank 2006). It is the synergies among these changes that are fueling a displacement epidemic for local people (Haque and Zaman 1989).

Between 1974 and 1981, Haque and Zaman (1989, 305) surveyed *char* residents who had left their homes because of river re-channeling and bank erosion (net of other floods and disasters) to learn what distances displacees traveled. Their findings are instructive in grasping *in situ* displacement: some 35 percent of the residents were forced to leave the area, while those who traveled only short distance were constrained for the lack of resources. In other cases, because friends and family nearby offered help, and displacees believed their lost lands would re-emerge, they too remained close to home. However, their hope often turned to despair, as only 5 percent of lost holdings actually did reemerge. For many *char* residents, the impact of land loss in the floodplain compromised their ability to reproduce their sustenance since losses included: 'homestead land, housing structures, crops, cattle, trees and household utensils. Thus, in contrast to floods and other pervasive types of riverine disasters, bank erosion generally leaves little hope for immediate recovery for marginal households.... [as] disruptions [often] bring enormous economic, social, and psychological costs' (Haque and Zaman 1989, 305). As for government assistance, the authors held out little hope:

> Although the incidence of destitute displacees is quite high in the region, relief and rehabilitation programmes are seldom undertaken by national government or nongovernment agencies or by local government offices... Among the limited assistance ...was that provided by relatives and friends... [as other] forms of assistance to hazard victims are almost non-existent in the study area. (Haque and Zaman 1989, 308)

This assistance vacuum and the limited options for displacees brought about an insidious dependence on local patronage systems and set in motion the multi-staged displacement alluded to earlier. Impoverished displacees become a cheap labor pool of sharecroppers and agricultural workers for local *talukdars* (bosses) and serve, as well, as violence enforcers of land grabbing for one or another local faction. That is, they are transformed into 'readily available *lathiyals* (clubmen, armed retainers) to organize violent fights when necessary to gain control over newly emergent *char* land' (Haque and Zaman 1989, 310–311).

This perverse process, turning displacees into displacers, is not an isolated incident. Elsewhere in the country, and more recently, Adnan (2011) investigated land struggles in the Noakhali coastal belt, where landless and displaced peasants squat on *char* and other coastal lands owned by the state. Once again, the lived experience of displacement occurs in stages and is not readily perceived as coercive, socially induced eviction at any given moment. The Noakhali *chars* and low-lying woodland areas were, in 2004, 'particularly attractive to wealthy and powerful interest groups interested in capital investment precisely because these were not encumbered with any pre-existing private property rights' (Adnan 2011, 6). Not only is Adnan suggesting that the local factionalism observed by Haque and colleagues occurs elsewhere in the *chars*, but also that local bosses themselves fall victim to the predations of larger external forces, often in league with the government or benefitting from its benign neglect (Haque and Zaman 1989, Shafi and Payne 2007, *New Age* 2011d). According to Adnan (2011, 12), squatter families displaced in the Noakhali *chars*

> faced recurrent violence from rival power-holders in the form of arson, harassment, assault, rape and killings, which were usually linked to threats of eviction from the lands under their occupation. Furthermore, these migrant households were subject to the absolute social power of the *banadasyu* leaders within their respective domains, who 'promulgated local laws' and used violence and intimidation to exploit and repress them. They had little option but to suffer in silence since they needed the protection of their own *banadasyu* leader against attacks from rival power-holders seeking to grab their tenuous landholdings.

Adnan (2011) is emphatic that Noakhali squatters lack fundamental tenure security as their survival requires loyalty from one forest bandit-boss or another. He also emphasizes the predation superimposed by large interests from beyond the region committed to converting *char* and other coastal land into shrimp and fish export operations. Citing several additional studies, Adnan reports that the removal of the *banadasyus*, often with unsparing violence, leaves originally displaced cultivators with little or no patronage defense. He notes, 'the poor peasants and squatters of the Noakhali *chars* had hoped that the elimination of the *banadasyus* would facilitate them in getting formal rights on state lands. Mob violence to which they fell prey as practitioners, now systematically turns against them by newly dominant interest groups' (Adnan 2011, 18). The displacement depicted here is complex and multi-stage and involves coercion and violence that terrorizes anyone who resist the dominant political and commercial interest groups positioned to seize their land. The same author astutely notes that the combination of land seizure and land denial (inaccessibility) is possible because 'influential others' are embedded in local and national power structures that enable them to draw on security forces of the central government when they choose.

Resistance to dispossession of this kind, and on this scale, is tempered not only by the fear of challenging the government on human rights or other grounds but also, we hasten to add, because of the illegibility of *in situ* displacement. As noted, *char* displacees appear to move short distances and, on occasion, return to their homes in the face of adversity, thus seeming not to be irrevocably displaced. But, less transparent, are the stories of multi-stage displacement that involves prolonged land grabbing with delayed, but significant, cumulative effects. In sum, life for all but a minority in the *char* is a life dependent on 'influential others' who dispatch the vulnerable to assemble *char* lands in a culturally, politically, and meteorologically tumultuous environment. The complicity of rural bureaucrats, political parties, and officials can be passive or active (Sarker *et al.* 2003), especially where accountability is low due to the frontier qualities of the zone and the often gradual rather than catastrophic nature of events. Gradualism, in other words, obscures land loss and displacement because, on a day-to-day basis, residents are characterized as resilient and enduring. Little thought is given to the displacement of the conditions of social reproduction itself, increasingly corrupted and dismembered by elite profiteering from the region and beyond.[15]

Per-urban land predation

We include, in the broader sensibility of land predation, inter-class forms of violence, plunder, and marauding. Such predation often operates with state complicity and includes actions by unscrupulous bureaucrats, political party members, judiciary and other law enforcement agents,[16] large landowners and investors, members of the *nouveau riche* business community and the military. These predations seem to be a regular occurrence in Bangladesh, in and around urban centers as well as in the *char*. Collaboration among these elites emerged in the post-Independence period with the ascendency of neoliberalism – privatization, deregulation, and free trade – under General Ershad's 1982 New Industrialization Policy.[17] The Policy, and those who benefited directly from it, was part and parcel of the rising Bangladesh garment

[15]At least indirectly, exceptions to this assertion do exist, e.g. Lynch and Alcorn (1994) and others who document the squeezing of the bio-physical resource base that enabled the 'historically dominant livelihoods' that depended on it (Miah *et al.* 2010). The same observers tersely state: 'presently, the whole region is on the verge of degradation largely because of frequent natural disasters, human activities and global climate change' (p. 427).

[16]As one newspaper article acknowledged, the DC [District Commissioner who is the highest ranking regional officer] had to be removed for failing to protect an archeological site – which finds its parallel in care about land as well as residents (see *Daily Star* 2011f). In another instance, the Supreme Court summoned a general secretary of the Awami League who sought to build a private hospital on the ground currently home to a temple (*New Age* 2011a).

[17]Collaboration among members of government, political parties and large landowners is not a completely new strategy of accumulation; but during this time it came to enjoy a degree of autonomy whose roots lie in how political party formation and a new policy environment reshaped by earlier forms of inequality.

sector that often played at the margins of legality.[18] The offshore garment boom in Bangladesh transformed export production from agricultural commodities to manufacturing goods and displaced agriculture as a central focus of investment. It also ushered in the end of subsidies for the agricultural sector and the shift from food self-sufficiency to food policies driven by comparative advantage. As key elements in the turn to neoliberal reform, these changes created a new context for state action that restructured state institutions. At this time, too, Ershad joined in an alliance with the Jamaat-i-Islami and gave greater voice to civilian and military elites in national decision-making as a way to secure his authority.

In this new policy environment, a phalanx of elites have been successful in manipulating laws, courts and the more banal activities of the state in securing private property through the control and manipulation of land titles, registries and maps on the expanding fringe of Dhaka and other urban areas. This has been vital in aiding the illegal expropriation of urban and rural lands held by small agricultural producers as owner-operators, contractors, or sharecroppers (Roy 2009). Such control also extends to the appropriation of government properties and embankments (dikes and levies) with no or only limited redress. This has had dire consequences for urban flooding and water quality, as well as for the lives and livelihoods of small farmers as agricultural lands shift to sites for housing and industry.

Corruption and government laxity that bear on urban land and water availability and quality spiral down heavily on the urban poor (Iftekharuzzaman 2005). While it often is difficult to access and compare illegal practices and their security impact on low income groups, international comparisons of corruption by Transparency International are suggestive since they placed Bangladesh 134th out of 178 countries in 2010 (TI 2010, 18). This ignominious status suggests that impoverished rural and urban dwellers, who generally lack voice and access to public resources – a measure of their lack of economic security – are vulnerable to losing control of their land as well as access to government or *khas* land. This suggestion garners support from micro-studies and program assessments that reveal that the judiciary is among the most corrupt among all the sectors, followed by law enforcement agencies and land administration (TIB 2011). Iftekharuzzaman (2005) in his household survey also notes that the poor, those that live on less than $1 a day, are those most likely to pay a bribe for government services. These findings point to failed government resolve in sustaining its commitment to good governance practices or, as we have signaled in the first case, redistributing *khas* land to the poor (e.g. Zaman 1991, Barkat *et al.* 2001, Blair 2003, EJF 2003, Adnan 2011).

Adding to the insecurity of those who depend on the land in the urban shadow is the role of gangs, often hired by elite encroachers, who use violence to appropriate land illegally. According to media reports, some of which are shared below, this includes threat and intimidation, robbery and physical assaults, rape and murder and a welter of legal cases against owners who have heretofore held their land undisturbed. Tellingly, the Government has a relatively poor record of monitoring

[18]The country's need for foreign exchange shaped government support for the garment sector, perhaps one reason why they ignored some of the questionable practices of factory owners. Support for the sector also came from the Multi-fiber Agreement the country enjoyed for preferential treatment of their exports. Together, these practices aided in legitimating the significance of the new emergent class of entrepreneurs.

these violent appropriations, particularly when it involves small-scale producers and non-elites[19]; in some cases, government staff and political appointees may even be the perpetrators of such undertakings. In other cases, such appropriations are either ignored by enforcement officers or entail partnership with land grabbers, including with the large realtors who, since the 1990s, are members of the urban *nouveau riche*.

These land grabs and the displacements they cause in cities and surrounding rural communities have led to mobilization among the poor and are widely discussed in newspapers and on radio and television – we offer a sampling of written media accounts below. Not only do they feature in the English and Bangla press, but, importantly, they affect the availability and price of food and fuel for the society as a whole, and of the stability of social order in the country. In what follows, we turn our attention to peri-urban land grabbing. Here and elsewhere, we surmise land grabbing to be a relatively recent problem, one that jeopardizes sustainable productive agriculture, the security of landholdings and of small landholders on a large scale. What is new in this context is the money that can be made by entrepreneurs who appropriate lands in and around urban centers where land values, rents and predatory behavior seem to be growing exponentially amidst the crush of dramatic migration increases.

One reason for the growth of urban land grabs is the expansion of industrial production that created an enormous demand for new laborers in both Dhaka and Chittagong, the two largest cities in the country. This demand has created new opportunities to extract profits from ground rent as land was transformed from agricultural to industrial holdings (World Bank 2007). Dhaka is now the eleventh largest megacity in the world and is estimated to reach 17.8 million by 2015. Assuming a current population of 12 million, Dhaka has increased fourfold in the last 25 years, with a population density of 27,700 persons per square mile. The land that is being appropriated for industry and housing, however, does not fulfill the needs of the migrant poor who enter the city in increasing numbers and are forced to live in slums. Rather, new housing is for a growing middle and professional class whose numbers have been increasing in and around the perimeters of the city, where housing and business sites are more affordable than in the city center.

As revealed in the press almost daily, those able to take control of this increasingly valuable land include the city's elite as well as members of the bureaucratic state. One recent headline captures the range of practices entailed in these efforts: 'Land price skyrockets as Buriganga bridge connects Keraniganj, prompts ruling party men, local influentials to gobble up *khas* land, canals...' (Roy 2011). The outright arrogation of right here is worthy of quoting at greater length:

> Some ruling Awami League men and influential locals have grabbed a vast swathe of khas land and a canal in Keraniganj near the recently built Shahid Buddhijeebi Bridge on the Buriganga. The bridge... had acted as a trigger for the rampant encroachment by causing the land price there to rise over fivefold in a span of nearly two years. The grabbing of wetlands and the canal – a branch of Aati Khal – is hard to escape one's notice. Then there are hoardings advertising plots on the encroached lands. ... Locals fear inundation of a large area during the monsoon, as most of the wetlands and the

---

[19]Some members of the expatriate community, particularly those from well-off families, also face land grabbers and poachers (Siddiqui 2009), often of their familiy members left behind such as the elderly who are unable to fend for themselves in protecting their landed property and the tenants who may reside on it.

> canal have already been filled up to make way for a sprawl of buildings. ... The grabbers are ... also launching a housing project – Modhu City – on the grabbed part of a branch of Aati Canal. The price of land has significantly shot up since then. One decimal of land was around Tk 1 lakh last year that now sells at Tk 7-8 lakh. Within days of encroachment, the developers have put up signboards there that read, 'Book your plot today under attractive packages'. (Roy 2011)

This news excerpt reveals the potent synergy shaping current urban encroachments on land and waterways. Important hydraulic public goods (canals, river and wetlands) are being privatized by those capable of dredging and filling them for the development of middle-class housing and industrial sites (see Figure 2).[20] These grabs, usually of public or government land, transform the City's waterways and open spaces, alter river flow, increase flooding probabilities and downgrade water quality. A number of newspaper bylines capture these practices and their costs well: 'Mindless sand lifting triggers serious erosion in Habiganj' (*Daily Star* 2011d), 'Conserve lakes, wetlands for Dhaka's survival' (*Daily Star* 2011c), 'Stop earth filling in Meghna' (*Daily Star* 2011a), and 'River land grabbed by agencies, Gatekeeper turning poacher' (*Daily Star* 2011b).

The incidence of these activities, and the government's facilitating role in them, is apparent in a recent editorial (*Daily Star* 2011b) stating that the land appropriations assume '... a comical quality since some of the violators are also part of the Task Force [asked to monitor violations]'. The same article characterizes such behavior as one of 'the most blatant violations of government rules and environmental laws by agencies of the government'. Those accused in the article, including the Bangladesh Army, carry the responsibility of serving the very custodians of public property and entrusted with the protection of Bangladesh's society and the environment. As the *Daily Star* (2011c) reports regarding an Army housing project that breaches numerous domestic laws,

> the Rajuk, the DCC and the Rapid Action Battalion have been filling up portions of Ramchandrapur, Baunia and a channel of the Buriganga. The Baunia filling up is to build an approach road for Rajuk's Purbachal project, while the filling up of Ramchandrapur canal is for building a road in Mohammadpur area. The Buriganga is being filled up to make room for RAB-10 office and residential quarters. And there is also the BIWTA, which is filling up the Buriganga to construct a market at Sadrghat.

The range of water encroachments indicated in this example signals the role of the military and government in the flagrant abuse of their authority and the intimidation that they employ to capture and secure urban and peri-urban water bodies without consideration of their environmental impacts.

The second practice, of land expropriations, occurs in the peri-urban areas of Bangladesh cities and towns, where agricultural land is being seized for lucrative housing developments and industrial sites, and removing it from agricultural production.[21] Such removals have consequences for environmental quality and food security since what replaces them create pollution, health risks and either pave over

---

[20]The cost of undertaking land preparation and control is extremely high so that only those with the resources able to do so can participate in such endeavors. Note too the housing structures that are proximate to the canal suggesting the ways that some buildings may fail to meet building codes.

[21]In some cases, struggles occur when people try to build brick factories on agricultural land creating major ecological problems for residents in the area (*Daily Star* 2011e).

Figure 2. Demolition of an unauthorised structure on the Kalyanpur Canal in Dhaka. *Source*: Rahman and Islam (2011).

or reduce soil fertility over the long term (*Daily Star* 2011e). Importantly, these land grabs often are accompanied by violence, as the following examples suggest. In the first, Knit Concern, a garment manufacturer whose structure displaced local residents from their holdings, responded to protests over their claim to land with an orchestrated gang of hoodlums. The violence that ensued was not atypical: 'a gang of 100 to 110 goons led by Jahangir, director of Knit Concern, and armed with lethal and sharp weapons went to Southern Sastapur to grab a piece of land... and faced villagers' protests' (*New Age* 2011b). In this incident, 'at least 25 people, including women, were wounded by the hired gang of an extortionist company in Southern Sastapur area of Fatullah thana of Narayangaj ... after they protested against their land being grabbed. The gang ... also vandalised 25 houses ... and two motorbikes' (*Daily Star* 2011g).

A second instance is that of Bangladesh Garment Manufacturers and Export Association (BGMEA), an influential institution currently under investigation regarding the legitimacy of the title they claim for their building site. At issue is whether and who had approved the construction project (*New Age* 27 April 2011). What is more, many suggest that there may have been collaboration with the Export Promotion Board, the government oversight institution charged with reviewing and monitoring such projects. The example also reveals the ambiguous role of government institutions in processes of land grabbing, especially when conflicts of interest emerge or when influential private sector constituencies require adjudication. Since members of the BGMEA are viewed by the urban middle classes and state bureaucrats as icons of the *nouveau riche,* and central to securing export earnings for Bangladesh, they seem to be untouchable despite operating on the margins of legality.[22]

[22]Notwithstanding often violent struggles and strikes over wages, rights, and worker safety, the BGMEA has been able to keep wages below the national minimum and to secure for Bangladesh the status as the country with the lowest wages in the world, despite thirty years of policy reform intended to improve living standards and economic growth.

It also is the case that government officials have been directly charged with appropriating both *khas* land for private use and privately held land for elite private development.[23] For the argument at hand, what is noteworthy is that the BGMEA completed their building project and, only subsequently, were they charged with the structure's lack of compliance and questioned about their right to the land. Once again, this is not unusual in Dhaka, where urban planning rules are regularly overridden by the interests of builder-developers who encroach on public or private space, whether by building beyond the boundaries of their legal holding; usurping public parks, roadways and embankments; accessing land from private owners; or moving onto lands recently cleared of slums and slum dwellers by the government (World Bank 2007). Rarely are these elite or government bureaucrats held accountable for doing so; as the latter, through either collaboration or turning a blind-eye to breaking the law, fail in their responsibility as overseers of public resources.

Beside media attention, there is significant public reaction to these encroachments. Both in the capital and in the provincial towns, people do not readily acquiesce to the grabbing of public and private land, especially when it is government institutions that are the expropriators. Rajshahi, a city in the north of Bangladesh, was witness to more than 200 farmers protesting against the Rajshahi City Corporation's acquisition of 136 acres of their farmland for a housing project. This is not the first time such appropriations have occurred, as the state previously removed agricultural land from production to build Rajshahi University (*New Age* 2011b, 2011c). What is telling is that opposition is growing in the countryside and peri-urban communities as people mobilize across the country against such practices. And, despite the fact that the High Court of Bangladesh continues to acknowledge the rights of farmers, government institutions have been slow to respond to farmer demands and to the contradictions revealed by policy rhetoric and ongoing practices.

A final example of government land acquisitions concerns the decision to acquire about 5,000 *bighas* or approximately 1,653 acres of land in Rupganj Upazilla, an area near the capital, for an army housing project, part of the acquisition of embankments, lakes and rivers noted in the lengthy quote above. The proposed site includes a designated flood flow zone (65 percent), water bodies (12 percent), and rural settlements (13 percent), where construction or earth filling is not permitted by the Dhaka Metropolitan Development Plan (Ali 2010). Public protests in the community began following claims that the army had set up camps over a period of six months to force farmers to sell off their land below market value. Ongoing confrontations turned 'deadly as more than 50 people were injured and one person died (three more reportedly missing) from clashes with the security forces' (Rezwan 2010). One blogger writing of the incident claimed: 'Those who have the faintest idea of buying and selling land knows [sic], one can never purchase all that land together without force. Not everyone will want to sell their ancestral house, family graveyard, and their means of livelihood' (Rezwan 2010).

[23]Notwithstanding often violent struggles and strikes over wages, rights, and worker safety, the BGMEA has been able to keep wages below the national minimum and to secure for Bangladesh the status as the country with the lowest wages in the world, despite thirty years of policy reform intended to improve living standards and economic growth.

Once again, there appears to be sinister collaborations between state authorities, security forces, private militia and the military itself. According to a BBC report, this network of collaboration can best be described as a conglomerate. Against these interests, small producers do not enjoy rights or the economic and social security of citizenship, nor are they protected by the property entitlements of the Constitution. Combined, these powerful interests trump environmental rules and other government regulations, changing to the landscape and unsettling agricultural production in the areas surrounding large cities in ways that alter the availability and proximate access of rice, fruits and vegetables for urban dwellers. That is, large numbers of people are deprived of familiar goods and services and the employment that comes with them. Some may remain functionally 'displaced in place', while many of their neighbors, when forced off their land, may be forced into the urban slum. These land use shifts have significant social and environmental repercussions, given the paucity of opportunities for secure alternative livelihoods. Importantly as well, legal and illegal encroachment policies remove arable land from production, alter the conditions under which the poor secure their own reproduction and heighten dependency on imports and foreign investment to meet the country's food needs. These consequences further instantiate a neoliberal agenda, whereby subsistence production is impaired if not destroyed, and people are forced to either secure wage or self-employment bereft of the resources to enable them to do so.

Conclusion

In this brief paper we have sought to open for scrutiny the myriad ways in which land grabbing unfolds in two areas of Bangladesh, in the *chars* and in peri-urban and urban centers. Our emphasis has been on the human displacement effects of encroachments that we differentiate into two dominant forms, *in situ* and *ex situ*. Using this displacement analytic reveals a much longer history to land grabbing, in Bangladesh as elsewhere, than current media and scholarly attention often supposes. We draw attention to routine domestic, rather than foreign, land grabs, the latter often being limited to episodic moments characterized by large-scale foreign investments. Domestic land grabs, in contrast, reveal how mass movements out of agriculture, spiraling land values, and bare-fisted predation are part of the encroachment process, and how perpetrators from the public and private sector often collaborate, blurring the distinction between legal and illegal takings. We argue here that banal domestic land grabbing and related displacement must enter conversations about global food security, the new green revolution, and arguments for increased economies-of-scale in agriculture to address global food insecurity, relationships we unfortunately have no space to fully attend to here.

Crucially, the displacement distinction we have drawn here is not an exercise in hair splitting. Once applied, a focus on *in situ* displacement reveals the growing magnitude of dispossessed peoples that in fact now exist, and how spatial referencing and human movement e.g. eviction and coerced relocation, are neither the only nor necessarily the best lens for exposing land grabbing and displacement processes, relations, and effects. The *in situ* perspective, moreover, suggests that depictions of adaptation and coping among displaced persons may simplify and understate a larger and longer unsettling of people, their communities, and the conditions of social reproduction. Recognizing the salience and subtlety of *in situ* displacement also casts a new light on the role of political, economic, and military elites who deploy violent rent-seeking behaviors that are fully capable of ruining lives in ways easily missed in framings of conventional, land grabbing and *ex situ* displacement.

There are similarities and differences we might highlight in the displacement processes at work in the two Bangladesh cases. In the interest of sharpening the *in situ/ex situ* distinction, we remind readers that these dual processes can overlap and complicate each other. For example, in both cases, we witness encroachments that can be 'explained away' by impersonal, complex forces and externalities, or by individual choices, as opposed to the predatory behaviors of elites driven by the lure of windfall profits and extraordinary forms of accumulation. Often, and even when such practices are evident, researchers identify natural disasters and incremental climate change as chiefly responsible for displacement in the *char*. In urban areas, the parallel explanation rests on population growth, growing urban densities and gentrification that drive land markets skyward and 'naturally' displace people.

Conveniently, displacement in both narratives is reduced to the collateral damage of development, with little attention to how influential others collaborate, coerce and benefit from land expropriation imposed on vulnerable smallholders. In both cases as well, the government has advanced plans to control *khas* and *char* land development, though these often serve as cover for the venal behaviors and entitlements of collaborating elites. Another cover narrative in both landscapes is that of functioning land 'markets,' which helps to obscure the violence and coercion that oblige farmers to relinquish their land as prices rise. In contrast, both cases situate displacement in a larger context of economic boom, a shrinking reliance on agriculture for food and employment, and severe downward pressure on the social wage that makes Bangladesh so attractive to global investors.

Important differences surface in these cases as well. In the *char*, central government presence is lax, despite government ownership of *char*-turned-*khas* lands ostensibly available for redistribution to the poor. An open-access situation results, inviting local intervention, rivalries, and the routine use of force by both public and private semi-militarized gangs. Another difference in the examples is the exorbitant land rents in and around Dhaka, which ignites brutal instances of encroachment and displacement. It is tempting to draw distinctions between the actions undertaken by the military and other security forces in and around urban and non-urban areas, but this may be an artifact of differential media coverage in the two zones. More certain are divergent uses of seized land in urban and remote coastal areas where conditions of reproduction are markedly dissimilar. As we have stressed throughout, *in situ* displacement, as exemplified in the *chars*, reveals a more languid if no less deadly form of displacement. Its progression, through a series of stages, challenges simple notions of coping, human adaption and resilience – understandings that ignore the violence and coercion experienced by even those who negotiate ways of remaining in the places they call home.

The multiple articulations of land grabbing – historically specific and contingent – thus force us, and we hope others, to examine displacement and its relationship to land grabbing as a large-scale, long-term process, rather than only as cataclysmic events that decisively expel people. The phenomenon we report is a daily occurrence and widespread in Bangladesh, yet it is beset by illegibility despite the lived experience of its victims that is severe, extensive, and rooted in a political ecology only recently submitting to reinterpretation.

References

Adnan, S. 2011. Resistance to accumulation by dispossession in the context of neoliberal capitalism and globalization: Struggles for defending and gaining land rights by the poor peasantry in the Noakhali Chars of Bangladesh. Paper presented at the International Conference on Global Land Grabbing, Institute of Development Studies, IDS, Brighton, UK, 6–8 April.

Adnan, S. *et al.* 1991. *Floods, people, and the environment: Institutional aspects of flood protection programmes in Bangladesh, 1990.* Dhaka, Bangladesh: Research & Advisory Services.

Agrawala, S., T. Ota, A.U. Ahmed, J. Smith and M. van Aalst. 2003. *Development and climate change in Bangladesh: Focus on coastal flooding and the Sundarbans.* Paris: OECD.

Akter, T. 2009. *Climate change and flow of environmental displacement in Bangladesh, Unnayan Onneshan – The innovators.* Dhaka: Center for Research and Action on Development.

Ali, A.S. 2003. SOS-arsenic.net [Accessed 13 August 2011].

Ali, S.M. 1980. Administration of Char Lands in Bangladesh. *Asian Affairs*, 2, 295–303.

Ali, T. 2010. Army Housing Scheme in Rupganj, Project area includes flood flow zone, *Daily Star*, 29 October.

Arendt, H. 1973. *The origins of totalitarianism.* New York: Harcourt, Brace and Co.

Barkat, A., G.M. Suhrawardy and P.S. Ghosh. 2011. *Commercialization of agricultural land and waterbodies and disempowerment of the poor in Bangladesh: An exploratory study.* Dhaka: Association for Land Reform and Development, 16 January.

Barkat, A., S. Zaman and S. Raihan. 2001. *Political economy of khas land in Bangladesh.* Dhaka: Association for Land Reform and Development.

Blaikie, P.M. and Z. Sadeque. 2000. *Policy in high places: Environment and development in the Himalayan region.* Kathmandu, Nepal: International Center for Integrated Mountain Development.

Blair, H. 2003. Civil society, dispute resolution and local governance in Bangladesh: Ideas for pro-poor programme support, Report prepared for the Department for International Development U.K. High Commission, Dhaka, Bangladesh, 22 February.

Borras, S.M. Jr., R. Hall, I. Scoones, B. White and W. Wolford. 2011. Towards a better understanding of global land grabbing: an editorial introduction. *Journal of Peasant Studies*, 38(2), 209–16.

Borras, S. Jr. and J. Franco. 2010. Towards a broader view of the politics of global land grab: rethinking land issues, reframing resistance. ICAS Working Paper. Series Number 001. Initiatives in Critical Agrarian Studies, May.

Brookings Institution. 2005. *Addressing internal displacement: a framework for national responsibility.* Washington, D.C. and Bern, Switzerland: Brookings Institution.

Buerk, R. 2004. Flooded future looms for Bangladesh, *BBC News in Bangladesh.* Available from: http://news.bbc.co.uk/2/hi/science/nature/4056755.stm [Accessed 25 February 2011].

Cernea, M.M. 2003. For a new economics of resettlement. *International Social Science Journal*, 55(175), 37–45.

Cernea, M.M. and C. McDowell, eds. 2000. *Risks and reconstruction: Experiences of resettlers and refugees.* Washington, DC: The World Bank.

Chars Livelihood Program. 2004. The voice-responsiveness framework: creating political space for the extreme poor. Chars Organisational Learning Paper 3, University of Manchester.

Center for Policy Dialogue (CPD). 2003. Strengthening the role of private sector housing in Bangladesh economy: The policy challenges. CPD Dialogue Report No. 64. Available from: http://www.cpd-bangladesh.org/publications/dr/DR-64.pdf [Accessed 11 October 2011].

Cohen, R. and F.M. Deng. 1998. Introduction. *In: Masses in flight: the global crisis of internal displacement.* Washington, DC: Brookings Institution Press, pp. 1–14.

Custers, P. 1993. Bangladesh's flood action plan: a critique. *Economic and Political Weekly*, 28, 29/30 (July 17–24), 1501–1503.

Daily Star. 2009. Bangladesh needs $4.2b to save coastal people, *Daily Star*, December 12.

Daily Star. 2011a. Stop earth filling in Megna, *Daily Star*, 15 August.

Daily Star. 2011b. River land grabbed by agencies, Gatekeeper turning poacher, *Daily Star*, 9 August.

Daily Star. 2011c. Rajuk, DCC among grabbers, *Daily Star*, 8 August.

Daily Star. 2011d. Mindless sand lifting triggers serious erosion in Habiganj, *Daily Star*, 7 August.

Daily Star. 2011e. Brick fields on crop land protested, *Daily Star*, 2 August.

Daily Star. 2011f. Mahasthangarh HC asks govt to ensure preservation Bogra, DC removed from site protection committee, *Daily Star*, 3 March.

Daily Star. 2011g. Land grabbers injure 25 in N'ganj, *Daily Star*, 27 January.

De Schutter, O. 2011. How not to think of land-grabbing: three critiques of large-scale investments in farmland. *Journal of Peasant Studies*, 38(2), 249–279.

Doutriaux, S, C. Geisler and G. Shively. 2008. Competing for coffee space: Development-induced displacement in the central highlands of Vietnam. *Rural Sociology*, 73(4), 528–554.

Environmental Justice Foundation (EJF). 2003. *Smash & grab: Conflict, corruption and human rights abuses in the shrimp farming industry*. London, UK: Justice Foundation.

Fagan, A. 2010. *The atlas of human rights*. London: Earthscan.

Feldman, S., C. Geisler and L. Silberling. 2003. Moving targets: displacement, impoverishment, and development. *International Social Science Journal*, 175, 7–13.

Feldman, S., G. Menon and C. Geisler, eds. 2011. Introduction: a new politics of containment. *In*: S. Feldman, C. Geisler and G. Menon, eds. *Accumulating insecurity: violence and dispossession in the making of everyday life*. Athens: University of Georgia Press, pp. 1–23.

Germanwatch. 2004. *Sea level rise in Bangladesh and the Netherlands: One phenomenon, many consequences*. Berlin: Germanwatch.

Global Database: Guiding principles on internal displacement. 2007. Internal displacement: frequently asked questions. Available from: http://www.law.georgetown.edu/idp/english/id_faq.html [Accessed 16 January 2011].

Goodwin-Gill, G. 2008. Convention relating to the Status of Refugees and Protocol relating to the Status of Refugees: Introduction. Audiovisual Library of International Law. Available from: http://untreaty.un.org/cod/avl/ha/prsr/prsr.html [Accessed 29 August 2011].

GOB (Government of Bangladesh). 2008. Agricultural Census. Bureau of Statistics. Available from: www.bbs.gov.bd/dataindex/Pre-report-Agri-census-2008-Final.pdf [Accessed 10 September 2011].

GOB (Government of Bangladesh). 2010. Department of Forests, Ministry of Environment and Forests. Available from: http://www.bforest.gov.bd/land.php [Accessed 10 September 2011].

Haque, C.E. 1988. Human adjustments to river bank erosion hazard in the Jamuna floodplain, Bangladesh. *Human Ecology*, 16(4), 421–437.

Haque, C.E. and M.Q. Zaman. 1989. Coping with riverbank erosion hazard and displacement in Bangladesh: Survival strategies and adjustments. *Disasters*, 13(4), 300–314.

Hutton, D. and E.E. Haque. 2003. Patterns of coping and adaption among erosion-induced displacees in Bangladesh: implications for hazard analysis and mitigation. *Natural Hazards*, 29, 405–421.

Hutton, D. and E.E. Haque. 2004. Human vulnerability, dislocation and resettlement: adaptation processes of river-bank erosion-induced displacees in Bangladesh. *Disorder*, 28(1), 41–62.

Iftekhar, M.S. and T. Takama. 2008. Perceptions of biodiversity, environmental services, and conservation of planted mangroves: A case study on Nijhum Dwip Island, Bangladesh. *Wetlands Ecology Management*, 16, 119–137.

Iftekharuzzaman. 2005. Corruption and human insecurity in Bangladesh, Paper presented at the seminar organized by Transparency International Bangladesh to mark International Anti-corruption Day, 9 December. Accessed at http://unpan1.un.org.

Islam, M.S. 2009. In search of 'white gold': Environmental and agrarian changes in rural Bangladesh. *Society and Natural Resources*, 22(1), 66–78.

Institute of Water Monitoring (IWM), 2009. *Institute of Water Modeling Report*. Bangladesh: IWM.

International Monetary Fund. 2005. Bangladesh: Poverty Reduction Strategy Paper. IMF Country Report, 05/410. Available from: http://www.imf.org/external/pubs/ft/scr/2005/cr05410.pdf [Accessed 10 September 2011].

Kamal, A., M. Kamaluddin and M. Ullah. 1999. *Land policies, land management and land degradation in the Hindu Kush Himalaya: Bangladesh Study Report*. Kathmandu, Nepal: International Centre for Integrated Mountain Development.

Leckie, S. 2010. *Climate change, displace persons, and housing, land and property rights: Preliminary strategies for rights-based planning and programming to resolve climate-induced displacement*. Geneva: Displacement Solutions.
Lein, H. 2008. The poorest and most vulnerable? On hazards, livelihoods and labeling of riverine communities in Bangladesh. *Singapore Journal of Tropical Geography*, 30, 98–113.
Li, T.M. 2011. Centering labor in the land grab debate, *Journal of Peasant Studies*, 38(2), 281–298.
Lynch, O.J. and J.B. Alcorn. 1994. Tenurial rights and community-based conservation, D. Western and R.M. Wright, eds. *Natural connections: perspectives in community-based conservation*. Washington, DC: Island Press, pp. 373–392.
Martin, A.J. 2010. After foreclosure: The displacement crisis and the social and spatial reproduction of inequality. ISSC Working Paper Series, 2009–10, No. 48. Berkeley: Department of City and Regional Planning.
Miah, G, M. Bari and A. Rahman. 2010. Resource degradation and livelihood in the coastal region of Bangladesh. *Frontier Earth Science China*, 4(4), 427–437.
Mongabay. 2008. List of countries by land mass. Last updated 8 December 2008. Available from: http://www.mongabay.com/igapo/world_statistics_by_area.htm [Accessed 20 June 2011].
Mukherji, I.N. 1976. Agrarian reforms in Bangladesh, *Asian Survey*, 16(5), 452–464.
New Age. 2011a. SC summons Jalil to explain alleged temple land grab, 11 August.
New Age. 2011b. Land grabbers injure 25 in N'ganj, 22 March.
New Age. 2011c. Protesting against a move of Rajshahi City Corporation to acquire 136 acres of their farmland for a housing project, 1 March.
New Age. 2011d. Landless people pass sleepless night fearing eviction, 27 February.
New Age. 2011e. Landless people pass sleepless night fearing eviction, 27 February.
Nishat, A. 2008. Climate change and water management in Bangladesh, Paper presented at the International Union for Conservation of Nature, Dhaka, Bangladesh, August 25.
Pearce, F. 1991. The rivers that won't be tamed, *New Scientist*, 13 April.
Peluso, N. and C. Lund. 2011. Special issue on New Frontiers of Land Control. *Journal of Peasant Studies*, 38(4), 667–81.
Plant, R. 1994. *Land rights and minorities*. London: Minority Rights Group.
Population and Sustainability Network. 2011. Climate change imperils global prosperity, UN warns, 2 November. *The New York Times Green Blog*, Available from: http://www.populationandsustainability.org/2014/news/climate-change-imperils-global-prosperity-un-warns.html [Accessed 8 January 2012].
Porter, K. 2009. Fragile states world's weakest nations pose great threats, *The Courier*, 14 September. Available from: http://www.stanleyfoundation.org/articles.cfm?id=593 [Accessed 10 September 2011].
Rahman, M.M. and S. Islam. 2011. Land grabbing: who will bell the cat? *New Age*, 14 March.
Rezwan. 2010. Land grab and protests in Bangladesh, *e*-Bangladesh, Available from: http://www/ebangladesh.com/2010/10/30/land-grab-and-protests-in-bangladesh [Accessed 30 October 2011].
Robertson, B. and P.P. Andersen. 2010. Global land acquisition: neo-colonialism or development opportunity? *Food Security*, 2, 271–283.
Roy, P. 2011. Grabbers arrive with the bridge Land price skyrockets as Buriganga bridge connects Keraniganj, prompts ruling party men, local influentials to gobble up *khas* land, canal. *Daily Star*, February 24.
Roy, P. 2009. Ajuk fills up Shitalakhya bank for housing scheme. *Daily Star*, September 17. Available from: http://www.thedailystar.net/newDesign/print_news.php?nid=106172 [Accessed 22 January 2011].
SAMATA. 2006. *Land conflict in Bangladesh: A resource based on news clippings published in national dailies in 2004*. Dhaka: National Policy Advocacy Cell, SAMATA.
Sarker, M.H., I. Huque, M. Alam and R. Koudstaal. 2003. Rivers, chars and char dwellers of Bangladesh. *International Journal of River Basin Management*, 1(1), 61–80.
Sarwar, M.G.M. 2005. Impacts of sea level rise on the coastal zone of Bangladesh. MA Thesis, Sweden, Lund University, International Masters Program in Environmental Science.
Shafi, S.A. and G. Payne. 2007. Local partnerships for urban poverty alleviation: land tenure security and land administration in Bangladesh. Final Report to LGED, UNDP and UN-Habitat, Project BGD/98/006, Dhaka, June.

Siddiqui, H. 2009. Land-grabbing and its implication on the expatriate community, Bangladesh Expatriate Council, 12 April.

Stavropoulou, M. 1998. Displacement and human rights: Reflections on UN practice. *Human Rights Quarterly*, 20(3), 515–554.

Titumir, R.A.M. 2007. Politics of confrontation, accumulation. *New Age*, Dhaka, 4th Anniversary Special. Available from: http://www.newagebd.com/store/anni07/10.html [Accessed 24 January 2011].

Transparency International. 2011. Global Corruption Report: climate change. Berlin. Available from: http://www.transparency.org/publications/gcr/gcr_climate_change2 [Accessed 5 September 2011].

Uddin, A.M.F. and J.T. Haque. 2009. Agrarian transition and livelihoods of the rural poor: The agricultural land market. Publication of Unnayan-Onneshan, The Innovators. Available from: http://unnayan.org/reports/ agri/Landmarket.pdf [Accessed 10 September 2011].

United Nations-Habitat. 2007. Enhancing urban safety and security: Global Report on Human Settlements. Available from: http://www.unhabitat.org/downloads/docs/GRHS.2007.0.pdf [Accessed 10 September 2011].

United Nations Environment Programme, Dhaka City State of the Environment. 2005. Available from: http://www.rrcap.unep.org/ pub/soe/dhakasoe05.cfm [Accessed 10 September 2011].

USAID. 2010. Country profile for Bangladesh: Property rights and resource governance. U.S. Agency for International Development. Available from: http://usaidlandtenure.net/usaidltprproducts/country-profiles/bangladesh/bangladesh-country-profile [Accessed 10 September 2011].

Vandergeest, P.P. Idahosa, and P.S. Bose, eds. 2007. *Development's displacements: ecologies, economies, and cultures at risk*. Vancouver, Canada: University of British Columbia Press.

Wang, S., J. Modvig and E. Montgomery. 2009. Household exposure to violence and human rights violations in Western Bangladesh (I): prevalence, risk factors and consequence. *BMC International Health and Human Rights*, 9(29), 1–18.

Wisner, B., P. Blaikie, T. Cannon and I. Davis. 2004. *At risk: natural hazards, people's vulnerability and disasters*. 2nd edn. London: Routledge.

Woodhouse, P. 2003. African enclosures: a default mode of development. *World Development*, 31(10), 1705–1720.

World Bank. 2006. *Bangladesh: Country environmental analysis, Vol. II. Bangladesh Development Series Paper No. 12*. Washington, DC: The World Bank.

World Bank. 2007. *Dhaka: improving the living conditions for the urban poor. Bangladesh Development Series Paper 17 June*. Dhaka: World Bank.

World Bank. 2010. Rising global interest in farmland: can it yield sustainable and equitable benefits? Washington, DC: The World Bank, Available from: http://siteresources.worldbank.org/INTARD/Resources/ESW_Sept7_final_final.pdf [Accessed 10 September 2011].

Yogeshsaxsena. 2010. Bangladesh's enemy property or vested property law is a black law, instrumental only in the persecution of the minority community. Available from: http://cplash.com/post/Bangladeshs-Enemy-Property-or-Vested-Property-Act-is-a-black-law,-instrumental-only-in-the-persecution-of-the-minority-community299.html [Accessed 5 September 2011].

Zaman, M.Q. 1991. The displaced poor and resettlement policies in Bangladesh. *Disasters*, 15(2), 117–125.

Shelley Feldman is International Professor of Development Sociology and Director of Feminist, Gender, and Sexuality Studies at Cornell University, and Visiting Professor of Sociology at Binghamton University. Her research explores agricultural and agrarian change, gender relations, state formation and questions of social reproduction and precarity. Recent publications include, with Charles Geisler and Gayatri Menon, Accumulating insecurity, violence and dispossession in the making of everyday life (Athens: University of Georgia Press, 2011), and with Stephen Biggs, International shifts in agricultural debates and practice: an historical view of analyses of global agriculture (In Issues in Agroecology: Present Status and Future Prospectus, Volume 2: Integrating Agriculture, Conservation, and Ecotourism: Societal Influences, Bruce Campbell, editor. Springer Publishers, forthcoming 2012), and The politics of international assessments: the IAASTD process, reception, and significance (Journal of Agrarian Change, 12, 1: 144–169, 2012). Much of her empirical research focuses on Bangladesh. Email: rf12@cornell.edu

Charles Geisler is a Professor of Development Sociology at Cornell University and focuses on displacement and dispossession, particularly under conditions of homeland insecurity in the US and abroad. He is currently engaged in research on changing police powers and appropriations in states of emergency. He co-directs the Cornell-based New Enclosures Research Working Group and recently co-edited *Accumulating insecurity, securing accumulation: violence and dispossession in the making of everyday life* with Shelley Feldman and Gayatri Menon. (Athens: University of Georgia Press, 2011).

Gendered experiences of dispossession: oil palm expansion in a Dayak Hibun community in West Kalimantan

Julia, with Ben White

This article explores the gendered experience of monocrop oil-palm expansion in a Hibun Dayak community in Sanggau District, West Kalimantan (Indonesia). It shows how the expanding corporate plantation and contract farming system has undermined the position and livelihood of indigenous women in this already patriarchal community. The shifting of land tenure from the community to the state and the practice of the 'family head' system of smallholder plot registration has eroded women's rights to land, and women are becoming a class of plantation labour. At the same time, as in other cases of expansion of agrarian corporate commodity production, we can discern a familiar pattern of ambivalence between, on the one hand, the attractions of regular cash income and, on the other, the loss of resource tenure and autonomy, which helps to explain the community's gendered experience of coercion, exploitation, intimidation, consent and resistance.

Indonesia is the world's biggest palm-oil producer. Oil palm plantations are found mainly in the four large islands of Sumatra, Sulawesi, Kalimantan and Papua, with further expansion planned mainly for the latter two islands. Meanwhile, the provincial government of West Kalimantan has recently raised its target for expansion of oil palm plantation from 1.5 million to 4.5 million hectares (ha) by 2025. *Ijin Usaha Perkebunan* (Plantation Business Permits) have been issued for approximately 3.6 million ha, although so far, the area actually cleared and planted with oil palm is only 0.75 million ha,[1] increasing rapidly from 0.4 million ha in 2008.

Aside from national companies, oil palm expansion in West Kalimantan involves foreign investors from Malaysia (the majority), Singapore, and the United States. Another crucial stakeholder is the creditors. Besides multilateral financial bodies

Earlier versions of this article were presented at the ICAS-JPS Workshop on Biofuels, Land and Agrarian Change, Halifax, October 2009, and the LDPI-JPS-FAC International Conference on Global Land Grabbing, at IDS, University of Sussex, April 2011. The authors are grateful to discussants at both sessions, and the two anonymous referees of this journal, for their helpful feedback and suggestions.

[1]Plantation Agriculture Office of West Kalimantan, http://disbun-kalbar.go.id/Disbun/index.php/statistik/menu-rekapitulasi-komoditi-perkebunan/152-komoditi-perkebunan-kalbar-2010 [accessed 25 Dec 2011]. See also '*Sawit Kalbar Makin Meluas*' (West Kalimantan oil palm expanding), *Harian Equator*, 12 April 2010.

such as the World Bank and Asian Development Bank, oil palm corporations also obtain credit from Indonesian national banks, as well as private international banks such as Rabobank, Citicorp International Ltd., Citibank, Shanghai Banking Corporation, Union Bank of Switzerland, Sumitomo Bank Ltd., Bank of Taiwan, Indosuez Bank-France, ABN-Amro Bank N.V, Japan Asia Investment, the Hong Kong Shanghai Banking Corporation, and Bank of Tokyo-Mitsubishi Ltd. (Prasetyantoko and Setiawan 2011).

Access to land, a fundamental issue for many indigenous and other forest dependent peoples, is a key issue in the debate on oil palm expansion. The complex regulations for the management of Indonesia's natural resources are all formally derived from one ultimate source, Article 33 of the 1945 Constitution, which states: 'Land, water and all natural resources that belong to common pools and public goods are under state control and will be utilized for the maximum welfare of the people' and thus formally fails to acknowledge customary rights. Article 33, and related forestry legislation passed in the Suharto period,[2] are the main reasons why some 70 percent of Indonesia's enormous territory – including virtually all land used or claimed under customary tenure arrangements by indigenous peoples – continues to be classified as 'state-owned forest' under the jurisdiction of the Ministry of Forestry. Local communities' rights to this land are not clear under Indonesian law, and are therefore structurally insecure (Marti 2008). Stephanus Djuweng has argued that the most obvious aspect of the conflict-ridden 'dialogue'[3] between indigenous peoples and the Indonesian state is that 'the state does not recognize the existence of indigenous peoples, whilst the indigenous peoples do not recognize the state claims to their ancestral lands' (Djuweng 1997, 3), as encapsulated by this Dayak leader:

> The government official asked me if I have a land ownership certificate and I answered that every single durian tree, every single *tengkawang* tree, and every single rubber tree that we or our ancestors have planted are certificates. I am an indigenous person born here. My ancestors have already defended this land for generations (Marti 2008, viii).

The establishment of oil plantations on land formerly held and cultivated under customary tenure brings a swift and radical transformation both to the environment and to the social, economic and political structure of local communities. Studies of these forms of plantation organization in Indonesia have shown that there is a large gap between what is promised to local people and their actual experience, which has often led to the deterioration of local livelihoods and triggered local conflicts (Colchester *et al.* 2006, Sirait 2009). Previous research has shown that incorporation into oil palm production may generate both wealth and poverty among local people, depending on the terms of incorporation: 'while palm oil appears to be a way to become rich, the way it is taken up actually reworks social relations and landownership in the rural domain in ways that may ultimately work against the poor' (McCarthy 2010, 847).

Behrman, Meinzen-Dick and Quisumbing have recently pointed out the dearth of information and analysis on the gender effects of large-scale corporate land deals. Gender is critical to understanding the impact of land deals because 'men and

[2]Law 5/1967 on Basic Provisions on Forestry, Law 41/1999 on Forestry.
[3]Writing in the last years of the Suharto regime, Djuweng uses the term 'dialogue' as euphemism for 'conflict'.

women have different social roles, rights, and opportunities and will be differentially affected by any major change in tenurial regimes, especially land transfers to extralocal investors' (Behrman *et al.* 2012, 51). In-depth studies are required to better understand how the expansion of industrial monoculture plantations has affected gender roles and relations in the emerging agrarian structures, in particular in access and control of land and other natural resources, the shifting gender division of labour and control over incomes, and strategies of survival and resistance in the face of corporate and patriarchal pressures.

In this paper we explore the gendered experience of corporate land acquisition and oil palm expansion in a Hibun Dayak community in Sanggau District, West Kalimantan.[4] The study involved a short period of mainly qualitative field research based on interviews with local informants.[5] We aim to show how the formalized procedures and relationships established with the expansion of corporate oil-palm ventures interact with existing local patriarchal structures to generate changes in gendered patterns of land rights, division of labour, livelihoods, voice in community affairs and resistance to some of these developments.

Our study is guided by the basic ideas of feminist political ecology. Political ecology, 'a field that seeks to unravel the political forces at work in environmental access, management and transformation' (Robbins 2004, xvi) applies the basic political economy questions of who has what, who does what, who gets what, and what do they do with it (Bernstein 2010) to the production of environmental change, inspired by a notion of environmental justice. While political ecologists focus largely on the uneven distribution of access to and control over resources on the basis of class and ethnicity, feminist political ecology treats gender as a critical variable in political ecology. Understanding gender as referring to both the social meanings given to male and female, and the socially constructed form of relations between men and women (Young 2002, 323) – 'the ways in which the social categories of men and women, male and female, relate over the whole range of social organization' (Pearson 2000, 385) – feminist political ecology focuses particularly on the interaction of gender with class, caste, race, culture, and ethnicity in shaping processes of resource access and control and ecological change, and the struggles of men and women to sustain ecologically viable livelihoods (Rocheleau *et al.* 1996). It draws particular attention to the nature of gendered knowledge, gendered resource access and control, and the engagement between local struggles and more global issues, while avoiding the seductions of 'earth mother myths' (Leach 2007).

Following a brief introduction to the study location, the next section of the paper will first outline the process of land expropriation and dispossession. We then turn to issues specifically relating to the experience of women in the establishment and operation of the oil palm plantation, focusing in turn on: the loss of women's tenure rights, the new gendered division of labour and feminisation of agricultural work, the loss of women's other sources of income, their new involvement in (illegal) scavenging of fallen palm kernels, and women's responses and resistance.

---

[4]One of the illustrative cases summarized in Behrman *et al.* (2012, 68–70) is drawn from an earlier version of this paper (Julia and White 2010).

[5]Fieldwork was carried out for a total of one month during July–September 2008 by Julia, in the context of the Walhi-ISS Short Course in Political Ecology (May–September 2008) coordinated by Ben White. More recent information based on further contacts with informants during 2010/11 has been incorporated in this version of the paper.

Sanggau District and Anbera Hamlet

Our study is located in Anbera Hamlet of Sanggau District, a sparsely populated region of West Kalimantan covering some 13,000 km^2. The first large-scale oil palm plantation in Sanggau District was established in the 1980s with an area of 14,000 ha, operated by the state-owned company PTP VII (now PTPN XIII). Since then, oil palm plantations have expanded until by 2008 there were as many as 20 oil palm companies operating in all of Sanggau's 15 sub-districts. Four of these corporations are based on foreign investment, 15 on domestic private investment, and one (PTPN XIII) is state-owned. Oil palm now dominates the export crop sector in Sanggau District with 145,477 ha (about 11 percent of Sanggau's total land area), having now exceeded the area under rubber (99,059 ha).[6]

The UNDP Human Development Index places West Kalimantan province 28th in rank among 33 Indonesian provinces. The 2007 Human Development Index of Sanggau District shows marked gender disparities in educational attainment at levels beyond primary school. While the proportions of the population aged 10 years and above who have attended or completed primary school are roughly equal by gender, there are about 30 percent more male junior secondary and 40 percent more male upper secondary graduates than female, while graduates of tertiary education are minimal (at less than one percent) for both genders.[7] The gender gap thus emerges at secondary level. This in turn has other consequences, such as early marriage among women; presently, according to informants, the marriage age of female community members ranges from 13 to 20 years.

Anbera Hamlet *(Dusun)*, a community of some 240 households, is part of the larger village agglomeration *(Desa)* called Dabat Village. Since the 1970s, when the trans-national highway to the Malaysian border was built, more and more people have moved to the area along both sides of the highway and formed hamlets.[8] Anbera Hamlet is therefore ethnically mixed but the Hibun Dayak peoples form the largest ethnic group, this being part of the customary territory of the indigenous Hibun Dayak peoples. The revitalized Hibun customary structure of Dabat Village falls under a bigger structure of Hibun customary organization called *Dewan Adat Hibun* (Hibun Customary Law Council) that consists of 91 member villages.[9]

One third of the village's total area is now occupied by oil palm plantations, with only seven percent remaining under smallholder mixed farming of food crops, rubber and fruit trees and just over half under forest cover. Anbera straddles the transnational highway and the hamlet is easy to reach, with frequent public transport. This, and the expansion of industrial monocrop plantation, have been main factors in the shift towards the kinds of urban lifestyles that villagers see, for example, in soap operas on national TV. Cellular phones and motorcycles are the typical indicators of modernity and prosperity of the villagers, although not all households in the hamlet have the luxury of electricity, good education or health care.

---

[6]2007 data provided by Sanggau District Office of Forestry and Plantation Crops.
[7]2007 data provided by Sanggau District Statistical Office.
[8]Interview with customary leaders of the Hibun Dayak community.
[9]Ethno-Linguistics Research Data of Institut Dayakologi, consulted in 2008.

Land expropriation through corporate land deals

The establishment of corporate estate zones in Indonesia is formally regulated by the issuance of long-term land use concessions (HGU)[10] by the government. Under the 2004 Plantation Act, HGU concessions may be granted for 35 years and extended to another 25 years. This law later on was overlapped by Investment Law No. 25/2007, which allows a single concession to be valid up to 95 years. Once the company has obtained its HGU permit, the local/indigenous community who formerly managed and controlled the land under customary law lose their rights over their territory. After the expiry of HGU concessions, the land reverts to State Land; customary rights are thus permanently abolished by the granting of an HGU concession, although local communities are often unaware of this.

Since the opening of the first oil palm plantation in West Kalimantan in the 1980s, several systems of production organization and control have been adopted by the oil palm companies, usually some combination of 'nucleus' plantation operated by hired workers and contract farming by smallholders, established on land previously held under customary tenure by local communities and now leased by companies from the Indonesian government through long-term HGU concessions. Common models are the PIR-BUN/PIR/PIR-Trans (Nucleus Estate – Smallholder Systems, with or without new settler farmers recruited under the transmigration program), KKPA (Membership-based Primary Credit Cooperatives), *Pola Kemitraan* ('Partnership Model', a kind of profit-sharing system), and more recently the *Pola Akuan* ('Acknowledgement' Model) which shares some features of the Partnership Model but leaves smallholders in some confusion in regard to their rights as it is not formally known in Indonesian plantation regulations.

The various plantation models or labour regimes share some common features including the division into the 'nucleus' (*inti*) area that belongs to the company and the smallholder (*plasma*) area that is allocated to smallholder contract growers. Local people who agree to be smallholder oil palm farmers have to surrender a certain amount of their (customary tenure) land in order to obtain a smaller area planted with oil palm trees, such as the 5:2 or 7:2 system which means the smallholder farmer surrenders five or seven hectares in return for two hectares planted with oil palm, the remaining three or five hectares coming under the company's direct control as the 'nucleus' zone. Although they have surrendered a larger area of their own land in exchange for their smallholder plot, the farmers still have to repay (amortization) credit for the plot(s) they obtained, through deductions from their monthly harvest income. Aside from these deductions for the (re)purchase of their land, the contract farmers' harvest incomes are further reduced by deductions for infrastructure maintenance, transportation, fertilizers and other inputs, etc. As smallholders may only sell their palm fruit to the company, the model involves both monopoly and monopsony practices. In the KKPA system, this control is exercised by the oil palm farmer cooperative formed by the company.

The latest system, the so-called 'Partnership *(Kemitraan)* System', applies a slightly different model where instead of having the smallholders work on their own plots, the company will hire people to work both in nucleus and smallholder zones, and the 'smallholder' becomes a non-farming, profit-sharing 'partner'. Agreement is

[10]See Sirait (2009) for details of the administrative processes that a plantation company should go through to obtains a HGU permit.

made on the mode of income sharing; usually the 70:30 system is applied where the company will get 70 percent of the value of the smallholder harvest, and the smallholder 30 percent (or, what is left of this 30 percent after the various deductions mentioned above).

When establishing the plantation, the company approached community leaders, customary leaders and other figures (teachers, religious leaders, etc.) who were all male, to do the public relations or information dissemination to the other community members. Usually, these formal and informal leaders received incentives (cash, promise of a smallholder plot, etc.) for this work, or for the number of community members who sign up as smallholders. The HGU lease conditions require the company to take over the full amount of land given in the Location Permit, and if the target is not fulfilled the permit can be revoked by the District Government.

In 1985–1986, 80 Anbera families handed over their customary land to PTPN XIII (then PTP VII). During the initial community consultation process, the company had promised to hand over the smallholder plots, already planted with oil palm seedlings (two ha for each five ha of released land) within 48 months after they received the local community's land. However, by 1999 (14–15 years after the loss of their land) none of the community members had received the promised oil palm plots. Finally, in 2001 the first smallholder plots were handed over, but by 2006 still as many as 10 out of 80 families had not yet received their promised plots.

In the case chronology compiled by the Anbera community we find many expressions of dissatisfaction with the allotted smallholder plots which were below the required standard in infrastructure (roadways, bridges, drainage) or in maintenance. The location of the allocated smallholder plots, in other hamlets far away from the farmers' residential area, has also created conflicts between Anbera smallholders and the villagers where their new oil palm farms are located. They also found that the number of oil palm trees planted in their plots was less than the number promised. The smallholders also filed a complaint about the high amount of credit installment charged to their smallholder areas since 2001, 15 years after they handed over their customary land to the company. Community documentation also revealed that PTPN XIII has actually operated without any HGU permit since 1985.[11] The company only obtained its HGU permit in 2003, 18 years after its first operation.

When problems arise with the local community, the company relies on community leaders, local government officials, and security officials (military and police). The latest method applied is the formation of a community-level 'Task Force', which is supposed to respond to problems arising between the community and the company. The members are representatives of local government, military, police, and local leaders (village and customary/*adat* leaders). The task force members are given monthly payments by the company.[12] In Anbera, the village and customary leaders are part of the task force, but they themselves have been involved in the long and tedious process of fighting to obtain their promised smallholder plots

[11]See '*Kebun Inti PTPN XIII Tanpa HGU Mesti Diusut*' (Nucleus Plantation of PTPN XIII that was without HGU should be Investigated), *Kapuas Post*, 3 February 2001; '*Tak Punya HGU, Belum Sah Kepemilikannya*' (Illegal Ownership Without HGU) *Pontianak Post*, 9 December 2000.
[12]See Sirait (2009).

and they acknowledge the ineffectiveness of the task force in ensuring that the company fulfills its obligations to the local community.

Women and oil palm: mothers, workers and scavengers

> The women were catching fish in the river in the middle of the oil palm plantation when suddenly a security guard came and shouted to them, 'Return all the fish to the river! None of you has the right to catch the fish here. This river does not belong to you anymore but to the company.' Then the women returned their fish to the river and with a heavy heart, walked home in silence (Stephanus Djuweng,[13] quoted in AIWN-AMAN 2007).

Aside from the social, cultural, economic and environmental problems and conflicts that other studies have identified (Marti 2008, Sirait 2009), oil palm expansion has had several gendered impacts on Anbera women.

The loss of women's tenure rights

The community's previous subsistence was from rubber, upland rice farming and other local agricultural products grown in sustainable mixed farming systems. According to Rona (2003, 11) the customary tenure system of Hibun people consists of three types: (1) collective ownership (*poyotono* in Hibun language) in which all community members have use rights and the land is managed collectively by the community; areas that fall into collective ownership category include *pengahieng* (cemetery), sacred places, *temawong* (mixed-crop agricultural areas that are planted with various fruit trees and cultivated for dry rice field), *sungi* (river), *himbo* (forest/ jungle), and *tapoakng madu* (honey tree); (2) ownership based on descent (*poyotiant*) where forest and other land that falls under this category is owned, managed and used by families claiming descent from the same ancestors, and (3) individual, heritable ownership that covers area such as *homing* (house), *kobon* (rubber orchard), and *muh* (rice field). Individual ownership has only been practiced since the introduction of the government's individual land certification programme.

In the Hibun Dayak community, although women do not play a part in communal formal politics, they have access rights to land. Hibun women inherit land rights from their parents, and according to customary leaders there is no gender differentiation in land inheritance.[14] The amount of inheritance usually depends on which child will take care of the parents. The one who will be in direct care of the parents will inherit more of the parents' properties. For example, in the case of the customary leader interviewed, it is his fifth daughter who will get a bigger share of her parents' properties (house, land, etc.) than her other male or female siblings, for this reason.

Moser (1993), underlining the importance of land rights for rural women as an essential source of protection and security for them and their children, has pointed out that state-codified, individual forms of land allocation and resettlement often annihilate women's rights to land. Recent studies of the gendered dynamics of land tenure have further explored the systematic discrimination that women have experienced in the process of globalization (Tsikata and Golah 2010), and in land

[13]Stephanus Djuweng is an influential Simpakng Dayak community leader.
[14]Interview with Mr. Antonius Lidos, 16 August 2008.

reforms (Jacobs 2010). One aspect of oil palm expansion with far reaching impact on the gendering of land tenure is the formal system of smallholder registration based on 'Family Heads' *(Kepala Keluarga)*, in both PIR-SUS/PIR-TRANS and KKPA schemes. The concept of 'Family Head' practiced in Indonesia designates the husband as the head of the family, except in a household where the woman is widowed or divorced. Although there is no formal gender restriction on who should be the smallholder participant, or what should be the gender of the household head, in practice the (male) Head of Family is registered as the smallholder. Hence, for a family which only obtains one 2-ha smallholder plot, the registered owner will usually be male. In cases where the family surrenders a big amount of land, such as the family of Mrs. Ayokng which handed over more than 43 ha of their customary inherited land to PTPN XIII, they were able to include their daughters among the smallholder recipients although, in the end, not all of them obtained the plot that had been promised.[15]

In the case chronology compiled by Anbera community leaders, among the 98 names registered as smallholder candidates who surrendered their customary land plots in 1985, only six are female candidates. In a case reported in neighbouring Kampuh Hamlet, a woman who had inherited customary land from her mother had to register this land under her husband's name so that he became the owner of the new oil palm plot, and controlled the income. She had to work as rubber tapper for her neighbour in order to have income to meet her own needs, and to provide her school-going children's daily lunch money.[16] Fortin (2011) records similar experiences of women in nearby Cempaka Village, who lost rights to their inherited land when their husbands registered themselves as smallholder farmer candidates using this land. Loss of these rights also bars women from access to formal credit sources, as they do not have collateral with legal title in their own name (Fortin 2011, 118).

Towards feminization of smallholder agriculture

Prior to the coming of oil palm there was a clear division of responsibilities between men and women. In communal agricultural work, for instance in the *perladangan* or shifting rain fed rice farming system which is the centre of Dayak cultural life, both men and women were involved in land clearing. Usually the women would clear the grass and other smaller brush, while the men would clear bigger bushes and trees. Burning the cleared fields would usually be done by men. In planting, again both men and women worked together, and the maintenance (weeding) would be the work of the women, until the time came for harvesting where again men and women would work together.

Dayak upland rice cultivation does not bring cash income as there is a taboo in the community, shared by all Dayak ethnic groups in West Kalimantan, on the sale of self-produced rice. Rice can only be given away for free in response to someone's request. Thus, the self-produced rice is only for the family's consumption or non-market exchange. Meanwhile, in the past, the community's main source of cash

[15]Mrs. Ayokng's son was one of the 10 remaining smallholder farmers who did not get their promised plot.
[16]Information obtained during indigenous women workshop organized by Indigenous Peoples Alliance of the Archipelago (AMAN), in Kampuh Hamlet, from 23–25 May 2009.

income was rubber. The rubber tapping was done by both men and women, and usually the man will be in charge of the marketing of the latex. Other income was derived from forest products such as rattan.

In contrast to the relatively balanced gender division of labour in Hibun Dayak subsistence and cash crop production, in oil palm cultivation women spend more time in agricultural work than men, especially in maintenance of the trees which requires much time and energy. The women are the ones who are mostly seen going to the field (either oil palm or their other agricultural fields) early in the morning between 6 and 7 am, and usually arriving back home at about 4 pm, with their hoe, machete and *jarai* basket or sprayer tool. Only in particular conditions, for instance where the women are unable to harvest the taller oil palm trees, will the men carry out the work.

Presently, traditional agricultural work that does not bring cash income is perceived by the Anbera community (particularly the male members) as female work, while income-earning work is socially considered the domain of men. In practice, however, women also work as oil palm farmers, labourers, or scattered oil palm fruit scavengers to generate cash income for their family, but this is normatively considered only as supplement to their husband's income. Thus, besides being now the main farmers on their remaining subsistence plots, women are also the main caretakers of the smallholder plots, although the holdings often do not formally belong to them, but to their husbands or male children. They also join in their Oil Palm Working Groups activities, although they are not formally registered as members. Smallholder farmers are organized in working groups, and are registered as member of one of the oil palm cooperatives established by the company and farmers themselves. An oil palm smallholder working group consists of 20 to 30 smallholder farmers whose plots are within the same contiguous location, thus enabling them to work collaboratively in area, facility and access maintenance, and also in harvesting. The oil palm harvest is sold by the group to the cooperative, and each month the proceeds are paid through the cooperative.

Subsistence agricultural activities are now also becoming the women's domain, with women taking the initiative to cultivate dry rice fields and also being more active in the *pangari* (reciprocal group labour) activities of the community. *Pangari is* a system of mutual collaboration in Dayak agriculture. *Pangari* groups are activated for agricultural activities that require much work and energy, such as *menugal* (seed planting with the dibble stick). The work is done in rotation from one member of the community's farm to another. Presently, in Anbera, the women have established their own *pangari* group, but with the establishment of oil palm plantation and other changes, *pangari* activity is no longer limited to agriculture; the women have extended the practice to many other activities, including house building.

> There is only *pangari* group, and usually it is only the women. The men are seldom involved in *pangari*. They only participate in heavy work, such as building the hut or other *pangari* work that cannot be performed by women. However, as long as the women can do the work, then the women will do it. (personal communication, Mrs. Tingaang)

Presently, besides reciprocating work with (the same or another kind of) work, *pangari* work contributions may be recompensed with cash. For instance Mrs. Bulan, whose family doesn't have a smallholder plot, was paid for her oil palm fruit bunch harvesting work in the *pangari* group. This suggests that *pangari* is losing its formerly strictly reciprocal and non-monetary character, at least in some activities connected to commercial agriculture.

Women and the emergence of a plantation labour class

According to informants none of the hamlet members have jobs at the PTPN XIII plantation which surrounds their village; at the time of interview, when they were still involved in pressing for the issue of rights to their smallholder plots, the company discriminated against Anbera hamlet members and had not recruited any worker or staff from Anbera. According to information obtained from female respondents and a village leader, some of the husbands had jobs as civil servants or at one of the five oil palm cooperatives located in the hamlet, running their own small businesses, working as rubber tappers or in village administration. But such jobs were very few in comparison to the total male population. Others were working at other plantation companies as staff, plantation security, daily labourer, or driver. There were also a number of jobless husbands. Another common occupation was migrant labourer, particularly to Malaysia, since the border is only three hours away from the hamlet. This applies to both women and men. In the case of women, however, there have been some 'trafficking' cases where the women later on were sold as sex workers or otherwise exploited. In the situation where the men do not have other employment and the family owns a smallholder plot, usually both husband and wife take care of their smallholder oil palm together.

The plantation companies use three types of labour in their nucleus zone: regular (contracted) labour, daily casual labour, and extra unpaid labour. Regular labourers are hired on contract and have a secure monthly salary, while daily (casual) labourers are recruited by a foreman for one specific task (spraying, area clearing, etc.) and paid based on the number of days they work. The third type (not formally recognized) is the unpaid family members (spouses, relatives, children) of a hired worker who may assist him/her in the work but are not included as parties in the labour transaction. Although contract-based labour may sound more secure in comparison to non-contract based labour in terms of income and labour rights, in many instances the terms of contract are violated by the employer (for example in pay deductions, failure to contribute to medical expenses, etc.).[17]

There are two groups in the community who become plantation workers. The first are villagers who are not of local origin, and do not have access to land. They include the Javanese and other in-migrants, or Dayak people from other villages/areas. This group takes up plantation labour jobs as their main source of income as they do not have land or a share of smallholder areas. They work as plantation labourers almost every day.

The second group are the smallholder farmers who have surrendered their land but did not obtain any smallholder plot. The plantation companies' failure to hand over smallholder plots as specified in the contract created the first plantation labourers. This is found not only in Anbera or West Kalimantan, but also in other parts of Indonesia where oil palm plantations are established.[18] An example from Anbera is GP Ltd.'s practices under the '*Pola Akuan*' system. While waiting for the handing over of their own smallholder areas, the farmers were allocated a certain

[17]As also reported in East Kalimantan (Darto *et al.* 2008).
[18]See Darto *et al.* (2008).

plot to work on as if it is their own smallholder plot, and paid for the work they do. Hence, they are no different to the daily labourers.

> I also don't understand the *akuan* system. The fertilizing work is paid daily, clearing around the trees is paid daily. I don't want to do this, but if we don't do the work, no payment is given. The promise was that even if we didn't work, the company would still pay us. It's Mr. Yun (the PR officer of GP Ltd.) himself who made the statement, he said that I need not work and I would still get payment, I could just sit at home and still get 30% of the harvest selling price. I told him to just give me my smallholder plot so I would know which and where it was. Although it was far, I could still pay somebody to work on it (personal communication, Mrs. Dom).

Women play a key role in the social reproduction of the oil palm plantation. This is closely related to their low level of formal education and also their (family's) tenure rights/access. In the case of Anbera almost all women in the hamlet have had experience as daily workers in one of the oil palm plantations in the area. According to informants at least two generations of Anbera women have worked as oil palm plantation labourers. Only one young woman from Anbera has been recruited to a structural company position. She is a graduate of one of Java's well-known universities and works in one of the foreign-owned plantation companies.

Mrs. Kusun had long years of experience as a casual daily labourer on one of the oil palm plantations. In 1988, her family gave up their land for conversion to oil palm smallholder plots but only got their plot 15 years later. She became a daily labourer in this period until her family got the plots (which were registered in the name of her husband, daughter and son). She became so experienced with the various types of daily labour work that she was entrusted to be a daily labour foreman. Meanwhile, her husband was working as contract-based staff of a plantation company.

Rural women are becoming a class of labour in oil palm plantations. They dominate the daily labour workforce; it has become an alternative source of income beside rubber tapping: '[I] joined the daily labour for seedling and area clearing, because if I don't go rubber tapping, it's the only other work available' (personal communication, Mrs. Bulan, 23 years old).

Many daily labourers are also young. They engage in daily labour particularly during the busy periods of rice cultivation, while the older women spend more of their time in the rice fields.

> It depends on when the work is carried out. During busy times in the rice fields such as right now, older women rarely join, there are many more young peoples, [it's for] anybody who wants to join the daily labour work...many women. I used to do daily labour and was paid Rp 6,000 – [about USD 0.60][19] to Rp 9,000 [USD 0.90] per day (personal communication, Mrs. Gi, 20 years old).

From the manager to field workers (field staff, foreman, checker, general worker), plantation security guards and the police, structural jobs in the plantation companies are very much men's space. The harvesting workers are divided into two groups, the group who harvest the fruit bunch from the trees and place them at the TPH (fruit bunch collection point), and those who transport the fruits from the TPH

[19]Rupiah exchange rates have fluctuated during the course of this study. For the sake of simplicity we have use a rough conversion rate of Rp. 10,000 = USD 1.00.

to the factory. In the company's own area, the harvesting workers are all male as the work is considered to be to heavy for women, despite the fact that women do the same work on smallholder plots.

The other oil palm plantation work which is mostly done by women is spraying and fertilizing. This work is considered to be easier for women, although in reality it is also heavy work, and no less hazardous as it involves working with highly toxic chemical substances (for pesticides, weedicides, and fungicides), such as Paraquat[20] and Roundup.

> All the daily labourers are women. We work in groups of 12 women who do the spraying, while the foreman, the driver, and the one who fills the sprayer tanks with the poison (herbicide) are men (personal communication, Mrs. Tingaang, 29 years old).

The spraying and fertilizing workers hardly ever wear safety equipment such as boots, mask, head cover, goggles or gloves. Aside from the hot weather that makes wearing protective clothing uncomfortable, the workers have to provide the safety items at their own cost, while their wages are far below the price of the items. Presently, the payment for spraying work is approximately Rp 26,000 (USD 2.60) per day. Meanwhile, a pair of rubber boots costs approximately Rp 150,000. (USD 15) and cloth mask costs Rp 25,000 (USD 2.50).

> The weather is too hot to wear a mask, we will die from it, sometimes I wonder how people manage to only have their eyes exposed, they are wearing veil, hat, all are covered, because it's hot. After spraying we have to drink milk and eat egg. . .usually we buy them ourselves and eat them at home, however the daily wage of only Rp 26,000 is not enough to buy milk and eggs, not to mention the distance we have to travel for the work, buying gasoline, food, sometimes we also bring our own food from home (personal communication, Mrs. Tingaang).

Berondol scavenging

Berondol is the local term for scattered oil palm fruits that fall out of the bunch during the harvesting process. All women of Anbera and many from nearby villages have experience as *berondol* scavengers. This (illegal) activity has become a part-time job for some women, particularly the women farmers, and full-time for others. Although there is no social restriction for men to join, *berondol* collecting has become almost exclusively women's space. *Berondol* collection provides women with a significant addition to the income their family gets from their oil palm smallholding, rubber tapping, mixed farming, small-scale businesses, etc. For some women, it has now become the main source of income.

Berondol are sold to traders or to other oil palm smallholders who add it to their own harvest sales. Women whose families have smallholder areas will usually add the *berondol* to their own oil palm harvest delivery.

[20]Brand name: Gramoxone. Paraquat (N,N′-Dimethyl-4,4′-bypyridinium dichloride) is one of the listed dangerous pesticides whose use should be prohibited according to WHO and the Pesticide Action Network (PAN). The Plantation Crops Office of West Kalimantan does not have any knowledge of the WHO prohibition on Paraquat. Also, none of our female respondents had any knowledge of the hazardous nature of the herbicide they are using.

> We start searching for *berondol* at 12 noon until evening, sometimes until 7 pm. The search becomes more fun as the time goes later. Usually, we can get 6 sacks in a day, and my husband will help to carry them. *Berondol* collection used to be freely allowed (personal communication, Mrs. Tingaang).

According to Mrs. Tingaang, *berondol* collecting started as the trees matured around 2003, when oil palm fruit bunch was priced at around Rp 300 per kg (USD 0.03). Mrs. Tingaang has been collecting *berondol* for approximately two years. At certain times the income that she gets through *berondol* collecting has been a determining factor in making make ends meet for her family.

> We do both, usually; after returning from working in our smallholder farm, we search for *berondol*. The amount is enough to cover the cost of the children's snacks, and my wardrobe and cupboard were also purchased with the income from *berondol*, and the bed mattresses, wood (for housing) and others. . .a lot, there are those who can buy [gold] necklace, such as Mrs. R, the wife of Mr. Sb, she bought a big necklace from *berondol* collecting, and is buying a motorcycle. Without collecting *berondol*, people won't be able to buy since they don't have smallholder plots. It will be impossible for us to eat, like this month, if we don't look for *berondol*, how can I pay my credit at the credit union also? We only got Rp. 107,000 [USD 10.70] [from the selling of oil palm fruit] last month (personal communication, Mrs. Tingaang)

Berondol kernels fetch a lower price than regularly harvested palm fruit bunches, usually Rp 800 per kg (USD 0.08) compared to Rp 1,200 per kg of fruit bunch (USD 0.12). Usually one woman can collect some 40–60 kg of *berondol* (worth Rp 32,000 [USD 3.20] to Rp 48,000 [USD 4.80]) in a day. *Berondol* collecting can be done every day whatever the weather and without any sophisticated tools. It requires only lots of energy, a large *jarai* basket, familiarity with the oil palm plantation area, lunch box, flashlight (if one plans to seek and collect *berondol* until after dark) and a good dose of courage, as the *berondol* collectors have to evade the plantation security officials or police who patrol the area. The security personnel now treat the *berondol* seekers as thieves, as they are considered to have trespassed on company territory and taken away company property. The women find this response exaggerated, as in their perspective, they only collect things that are discarded by the plantation, for the *berondol* are left to rot when no one collects them.

The security personnel's responses to the *berondol* women range from psychological intimidation and harassment to neutrality, depending on the individual. The women are often intimidated, being told they 'will be sent to Sanggau' or 'will be jailed in Sanggau' if they are caught. Such intimidation has created intense fear among some women and made them vulnerable to harassment. One informant told of a woman who was forced to have sex with a security guard who caught her collecting *berondol*. Children who collect *berondol* during school holidays also experience intimidation. Usually, women go in groups to collect *berondol*. This gives them a sense of security and courage to confront the security personnel. Despite the risks and anxiety involved in *berondol* collection, it has become one of the rare occasions when women still meet in groups and work together.

Sanggau Regional Regulation No. 3/2004 on the partnership management of oil palm plantation, Chapter VI, article 11 states that (1) The security management for the organization of the partnership model of oil palm production is done by the

District Head (Bupati) together with relevant government offices which are members of the TP3K (Task Force); (2) Security measures should use 'persuasive, preventive, and repressive methods'; (3) All agreements made should be read and signed by both parties; (4) Before the application of 'repressive' measures as mentioned in point (2), the matter should be dealt with in deliberation through the TP3K and/or the legal security officer, SATGAS/SATLAK, and Customary Council. The locals, including the *berondol* women, are not aware of this regulation and the restrictions on the repressive measures that the plantation can impose on them.

In May 2010 two women and one man from Anbera were arrested, tried and convicted for the theft of 60 kg of *berondol*, worth Rp 60,000 or approximately USD 7. This was the first court case against *berondol* scavengers. In an interview, one of the lawyers for the three villagers stated that the case should not have gone to trial as based on judicial perspective, the amount of trial 'material' (the *berondol*) was too small.[21] Despite the fact that the women were willing to return the *berondol* they collected, the company (PTPN XIII) insisted on having the case go to court, and the the police brought the charges to Sanggau District Court. Fortunately, the accused villagers were assisted by a local NGO who provided defence lawyers. The case also caught the attention of West Kalimantan's Regional Representatives (DPD) in the National People's Representative Council (DPR-RI), one of whom lodged a formal protest (*Nota Protes*) with the Ministry of State-Owned Enterprises (as PTPN XIII is a state-owned company).[22] Following this publicity the three Anbera villagers were released despite the fact that, in the court verdict, the two women were sentenced to one month in jail with two months' probation, while the man was sentenced to 20 days in jail with two months' probation. The case was also publicised in both local and national media; judging from the way the news was written, the media were in support of the three local villagers, especially the two women.[23] The release of the *berondol* scavengers was considered a rare case of successful local resistance against the big oil palm company, as more often the stories are about local people or NGO activists who support them ending up in prison.

Money, ambivalence and resistance: paradoxes of women's responses to oil palm

In previous sections we have seen how the coming of oil palm (in its corporate, monocrop form – we should not blame the crop itself) has led to the loss of women's tenure rights, a new gendered division of labour and feminisation of agricultural

---

[21]In stark contrast, on 12 March 2011, when PTPN XIII celebrated its 15th anniversary in one of the the provincial capital Pontianak's luxury hotels, the CEO stated that PTPN XIII profits had reached Rp 200 billion or approximately USD 20 million in 2010, and the company was projecting annual profits of Rp 1 trillion (or approximately USD 100 million) by 2014. '*PTPN XIII Optimis Bukukan Laba Rp 1 T*' [PTPN XIII is optimistic to book profits of Rp 1 trillion], Pontianak Post, 15 March 2011.

[22]See '*DPD-RI Kirim Nota Protes Kepada Kementrian BUMN*' (DPD-RI Sent Note of Protest to the Ministry of BUMN), *Borneo Tribune*, 9 August 2010.

[23]See '*Petani Miskin Cari Keadilan*' (Poor Farmers in Search of Justice), *Pontianak Post*, 9 August 2010; '*Curi Sawit untuk Bayar SPP*' (Stealing Oil Palm Fruits to Pay School Fees) *Tribun Pontianak*, 9 August 2010; '*Maling Sawit Terancam Tujuh Tahun*' (Oil Palm Thief could be Convicted for Seven Years), *Pontianak Post*, 1 September 2010; '*Hanya Merugi Rp 60 Ribu, PTPN 13 Tetap Proses Hukum*' (For a loss of only 60 thousand rupiah, PTPN 13 still pursues legal action), *Portal Metro Pontianak*, 1 September 2010.

work, the loss of women's other sources of income, and their new involvement in (illegal) *berondol* scavenging. In this section we explore further some of the bases of women's ambivalence to their oil-palm-dominated existence, and their involvement in resistance and protest.

Cash economy and being 'modern'

Money economy has become the main driving force of Anbera women's views and responses towards oil palm. The increasing variety of family cash needs, including purchased food, children's education and health services, has been the main rationale for the women of Anbera to see a positive side to the coming of oil palm. Older women particularly experienced the effort to support their families when cash economy was very limited in the area as 'a truly difficult time'. This was particularly evident among women who are not originally from the hamlet and who initially had no land in the village, such as Mrs. Kusun and Mrs. Dom, who later on were able to buy smallholder plots in the hamlet from other villagers.

> We used to eat rice mixed with cassava, we searched for vegetables in the forest, [we searched for] anything in the forest that time, we ate our own chickens, we obtained fish from the river, before the oil palm plantation came, we truly suffered. We never bought meat, we had to search for it in the forest, by fishing, [if we wanted chicken] we had to raise one ourselves. For pork, if there were people who had it for sale, then we could buy, it was cheap then (personal communication, Mrs. Kusun).

This respondent's view raises questions about romantic views of indigenous 'self-sufficiency'. In the 1980s, smallholder rubber did not provide enough cash to meet most families' needs. Aside from the low price, the women also identified a drawback of rubber: that it could not be tapped during the rainy season. Presently, the farmers find it difficult to use their rubber trees as collateral when they apply for loans to the bank or credit union (cooperative) due to the absence of land certificates, as it is planted on land in customary tenure. The oil palm plantation has brought in a new and more regular source of cash income, for both daily labourers and smallholder farmers. Mrs. Ayonkng also recalled the initial support given for the establishment of oil palm plantation due to the promises of higher income.

Anbera people seem proud to use the term *gaji* to refer to what they receive every month from the company either for sale of their oil palm fruit or in payment for daily labour, although these do not fall within the lexical or popularly understood meaning of *gaji* ('salary'). The villagers find it exciting to receive the payment for their casual labour or their harvest on a fixed day of the month – like a modern-day office or factory worker getting a monthly salary. Hence, local peoples' evolving view of 'work', influenced by ideas of 'development', sees 'work' as a transactional activity (labour) that results in a salary paid by another party. Thus, self-generating income activities such as subsistence agricultural work no longer fall into the category 'work'.

The past 24 years have provided women with sufficient knowledge and experience (both positive and negative) of oil palm expansion in the area and also of the volatility of oil palm fruit price. Anbera women of are now more aware of the importance of women's land rights and the negative (environmental and other) impacts of the plantation, and have identified alternatives to involvement in corporate oil palm production, including rubber (which presently is well-priced and easy to market) and also self-planted oil palm. Rubber is becoming more popular, as they find the rubber tapping easier than oil palm work, and a more sustainable source of income for the long term due to its longer-lasting age if it is tapped appropriately compared to oil palm (that has a productive age limited to approximately 18 years only). Cacao is seen as another potentially profitable cash crop due to its high price. Mrs. Dom made it clear that she would not hand over her family's land to another oil palm plantation unless the oil palm company can guarantee fair practice. She is the main manager of her family's land and aside from working in the smallholder plots, she has planted her land with rubber, citrus trees, and also a self-planted oil palm plot to provide a variety of income-generating sources. Mrs. Ayokng commented:

> I do not want to have it if it is still like this [the plantation malpractices], no more giving away [of land], there is rubber (orchard) and *tembawang* as our land certificate. In the past, our land did not have written document, the plants were the original certificate. Sorry to say that I see this oil palm plantation is only to appropriate our ancestral control over natural resources (personal communication, Mrs. Ayokng).

Another issue that particularly troubles women is the mushrooming of 'café'[24] 'Café' in the area since the oil palm trees have matured and men control monthly cash incomes. 'Café' is the term used to refer to a food and snack stall where the female staff also engage in commercial sex with customers. Many café women in Sanggau District are local Dayak women from other villages, or other districts; they do not work in cafés near their own communities, as the locals may know them. There are also café women who are returned migrant workers, particularly from Malaysia. As may be expected, these cafés are an important cause both of family problems and sexually transmitted diseases, which are now quite frequent.[25] Anbera women report that the cafés are particularly crowded on the monthly company payday for smallholder farmers and male daily labourers. Usually, it is the men who have the oil palm fruit bunch transported and sold to the cooperative. They are also the ones who generally go to collect their farmer 'salary'.

> If the man does the harvesting, he is also the one who transports it and receives the salary; sometimes people are paid for doing the harvest, and this can be used as an excuse to explain to the wife where the money has gone. Well nowadays the money can be for [sexual] affairs...actually just like me and my husband who are both working (on the smallholder plot), those who understand will ask their wives to manage the money or have it given to them; if a man wants to go to café, he will just get the salary and spend it all by himself. Then how can the wife and children at home eat? So there is conflict at home, if he arrives from the café and hears the wife's complaint, then the wife will get punched and kicked for having no food at home (personal communication, Mrs. Ayonkng).

---

[24]According to one informant, 'café' appeared in Anbera some time in the 2000s, as an 'evolution' from the 'Karaoke tempel' which was an arena of hidden prostitution in the form of mobile stage dance entertainment using showgirls. 'Café' is the latest evolving name for the 'Kopi Pangku' ('coffee on your lap') phenomenon that has existed in West Kalimantan since the 1990s.

[25]Interview with one of the health officials in the area.

The assault on local biodiversity by expanding monoculture oil palm has caused the disappearance of various locally grown foods (such as those grown in *tomawokng*) or forest food: 'Yes, everything should be bought now, there are no more vegetables obtainable in the forest' (personal communication, Mrs. Bulan). Some Anbera women still are skilled in rattan handicrafts, producing *jarai* (local baskets/carriers) for sale. The destruction of the hamlet's surrounding forest has made it difficult for women to obtain the raw material. A *jarai* maker explained that in the past, rattan was plentiful only minutes' walk from her house. She now has to walk for a whole day to the forest near the community's protected mountain area[26] to find rattan. Her *jarai* income is declining and to make ends meet, she has now joined other women in *berondol* scavenging.

One serious issue experienced by residents of Anbera and the surrounding areas is clean water. Once-clear river water has become dirty and muddy since the conversion of upstream area into oil palm plantation. The community is now dependent on well water and also clean water supply from their protected sacred mountain.

> There are advantages and disadvantages of oil palm. The advantage is it generates cash income; the disadvantages are when the soil is destroyed, and water is polluted. Later on, this community will no longer have any source of clean water. We are happy when we get cash, but it also truly brings difficulty, especially the water problem (personal communication, Mrs. Tingaang).

Resistance and protest

For 15 years after handing over their land to the company, the community made efforts through its leaders to obtain their promised oil palm plots through peaceful approach to the plantation managers. No legal action was initiated as the community members did not understand the legal procedures, and they were also afraid of intimidation. In 1999 however, the Anbera villagers took their case to court with the help of a couple whose husband was a local Anbera villager returning to his village after finishing his military service, and who had contact with a Legal Aid organization based in Pontianak. The (male) community leaders have since been active in taking legal action, sending letters and formal statements to government leaders and institutions, documenting the case and compiling a case chronology, sending information to the media, and organizing regular coordination meetings among the smallholders. In response to the community's call, several negotiation meetings were organized by local government or the company, but these meetings generally ended without satisfying results/agreement. The process itself is tedious, time- and energy-consuming, and accompanied by intimidation.

In early 2001, an agreement was reached between PTPN XIII and Anbera community. After 15 years' delay, the first smallholder plots were handed over to 58 families (out of 80 families) and organized under the KKPA system. These families received smallholder plots of 1.8 to 1.9 ha, close to the promised 2 ha. The next batch of 12 families, however, received only 0.9 ha per farmer, and the remaining 10 families remained without any plot. In 2001, Anbera villagers carried out an open

[26]Anbera Hamlet community recently succeeded in obtaining recognition by Sanggau Forestry Office for 3000 ha of protected indigenous forest and mountainous terrain.

protest against PTPN XIII by blockading and harvesting a part of the company's nucleus zone. Only about five women were involved directly in the open protest. The military were brought in to confront the protesting villagers, who were all taken into police detention. After negotiation between the male villagers and the police, the women were released and sent back to Anbera hamlet, while the male villagers were jailed for a day.

In 2008, after approximately nine years of negotiations and meetings and another open protest, the remaining 10 smallholders were offered Rp 10 million (about USD 1,000) per farmer as compensation for the company's failure to deliver their smallholder plots. Although unsatisfied with the solution, they finally accepted the compensation for the loss of their land, as there was no other option. One farmer stated that they also did not get the full Rp 10 million, as they also had to pay some village leaders who had negotiated on their behalf.

Traditionally, in Hibun Dayak communities, there is a taboo on women's leadership roles and participation in local political process; women do not take part in meetings and community decision-making processes, and women's main space is normally defined as the domestic domain.[27] Hibun Dayak women's weak position in community politics put them at a disadvantage when the oil palm plantation company came. They did not participate in the initial community consultation process before the establishment of the plantation, had no clear knowledge of the plantation establishment plan, and no way to voice their opinion about it. The voice of the men was considered to be the unanimous voice of the villagers.

Many Anbera women saw their reproductive tasks as the main factor in their inability to participate in the consultation meetings, aside from seeing their husbands' participation in the meeting as adequate to represent their household, as told by Mrs. Kusun:

> It's not that it's not allowed, none of us participate, if we already have our husbands [participating] then we do not need to [attend the meeting], but if we wanted to, we could. Then, I still had small children, nobody was taking care of them. Then, it [the condition] wasn't such as today, it's free now, wherever the people are going, we can just follow and participate in the people's meeting or others (personal communication, Mrs. Kusun, 49 years old).

Today, men still dominate the village's formal governance structure from the Village Head to the Neighborhood (RT) Head, as well as in the revitalized Hibun customary structure where in the case of Dabat Village, the positions are all filled by men.[28]

[27]The oral history of the community records only one female leadership figure, called *Entulai Ndou' Labaa'*, who appealed to community leaders for the stopping of inter-tribal war. She was afraid to deliver her thoughts to the male leaders until she found a stone with supernatural power that gave her the courage to present herself in the leaders' meeting and call for peace.

[28]However, within the present larger structure of Hibun Customary Council, women have been incorporated in the board and a separate division for women's empowerment is established where the personnel responsible are also women. This suggests that Hibun women may have greater access to decision-making processes within their revitalized customary structure, although it may still require time and struggle, as the Hibun customary taboo on women leadership is still dominant in the perspective of the male leaders. One example of this concerns a woman who was elected as the Village Head of a Hibun village. When the village later on experienced many problems, both natural and social, customary leaders claimed that they were caused by the village's violation of the taboo, and that the woman Village Head should be replaced.

One strong voice of concern did come out from a woman of Anbera when she learnt that the land would be surrendered to the plantation company. Mrs. Ayokng retold what the woman said then:

> Mrs. Daum was the first person who didn't want to give up the land, she didn't want to give away all the land. Don't give up all the land as no one eats oil palm fruit, she said, '*Kidoh nyorapi kelapa sawit, masi borah lonyu nyorapi*' [no one cooks oil palm fruit, still one cooks rice]. She wanted to keep all of her rubber orchard. So, when her husband and father-in-law were persuaded [to give up the land], she moved to [a nearby village], so desperately not wanting to give up the land, she ran away (personal communication, Mrs. Ayokng).

Although women do not have their space in the formal community political domain, some women such as Mrs. Dom and Mrs. Ayokng showed negotiation skills and assertiveness when it came to defending their (or their families') rights in an informal context.

> It's like this sir; I don't want money, it's not that I have a lot, I don't have much but the most important thing for me is land, because I'm a newcomer, I don't have land, so I want to say that for my rubber orchard that has been burned down, you can just cut them down and replace with oil palm but with condition that the same area will still be mine (personal communication, Mrs. Dom).

> ...then we got two (smallholder plots), one registered in my name, the other in my husband's, I insisted on that [the plot registered under Mrs. Ayokng's name], just ask that lady's husband, they were the ones who supported me to demand my share of land, my husband's parents too, Mrs. B and Mrs. L, as I did not get the share, I asked the village head to make me an ID card as widow or if it's not possible, then as a widower (personal communication, Mrs. Ayokng).

Mrs. Ayokng joined the most recent open protest in a village meeting with the company. She also stated her objection to the violation of promises made by the company at the sub-district office.

> I protested in May-June this year [2008]. The company [GP Ltd] promised us that after it had been planted with oil palm trees, the land would be given to the farmers but 10 years later, it was not cleared, not maintained. So we re-occupied it. We did ask for compensation then, but the company said that oil palm plantation did not give compensation, only *derasah* for our previous activity there. Our rubber trees were felled without compensation, so I took hold of the plantation plot as compensation, because it belonged to us. The rubber trees were compensated with the oil palm trees, so don't say that we steal (personal communication, Mrs. Ayokng).

The gender-based allocation of tenure rights by the oil palm plantation system has also led to the exclusion of many women from the organization that might have offered a channel to voice their concerns and campaign for their rights. *Serikat Petani Kelapa Sawit* (SPKS) is an oil palm farmers' union formed by smallholders in Sanggau District with the support of NGOs in response to the many problems arising in the district. The union engages in advocacy and campaigns for just practices in the oil palm sector. It promotes observance of farmers' rights by the plantation companies and government policies. SPKS membership is restricted to registered smallholder farmers. Due to the system of smallholder registration in the name of household heads explained earlier, the SPKS branch of Sanggau District has only 200 female farmers among its 1025 registered members. Almost all Anbera women have no knowledge about SPKS or other farmer organizations.

Although Anbera women still rarely participate in open protest, at Sanggau District level, particularly the women who are organized by the SPKS (Oil Palm Farmers Union) do engage in open protest against the malpractices of the oil palm plantation companies in the district. Mrs. Rini, a Hibun Dayak woman teacher who lives in a neighbouring village, is known for her strong resistance against the oil palm company in defending her family's and community's right for their smallholder areas.[29]

> As a teacher working in the village and as indigenous Hibun, my husband and I feel that it is our obligation to react collectively. We and other teachers in the area as well as the other communities here are members of the cooperative. We were supposed to receive and own an oil palm plot but we still do not know which is our plot. Our land was taken with empty promises, but none of our *adat* leaders have reacted. No wonder, they are members of the Satlak [task force] of the company. They have become the guardians of the company, not any more our leaders. By joining SPKS, step by step we will gain back our cooperative and we can negotiate with the company to gain a better position (Mrs. Rini, qtd. in Sirait [2009, 45]).

Conclusions

Prior to the coming of oil palm, Anbera villagers had broader access to and control over land and forest resources managed under Hibun Dayak customary law, including women whose rights to inherit land were recognized in customary law. Mainly driven by their increasing need for cash income, Anbera men agreed to surrender parcels of their land in order to join the oil palm plantation scheme that, they were told, would bring them prosperity. Oil palm plantation expansion has extended and strengthened the patriarchal system of the state and the Hibun Dayak community.

The shifting of land tenure from the community to the state via the plantation company and the practice of the 'household head' system of smallholder plot registration has narrowed women's tenure access. Furthermore, the irregular practices and broken promises of the plantation companies have increased the area of appropriated land, reducing opportunities for traditional mixed farming.

Greater involvement in cash economy, the decrease in available agricultural land and the depletion of local food products as the result of forest destruction have shifted the pattern of food acquisition and consumption. Most foods and other household goods are now purchased from local and mobile sellers, with cash or credit. This is one factor influencing women to seek income additional to whatever the family obtains from smallholder plots or the husband's income, by working as plantation labourers or *berondol* seekers, or as migrant workers.

In terms of gender relations, as in many other cases of agricultural 'modernisation', formalisation has been accompanied by masculinisation (of oil palm plot ownership, of membership in producers' organizations, and of access to

---

[29]Mrs. Rini attended a conference on the Roundtable on Sustainable Palm Oil (RSPO) in Singapore in 2005, and a meeting of RSPO in Kuala Lumpur in 2007, where she met the CEO of the Malaysian company Synergy Drive (owner of one of the companies active in Sanggau) and received an apology for the company's intimidation of SPKS members.

credit sources linked to land titles) undermining the position and livelihoods of women in this already patriarchal society. Women were largely absent from community discussions and negotiations with the oil palm corporations. Further indirect gendered impacts of the plantation industry include increasing social problems, including prostitution, STDs and gendered violence. At the same time, as we have seen in the preceding section, some women have become active in open protest.

As in many other cases of expansion of agrarian commodity production, we can discern a familiar pattern of ambivalence between the attractions of regular cash income on the one hand and on the other the loss of resource tenure and autonomy, exploitation and intimidation, which helps to explain the Anbera community's gendered experience of coercion, intimidation, consent and resistance.

References

AIWN-AMAN. 2007. Information kit – portrait of the indigenous women of Asia. Asian Indigenous Women's Network (AIWN), and Aliansi Masyarakat Adat Nusantara (AMAN), in partnership with Rights & Democracy. Available from: http://www.dd-rd.ca/site/publications/index.php?id=2208&page=8&subsection=catalogue [accessed 17 Feb 2012].

Behrman, J., R. Meinzen-Dick and A. Quisumbing. 2012. The gender implications of large-scale land deals. *Journal of Peasant Studies*, 39(1), 49–79.

Bernstein, H. 2010. *Class analysis of agrarian change*. ICAS Agrarian Change and Peasant Studies Series. Halifax: Fernwood Publishing and Sterling, VA: Kumarian Press.

Colchester, Marcus, Norman Jiwan, Andiko, Martua Sirait, Asep Y. Firdaus, A. Surambo and Robert Pane. 2006. *Promised land: palm oil and land acquisition in Indonesia*. Moreton-in-Marsh: Forest Peoples Programme and Bogor: Sawit Watch.

Darto, Chica and Jefri. 2008. Buruh yang Menderita Tak Kunjung Sejahtera Ditengah Meroketnya Harga Minyak Sawit Dunia. *Tandan Sawit*, 8(1), 12–8.

Djuweng, S. 1997. *Indigenous peoples and land-use policy in Indonesia: a Dayak showcase*. Pontianak: Institute of Dayakology Research and Development.

Fortin, C.J. 2011. The biofuel boom and Indonesia's oil palm industry: the twin processes of peasant dispossession and adverse incorporation in West Kalimantan. MA Thesis, Saint Mary's University, Halifax, Nova Scotia.

Jacobs, S. 2010. *Gender and agrarian reforms*. London: Routledge.

Julia, and B. White. 2010. The gendered politics of dispossession: oil palm expansion in a Dayak Hibun community in West Kalimantan, Indonesia. Paper presented at the LDPI-JPS-FAC International Conference on Global Land Grabbing, 6–8 April 2011 at IDS, University of Sussex.

Leach, M. 2007. Earth mother myths and other ecofeminist fables: how a strategic notion rose and fell. *Development and Change*, 38(1), 67–85.

Marti, S. 2008. *Losing ground: the human rights impacts of oil palm expansions in Indonesia*. Bogor: Friends of the Earth Indonesia, Sawit Watch, LifeMosaic.

McCarthy, J.F. 2010. Processes of inclusion and adverse incorporation: oil palm and agrarian change in Sumatra, Indonesia. *Journal of Peasant Studies*, 37(4), 821–50.

Moser, C. 1993. *Gender planning and development*. London and New York: Routledge.

Prasetyantoko, A. and D. Setiawan. 2011. *Climate finance: between people's needs and safety*. Jakarta: Wahana Lingkungan Hidup Indonesia.

Pearson, R. 2000. Rethinking gender matters in development. *In*: T. Allen and A. Thomas, eds. *Poverty and development into the 21st century*. Oxford: Open University/Oxford University Press, pp. 383–402.

Robbins, P. 2004. *Political ecology*. Oxford: Blackwell Publishing.

Rocheleau, D., B. Thomas-Slayter and E. Wangari. 1996. *Feminist political ecology: global issues and local experiences*. New York: Routledge.

Rona, M. 2003. '*Onok Adak Onok Nusantara*' in *Otonomi Masyarakat Adat, Kedai V, Sarasehan KMAN II*, Kumpulan Diskusi dan Presentasi Kedai (Kelompok Diskusi Adat Indonesia). Bogor: International Centre for Research in Agroforestry (ICRAF) and AMAN.

Sirait, Martua T. 2009. *Indigenous peoples of West Kalimantan and oil palm plantation expansion in West Kalimantan, Indonesia.* Amsterdam: University of Amsterdam and Cordaid.

Tsikata, D. and P. Golah, eds. 2010. *Land tenure, gender and globalisation: research and analysis from Africa, Asia and Latin America.* New Delhi and Ottawa: Kali for Women and IDRC.

Young, K. 2002. WID, GAD and WAD. *In*: V. Desai and R. Potter, eds. *The companion to development studies.* London: Arnold, pp. 321–5.

Julia (this is her full name) is a board member of Kalimantan Women's Alliance for Peace and Gender Justice (AIPeKaJe), former staff of The Dayakology Institute, Pontianak, and former chairperson of West Kalimantan Regional Council of Indonesian Friends of the Earth (WALHI). She has been engaged in several research projects on the local impacts of oil palm expansion.

Ben White is Emeritus Professor of Rural Sociology at the International Institute of Social Studies, Erasmus University Rotterdam, The Hague. His research interests focus on processes of change in rural societies and the anthropology and history of childhood and youth, particularly in the area of child work and education. He has been engaged in research on these issues in Indonesia since the early 1970s. He is currently engaged in research on agrarian dispossession, rural youth and the future of farming. His most recent book is *Rights and wrongs of children's work*, Rutgers University Press, and his most recent article is 'Agrofuels capitalism: a view from political economy', *Journal of Peasant Studies*, 37(4), 2010.

Plantation rubber, land grabbing and social-property transformation in southern Laos

Miles Kenney-Lazar

This paper critically examines theories of accumulation, dispossession and exclusion for analyzing the agrarian transformations that result from contemporary large-scale land acquisitions across the Global South. Building upon Marx's primitive accumulation, Harvey's accumulation by dispossession and Hall *et al.*'s *Powers of Exclusion*, conceptual lenses are developed through which to examine how land grabs transform property and social relationships of resource-based production. I examine the concession of 10,000 hectares by the central government of Laos to a Vietnamese corporation for extracting timber and planting rubber in the southern province of Attapeu. This acquisition has excluded farmers from land and resources that constituted their primary sources of (re)production, reconfigured rural property relations, altered the peasant relationship to land and produced new exploitative forms of wage labor.

Introduction

Following the food and financial crises of 2007–2008, a diverse range of global investors has rapidly acquired vast tracts of rural land across the Global South for resource commodity production. Investors acquired 46.6 million hectares (ha) globally between 2007 and 2009 through both direct purchases and concessions, which was a significant increase in comparison to an average annual global farmland expansion of four million ha before 2008 (World Bank 2010). The rapid rise of such land deals has been attributed to a variety of causes: food security fears by wealthy countries with little arable land, commodity price hikes increasing the profitability of directly controlling agricultural production, and the use of land for biofuels and carbon sinks (Zoomers 2010, Borras *et al.* 2011). These acquisitions have been condemned by Jacques Diouf, the head of the UN Food and Agriculture Organization (FAO), as a form of neo-colonialism–poor countries are producing food for wealthy countries at the expense of their own hungry people (Borger 2008).

I would first like to thank all of my friends family and colleagues–in and outside of Laos–who supported this research. I would also like to thank the following people for their comments on previous drafts of the paper: Mazen Labban, Jody Emel, Tony Bebbington, Wendy Wolford, James McCarthy, Bill Kutz, John Lauermann, Mark O'Keefe, Adrienne Johnson, and four anonymous reviewers. Any remaining errors or omissions are solely the responsibility of the author. Funding from the Fulbright Program and a National Science Foundation Graduate Research Fellowship to support this research is sincerely appreciated.

These acquisitions have also been referred to as 'land grabbing', a term which captures the rapid, abrupt and unjust nature of the deals (GRAIN 2008, Daniel and Mittal 2009). International development and research organizations–such as the FAO, World Bank, International Food Policy Research Institute (IFPRI) and International Fund for Agricultural Development (IFAD)–have recognized the potentially negative impacts of large-scale land acquisitions upon poor, marginalized and land insecure peoples of the Global South, but they have mainly focused on the positive potential, asking whether the acquisitions can lead to development opportunities (Cotula *et al.* 2009) or 'win-win' situations for both investors and small farmers (von Braun and Meinzen-Dick 2009, Haralambous *et al.* 2009). Meanwhile, the World Bank euphemistically referred to the dynamic as 'rising global interest in farmland' (World Bank 2010). These institutions have further attempted to legitimize land grabbing by arguing that if they follow voluntary codes of conduct and responsible investment principles then they can lead to rural development (see FAO *et al.* 2010).

Subsequent to the deluge of media and development reports on land grabbing, a critical academic literature on the topic has begun to emerge,[1] providing a scholarly basis for political action. Zoomers (2010) has argued that land grabbing amounts to the foreignization of space, whereby land is increasingly owned and used extraterritorially by foreigners, who act as long-distance absentee landowners, while locals are excluded. McMichael (2012) has placed land grabbing within the historical context of a shifting food regime that is no longer characterized by heavily subsidized, surplus-producing Northern agro-industries, but by outsourced, financialized investments in cheap land, water and labor across the Global South in order to overcome food deficits within the global economy. Other recent case studies have analyzed land grabbing as a process of dispossession, denying small farmers, forest dwellers and pastoralists access to the land upon which they depend for their livelihoods (see Ballvé 2011 on narcotics production and territorial dispossession in Colombia and Levien 2011 on special economic zones in Rajasthan, India). Borras and Franco (2010a) contend that codes of conduct and responsible investment principles are fundamentally problematic, because they are not driven by a social justice-based analysis of the causes of rural poverty and do not advance solutions that protect and improve access to land and property by the rural poor.

International development and research institutions have attempted to provide a balanced analysis of the socio-economic impacts of land acquisitions, particularly due to the dearth of empirical data available (Cotula *et al.* 2009, von Braun and Meinzen-Dick 2009, World Bank 2010). Their approach analyzes the costs and benefits for those affected by the projects; although farmers may lose access to land and resources, they may gain access to employment or benefit from technological spillovers and nucleus-outgrower schemes. Such institutions have also recognized the negative distributional impacts of land acquisitions upon small-scale land users (see Cotula *et al.* 2009, World Bank 2010), and therefore have been challenged to frame the phenomenon positively. Yet, it is precisely such analyses of the relationship between land deals and rural poverty that ought to direct the FAO and World Bank to institute policy approaches that deal with the crux of the problem rather than papering it over with voluntary guidelines for investment. They ought to fully embrace a broadly defined agrarian political economy approach, which Borras *et al.*

[1]Such work has been featured in this journal and the 2011 International Conference on Land Grabbing; papers from which can be accessed at www.iss.nl/ldpi

(2011) frame as six simple, but key critical questions, four of which come from Bernstein (2010): (1) who owns what, (2) who does what, (3) who gets what, (4) what do they do with the surplus wealth that has been created, (5) what do they do to each other and (6) how are political changes shaped by and shape dynamic ecologies. Li's call to center labor in the land grabbing debate (2011) is also welcome as a focus on the conditions under which people appropriate nature and relate to land through various forms of property relations and social relations of production.

The agrarian political economy framework that I use in this paper builds upon the theoretical edifices of Marx's critique of primitive accumulation, Harvey's framework of accumulation by dispossession and the 'powers of exclusion' of Hall *et al.* (2011). Various aspects of these three frameworks are combined in order to analyze social-property transformations. Primitive accumulation is emphasized in order to examine the transformations of property and social relationships of production in their entirety, rather than solely the process through which assets are acquired. The term 'exclusion' is used to critically examine the terms 'accumulation' and 'dispossession' so as not to mask a detailed analysis of the agrarian and social-property transformations resulting from land grabbing. These transformations can be separated into those of (1) property relationships, through processes of reconfiguration, formalization, concentration and abstraction, and (2) social relations of production, through (semi-)proletarianization.

These theoretical interventions are used as conceptual lenses to critically examine a 10,000-ha land concession to a Vietnamese multinational corporation, Hoàng Anh Gia Lai Joint Stock Company (HAGL), in southern Laos. In this case study, a precarious alignment of state and capitalist interests facilitated the exclusion of peasants from the land and resources that constituted their primary means of (re)production. This led to the abstraction of a heterogeneous landscape of land uses and property relationships into a singularly and privately owned rubber plantation. Government officials permitted exclusion in part due to developmental promises of gainful employment, rural stability and infrastructure construction. This resulted in a radical transformation of property relations and of the peasant relationship to land and nature. Semi-proletarianized peasants continued to use the same land but under reconstituted social relationships that precluded them from capturing the full value of their labor.

In the next section, I critically examine and combine elements of Marx's primitive accumulation (PA), Harvey's accumulation by dispossession (ABD) and Hall *et al.*'s powers of exclusion to examine social-property transformations resulting from land acquisitions. The third section functions as a bridge between the global and abstract literatures and the case study by examining how the Lao state has used land concessions as a strategy of resource-led development and how concessions relate to the country's existing land management system. In the fourth section I describe how HAGL acquired its land and how the project was implemented, and detail the resulting transformations of land uses and livelihoods. In the fifth section I analyze how the impacts of HAGL's concession can be understood as a transformation of agrarian property relations and social relations of production. I conclude the paper by hypothesizing potential opportunities that land grabbing presents for uniting peasant and worker struggles under contemporary capitalism.

The fieldwork for this research was supported by a Fulbright fellowship and conducted between October 2009 and August 2010. Most of the empirical data were collected in semi-structured interviews with government officials, company representatives, village leaders and village households. At the provincial level, interviews were conducted with the Department of Planning and Investment, the Provincial Agriculture and Forestry Office, the Provincial Land Management Authority (PLMA) and the Water Resources and Environment Administration; at the district level interviews were conducted with the District Agriculture and Forestry Office. At the village level, interviews were held with village leaders at all seven villages *(ban)* from which the company acquired land: *Ban* Hatxane, *Ban* Nyai Oudom, *Ban* Keng Nyai, *Ban* Datkoum, *Ban* Mai, *Ban* Khamvongsa and *Ban* Kang. Within each of the first three villages, 10 in-depth semi-structured household interviews were conducted. A snowball sampling technique was used to select the three focus villages and the interviewed households. Archival documents and spatial data were also collected, including investment contracts, feasibility and impact surveys, GPS coordinates, government and company maps and land and forest allocation signboard maps.

Conceptualizing accumulation, dispossession and exclusion

Part VIII of Marx's *Capital, Volume I* (1976), entitled 'The so-called primitive accumulation', is a critique of Adam Smith's account of how the division of labor and the unequal accumulation of wealth were initially established under capitalism. Smith argued that those with more 'stock', or capital, had simply worked harder over time. Marx, however, turned this thesis on its head by showing that such wealth was accumulated through violence and theft that separated owners from their means of production in the countryside, expelling a rural population to the urban factories as a newly created proletariat (Marx 1976). He analyzed a process whereby the capital relationship – the social relations between capital and labor – was established and the capitalist mode of production emerged. His argument is often understood as historical, concerned with a specific sequence of events that led to the birth of capitalism. As recent theorists have noted, however, his is also a logical argument that explains how and why capitalist property and social relations of production are established and intensified via dispossessing processes (De Angelis 1999, 2004, Perelman 2000, Read 2002, Harvey 2003, 2006, McCarthy 2004, Glassman 2006). PA cannot be relegated to a pre-industrial past, but is instead an integral component of the contemporary expansion and entrenchment of capitalism.

De Angelis (1999) distinguishes between Marx's understanding of regular accumulation, which is based upon pre-established capitalist social relations, and PA, which establishes such social relations where they do not previously exist. De Angelis argues that the capitalist mode of production operates by expanded reproduction under the social and property conditions of its existence, until it encounters producers as obstacles to such expansion who must be separated from their means of production. Harvey (2003) similarly distinguishes between accumulation by exploitation under the expanded reproduction of capitalism and accumulation by dispossession, whereby non-capitalist or semi-capitalist aspects of society are reconfigured by privatization and commodification into more exclusively capitalist forms. For Harvey, ABD is a strategy of overcoming capitalist crises of overaccumulation by releasing assets at little to no cost, opening up new areas of investment and increasing effective demand. Glassman uses the phrase accumulation by 'extra-economic' means (2006) to point out that PA and ABD are strategies of accumulation that lie outside of capitalist production despite being intimately bound up with it. Such accumulation is of alienated assets acquired by extra-economic force, the value of which can only be realized through further capitalist commodity production and sale. Thus, PA is external to but contemporaneously articulated with the dynamics of capitalist accumulation (McCarthy 2004). It is simultaneously a form of accumulation and the means for establishing the conditions necessary for further capitalist accumulation (Read 2002).

Harvey's ABD framing has become popular for analyzing acts of dispossession, but it has often been used in scholarly work on land and natural resources in such a way that largely examines the unjust acquisition of assets without analyzing the resulting transformations of social-property relations (see Sneddon 2007, Spronk and Webber 2007, Roberts 2008). The missing element, especially for analyses of agrarian transformations resulting from land acquisitions, is Marx's original focus on how the act of separation leads to fundamental transformations of property, social and class relationships, as well as modes and forms of production. Second, the terms 'accumulation' and 'dispossession' can be misleading and can close off careful analyses of the relationships between land grabbing and capitalist production as well as the distinctions between property, access and ownership. Marx defined capitalist accumulation as the expansion of capital as value circulates through spheres of production, distribution, exchange and reinvestment, continually producing and realizing surplus value as a function of labor exploitation. Marx used the term 'accumulation' in Part VIII to critique Smith's approach to 'primitive' or 'original' versus 'regular' accumulation under normalized conditions of capitalist production (De Angelis 1999). The use of the term 'accumulation' can be easily misinterpreted when discussed in relation to land dispossessing processes. Land grabbing is primarily a process of acquisition rather than accumulation, which may or may not lead to future accumulation, depending on how such assets are incorporated into capitalist production.

The concept of 'dispossession' also requires careful treatment. It is important to identify who is being dispossessed of what and the types of rights and power they had to access property prior to dispossession. While property is often understood as the legal *right* to benefit from things, or a 'bundle of rights', Ribot and Peluso (2003) define access as the actual *ability* to derive benefit from things, or a 'bundle of powers'. In cases of dispossession, one must consider the types of legal rights land users maintained prior to dispossession and in what ways those rights were actually recognized, legitimated and enforced through power-laden social relationships. Paying special attention to the process of dispossession is telling of the property regime that was actually in place – the interrelated rights and powers that land users actually maintained. Cases of dispossession provide the opportunity to examine peasants' rights to land and their empowerment to protect those rights. It is helpful to consider the term 'exclusion' in conjunction with dispossession, because it powerfully captures the process of denying access to land, resources and assets across a diverse range of property regimes. Rather than a condition of lacking access to land, Hall *et al.* (2011) define exclusion as a process whereby poor people are evicted from land by or on behalf of powerful social actors. However, they also show that exclusion is a double-edged sword that can be wielded by the weak against the powerful with the right strategies and under the right conditions.

Acts of exclusion and dispossession resulting from land grabbing can lead to transformations of property relationships and social relationships of production, which are intimately connected as transformations of social-property relations

(Brenner 1982). Changing property relationships due to exclusion are commonly termed 'enclosure'–understood as the conversion of common lands into private property or the increasing commodification and exclusion of access to land and water (Woodhouse 2003, Blomley 2007, Vaseduvan *et al.* 2008). Like the term 'dispossession', however, the idea of enclosure can be misleading because it does not capture the variegated transformations of property relations that may result from capitalist resource extraction and production. Common lands have complex, varied and particular property regimes. Moreover, land types can be enclosed that are not easily categorized as communal or private, such as land belonging to the state that is legally tenured by individual households. Therefore it is important to simultaneously consider concepts of reconfiguration, concentration, formalization, and abstraction. Reconfiguration captures how property boundaries are re-drawn and new spatial boundaries are established. This may lead to property concentration, whereby smaller plots of land belonging to multiple actors are amassed into a larger piece of property owned by a single social actor. Formalization entails strictly defining land tenure regimes and property relations that are ambiguous, fluid and even contradictory – where multiple claims and understandings may overlap. Finally, the concept of abstraction captures the totality of this process – how particular and local regimes of property used for subsistence and simple commodity production are replaced by universal, abstract and capitalist delineations of spatial boundaries. Previous to capitalist commodity production, property regimes are often characterized by particular and concrete practices of land uses rather than abstract plans conceptualized by elite entities such as the state and capital. As land is acquired for capitalist production, however, a process of abstraction occurs whereby capitalist conceptions of spatial boundaries, land use and property relations determine how the land is actually used in practice.

This transformation of property relations is concomitant with social transformations of production and labor, particularly processes of proletarianization. Without the means to independent social reproduction, producers must work as wage laborers with the means of production belonging to others (Marx 1976). While some may lose all of their land and become fully proletarianized, many of those impacted by land grabbing might find themselves in the awkward predicament of semi-proletarianization. Lenin defined semi-proletarians as peasants who retain a small plot of land, but one that is not large enough for subsistence and requires them to engage in some form of wage labor (Lenin 1966). Semi-proletarianization may result from a variety of agrarian problems such as land degradation, larger firms out-pricing smallholder crops on the market, or the availability of more lucrative off-farm employment. This paper, however, examines the semi-proletarianization of peasants who have been dispossessed of part, but not all, of their land. Semi-proletarianization can actually be of benefit to capital. Immediate producers, who retain some form of subsistence production, but not enough for them to evade wage labor, have cheaper consumption costs and thus can be paid lower wages (Perelman 2007, Roberts 2008).

Land concessions and land management in Laos

The Lao People's Democratic Republic (hereafter, Lao PDR or Laos), along with the Philippines, Cambodia and Indonesia, has been identified by development and media reports as one of the Southeast Asian countries that has conceded significant amounts of land to foreign investors in conjunction with the rise of global land acquisitions (GRAIN 2008, MacKinnon 2008, World Bank 2010). Land concessions began to accelerate during the early to mid-2000s as investors connected to global flows of capital and world commodity markets rapidly acquired large tracts of land in the country. Currently, as much as 3.5 million ha of land are under agri-business concession agreements and contracts, which constitutes over 15 percent of the country's total land area (UNDP 2010 cited in Barney 2011). Many of these concessions are in the central and southern provinces, while a contracting model is predominant in northern Laos (see Alton *et al.* 2005, Shi 2008, Dwyer 2011).

The government of Laos (GoL) has increasingly recognized land concessions as a problematic mode of investment. By 2007, enough evidence had mounted through media, development and government reports to show that land concessions as a system of resource acquisition was detrimental to the country for a number of reasons (Schumann *et al.* 2006, Barney 2007, Dwyer 2007, Hunt 2007). Concessions led to an uncompensated loss of assets at both the village and state level. The former Prime Minister, Bouasone Bouphavane, recognized these losses and instituted a moratorium on all land concessions over 100 ha for industrial trees, perennial plants and mining (Dwyer 2007, Vientiane Times 2007). However, it has been widely recognized that this effort failed to put an end to concessions, which continued to occur by more corrupt and illegal means, often at the provincial level (Schoenweger and Üllenberg 2009). Concessions continued in part due to the government's ambivalent policies. Although the moratorium was put in place in 2007, it was repealed in 2009, and then reinstated one month later with weaker regulation, only prohibiting concessions over 1,000 ha (Vientiane Times 2009a, 2009b). Moreover, central level government policies regulating land concessions are not often implemented at the provincial level due to policy misunderstanding and miscommunication, a lack of resources to implement policies, and unrealistically low ceilings for the amount of land that provincial authorities can concede to foreign investors (Schumann *et al.* 2006, Hanssen 2007).

The policy debate over land concessions reflects a longer-standing evolution of the Lao state's approach to regulating land and resources ever since the Lao People's Revolutionary Party came to power in 1975 and created a centrally planned economy based upon Soviet and Vietnamese models. In 1986, the economy began to transform as the government initiated economic reforms under a policy known as the New Economic Mechanism that gradually incorporated market-based elements into the economy. These reforms raised governance questions of who to allocate land use rights to and for what purpose. The central government was responsible for allocating land use rights to villages and individual households, which was principally accomplished through the Land and Forest Allocation (LFA) program. The LFA program was originally intended to manage forest resources, delimit village boundaries, reduce swidden cultivation and allocate agricultural plots for individual households. It was not primarily intended to allocate large plots of land to foreign investors, although there was room within the program for corporate investment. Concessions were envisioned as a way to develop extra land that could not be developed by local farmers. Villages gained the right to legally use some portions of their community forestland while they were required to conserve and protect other areas. Although LFA decentralized land management decisions to the village and district level, it also enrolled villagers to manage and protect forest categories that were created at the national level (Fujita and Phanvilay 2008).

LFA was initiated to create the types of legal reforms that would set the preconditions for establishing land markets and permanent land titles in rural areas, which would allow for market-led development. The government envisioned that this would be achieved by individual farming households increasing the productivity of their land rather than by conceding large portions of village-managed land to foreign companies (Ducourtieux *et al.* 2005). Therefore, there is somewhat of a contradiction between the country's two land management systems, if not historically then at least in the abstract. This contradiction may help explain why land types have been allocated to companies that by law cannot be conceded, such as conservation and protection forest (see following section) and individual agricultural land (Obein 2007, NLMA and GTZ 2009). Agribusiness companies have even funded the LFA process in order to allocate village land for their own plantations (Barney 2007).

Although LFA was initially focused on smallholders and local producers, the allocation of land to foreign investors has become a critical component of the state's resource-led development strategy. Laos has often been discursively framed as an under-populated but resource-rich country, dense with natural resources such as timber, minerals, hydropower and rich soils (Barney 2009). Yet the country's public and private sectors do not have enough capital to extract and develop these resources, and thus there is a dependence upon foreign capital. Currently, 80 percent of foreign investment flows in to the resource sector, mostly consisting of hydropower and mining, but also timber extraction and industrial tree plantations (Schoenweger and Üllenberg 2009). The central Lao government has made resource investments attractive to foreign capital through low land lease rates, while also trying to capture significant benefits in the form of government revenue and local economic development. To achieve this task, they have pursued a policy of state land leases and land concessions, officially known as 'turning land into capital' (Dwyer 2007). Under this strategy state land is leased to foreign investors in return for rental fees and production taxes. However, the central government has not been able to fully take advantage of this income source. From 2004 to 2005, revenues from state land assets totaled USD 7.2 million, or only 0.24 percent of GDP (Schumann *et al.* 2006). This is largely because leasing rates, at USD 3 to 9 annually per hectare, are incredibly low in comparison to rates between USD 30 and 70 for the rest of Southeast Asia (Schoenweger and Üllenberg 2009). Land concessions are intended to bring benefits to the local level through infrastructure development and wage employment. However, a number of reports have shown the social and ecological impacts to be devastating due to polluted environments, meager employment opportunities and lost access to essential land and resources, (Obein 2007, NLMA and GTZ 2009, Baird 2010). In order to address this wide range of problems related to land concessions, the National Land Management Agency (NLMA) was created in 2004 as a central agency in charge of all issues related to land management and administration, particularly the regulation of land concessions.

As the Lao PDR has opened its doors to foreign investment for resource production, the national government has sought to use land concessions as a strategy for capturing the development benefits of such projects. Despite investor imaginaries of Laos as an empty and little-populated resource frontier (Barney 2009), rural Laotians use most of the country's land for cultivation, foraging and raising livestock. Not only do villagers engage in customary land tenure practices, the national government has increasingly formalized their statutory tenure rights to such land via land reforms since the 1990s. By law, all land within the Lao PDR is under the ownership of the national community, the representative of which is the national government (GoL 2003). However, the manifestations of this law in practice are much more ambiguous, depending upon the state's ability to enforce eminent domain over the land use claims of non-government actors and overcome competing attempts to access land.

HAGL's 10,000 hectares

In 2008, the Lao PDR Ministry of Planning and Investment (MPI) and Ministry of Agriculture and Forestry (MAF) gave HAGL the rights to extract timber from and plant rubber on 10,000 ha of land for an initial period of 35 years in Attapeu province, southern Laos.[2] Although the project contract stipulated that HAGL was limited to only using state land, they acquired land that the Attapeu government had previously allocated to seven villages (*ban*) through land formalization programs for growing agricultural and tree crops, collecting forest products and raising livestock. By acquiring such land, HAGL excluded villagers from accessing land that constituted their primary sources of resources, food and income, providing little to no monetary compensation to land users in return. Meanwhile, they promised to build village-level infrastructure such as electric lines, roads, bridges, schools and health clinics. With minimal self-governed land left to rely on, many villagers began working for HAGL as wage laborers on the land that was swept from beneath their feet.

Before acquiring land in Attapeu, HAGL had grown to become a giant multinational corporation over the past two decades. Its current CEO, Đoàn Nguyên Đú'c, founded HAGL in 1993 as a small carpentry business in Gia Lai province of Vietnam's Central Highlands. The business grew rapidly as it entered Vietnam's timber sector in the early 1990s and Ho Chi Minh City real estate in 2002. By the end of 2009, 80 percent of the company's revenue originated from real estate, its stock had tripled since first being publicly listed in 2008 and its total assets had passed USD 1 billion (Forbes Asia 2009). HAGL has since become the third largest corporation on the Ho Chi Minh Stock Exchange (Than Nhien News 2009). More recently, the company has focused on primary resource commodity production, with expectations that rubber, mining and hydropower will make up the core of the group's revenue stream by 2012. As Đú'c explained in an interview with Forbes Asia, 'I think natural resources are limited, and I need to take them before they're gone' (Forbes Asia 2009). The company has increasingly set its sights on Vietnam's less-developed and more resource-rich neighbors, Cambodia, Laos and Myanmar. Laos, however, will host the bulk of the investment, up to USD 450 million in the coming years (Vietnam News Service 2010).

HAGL acquired the 10,000 ha in Attapeu as a concession in return for helping to finance the 25th Southeast Asian (SEA) Games hosted by the central Lao government in 2009. The corporation gave USD 19 million to the government to construct an athletes' village in the nation's capital, Vientiane, for the games – four million dollars in the form of a cash grant and the other 15 million as an interest-free loan to partly be paid back by voiding the leasing fees on the 10,000 ha.[3] This means that the annual leasing rate would have to be about USD 43 per ha in order to pay

[2]Investment contract signed by representatives of the MPI, MAF, Vietnamese National Assembly, and HAGL, 2008.

back the USD 15 million interest-free loan over 35 years, which is significantly higher than the national average concession rate. However, since the specifics of the deal lie behind closed doors, the actual 'interest' on the loaned capital may come from a lower concession rate, additional resource rights given to the company or simply the purchase of sway and influence. For example, the deal paved the way for HAGL's additional investments in Laos, including the construction of a real estate complex in Vientiane, mineral exploration rights in two southern provinces and the construction of a rubber processing factory in Attapeu (VietnamNet Bridge 2008).

In Attapeu, HAGL rapidly clear-cut forest, logged timber and grew rubber in and outside of the allocated concession boundary.[4] By February 2010, 8,078 ha of the 10,000-ha quota had been allocated to HAGL (Attapeu PLMA 2009). The provincial government was unable to find a continuous tract for such a large amount of land, and therefore it was separated into three different plots in Saysettha, Sanxay and Phouvong districts (Table 1, Figures 1 and 2).

HAGL rapidly developed the land that it had been allocated, planting 3,218 ha with rubber and clear-cutting 2,005 ha of land between May 2007 and August 2009.[5] While the provincial line agencies did not know the number of hectares of different land use categories that HAGL acquired, a rough estimate can be calculated by overlaying company maps with digitized LFA signboards for two villages (Table 2).

Most of the land that HAGL acquired was previously agricultural and forest land and had been allocated to most of the villages through the LFA process between 2004 and 2008. Although the official LFA program ended in 2003, it was carried out in these villages as part of development projects sponsored by IFAD. As determined by LFA regulations, village land was divided into village and household land types. For village land, six different forest types were allocated. Villages were restricted from using conservation, rehabilitation and protection forest, but they could collect non-timber forest products (NTFPs) from utilization and degraded forests, while production forest could be used for future agricultural plots. Village forestland, and especially land types specifically identified as grazing land, could be used for raising livestock. Household agricultural land was mostly used for rice production, whether as wet, paddy rice or some form of dry rice, mixing shifting

Table 1. Plantation plot sizes (2010).

Plot number	Districts	Total area (ha)	Area planted (ha)	Area cleared (ha)	Area not yet cleared (ha)
1	Saysettha, Sanxay	4419	2264	1645	510
2	Saysettha	2100	0	0	2100
3	Phouvong	1559	964	360	245
	Total:	**8078**	**3218**	**2005**	**2855**

Source: HAGL plantation maps.

[3]Interview with HAGL field operations manager, Phouvong district, Attapeu province, February 2010.

[4]The provincial government recognized that the company had clear-cut 33 ha beyond the concession boundary (Attapeu PLMA 2009). By overlaying GPS coordinates from plantation boundaries with company concession maps I was able to conservatively estimate that HAGL had cleared at least 159 ha beyond the boundary.

[5]Due to the recent nature of the project, not all of the land had been cleared.

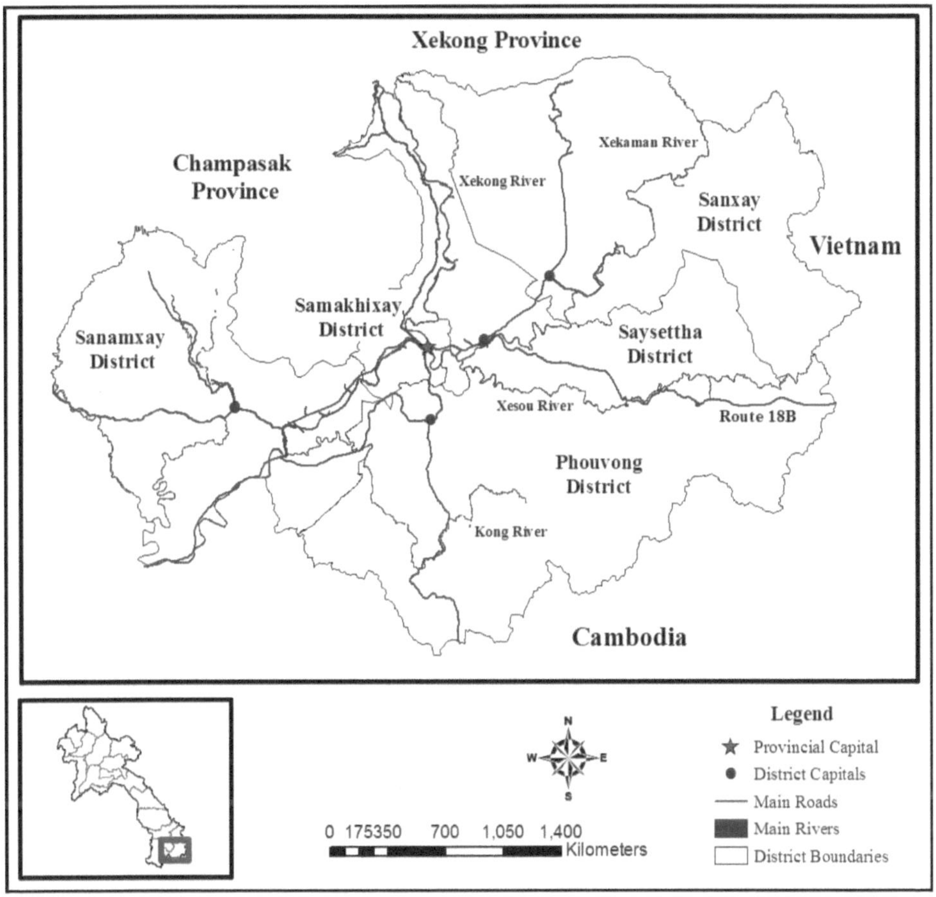

Figure 1. Attapeu province (2010).
Source: Author's drawing based upon Lao PDR National Geographic Department database and own GPS points.

cultivation with vegetable crops. Household agricultural plots were also used for small vegetable gardens or producing cash crops or commercial trees, such as teak. LFA categories were not wholly accurate because they were not strictly followed. Although all seven villages were located in flat-lying, lowland areas, most of the households in these villages came from ethnic groups that historically lived in upland areas[6] where agricultural practices differed significantly, leading to a mix of upland and lowland practices after migration.

Much of the conceded village land, as defined by formal LFA categories, was village forest and grazing land. Although not individually tenured, such land is critically important for village households, especially for the collection of NTFPs, which constitutes a significant portion of many households' monetary income.

[6]Brao and Jeng ethnic minorities are the predominant ethnic groups in the seven studied villages. Laos has a significant non-Lao ethnic population constituting almost half of the country's 6.5 million people (World Factbook 2011). Four major ethno-linguistic groups (Tai, Austroasiatic, Miao-Yao and Sino-Tibetan) and 132 ethnic groups have been identified (Chazée 1999).

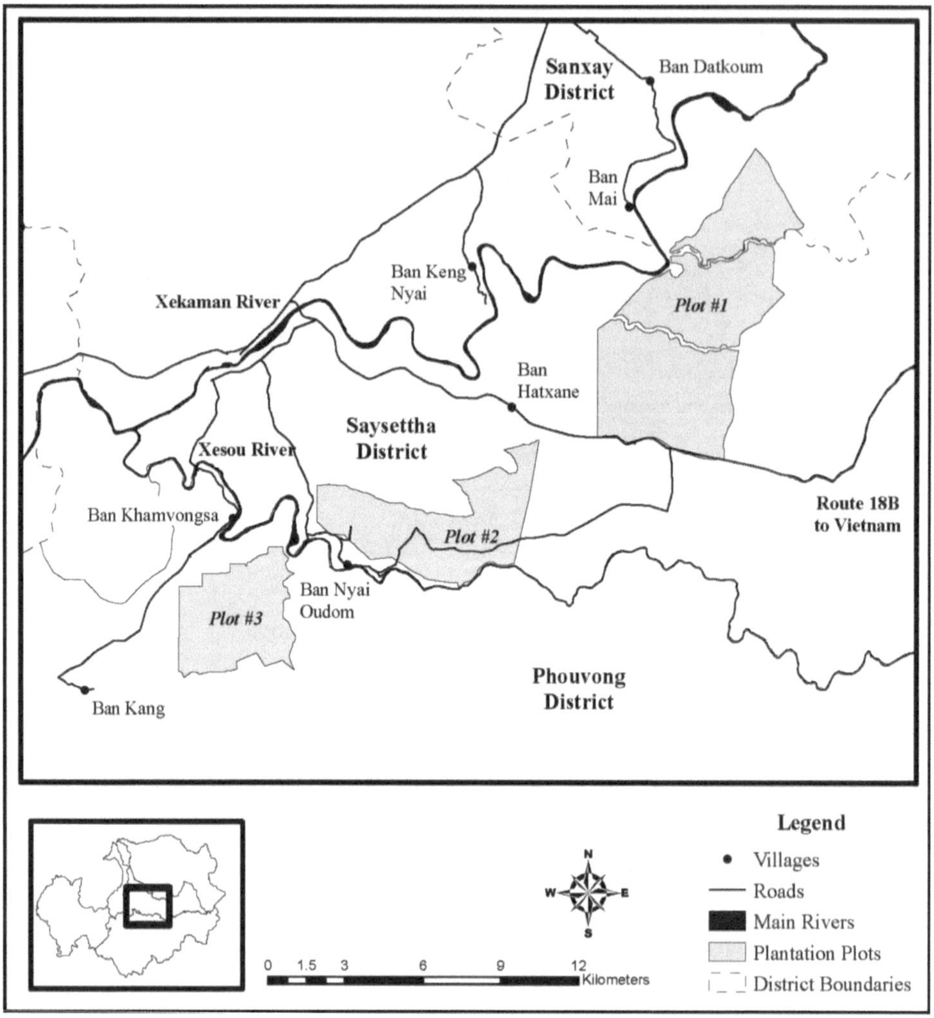

Figure 2. Map of plantation field locations (2010).
Source: Author's drawing based upon HAGL plantation maps, Lao PDR National Geographic Department database and own GPS points.

NTFPs play an important part in rural livelihoods throughout the country, averaging 50 percent of cash income in rural villages (Foppes and Ketphanh 2000). Interviewed villagers collected a wide variety of NTFPs for consumption and sale, such as small animals, vegetables, medicinal plants, firewood, and construction materials. Village land was also important for raising livestock for consumption and sale. The loss of village land used for foraging and raising livestock had widespread impacts since most villagers had incorporated such land into their livelihood changes. Interviewed households earned between USD 24 and USD 473[7] per year

[7]These monetary amounts are based upon an exchange rate of 8,470 Lao Kip per USD 1 as reported in the Vientiane Times on 1 February 2010.

Table 2. Amount and type of land appropriated (ha) (2010).

Village:	*Ban* Hatxane	*Ban* Nyai Oudom
Agricultural land	962	357
Paddy land	40	71
Industrial tree land	0	41
Grazing land	0	496
Rehabilitation forest	0	32
Utilization forest	791	581
Conservation forest	0	12
Protection forest	0	89
Cemetery forest	0	14
Unknown land use	0	107
Total:	1793	1801

Source: Author's tabulation based upon overlay of HAGL plantation maps with LFA signboard maps.

from the sale of forest products.[8] In the three focus villages, the main areas for collecting these products were forests that had already been cleared or were located in the concession area. For a number of poorer households, the loss of forest resources was devastating because one of their only sources of income was logging timber from forested areas to build houses for wealthier families in the village. The loss of village grazing land was also detrimental. Many households reported that the clearing of forestland for the plantation made it difficult to find livestock grazing land.

The most important source of livelihoods, however, was the cultivation of wet paddy rice, which villagers relied on both for food and income. Poorer households sold their labor in a variety of forms, which made up an important part of their income. They either expanded and prepared others' paddy land or built houses for wealthier villagers. The loss of individual agricultural plots was less common than the loss of communal lands, yet the impacts were more acute and intense since they led to a direct loss of agricultural production, income generation and food production. Provincial government agencies did not know the total number of households that lost land and the total amount of land that was lost. They did know that 115 families lost land, but this only included four of the six villages where agricultural land was appropriated (Attapeu PLMA 2009).

The most contentious aspect of HAGL's work to transform the landscape was the separation of farmers from their land. HAGL forcefully clear-cut village and household land without warning at three of the villages. In many instances, company employees may not have been aware that the land was used or tenured, especially forest and grazing land not visibly under production. This was in part due to the country's unclear land management system, composed of conflicting land use and land tenure categories. Furthermore, it is often difficult to find maps that clearly show how the land is managed, which often only exist as signboards at the entrances of villages – provincial line agency offices did not have paper or digital copies while conducting fieldwork for this research. The clearest and easiest maps to locate were those generated by HAGL of their plantation plots, only one of which acknowledged

[8]Household interviews, *Ban* Hatxane, *Ban* Keng Nyai, *Ban* Nyai Oudom, Saysettha district, Attapeu province, February 2010.

the location of village land. Additionally, at two of the villages the signboards had washed away during a 2009 flood. This confusion worked to the advantage of HAGL, making it easier for them to claim the land they acquired as unoccupied or unproductive. Such claims, however, are not fully tenable because HAGL also cleared land that was visibly under production, including wet rice fields, vegetable gardens and commercial tree plots.

In four of the villages, HAGL and provincial government officials pre-informed village leaders that part of the village land had been allocated to HAGL. In these cases, officials portrayed the concession as both inevitable and beneficial. When provincial government officials arrived at *Ban* Nyai Oudom to inform the village about the plantation project, they told village leaders that the provincial government had allocated 2,100 ha in the area, part of which was village land. Village leaders felt that they did not have recourse because the decision was made at the national level.[9] One of the village leaders at *Ban* Hatxane explained that a Lao company representative told him that land in the village would be appropriated as stipulated by the project contract without asking permission from the village. He stated, 'If the central government, province and district have agreed to this, how can a small village disagree?'[10]

HAGL's project was also presented as an opportunity to develop and modernize villages that were characterized as unable to escape poverty on their own. When the same Lao representative of HAGL visited *Ban* Kang, he told them that although the village had used their land for many years, they remained poor because they had not used the land productively. He then argued 'if you give up your land you will be rich like us, with cars and planes[11], but if you do not accept then you will remain poor.'[12] Provincial government officials and company representatives both argued that if HAGL was allowed to used the land productively then the villages could become modern and wealthy. The company would build infrastructure such as roads, bridges, electric lines, schools and health dispensaries. They would also provide stable employment on the plantation. The HAGL manager of the plantation operations even claimed that 99 percent of the employees would be Laotian.[13] Such claims were backed up by the legitimacy of the Lao state, which village leaders viewed as a guarantor of their success: 'The government is like our father. Trusting the government will lead to improvement'.[14]

After HAGL had cleared the land, they compensated villagers in exchange for their temporary land use certificates and tax receipts to such land. Compensation was a pre-emptive tactic to reduce resentment and resistance, but one that occurred retroactively, after the land had been wrested from household control. Additionally, compensation was a strategy for lessening possible civil unrest, sabotage to

[9]Interview with village leader, *Ban* Nyai Oudom, Saysettha district, Attapeu province, February 2010.

[10]Interview with village leader, *Ban* Hatxane, Saysettha district, Attapeu province, February 2010.

[11]He is referring to the HAGL CEO's private jet.

[12]Interview with village leader, *Ban* Kang, Phouvong district, Attapeu province, February 2010.

[13]Interview with HAGL plantation manager, Saysettha district, Attapeu province, February 2010.

[14]Interview with village leader, *Ban* Nyai Oudom, Saysettha district, Attapeu province, February 2010.

plantations and other tactics of resistance. The compensation was characterized by an exceptionally unequal power relationship, as a result of negotiations held solely between Laotian peasants and HAGL representatives without any regulation by a third party, such as the PLMA. Whether intentional or not, the provincial government was conveniently absent when it came time to act on the villagers' behalf. Thus, compensation rates were determined by HAGL's rough under-estimation of the economic productivity of the land and the results of ensuing negotiations with the farmers. Compensation was mostly given when villagers confronted the company directly. HAGL only offered compensation when it was obvious that they had cleared economically productive land, such as commercial tree plots, or when farmers were working in their fields as their land was being cleared. Villagers only received compensation to land to which they held some form of documentation, such as a temporary land use certificate or tax receipt.[15] Any customary or village land without title was not compensated. The result of these negotiations was that farmers were compensated at rates far below the market value of their land and the amount that they requested. HAGL forced villagers to accept low prices by threatening not to award compensation at all, a threat against which villagers felt they had no recourse.

There were a few cases, however, in which villagers used exclusion to their own advantage to prevent HAGL from acquiring their land. In one of these cases, the village leaders of *Ban* Kang repeatedly denied HAGL and the provincial government's requests to survey and acquire land. HAGL employees still attempted to clear-cut village land, but village leaders and armed village police stopped bulldozer operators and forced them to pay a fine of USD 355 for breaking the law and destroying village land. Provincial government officials and company representatives continued to unsuccessfully attempt to acquire village land. The village leaders had refused because of the land's importance for village livelihoods. Additionally, they knew the results of plantation development in advance because they had seen what happened to their relatives in Bachieng district of neighboring Champasak province who lost their land to a different Vietnamese rubber company. Based upon the village leaders' responses, I find that the ability for the village to exclude rather than to be excluded was predicated upon the following factors: (1) village leaders had knowledge of the potential negative impacts of land concessions and plantation development upon their livelihoods, and also of their legal rights under Lao law,[16] (2) village leaders made their decision to oppose the concession only after democratically consulting the rest of the village and (3) village leaders were unwavering and forceful in their resistance.

Despite such acts of resistance, much of the area's diverse landscape of land use practices and property relationships had been cleared by the end of the fieldwork period. Trees of significant size in forested areas were felled and lay in the cleared fields waiting to be transported to HAGL's wood processing and furniture construction factory 15 kilometers down the road.[17] The enclosure of the land was clearly etched into the physical landscape, marked not only by the bulldozed land and

[15]Household interviews, *Ban* Hatxane, *Ban* Keng Nyai, *Ban* Nyai Oudom, Saysettha district, Attapeu province, February 2010.

[16]The interviewed village leaders did not refer to any specific law, only that village land could not be acquired without the approval of the village leadership.

[17]Interview with HAGL plantation manager, Saysettha district, Attapeu province, February 2010.

the burned leftover brush, but also by the construction of a meter-wide, meter-high ditch around each of the plantation plots to keep village cattle from grazing on the rubber trees and to distinctly defined the borders between village and HAGL land.

Not all households lost agricultural land and some of those that did retained some of their individual plots or only lost access to communal village land. However, as self-governed livelihood strategies were eroded through land loss, villagers were left with little choice but to work as wage-laborers on the plantation, and thus began a transition from land users to land laborers. Some even worked on the land that had been taken from beneath their feet. Since the plantation was still expanding, there was a significant amount of employment available, although this may change once production advances into later stages and the work of weeding and tree maintenance is less labor intensive. Workers were supposed to be paid USD 3.55 per day, but their pay could be reduced to between USD 2.36 to USD 2.96 if supervisors found that they did not work hard enough.[18]

While villagers could potentially make higher incomes from full-time plantation work than independent farming, most villagers found the work to be incredibly demanding with little pay and therefore worked an average of two to three times per year – five days each time – when they were in need of money. At the official wage of about USD 3.55 per day, workers could make up to about USD 53.25 if they worked five times per year. Yet this is even less than some of the poorest interviewed household's incomes of between USD 88.24 and USD 588.23 per year from logging timber and building houses for wealthier households.[19]

During household interviews, villagers were most angry when discussing labor conditions on the plantation. They were frustrated with the 12-hour workdays, prohibitions against smoking and conversation at work, and constant supervision. One of the village leaders was convinced that the company was purposely overworking Laotian employees so that they would stop showing up for work, thus giving HAGL an excuse to evade Lao labor laws and import Vietnamese labor. The most common complaint was that the low wages did not match the difficulty of the work. Additionally, HAGL did not pay workers directly, but instead paid the group supervisors who were charged with distributing wages to each individual. Such payments were not regulated and supervisors often smuggled part of the wages for themselves. Supervisors paid wages late, and sometimes at less than the contracted amount by making excuses that they did not have small banknotes; in other instances, workers were not paid at all. Such labor practices make plantation labor a questionable employment alternative to previous livelihood strategies.

Social-property relations under transformation

HAGL's land acquisition in Attapeu dramatically transformed agrarian property and social relations of production. First, methods of coercive exclusion were used to prevent peasants from accessing land they had previously used for food, income and resources. Second, various forms of household, village, and state land were abstracted and privatized. Finally, peasants entered an altered relation to the land as semi-proletarianized wage laborers. Ultimately, these interrelated processes led to

[18]Household interviews, *Ban* Hatxane, Saysettha district, Attapeu province, February 2010.
[19]Household interviews, *Ban* Nyai Oudom, Saysettha district, Attapeu province, February 2010.

the establishment of the property and social relationships necessary for capitalist resource extraction and production. Such a transformation resulted from a precarious interface of class interests and power. While HAGL was more interested in acquiring resources than proletarianizing peasants, the Lao state sought to provide benefits for the population through stable employment and infrastructure while also generating new revenue streams.

Acts of coercive exclusion did not lead to direct violence towards villagers, but the act of bulldozing village forests and crop fields was a form of indirect or environmental violence (Peluso and Watts 2001). Tactics of legitimation were also used to separate farmers from their land by portraying the concession as both inevitable and beneficial. By presenting the project as one that would lead to modernization and development, government officials and company representatives created a powerful illusion of progressive change. Furthermore, HAGL ensured that farmers remained excluded and separated from their land by providing them with limited monetary compensation. This created the illusion of legitimate exchange and an appearance of ownership that never existed. Retroactively, the villagers became landowners precisely at the moment they were excluded from what they never fully owned. The land was transformed into private property at the moment of its alienation.

The exclusion of farmers from their land allowed HAGL to spatially reorganize property relations in the Lao countryside to facilitate capitalist production of resource and agro-industrial commodities. While it was unnecessary for HAGL to remove any peasants from their residential land – unlike village relocations that are common for mining and hydropower projects (UNDP 2006, International Rivers 2008) – the company had to alter the relationship between peasants and their land. This relationship was previously characterized by both customary and statutory land tenure and use, managed at the household and village levels but ultimately owned by the state. However, the abstraction of household, village and state lands – the redistribution of spatial relations and the social relations tied to boundaries and categories of land – erased this social connection. An altered landscape was constructed that was characterized by monocropped rubber land held in ownership by foreign capital, managed by Vietnamese employees and labored by Laotians. Thus, a heterogeneous landscape of diverse livelihood practices and land tenure was transformed into a homogeneously owned and used plantation of monocropped rubber. This transformation was not only one of spatial arrangements, maps and landscapes – it was also of social relationships. Previous relationships between villagers and the state concerning their right to use land in combination with village-level interpretations and manipulations of these rules was erased in favor of a project of large-scale capital accumulation-HAGL expects to eventually export USD 120 million of latex and wooden products every year from Attapeu province once the project is fully operational (VN Business News 2008). Such profit generation is only possible through the erasure of linkages between people and land, the emptying of social attributes from polygons on a concession map.

In Laos, proletarianization proceeded differently from a functionalist reading of PA that views dispossession as a strategy of capital to create a proletarianized labor force. Such a labor force already exists across the border in Vietnam and China, one that foreign capital may even prefer. Foreign investors may be less interested in a Laotian labor force and more interested in the resources that can be produced from land, which makes dispossession merely a tactic for getting people out of the way.

Perceptions that Lao laborers are unproductive in comparison to foreign laborers, especially the Vietnamese, date back to views expressed by French colonial administrators (Hodgdon 2008, Ivarsson 2008). A recent survey of businesses in Laos found that most find foreign, rather than Lao, labor to be more productive, especially for skilled positions (Asian Development Bank 2010). Vietnamese and Chinese companies may prefer employing laborers from their own countries due to ease of communication and preconceived worker understandings of wage labor demands. Additionally, plantation managers may find it useful to employ foreign workers due to their lack of legal rights. However, Lao labor laws partially restrict Vietnamese labor migration by requiring that the work force for manual labor on all investment projects be 90 percent Laotian (GoL 2006). As a result, the issue of proletarianization can become a point of contention rather than consensus between the Lao government and foreign investors that has important implications for the types of politics that surround plantation labor.

Conclusion

This paper shows how land concessions in Laos, and global land grabbing more broadly, can be analyzed through various lenses of accumulation-, dispossession- and exclusion-oriented frameworks. These lenses help examine the holistic transformation of property relations and social relations of production emphasized in Marx's PA while avoiding the pitfalls of uncritically importing of the terms 'accumulation' and 'dispossession' when analyzing capitalist production of land and resources. This theoretical intervention can be used to analyze how land grabs lead to interrelated transformations of property relations and social relations of production, or social-property relations. These transformations are achieved by excluding the rural poor from land and resources that they had previously used for their livelihoods and wellbeing. Analyzing these transformations captures the ways in which variegated tenure and property regimes become abstracted into rigidly defined plots of land, previous users are excluded from accessing such land, and their livelihoods hang in the balance between individual subsistence production and wage labor.

Examining these transformations in totality helps address Harvey's call to unite political struggles against 'accumulation by dispossession' with those against 'accumulation by exploitation' (Harvey 2003). To reconcile such struggles, it is important to analyze the myriad ways in which they are connected through processes of capitalist acquisition, production, distribution, exchange, consumption and ultimately accumulation. The emergence of global land grabbing creates a formidable new challenge for peasant struggles, placing extraordinary pressure upon countless livelihoods worldwide. Yet, this threat has also created a renewed urgency to address critical agrarian issues, such as land tenure, land reform and food security (Borras and Franco 2010b).

Borras and Franco, for example, argue for struggles against land grabbing based upon land sovereignty, or the 'right of the working class people to have effective access to, control over and use of land and live on it as a resource and territory' (2010b, 35). This concept could be imported to other struggles against capital, such as in the workplace or the city, for the proletariat to have control over the means of production, or over housing and public spaces. As Borras and Franco mention, land sovereignty is a cross-class political project that also spans different strata of the working classes – rural and urban, and across borders. This type of rights-based approach presents opportunities for thinking through connections between the right to food, the right to land, the right to a fair wage and the right to the city. It also helps contemplate how land is not the only focal point of social justice struggles in rural areas; labor issues and control over other forms of property are as important. The global land grab presents an opportunity to rethink these relationships as much as it provides a platform for conceiving strategies to grab back what was unfairly taken with the aim of creating more socially just forms of production and distribution.

References

Alton, C., D. Blum and S. Sananikone. 2005. *Para rubber cultivation in Northern Laos: constraints and chances*. Vientiane, Laos: GTZ.

Asian Development Bank. 2010. *Final labor market assessment report*. Vientiane, Lao PDR.

Attapeu PLMA. 2009. *Environmental and social assessment evaluation report*. Attapeu, Lao PDR.

Baird, I.G. 2010. Land, rubber and people: rapid agrarian changes and responses in southern Laos. *The Journal of Lao Studies*, 1(1), 1–47.

Ballvé, T. 2011. Territory by dispossession: decentralization, statehood, and the narco land-grab in Colombia. Paper presented at the International Conference on Land Grabbing, Brighton, UK.

Barney, K. 2007. *Power, progress and impoverishment: plantations, hydropower, ecological change, and community transformation in Hinboun district, Lao PDR*. Field Report, Vientiane, Laos.

Barney, K. 2009. Laos and the making of a 'relational' resource frontier. *The Geographical Journal*, 175(2), 146–159.

Barney, K. 2011. Grounding global forest economies: resource governance and commodity power in rural Laos. PhD Thesis. Toronto, ON, York University.

Bernstein, H. 2010. *Class dynamics of agrarian change*. Halifax: Fernwood.

Blomley, N. 2007. Making private property: enclosure, common right and the work of hedges. *Rural History*, 18(1), 1–21.

Borger, J. 2008. Rich countries launch great land grab to safeguard food supply. *The Guardian*, 22 Nov.

Borras, S.M. and J. Franco. 2010a. From threat to opportunity? Problems with the idea of a 'code of conduct' for land-grabbing. *Yale Human Rights & Development Law Journal*, XIII(ii), 507–523.

Borras, S.M. and J. Franco. 2010b. Towards a broader view of the politics of global land grab: rethinking land issues, reframing resistance. ICAS Working Paper Series No. 1.

Borras, S.M., R. Hall, I. Scoones, B. White and W. Wolford. 2011. Towards a better understanding of global land grabbing: an editorial introduction. *Journal of Peasant Studies*, 38(2), 209–216.

Brenner, R. 1982. The agrarian roots of European capitalism. *Past & Present*, 97, 16–113.

Chazée, L. 1999. *The people of Laos: rural and ethnic diversities*. Bangkok: Orchid Press.

Cotula, L., S. Vermeulen, R. Leonard and J. Keeley. 2009. *Land grab or development opportunity? Agricultural investment and international land deals in Africa*. Rome, Italy: FAO, IIED, and IFAD.

Daniel, S. and A. Mittal. 2009. *The great land grab: rush for world's farmland threatens food security for the poor*. Oakland, CA: The Oakland Institute.

De Angelis, M. 1999. Marx's theory of primitive accumulation: a suggested reinterpretation. Available from: http://www.homepages.uel.ac.uk/M.DeAngelis/PIMACCA.htm [Accessed November 2010].

De Angelis, M. 2004. Separating the doing from the deed: capital and the continuous nature of enclosures. *Historical Materialism*, 12, 57–87.

Ducourtieux, O. 2005. Land policy and farming practices in Laos. *Development and Change*, 36(3), 499–526.

Dwyer, M. 2007. *Turning land into capital: a review of recent research on land concessions for investment in Lao PDR: Part 1 of 2 – Existing literature*. Vientiane, Laos: Land Issues Working Group.

Dwyer, M. 2011. *Territorial affairs: turning battlefields into marketplaces in postwar Laos*. PhD Thesis. Berkeley, CA, University of California, Berkeley.
FAO, IFAD, UNCTAD, and the World Bank Group. 2010. *Principles for responsible investment that respects rights, livelihoods and resources*. Rome, Italy.
Foppes, J. and S. Ketphanh. 2000. Forest extraction or cultivation? Local solutions from Lao PDR. Paper presented at the Workshop on the Evolution and Sustainability of 'Intermediate Systems' of Forest Management, Lofoten, Norway.
Forbes Asia. 2009. Condo Boss. 30 November.
Fujita, Y. and K. Phanvilay. 2008. Land and forest allocation in Lao People's Democratic Republic: comparison of case studies from community-based natural resource management research. *Society and Natural Resources*, 21, 120–133.
Glassman, J. 2006. Primitive accumulation, accumulation by dispossession, accumulation by extra-economic means. *Progress in Human Geography*, 30(5), 608–625.
GoL (Government of Laos). 2003. Land law. No. 04/NA. Vientiane, Lao: PDR.
GoL (Government of Laos). 2006. Labour law. No. 06/NA. Vientiane, Lao: PDR.
GRAIN. 2008. SEIZED! The 2008 land grab for food and financial security. Barcelona, Spain.
Hall, D., P. Hirsch, and T.M. Li. 2011. *Powers of exclusion: land dilemmas in Southeast Asia*. Singapore: National University of Singapore Press.
Hanssen, C.H. 2007. Lao land concessions: development for the people? Paper presented at the International Conference on Poverty Reduction and Forests. Bangkok, Thailand.
Haralambous, S., H. Liversage, and M. Romano. 2009. The growing demand for land: risks and opportunities for smallholder farmers. Rome, Italy: IFAD.
Harvey, D. 2003. *The new imperialism*. New York: Oxford University Press.
Harvey, D. 2006. Comment on commentaries. *Historical Materialism*, 14(4), 157–166.
Hodgdon, B. 2008. Frontier country: the political culture of logging and development on the periphery in Laos. *Kyoto Journal*, 69, 58–65.
Hunt, G. 2007. Regional linkages: plantations investment and land alienation in the Theun and Hinboun basins. Paper presented at the Greater Mekong Conference, Sydney, Australia.
International Rivers. 2008. Power surge: the impacts of rapid dam development in Laos. Berkeley, CA.
Ivarsson, S. 2008. *Creating Laos: the making of a Lao space between Indochina and Siam, 1860–1945*. Copenhagen: NIAS Press.
Lenin, V.I. 1966. *Collected works volume XXXI*. Moscow: Progress Publishers. First published in 1920.
Levien, M. 2011. The land question: special economic zones and the political economy of dispossession in India. Paper presented at the International Conference on Land Grabbing, Brighton, UK.
Li, T.M. 2011. Centering labor in the land grab debate. *Journal of Peasant Studies*, 38(2), 281–298.
MacKinnon, I. 2008. The resentment rises as villagers are stripped of holdings and livelihood. *The Guardian*, 22 Nov.
Marx, K. 1976. *Capital: a critique of political economy*, vol. 1. New York: Penguin Books. First published in 1867.
McCarthy, J. 2004. Privatizing conditions of production: Trade agreements as neoliberal environmental governance. *Geoforum*, 35(3), 327–341.
McMichael, P. 2012. The food regime in the land grab. *Journal of Peasant Studies*.
Nguyen, T.M.V. 2010. Labor migration in East-West economic corridor: a case study of Vietnam. Khon Kaen, Thailand: Mekong Institute.
NLMA and GTZ. 2009. Findings of the state land concession and lease inventory project in the pilot of Vientiane Province: phase II. Vientiane, Laos.
Obein, F. 2007. *Assessment of the environmental and social impacts created by the VLRC industrial rubber plantation and proposed environmental and social plans*. Vientiane, Laos: Agence Française de Développement.
Peluso, N.L. and M. Watts. 2001. Introduction: violent environments. In: N.L. Peluso and M. Watts, eds. *Violent environments*. Ithaca, NY: Cornell University Press, pp. 3–38.
Perelman, M. 2000. *The invention of capitalism: classical political economy and the secret history of primitive accumulation*. Durham, NC: Duke University Press.

Perelman. 2007. From feudalism to capitalism. *Capitalism, Nature, Socialism*, 18(2), 44–61.
Read, J. 2002. Primitive accumulation: the aleatory foundations of capitalism. *Rethinking Marxism* 14(2), 24–49.
Ribot, J.C., and N.L. Peluso. 2003. A theory of access. *Rural Sociology*, 68(2), 153–181.
Roberts, A. 2008. Privatizing social reproduction: the primitive accumulation of water in an era of neoliberalism. *Antipode*, 40(4), 535–560.
Schoenweger, O. and A. Üllenberg. 2009. Foreign direct investment (FDI) in land in the Lao PDR. Exchborn, Germany: GTZ.
Schumann, G., P. Ngaosrivathana, P. Soulivanh, S. Kenpraseuth, K. Onmanivong, K. Vongphansipraseuth and C. Bounkhong. 2006. *Study on land leases and concessions in Lao PDR. Land Policy Study No. 4 under LLTP II*. Vientiane, Laos: GTZ.
Shi, W. 2008. *Rubber boom in Luang Namtha: a transnational perspective*. Vientiane, Laos GTZ.
Sneddon, C. 2007. Nature's materiality and the circuitous paths of accumulation: dispossession of freshwater fisheries in Cambodia. *Antipode*, 39(1), 167–193.
Spronk, S. and J.R. Webber. 2007. Struggles against accumulation by dispossession in Bolivia: the political economy of natural resource contention. *Latin American Perspectives*, 34, 31–47.
Than Nhien News. 2009. Hoang Anh Gia Lai provides cautious profit guidance. 14 May.
UNDP. 2006. National human development report: international trade and human development. Vientiane, Lao PDR.
Vandergeest, P. 2003. Land to some tillers: development-induced displacement in Laos. *International Social Science Journal*, 55(175), 47–56.
Vasudevan, A., C. McFarlane and A. Jeffrey. 2008. Spaces of enclosure. *Geoforum*, 39, 1641–1646.
Vientiane Times. 2007. Govt suspends land concessions. 9 May.
Vientiane Times. 2009a. Govt resumes land concessions. 16 June.
Vientiane Times. 2009b. Govt again suspends land concessions. 2 July.
Vietnam News Service. 2010. HAGL chairman unveils regional expansion plans. 14 May.
VietnamNet Bridge. 2008. HAGL built athletes' village in Laos. 15 July.
VN Business News. 2008. HAGL plans over US$ 100m for projects in Laos. 25 June.
von Braun, J. and R. Meinzen-Dick. 2009. "Land grabbing" by foreign investors in developing countries: risks and opportunities. Washington, DC: International Food Policy Research Institute.
Woodhouse, P. 2003. African enclosures: a default mode of development. *World Development*, 31(10), 1704–1720.
World Bank. 2010. Rising global interest in farmland. Washington, DC.
World Factbook. 2011. Laos. CIA, Washington, DC. Available from: https://www.cia.gov/library/publications/the-world-factbook/geos/la.html [Accessed January 2012].
Zoomers A. 2010. Globalisation and the foreignisation of space: seven processes driving the current global land grab. *Journal of Peasant Studies*. 37(2), 429–447.

Miles Kenney-Lazar is a PhD student in the Graduate School of Geography at Clark University in Worcester, Massachusetts, USA. His research interests include the political economy of resource commodity production, the financialization of land and property investment, theoretical understandings of nature-society relations under capitalism, and contemporary agrarian transformations in Southeast Asia. His dissertation project focuses on understanding the motivations and strategies behind financial investments in agricultural land, throughout the Global South and specifically in Laos. More information about his research can be found at http://mileskenneylazar.com.

Large-scale land deals from the inside out: findings from Kenya's Tana Delta

Rebecca Smalley and Esteve Corbera

Although there is alarm over the global land rush, many plans for the large-scale transformation of land acquired by investors remain on the drawing board. Based on a study of two land deals in Kenya's Tana Delta, this paper considers the processes by which blueprint designs are amended or delayed through the involvement of local actors. It demonstrates that even top-down acquisition of land by powerful state-linked actors with the support of policy discourse can be stalled by the rural poor, particularly if the latter have strong customary claims and links to wider opposition. At the same time, large-scale land acquisition is not automatically opposed by local people, who may see land deals as an opportunity to safeguard access to resources and to support their development expectations. The paper also suggests that although consultation and the existence of recognised property rights appear to result in fairer project designs, land deals are likely to reflect the decision-making power of an elite that is not fully informed. The conclusion affirms the need for more nuanced, place-based analyses of large-scale land deals, taking into account tenure arrangements, resource access mechanisms, land management discourses and the role of cross-scale agency and alliances in building support for, or opposition to, such deals.

1. Introduction

The role that agriculture can play in sub-Saharan African rural development is a live topic of debate. After years of intellectual and financial neglect of the sector, the persistence of rural poverty and the failure of agrarian economies to take off have focused minds on the old question of whether agriculture can contribute to poverty reduction and, if so, what kind of agriculture (Akram-Lodhi 2008). Although smallholder farming dominates in sub-Saharan Africa, the average farm size being 2.4 hectares (ha) (World Bank 2011), there has been a modern development of large, export-oriented operations. These began with the settler farms and corporate

The authors would like to thank the interviewees who gave up their time and agreed to have their wishes and concerns documented. We would also like to thank Dr. Abdirizak Nunow of Moi University for his support, the research assistants in Tana Delta without whom this research could not have taken place, and three anonymous reviewers for commenting on an earlier version of this paper. Esteve Corbera acknowledges the financial support of the Spanish Ministry of Science and Innovation (RYC-2010-07183) and of a Marie Curie Career Integration Grant within the 7th European Community Framework Programme (PCIG09-GA-2011-294234). Any errors or omissions remain the authors' sole responsibility.

plantations of colonialism, and continued after independence as new African governments explored options for intensifying and commercialising their agricultural production through state farms, private estates and agro-food complexes (Hinderink and Sterkenburg 1985, Loewenson 1992). Large estates have been checked by the forces of land reform, among other factors, and today many African countries tend to have a dualistic structure of a smallholder majority with a few large commercial enterprises (OECD 2008, World Bank 2009). We understand large-scale commercial farms to be privately owned, capital-intensive, centrally managed estates or holdings 'with an acreage far above the national average' and which 'employ much more hired than family labour and practise clear labour division within a management hierarchy' (Von Blanckenburg 1994, 1).

Large-scale commercial farms have comparative advantage over smallholdings when it comes to mechanised or irrigated cultivation of certain plantation crops such as sugar (Poulton *et al.* 2008). Overall, though, small farms have been judged more efficient, and since the 1960s they have been the focus for rural development strategies (Ellis and Biggs 2001). However, observers now question whether the viability of small farms can continue. On one hand, technology and an increasingly globalised agri-food system are creating economies of scale for larger, vertically integrated farms; on the other, subdivision and environmental marginalisation are hindering small-farm productivity (Diao *et al.* 2010).

In this context, and particularly since 2008, rural development studies have reported a new phenomenon of 'land grabbing'. This describes the acquisition of large tracts of farmland or uncultivated land, typically for producing food or fuel crops (Cotula 2011, World Bank 2011). A more neutral term is 'large-scale land acquisition', which allows for the possibility of a just transfer of land. The land is being bought or, more often, leased by investors, food-insecure countries and agribusiness firms, spurred by the 2007–2008 food-price crisis, demand for biofuels, capital flight into land and an improved investment environment in developing countries (Cotula *et al.* 2009, Haralambous 2009, IFPRI 2009). The largest number of cases occurs in sub-Saharan Africa, where it is widely perceived that empty or under-used land is available (Haralambous 2009, World Bank 2011). The acquisitions range from tens to hundreds of thousands of hectares (IFPRI 2009, Cuffaro and Hallam 2011). However, many land-grab deals remain on the drawing board. Only 21 percent of deals analysed by the World Bank in 2011 had been operationalised (World Bank 2011). Explanations put forward for this include unrealistic project designs, technical obstacles, changing investment conditions and the land being hoarded by speculative investors. The Bank also noted that projects have been disrupted by opposition from local land users (see also IFPRI 2009). As Poulton *et al.* (2008, 17) indicated in their review of barriers to commercial agriculture in Africa, 'the problems arising from multiple and overlapping claims to land ... can be a particular problem for large-scale agribusiness ventures'.

We present here a case study of two large-scale deals in Kenya's Tana Delta, with the aim of identifying conditions that facilitate large-scale land acquisition or impede its progress. The Kenyan agricultural sector is known for a relatively high degree of export-oriented commercialisation and for the contribution that large farms and plantation estates have made to this since colonial times (Bratton 1977, Mwangi 1981, Deininger and Binswanger 1995, Winter-Nelson 1995, Bates 2005, Dever 2009). Nevertheless, smallholdings of 3 ha or less have also been instrumental to production (Jaffee 1994, Gow and Parton 1995, Poulton *et al.* 2008) and they remain the predominant farm size in Kenya, accounting for 75 percent of total agricultural output (GoK 2010, Mandere *et al.* 2011). The majority of farmers in Tana Delta have an average farm size of just 0.8 ha (RoK 2009).

In the following section we set out our chosen conceptual tools, drawn from access and discourse theory. The third section introduces the case-study area, and in the fourth section we present evidence for support and resistance towards both deals among affected communities, attributable to a range of institutional, discursive and historical drivers. We argue that one of the deals, a foreign acquisition, has proceeded faster than the other, a domestic development, owing to differences in the tenure arrangements, consultation processes and value of the targeted land. We go on to discuss the mechanisms by which actors gain or maintain access to the land affected by the deals and the implications of the findings for the land-grab debate. We conclude by noting that although Kenya's policy environment overwhelmingly supports the introduction of large top-down agricultural schemes, local visions of rural development still envisage a future for small-scale farming and pastoralism, and have not allowed the deals to go ahead without amendment or opposition.

2. Land deals through the lens of access and discourse theory

2.1. Contextualising land acquisition

Large-scale commercial agriculture has historically taken distinct forms depending on the country and location's tenure regime, existing agrarian law, the type of crop and the targeted market, resulting in distinct production arrangements and business models (Vermeulen and Cotula 2010b). In Kenya, rural people have been involved in agribusiness models since colonialism, through wage labour on tea, coffee and sisal estates, participation in outgrower schemes for sugarcane and cut flowers, and work on medium- and large-sized horticulture farms. Kenya has one of the largest contract farming sectors in sub-Saharan Africa (Jaffee 1994, UNCTAD 2011, Oya 2012). The socio-economic outcomes of its outgrower schemes have been well documented, and they are by no means universally positive for participating farmers or surrounding communities (Mulaa 1981, Glover and Kusterer 1990, Wegulo and Obulinji 2001). In some cases, farm enterprises have retreated from outgrowing back to a pure estate model (see Jaffee 1994 and Dever 2009 for examples).

Although the recent global land rush is feared to signal a renaissance of plantations, many initiatives include a component for outgrower farmers to supply a nucleus estate (Haralambous *et al.* 2009, Cotula 2011, GRAIN 2012, Oya 2012). The prospect that large-scale land acquisition could introduce more contract farming into rural economies gives cause for optimism to those who believe smallholders will benefit from technology transfer and incorporation into the global agri-food regime (see IFPRI 2009, Robertson and Pinstrup-Andersen 2010, World Bank 2011). Assessed from this perspective, contract farming offers a means of small-farm survival (Oya 2012). Agrarian political economists approach the issue in terms of transformed land and labour relations, and, along with certain activist organisations, are more deeply critical of the arguments for smallholder agribusiness models (GRAIN 2008, Friends of the Earth 2010, Cochet and Merlet 2011, Li 2011). Some critical commentators situate the land rush within the political economy of global agro-food/fuel networks and capital's search for frontiers of accumulation, often framing it as neo-colonialism (Huggins 2011, Land Research Action Network 2011, McMichael 2011). We believe that conceptualising events in this way understates the contribution of local actors in making or resisting land deals and the complicit role of domestic states and developers. Cases of domestic land acquisition have received comparatively little attention, even though most recorded land deals involve domestic individuals or companies (World Bank 2011). Moving the focus of research from structural core–periphery forces to the local context can reveal differences between individual cases of large-scale acquisition (Borras and Franco 2010) and dig down to the roots of local support and opposition.

2.2. Framing access

Borras, McMichael and Scoones (2010, 582) write of the 'shades of grey' in how proposed land deals play out in specific cases. The institutions and micro-processes that determine outcomes include the level of democracy among villages, the terms of the deals and ecological factors. We take a step further and adopt Ribot and Peluso's access theory (2003) to argue that benefiting from large-scale land deals, from their contractual terms to the actual commercialisation of any produced crop, is mediated by what Ribot and Peluso call rights-based and other structural and relational means of access. The former encompass legal mechanisms such as title, and illegal mechanisms such as coercion and theft. In the context of large-scale land deals, these can be critical means through which certain actors are able to secure and/or enforce property rights over land for agricultural development, or to lease and/or cede land for that purpose.

The dominant message on property rights in the development literature is that defined, registered tenure is essential for safeguarding local interests in land acquisitions (World Bank 2009, Deininger 2011). Yet evidence points to common failures of land titling programmes in Africa (Peters 2009), and initiatives to formalise customary arrangements in a bid to strengthen people's tenure can backfire by encouraging elite capture (Hodgson and Schroeder 2002, De Schutter 2011). A review of recent land deals in six African countries found that most of the acquired land was under government ownership but managed *de facto* by local users (Vermeulen and Cotula 2010a). According to the authors, even in cases where local land users had officially-recognised tenure, this would not have given them enough bargaining power to affect outcomes of deals in their favour, because of the weak procedural rights afforded to everyday people.

Illegal mechanisms can provide alternative means of gaining or retaining access to land. Examples of land claims being supported by force, from village-level arson to military-backed land grabbing, have been documented in southeast Asia and some Latin American countries (Grajales 2011, Hall *et al.* 2011). This is not surprising, as bribery, corruption and coercion have very often underpinned the management and conservation of natural resources (Robbins *et al.* 2006, Kolstad and Soreide 2009), as well as the divestiture or restitution of land (Dorondel 2009, World Bank 2011).

Ribot and Peluso's category of structural and relational mechanisms includes a variety of other mechanisms relevant to understanding the development of large-scale land deals, such as access to capital, technology, labour and markets. Because the deals analysed here are still embryonic, we do not pay in-depth attention to how all these mechanisms actually play out. We nonetheless reflect on how people's opinions on these issues influence their position in either supporting or rejecting the land deals. We also consider how access to knowledge and authority enables not only project developers to plan and implement deals, but also local land users to articulate their responses and participate in the deals' negotiations. As Vermeulen and Cotula (2010a) have shown, local people's ability to secure favourable terms in deals with outside investors depends on their capacity to access decision-makers and acquire information. Additionally, access to land resources and their flowing benefits is strongly mediated by cross-scale institutions, ranging from intra-household and -community relationships to national processes such as the justice system and impact assessment procedures that define whose voice counts (Ribot and Peluso 2003, Amechi 2009, Vermeulen and Cotula 2010a).

2.3. The importance of discourse

When land deals are not yet operational and actors are still competing over the right to benefit in future, the role that discourse plays as a mediating access mechanism is perhaps more important than, say, securing access to labour or markets. We define discourse as a certain way of seeing and giving meaning to the world. An example is the colonial discourse of 'the profligate native', which made certain assumptions about indigenous land-use practices and judged them against European values (Li 2007). Discourse is given expression through narratives, rhetoric and argumentation and, over time, becomes naturalised and accepted as truth (Gasper and Apthorpe 1996, Adger *et al.* 2001). As Ribot and Peluso (2003) also maintain, discourse is a key means to produce categories of knowledge and justify forms of intervention and control over natural resources. In colonial Kenya, for example, the idea that Maasai were 'profligate natives' who did not use land to its full potential provided justification for the appropriation of their land by Europeans (Collett 1987).

Authors such as McCarthy and Cramb (2009), Hunsberger (2010) and Da Via (2011) have found that policy and development narratives often support large-scale land acquisition. Project proponents present a certain vision of rural development that involves commercial agriculture, and depict targeted land as empty or underused. It is reminiscent of the colonial vision of the African landscape as a blank slate, simultaneously an Edenic wilderness and potential farmland 'ripe for settlement and colonisation' (Wylie 2007, 133, see also Ariza-Montobbio *et al.* 2010). Another characteristic of discursive strategies is the rhetoric of illogical persuasion (Gasper and Apthorpe 1996). For instance, several proposals for large-scale land deals promise to create local employment while at the same time heralding the labour-saving advances of irrigation technology (Li 2011). Conversely, research from other contexts provides evidence of marginalised groups mobilising narratives and of discourses being intentionally embedded in resistance strategies (e.g. Neumann 1995, Schwartz 2006). In order to understand and disaggregate local people's attitudes towards land deals as they coalesce around certain political positions, it might be helpful to broaden thinking from the preoccupation with what Wetherell (2007) calls 'big discourse' as a structural means of legitimisation or resistance. This creates more breathing space for the historical experiences and understandings of nature shaping personal life-worlds (Schutz and Luckmann 1973).

In summary, this paper is guided by the proposition that domestic players have a greater influence on large-scale land deals than they are accorded in the neo-colonialist argument. Institutions and a variety of access mechanisms – operating as part of or across institutions – may shape how residents respond to the deals. Additionally, the paper is informed by a particular interest in how local individuals frame their life experience and how evolving land deals fit into their development concerns and expectations.

3. Case study and methods

3.1. Two land deals in Tana Delta

The Tana River delta is located at the Kenyan coast (Figure 1). It begins near the town of Garsen, where the river floodplain widens into the characteristic fan shape of a delta and the Tana meanders for approximately 45 km until reaching the ocean (Hughes 1990). The area is sometimes referred to as the Lower Tana. Upstream rains flood the wetland twice a year, corresponding to the rainy seasons of April–June and November–December. The flooding fills oxbow lakes and supports grassland, bush and riverine forest, plus an exuberance of birds and mammals that led conservationists to declare the Lower Tana a biodiversity hotspot and an internationally important bird area (Hamerlynck *et al.* 2010, Nature Kenya 2011). The district overall is considered low-potential, arid and semi-arid land (ASAL), with low, erratic rainfall (averaging 611 mm a year in Garsen)[1] and a high evaporation rate. The Tana easily breaches its soft alluvial banks, and there are winners and losers when the river floods or changes course; unlucky communities find themselves cut off or crops and livestock swept away.

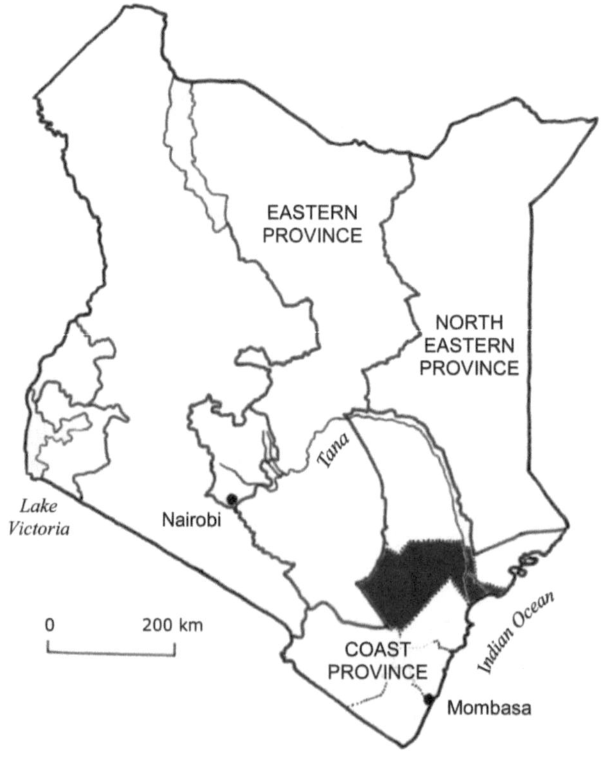

Figure 1. Tana Delta District.

[1]Data for 1994–2007. Source: Kenya Meteorological Department, Garsen station.

'Tana Delta' also refers to the administrative district, with a population of 96,664 and density of 6 people/km^2 (KNBS 2009).[2] Comparable census figures of 1 person/km^2 in 1969 and 3 people/km^2 in 1979 indicate the extent of population growth (NES 1985). Tana Delta remains overwhelmingly rural, the main livelihoods being crop farming, pastoralism and fishing (RoK 2009). These activities are often associated with the largest ethnic groups: Pokomo farmers, Orma, Wardei and Somali pastoralists, and immigrant Luo and Luhya fishers.[3] However, residents pursue a range of livelihood strategies that do not fit neat ethnic categories. Farmers and pastoralists alike engage in fishing, timber harvesting and charcoal making. Women might generate income by weaving mats or selling food; men pursue cattle broking, salaried jobs or casual labour. To cope with drought and livestock losses, Orma and Wardei pastoralists also grow crops. Their transition to agro-pastoralism is part of a process of agrarian change that has brought pastoralists into the money economy, changed their staple food from milk or meat to maize and increased elders' approval of education for children. Most Orma and Wardei that farm still keep livestock, however, which seems to provide more income options than are available to farmers, as women sell milk and men can 'cash in' animals when they need money. Thirty-four percent of the district's population aged 15 and over can read and write, compared with the national average of 71 percent (RoK 2009, UNDP 2010). There are 15,068 beneficiaries of food relief in the district (interview, Kenya Red Cross Society, 2 July 2011).

In the early twentieth century colonial officials established unsuccessful cotton plantations in the Lower Tana, and planners have wanted to reintroduce large-scale irrigated monoculture to the area since the 1950s (Ominde 1971, Miller 1981, NES 1985, Hughes 1987). Under the stewardship of the Tana and Athi Rivers Development Authority (TARDA) and the National Irrigation Board, a number of projects were introduced to grow rice, cotton and maize, notably schemes at Bura and Hola north of Tana Delta funded by the Kenyan government and the World Bank, the Lower Tana Village Irrigation Programme, funded by the World Bank and the Dutch government, and the Tana Delta Irrigation Project (TDIP), funded by the Japanese government. These projects were dogged by mismanagement, a lack of alignment with local practices and expectations, and extremely high capital costs (Adams 1990). Reliance on external funding and expertise came at the expense of local capacity and ownership (Osoro *et al.* 1983), and overlapping donor interests may have inhibited the emergence of cross-sectoral, regional planning (Little 1987, Rowntree 1990, KWF 2005). Managing projects was also hindered by remoteness and difficult environmental conditions (Ominde 1971, Miller 1981, Rowntree 1990, Van Eijk 1998). The TDIP scheme in particular was damaged by El Niño rains in 1997 (Hamerlynck *et al.* 2010) and lay dormant until being re-launched in 2009.

Despite advice to modify the development strategy, irrigating the Lower Tana remains a government objective. Encouraged by favourable research by the Kenya Sugar Research Foundation, TARDA has struck a deal with Kenya's Mumias Sugar

[2]In this article, we use 'Tana Delta' to refer to the wider geographical area demarcated by the district but excluding Assa location (see Figure 2), and 'the delta' to refer to the wetland.

[3]Smaller groups include Banjuni, Boni, Giriama, Watta and Wasanye. There are distinct Pokomo groups of Upper Pokomo (surveyed village 6), Lower Pokomo (villages 3 and 4), Malakote and Korokoro.

Company to cultivate sugarcane on a 16,000-ha estate east of the Tana. Farmers are expected to provide 4000 ha of their own land for outgrowing. The project, called the Tana Integrated Sugar Project (TISP), would extract water from the river for irrigation and includes an ethanol production plant and a livestock component for fattening beef cattle (HVA International and MA Consulting 2007).

Since 2008 there has been another wave of interest in the Lower Tana, particularly from foreign investors. Among them is the Canadian company Bedford Biofuels. It has acquired 93,337 ha to grow and process *Jatropha curcas*, a plant that has been used to produce biodiesel elsewhere in Kenya (Hunsberger 2010). Only 64,000 ha will be cultivated; the remainder will be kept for other uses. Bedford has sub-leased the land from six ranches, which are large tracts – typically several thousand hectares – leased for livestock production. The project is widely reported to be located in the delta, but most of the ranch land lies outside it see (Figure 2). Although this area is more arid, Bedford Biofuels does not anticipate the use of irrigation, planning instead to use dams and boreholes and rely on *Jatropha*'s reputation for requiring little water (ABF 2010).

In accordance with Kenyan regulation, the TISP and the *Jatropha* project were both subject to an environmental and social impact assessment (ESIA) for approval by the National Environment Management Authority (NEMA). The TISP ESIA (HVA International and MA Consulting 2007) identified numerous potential negative impacts, including water shortages, biodiversity loss and displacement of villagers from project land. The authors of the Bedford Biofuels ESIA forecasted a similarly long list of impacts (ABF 2010). One finding was that clearing land for *Jatropha* could disrupt wildlife migration and increase the incidence of human–wildlife conflict. However, both ESIAs concluded in favour of the projects owing to the projected benefits of rural development, improved fuel security and poverty alleviation.

While insisting on certain conditions to minimise its environmental impact, NEMA licensed Bedford Biofuels to grow *Jatropha* as a 10,000-ha pilot, and the process of conversion from ranch to plantation is under way. In contrast, the proposal by TARDA–Mumias was deemed too sensitive to approve wholly. In 2008, NEMA awarded a conditional licence for an initial rice-growing phase of 5000 ha, agreeing to consider expansion at a later date (NEMA 2008). At the time of writing, the TISP is in stasis (see Table 1 for key project characteristics).

3.2. Methodology

Fieldwork was conducted during July–September 2011. At national and regional levels, thirteen interviews were held in person, by phone or by email with stakeholder organisations based in Tana Delta, Nairobi and Nakuru. The organisations and individuals interviewed included Garsen market officials, a government planner, Kenya Land Alliance, Kenya Livestock Marketing Council (KLMC), Kenya Red Cross Society, Kituo cha Sheria (a Kenyan NGO that provides advocacy and legal aid on human rights), the Lower Tana Delta Conservation Trust (LTDCT), the Ministry of Livestock (MoL), the National Agriculture and Livestock Extension Programme, Nature Kenya, the Pastoralists Development Network of Kenya and two local activists. Unfortunately, Bedford Biofuels and TARDA declined to participate and it was not possible to interview conservationists and pastoralists who were part of the initial opposition to the TISP in 2004/5, or informal local activist

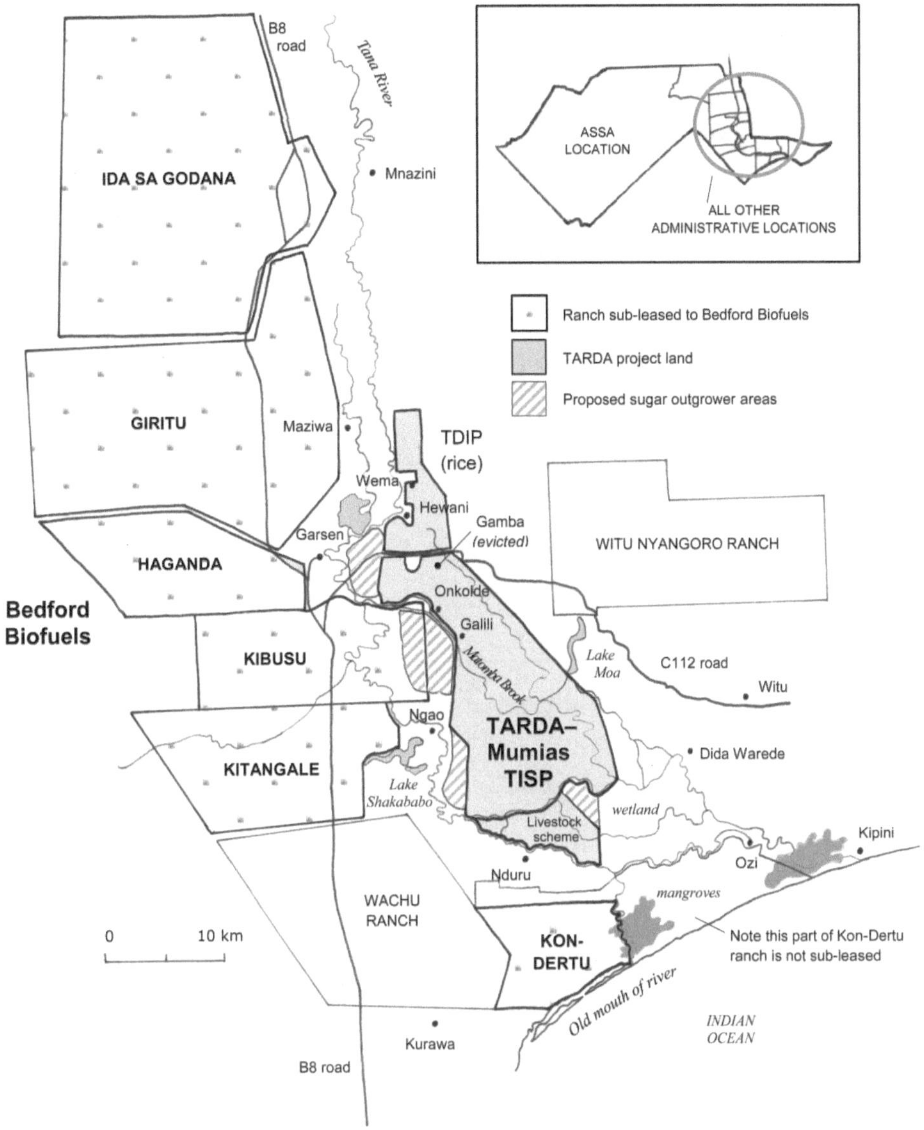

Figure 2. Project areas and position in Tana Delta District (inset).

groups who are apparently beginning to mobilise against Bedford Biofuels. It was difficult to obtain written information given the secretive and sensitive nature of the deals, but a number of secondary data sources, particularly policy and project documents and newspaper reports, were consulted.

At the local level, ESIAs and maps suggest there are at least 40 villages and settlements on or near the targeted project land. Of these, eight were selected for interviewing. To preserve anonymity the villages are not identified here, but their ethnic affiliation and location are shown in Table 2. Forty-eight interviews with villagers and another seven interviews with members and officials of Ida sa Godana, Kitangale and Witu Nyangoro ranches were conducted. Respondents were invited to participate through random, quota sampling to capture the views

Table 1. The two land deals.

	TARDA–Mumias Sugar (TISP)	Bedford Biofuels
Proponent's nationality	Domestic (Kenyan)	Foreign (Canadian)
Crop	Sugar (food and biofuel)	*Jatropha* (biofuel)
Size	33,000 ha	64,000 ha (total acquired: 93,337 ha)
Farm model	Plantation (16,000 ha), on-site processing, small-scale outgrowing (4000 ha) in future, livestock	Six plantations, nearby processing, small-scale outgrowing in future
Type of land	Mostly floodplain grassland	Mostly arid rangeland
Irrigation	Irrigation needed	No irrigation planned
EIA status	ESIA partly approved July 2008	ESIA partly approved May 2011

Table 2. Villages and people interviewed.

Village number	Dominant ethnic group	Location/characteristics	Number of people interviewed
1	Luo/Luhya and Orma	On or near TISP land	13
2	Orma	On or near TISP land	14
3	Pokomo	On or near TISP *and* on or near Kitangale Ranch	13
4	Pokomo	On or near Kibusu Ranch	6
5	Orma/Wardei	Herders who graze on project land	14
6	Pokomo	On or near Ida sa Godana Ranch	15
7	Wardei	On or near Ida sa Godana Ranch	9
8	Wardei	On or near Ida sa Godana Ranch	5

of men, women, elders and youth. Interviews at the local level were semi-structured and mostly one-on-one, though some were group interviews. Some questions about livelihoods and land use were clarified with transect walks and participatory mapping. Many of the respondents feared eviction and, furthermore, the land deals had become political and there were reports of intimidation on all sides. Therefore, although interviews were carried out under guarantee of anonymity, this is likely to have affected people's openness, and the research findings should be treated with caution.

4. Encountering resistance, engendering support: how land deals were made or broken

4.1. The local response

The interviews support the message from secondary data of widespread resistance to the TISP. The local opposition has linked to political and environmental movements in 'resistance coalitions' that have made it difficult for the TISP to progress. In contrast, Bedford Biofuels' proposal has been received more warmly. Our interviews recorded broad support at villages 3, 4 and 5, although a handful of respondents there expressed wariness over *Jatropha* and the deal's transparency. Others at villages 5, 6 and 7, along with local activists, oppose it.

The main discursive positions that affect local opinion on the TISP and Bedford Biofuels deals are shown in Table 3. Only one location, village 5, had a mixture of opinions for or against. Elsewhere, villagers were united, even if they varied in their reasons. Interviews revealed heterogeneity in respondents' livelihoods, education levels and life-worlds. But on the issue of the two deals, there is no significant divergence along gender, age or class lines. The limitations of small samples notwithstanding, this might suggest villagers' discursive positions are shaped collectively.

In what follows we present our findings structured around the four key differences between the two deals that explain the dynamics of opposition and support in Tana Delta: differences in the underlying tenure system, in the investors' reputation, in the perceived value of the targeted land and in the extent to which local concerns have been able to coalesce into activism.

4.2. The **de jure** *setting*

The status of land tenure in Tana Delta is complex and unclear. As is common in postcolonial countries, land relations are characterised by institutional pluralism involving overlapping – and contested – statutory and customary claims (Sikor and Lund 2009).

Officially, land in Kenya was, until a change to the national Constitution in 2010, categorised as either trust land, government land or private land.[4] Trust land was so named because the land was held 'in trust' by local county councils on behalf of its residents, allowing for collective, customary tenure. Government land, vested in the Kenyatta government after Independence in 1964, included townships, forests and national parks but also unalienated agricultural land (Owiro 2000, Kenya Land Alliance 2002). Land could become private through a process of adjudication introduced by the colonial authorities in 1954, whereby Kenyans were encouraged to register surveyed plots and receive individual freehold title (Thurston 1987). Although most of the intensively farmed Central and Western provinces have been subdivided and privatised in this way (Ministry of Lands 2009, Syagga and Mwenda 2010), adjudication has been slow in other parts of the country (Kanyinga 1998). In Tana Delta, attempts to demarcate and register land are reported to have triggered conflict between farmers and pastoralists in 2001–2002, and the adjudication programme was suspended (Global IDP 2004). Today only 4.3 percent of farmers in the district have title deeds for their farm plots or *shambas* (RoK 2009). Several respondents said they applied for title years ago and are still waiting.

In addition to this process of registration for everyday Kenyans, parts of trust land and government land could be acquired – or 'set apart' – through the authorities for public or private purposes. Although there were checks in the legislation designed to protect the public interest (Owiro 2000), Kenya's postcolonial tenure system failed to safeguard community land and valuable ecosystems, and there was systematic divestiture of public land to wealthy individuals via parastatals and land-buying companies (Klopp 2000, Lukudu *et al.* 2000, Southall 2005,

[4]The new categories in the 2010 Constitution are community land, public land and private land.

Table 3. Discourses of local support and resistance.

		Village[a]						
Proposal	Reasoning behind support (✓) or opposition (×)	1 Orma, Luo/Luhya Agropast, fishing	2 Orma Agropast	3 Lower Pokomo Farming	4 Lower Pokomo Farming	5 Orma, Wardei Agropast, broking	6 Upper Pokomo Farming	7 Wardei Agropast, trade
TISP	*Reasons to oppose*							
	Belief in need for delta to remain grazing	×	×		–	×	–	×
	Community claim to delta		×		–		–	
	Fear of eviction or lost access	×	×		–		–	
	Vision of development through self-employed farming	×	×		–		–	
	Resentment towards TARDA	×	×		–	×	–	×
	Reasons to support							
	Antipathy towards pastoralists			✓	–		–	
	Vision of development through jobs			✓	–		–	
	Desire for agricultural development projects			✓	–		–	
Bedford Biofuels	*Reasons to oppose*							
	Belief in need for ranch to remain grazing	–	–			×		×
	Community claim to ranch land	–	–			×	×	×
	Fear of eviction or lost access	–	–				×	×
	Resentment towards ranch	–	–				×	×
	Reasons to support							
	Vision of development through jobs	–	–	✓	✓	✓		
	Desire for agricultural development projects	–	–	✓	✓			

– deal not discussed in interviews
[a]Respondents at Village 8 requested not to comment on specific deals

Kanyinga *et al.* 2008). Authorities helped create a culture of patronage whereby land became a tool of securing political support, while customary residents legally became squatters.

At the time of the 2010 Constitution, approximately 70 percent of Tana Delta District was government land, versus 20 percent of trust land (RoK 2009, see also Ogolla 1996, 86, fig. 2). The precise tenure categories of the land acquired by TARDA–Mumias and Bedford Biofuels are unclear. We did not see the land deeds, and secondary sources give conflicting information, reflecting the lack of transparency in Kenya's land system (Ogolla 1996). Some accounts suggest the TISP area was trust land (Kithi 2004, Glenday 2005, Nature Kenya 2008, Omondi 2008). Others, including the TISP ESIA, report that it includes government land also. When asked, respondents living near the TISP area in villages 2 and 3 said they believed their *shambas* were located on trust or 'reserve' land.[5] The ranches that Bedford Biofuels has sub-leased were apparently located on government land (ILCA 1979, 97, NES 1985, 45, Little 1987), although the ESIA defines it as trust land. Eriksen *et al.* (1996, 217) write that when one of the ranches, Kon-Dertu, was registered, 'the fact as to whether the land was public (government-owned) or trust land was never clearly established'.

The process by which TARDA acquired its land in Tana Delta, be it government or trust, was long and protracted. Following an informal grant by the authorities in the 1970s, TARDA received an official letter of allotment for 28,680 ha in 1995 (Glenday 2005, HVA International and MA Consulting 2007). This grants the potential landowner leasehold title, providing they comply with certain conditions and fees. According to the ESIA, TARDA was also instructed to compensate displaced 'squatters'. Letters of allotment are valid within a certain period, usually 30 days and in this case three months (Lebrun *et al.* 2010). However, TARDA was not awarded title until 2009. Opponents therefore claim that its 45-year lease is void. They argue the land was, or should be, trust land and should not have been alienated for commercial development (KWF 2005, Kweyu 2008). Bedford Biofuels, meanwhile, is in the process of acquiring title for its ranches that lack the paperwork.

Thus, the inefficiencies and abuses of Kenya's land system have created an environment in which it is possible, and even accepted, for developments to involve improper acquisition of land or to infringe on people's customary rights. The ease with which trust or government land could be set apart for parastatals and ranches laid the foundations for the current deals in Tana Delta. It also fostered resentment among local land users, affecting TARDA in particular, as we explain below.

4.3. Livelihoods and contested tenure rights

Both of the Tana Delta ESIAs recognise that parts of the newly acquired land are already put to small-scale use by communities with a stake in the area's future development. The diversity of livelihoods in Tana Delta means that people rely on a range of natural resources. In addition to fishing grounds, *shambas*, pasture and forest, water sources are crucial. During the year, livestock is moved through the landscape to find pasture and water. The ranches and TISP land are both used by

[5]'Native reserves' were one of the land categories amalgamated under the 1963 Trust Lands Act.

residents and visitors, mostly for grazing, fishing and charcoal burning. This complex pattern of land use is managed by customary arrangements, since so few people have private land rights. The customary Pokomo system of patrilineal inheritance of *shambas* has evolved to allow unofficial buying and selling of *shambas* outside the family and inheritance by daughters. Among Orma and Wardei, land is not owned or inherited. Rather, 'bylaws' are used to allocate grazing areas, farming areas and water rights. Wooded areas where people harvest timber and firewood are open access, though communities sometimes claim ownership. The few private claims to land are made by TARDA and the ranches, plus government irrigation schemes and individual investors.

The co-existence of customary and legal forms of ownership has led to overlapping legitimacy claims. *De jure* ownership of alienated government and trust land is not necessarily respected, for three reasons. First, customary claims are viewed as a legitimate alternative. Second, many people are not aware of the boundaries of private land, and trespass accidentally. Respondents at village 2, for example, said the boundaries of TARDA's property are not clearly marked; one man relies on a neighbour who has documents and knows where the beacons are. Third, some users believe private land was acquired illegally, and rightfully belongs to their community. When asked why his community did not pay rent to ranch owners for the use of ranch land, a respondent from another village said, 'Why should we pay? They're not the government.' The ambiguity of *de jure* claims stems partly from the practice in Kenya of awarding landowners a provisional letter of allotment separately from registered title. In practice many land acquirers, including until recently TARDA and several ranches, have only the letter yet still claim ownership. This leads to bureaucratic confusion, encourages double allotment of the same plot and undermines the official tenure system.

Among informal, *de facto* systems there are also multiple claims of legitimacy. This stems from, firstly, the presence of several ethnic groups with various institutional norms for land use – Tana Delta is one of the most ethnically heterogeneous districts in Kenya (Kimenyi and Ndungu 2005, see Section 3.1 above). There is obvious overlap, for example, between individual Pokomo *shamba* ownership and the commons system of Orma and Wardei. Contested areas include land adjacent to the river: traditional Pokomo territory which is also used for watering livestock (Miller 1981, Hughes 1987, Irungu 2000). There are also differences within ethnic groups. One elder explained that the Lower Pokomo are more concerned with jobs and education than Upper Pokomo villagers, a stereotype that might nevertheless help explain the differences between village 6 and villages 3 and 4 (see Table 3).

Secondly, both farming and pastoralist groups claim ancestral links to the area. Long-term residents contrast this indigeneity with the apparently weaker rights of more recent settlers, who, nevertheless, might in turn claim legitimacy if they were given permission to settle by village elders. In Tana Delta, land relations are dynamic and have changed over time. Decades of movement, transhumance and in-migration, as well as changes in economic livelihoods and the riverine landscape itself, have continuously opened up new spaces for negotiation of access to land, water and associated resources by various indigenous and incoming groups.[6] During the 1970–80s, the government created

[6]For a historical review see Miller 1981.

several settlement schemes for incoming workers in neighbouring areas (ILCA 1979, RoK 1989, ILO 2000). Near the village of Dida Waride, we visited a scheme reportedly launched by Mwai Kibaki during his 2007 Presidential election campaign, whereby participants paid a fee for a plot with title deed. Respondents told us the scheme attracted Luo and Kikuyu settlers from western Kenya. But in general we came across more instances of self-organised in-migrants who were motivated by economic opportunity or climate shocks, such as Giriama smallholders or the fishers at village 1.

To respond to perceived pressure on resources, one Pokomo respondent mentioned the establishment of new villages to protect 'ancestral' land – a strategy of territorialism (Sikor and Lund 2009). In another form of asserting access, several Pokomo respondents described Orma and Wardei neighbours as Cushites or *wageni* (strangers or guests), discursively framing them as newcomers with weaker claims to land. Although ethnicity is only one possible dimension of differentiation (Peters 2004), it does appear to be a factor that can ease or obstruct access in Tana Delta. Some Orma respondents reported a sense of kinship with Somali and Wardei, for example, which obliges them to share grazing grounds and offer a temporary home. The ranches offer membership of a different kind, as we will discuss below, while membership of government- and donor-led irrigation schemes can provide a villager with a *shamba*, pumped water and access to extension services. Evidently Ribot and Peluso's access mechanisms of social relations and identity are in play in Tana Delta.

Acknowledging such diversity and understanding complex processes of land access is important for the land deals. On the one hand, it means that investors do not face a single polity with unified interests, and can benefit from tensions between user groups. On the other, managing and compensating messy grievances over land presents a costly complication for investors (Poulton *et al.* 2008, McCarthy and Cramb 2009). We have seen that, without official rights, communities on government and trust land have historically been vulnerable to dispossession and displacement in a culture of land grabbing by the political elite. However, the very persistence of customary tenure systems in Tana Delta *in the absence of private ownership* has created a legal platform for resistance. Since the customary claims are cultural institutions, it can be argued that they take precedence over more recent *de jure* claims and should be upheld as an indigenous or human right.

Thus, the strength of customary claims and lack of respect for official ownership provide scope for both deals to be opposed. Nonetheless, differences in the underlying tenure system help explain the contrasting approaches taken by TARDA–Mumias and Bedford Biofuels, and why their proposals have not been responded to in the same way. Although the TISP area has been private land unofficially since TARDA was granted the land in the 1970s and officially since it received title in 2009, the land has not yet been developed and has been *de facto* operating as commons. The TISP proposal could therefore be considered a form of enclosure, involving a physical and symbolic wrenching of land away from community users. This is one of the reasons why we encountered strong resistance to the project among fishing and agro-pastoralist villagers who live on or near TARDA property. A Wardei respondent from village 7 claimed, 'TARDA got our elders to sign away land for sugar, and because the elders were illiterate they just signed with their thumbprint.' Pokomo villagers have also opposed enclosure: in

1994, members of Salama location in the TDIP area filed a civil case to protest against TARDA's allocation of land (Lebrun 2009). In contrast, the ranches sub-leased by Bedford Biofuels were already private land and, significantly, they are owned by local people who constitute the ranch management and membership. Consequently, the *Jatropha* deal is generally perceived not as a land grab but as an agreement or transfer of legal ownership. A respondent from village 1 near the TISP area told us, 'We want to be able to negotiate with developers like they did with the *Jatropha*'. This support for the deal can be explained in the light of the ranches' history.

4.4. Ranches: the incentive to sell

The policy of establishing livestock ranches in Kenya's ASALs emerged in the 1960s, influenced by ideas on range management from the United States (Von Kaufmann 1976, Umar 1997) and the perception that pastoralism is an underproductive activity (Eriksen and Lind 2009). The hope was to sedentise nomads and commercialise subsistence pastoralism, thereby making pastoralists 'economy-minded' (Abercrombie 1974, 13). The best known are the group ranches established by Maasai, Samburu and others (Rutten 1992). In this model, groups would come together under the Land (Group Representatives) Act and register an area of trust land for keeping livestock, as a compromise between traditional grazing systems and the policy imperatives of increasing productivity, avoiding overgrazing and introducing private property (Simpson 1973, Lenaola *et al.* 1996). Other ranching models were designed for different tenure situations and ethnic groups who were thought to be more entrepreneurial (see Ominde 1971, Abercrombie 1974). Relevant to our case study are company ranches and cooperative ranches. The former bought or leased government or trust land, running the business with capital provided by shareholders. A modified form known as Directed Agricultural Companies was designed with the presence of a ranch manager from the government, and these DAC ranches were similar to cooperative ranches, in which members paid share capital to join a ranch cooperative, run by a management committee and, typically, a hired ranch manager (Simpson 1973).

Despite their lack of experience in livestock management, Pokomo communities were encouraged to form cooperative or company ranches in Tana Delta, as were local Orma. By 1978, ranches flanked the west bank of the Tana, including five of those acquired by Bedford Biofuels (the sixth, Kibusu, was founded in 1990). In contrast with group ranches, participants did not live on site but contributed livestock as their share to be managed by a small number of staff. The government's objective for Lower Tana ranches was 'to organise the pastoralists into groups and then settle them' (NES 1985, 46). Locally, however, the rationale was to acquire or safeguard land. Respondents recall that land at that time was threatened by appropriation by bandits, politicians and land-buying companies and, in the case of Kon-Dertu, by plans for a wildlife reserve (interview, LTDCT, 3 September 2011).[7] Interviews suggest that communities therefore identified ranches as a possible means

[7]In this paper we use the term 'interview' in brackets for data obtained through face-to-face conversations, while we use the term 'personal communication' for commentaries made by interviewees via email.

of exclusion, encouraged in some cases by their Member of Parliament (MP) who urged them not to become 'squatters on your own land' (interview, Chair of ranch forum, 22 July 2011). In the words of a Kitangale ranch member, 'idle land is not yours until you develop something'.

As commercial enterprises, the Tana Delta ranches have not succeeded and only Ida sa Godana, a cooperative ranch, is considered active today. Respondents attribute this failure to a combination of under-investment, mismanagement and corruption, environmental problems and cattle rustling. Though some ranches received state loans, others lacked the necessary paperwork. Again, Kenya's complex tenure system caused difficulties: there were delays in receiving letters of allotment and thereby title deeds (NES 1985, Little 1987). Giritu ranch was able to secure its letter of allotment in 1975, but four years later, an observer wrote that 'no development loan has been granted because the members' equity contributions are insufficient and because a Letter of Allotment ... cannot be used as security' (ILCA 1979, 97). Livestock ranching is highly capital-intensive (Von Kaufmann 1976), and funds to maintain cattle dips and other equipment dried up. This was exacerbated by the lack of market linkages and rural infrastructure needed to support commercial livestock production (KLMC, personal communication, received 25 September 2011). Furthermore, a hiring freeze at the Ministry of Livestock was introduced through structural adjustment in the 1990s, limiting the support available to ranches through extension (interview, MoL, 24 August 2011). Some members suspect funds were embezzled by ranch elites, or that the illiteracy of Orma owners, in particular, was exploited by government officials. Aside from these managerial and economic challenges, livestock was lost to disease, flooding and drought (EcoSystems Ltd 1985). Officials at Kitangale and Witu Nyangoro say bandits stole hundreds of cattle (ranches are not fenced). These findings chime with evidence in other areas of the country, where ranches failed to suit pastoralists' norms of land and cattle ownership, or the strategies of transhumance and large herd sizes developed by pastoralists to cope with Kenya's variable climate and pasture ecosystems (Ensminger and Rutten 1991, Mutiso 1995).

The decline of the Tana Delta ranches has given ranch owners a strong motivation to sub-lease the land to Bedford Biofuels. Firstly, the ranches fell into debt over land rates and other obligations. Bedford Biofuels offered to clear these and has apparently paid each ranch Ksh 2 million (USD 20,000) in rent. Secondly, as activity on the ranches ceased, farmers and seasonal herders who were not members began to use the land more heavily. Their *de facto* and illegal access mechanisms – squatting, trespass and the threat of violence – have proved more effective than the ranch owners' *de jure* ownership, and the ranches need to find a new purpose for the land in order to protect it. Thirdly, owners worried that because the ranches were dormant they could be classed as idle land and repossessed by the government. Owners were not sure that their 45-year leases would be renewed, even if they had the funds. In the absence of alternative investment offers, they decided to sign with Bedford Biofuels. At nearby Witu Nyangoro ranch, shareholders were able to turn down Bedford Biofuels partly because they had negotiated a more favourable deal with another investor (interview, ranch official, 17 July 2011).

In sum, despite both project areas having been privately owned by TARDA or the ranches for some years, people have continued to use the land as if it were trust land, following customary or *ad hoc* arrangements. But whereas the existence of *de facto* claims to TARDA's land is impeding the sugar deal's progress, the inability of the ranches to repel *de facto* claims to their land is creating motivation for the *Jatropha* deal to go ahead. Some of those claims are made by local residents who maintain that ranches were located on existing pasture or village land, leading in at least one case to forceful eviction (interview, Pokomo elder, 27 July 2011). Among our interview sample, such claims were made at villages 6 and 7 near Ida sa Godana ranch and by the LTDCT at Kon-Dertu. In addition, the requirement to pay a membership fee excluded the poorest community members from joining ranches, and there were also limits to the number of members allowed so as to avoid overstocking. The establishment of the ranches, then, has created some local resentment among people who contest access or see the ranches themselves as a form of enclosure.

4.5. Consultation, promises and dealing with dissent

Like Mali, Mozambique and several other African countries, Kenya requires project proponents to undertake public consultations, which typically provide an opportunity for affected groups to voice concerns but also for community elites to disproportionately benefit from discussions (Cotula *et al.* 2011). The issue of consultation was mentioned in 37 of our 48 village interviews (involving 59 people), and about a third of the respondents said they had attended public meetings or consultations regarding one of the deals, although they did not necessarily feel adequately informed. A slightly larger group did not attend any meetings and heard about the deals through neighbours, elders or the village chief, but for the remaining third it was not clear to what extent they had been consulted, and our findings are not conclusive on this point.

Consultations for the deals were undertaken by the developers, and by ESIA teams and NEMA as part of the mandatory impact assessments. Judged against best practice (Bisset 2000, Vermeulen and Cotula 2010a), there seem to have been some flaws in both consultation processes. Large public meetings were held, which was apparently not the best forum for imparting detailed information (judging by respondents' poor knowledge of *Jatropha* and how outgrower schemes work, for example) or eliciting opinions from women, who are not used to speaking out in public (A Rocha 2011). There are allegations from interview respondents and conservationists (EAWLS 2008, RSPB 2011c) of people being hired to give support at meetings, and of village elders being bribed to endorse deals. Both projects promise local employment; several respondents said they were told that the Bedford Biofuels project will generate 10,000–15,000 jobs. However, most large-scale land deals create less employment than promised (Franco *et al.* 2010, Li 2011, World Bank 2011) and jobs tend to be 'unskilled, short-term and small in number relative to the size of the investment' (Cotula and Vermuelen 2011, 45).

As a form of enclosure, the TARDA–Mumias proposal (and its earlier incarnations) did not need to be negotiated with locals. Pastoralist spokespeople who oppose the project claimed communities were not consulted, as did NEMA (Kithi 2005, Opala 2006). However, it is not accurate to say that residents were ignored entirely. In the early stages TARDA attempted to bring pastoralists on board and dampen opposition by convening talks and carrying out a process of sensitisation (Anyanzwa 2004, Ndunda 2004, Mango 2005). The sense that the project was treated as a foregone conclusion during these consultations is encapsulated in this quotation from the former managing director of TARDA:

'We shall not force the proposed sugar project on Tana River people, but we will hold consultations with all those concerned before we go ahead' (Daily Nation 2004).[8] In 2007 and 2008, with Mumias Sugar now on board and pastoralists remaining unconvinced, the number of public hearings about the sugar project increased in a new wave of sensitisation-*cum*-consultation, partly carried out by NEMA and the ESIA team (The Standard 2007, Ringa 2008). The government revised the shareholding structure of the venture to give locals a larger stake, and the TISP's livestock component was another attempt to placate pastoralists (Henry 2008). Unfortunately for TARDA, its chances of winning local support were always threatened by its association with corrupt government and previous top-down development projects such as the TDIP. These are perceived to have taken grazing and farming land and disrupted the natural flooding regime. According to the KLMC representative, pastoralists are automatically hostile as a result: 'You see, anything associated with TARDA is opposed because of past historical injustice TARDA did to the livestock community on acquisition of grazing land and displacement of people' (personal communication, received 25 September 2011). The low wages reportedly paid to TDIP labourers, meanwhile, do not make the promise of jobs on the sugar plantation an attractive proposition either, and several respondents believe they lack the necessary certification for employment.

Whereas TARDA was granted its land from the government, Bedford Biofuels needed to secure leases for the ranches from local owners, and consequently it was obliged to consult and negotiate with local authorities and ranch officials (i.e. the chairman, treasurer and other officials of the six ranches). Accordingly, the process appears to have been more of a partnership. The ESIA report states that Bedford Biofuels made several concessions, such as leaving 60 percent of the ranches *Jatropha*-free. Its promise to allow herders to graze alongside *Jatropha* for a small fee and to protect livestock routes is a key reason why some respondents in the Orma/Wardei village 5 support the deal. As well as clearing debts, the company pledged millions of dollars for a humanitarian programme called EMPOWER, through which ranch companies can nominate and establish community projects (Bedford Biofuels 2011).

Via negotiations, then, the Bedford Biofuels proposal has been brought closer than the TISP to local people's vision of development. Neither deal fully addresses the livelihood aspirations of the interviewed villagers. The villagers' priorities – for more teachers, better healthcare, credit, facilities for processing fish, mangoes and milk, and, above all, support for small-scale irrigation – are actually already being addressed by government initiatives such as the Arid Lands Resource Management Programme and donors such as the World Food Programme. However, these initiatives lack funding and do not always get off the ground (RoK 2006, 2009). Bedford Biofuels benefits from a widespread narrative of Tana Delta having been neglected by the government and, unlike TARDA, it is not associated with a corrupt political regime.

Bedford Biofuels and the ranches hope the *Jatropha* project will bring community benefits. It is not always clear who is included in this notion of 'community'. Establishing the ranches seems to have created an access hierarchy, with ranch

[8]As another example, in a December 2004 visit to Garsen, President Kibaki is reported to have told resident pastoralists to stop being 'selfish' and that the government had already approved the scheme (Ringa 2004).

officials at the top, followed by ranch members, landless *de facto* rights holders, and recent squatters, in-migrants and transient herders at the bottom. When the ranches were inactive, the officials' *de jure* ownership could be ignored by herders, charcoal-makers and other resource users. But as Sikor and Lund (2009, 6) note, though property rights might not guarantee access they can 'come in handy if circumstances change'. With the arrival of Bedford Biofuels, the ranches can again be used as a means of exclusion. Although some concerns of marginalised users were incorporated into the *Jatropha* deal agreement, other concerns, particularly the potential loss of grazing areas, relocation of squatters and a general lack of information about the project, were recorded by the ESIA team but not regarded as a reason to stop the development (ABF 2010, 117, 104 and 70). However, some of the squatters have the backing of a politician who encouraged them to settle. Meanwhile, Wardei and Pokomo squatters have brought lawsuits against Ida sa Godana ranch and Giriama farmers have written to the Ministry of Lands to protest against Kibusu ranch being established without their knowledge (Aluanga 2011). Since Bedford Biofuels' strategy involves development in phases, ranch by ranch, its response to opposition can be more flexible than the TISP. An elder at village 6 believes Bedford Biofuels has postponed its interest in Ida sa Godana, while the LTDCT have been told that *Jatropha* will not be grown at Kon-Dertu, something we could not confirm with the company.

Finally, it is worth noting that some respondents from higher up the hierarchy – ranch members and others from nearby, well-established villages – said they felt involved in the deal discussions. According to one member, in the case of Ida sa Godana, members gathered, discussed the company and decided to accept the deal. 'When the ranch was formed,' he said, 'the elders were illiterate. But now new shareholders, and those who have inherited shares from their parents, are learned and are not being exploited.' But not all of the nine ranch members interviewed were supportive of, interested in or even aware of the deal. A more critical Ida sa Godana member cast doubt on *Jatropha*'s viability and the ranch leadership. Ultimately, it is the ranch elite who have decision-making power. They established a forum committee to deal with Bedford Biofuels as a single executive representing all six ranches, and were able to override misgivings from ranch members and other villagers. Some respondents tactfully accept their authority: 'Since the ranch owners accepted [the deal], we the community have no reason to oppose it,' said one Orma elder. Others are outspoken, one pastoralist at village 5 describing the *Jatropha* deal as another example of grazing land being taken 'through pen and paper'. Reports suggest members of Kitangale and Kon-Dertu ranches were openly divided over the project (Ndurya 2009 and interviews), providing tentative evidence for the deal being made without full ranch membership consent. In short, participation in consultations was unlikely to be wholly satisfactory for all local residents. Aware that they do not have perfect information, informants assess the proposals against past experience with the owners and with previous development projects, through the lens of their own social status and land-use related interests.

4.6. The value of the land

So far we have argued that differences in how the land was acquired and in the reputation of the project proponents have influenced the design and progress of the two deals. A third difference is how the project land is valued economically and culturally. The land targeted by TARDA–Mumias plays a key role in the regional pastoralist economy. During the dry season, livestock from local villages is taken into the area to graze, where it is joined by animals trekked or trucked into the delta to be fed until the grassland floods and they are strong enough to return to the hinterland. The TISP ESIA estimated that 60,000 cattle are grazed in the delta, of which 4500–8000 live permanently on or near TISP land. During the 2011 drought animals came from as far as Masarbit (Figure 3). Garsen has a weekly livestock market that draws buyers and sellers from neighbouring districts. Farming and fishing also take place in the project area, and natural flooding is relied upon downstream. Widespread resistance suggests that TARDA and Mumias Sugar have underestimated the delta's importance to local people, as have the government planners that have encouraged its conversion to crop agriculture. Villagers and herders maintain that pasture is already becoming scarce through overgrazing by

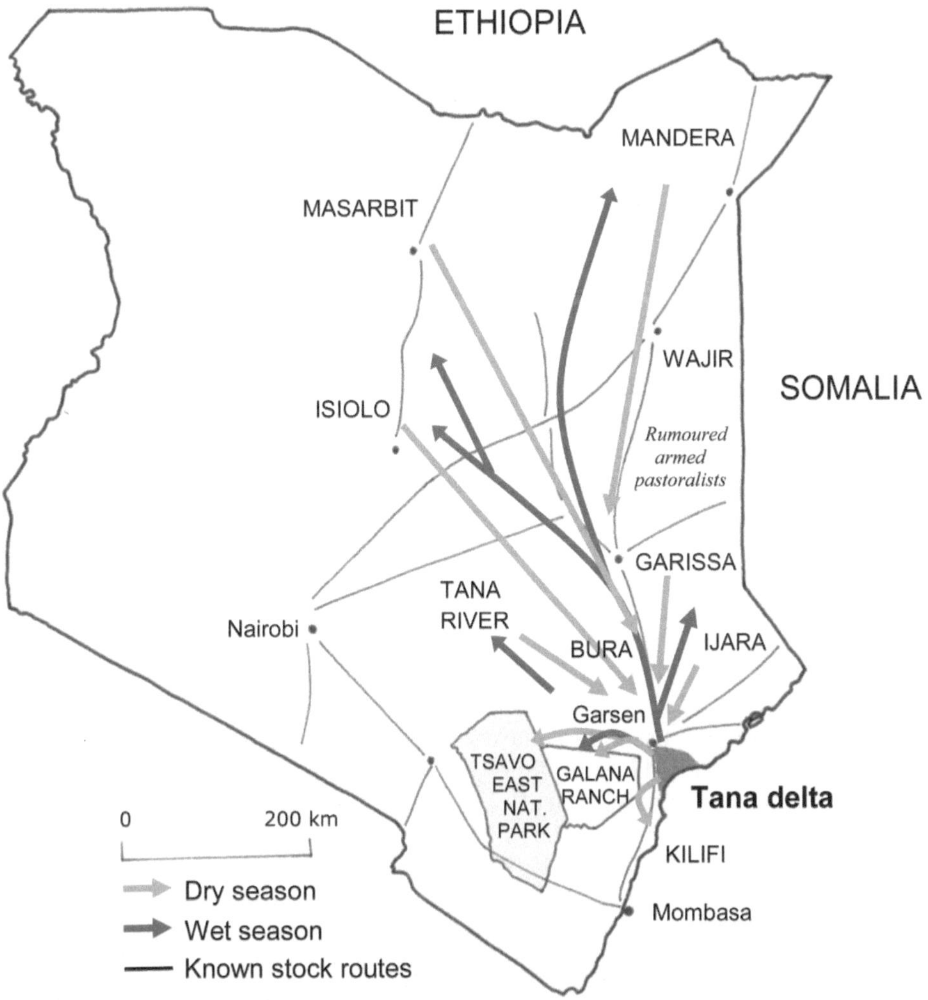

Figure 3. Stylised representation of seasonal livestock influx into the delta, superimposed over stock routes (compiled from Ominde 1971, Muthee 2006, Mahmoud 2010).

outsiders, conversion to irrigation schemes and a drying of the climate that they attribute to deforestation and changes in rain and river flooding patterns. Repeatedly, respondents asked, 'Where will the animals go?' The delta also has symbolic value as a haven. Fisherfolk have moved from Lake Victoria in western Kenya to fish the oxbow lakes, and several Orma and Wardei respondents described moving to the area to escape flooding and drought elsewhere. Initially, the government suggested that part of Galana Ranch to the west could be set aside for pastoralists to use instead of the delta in dry season. Pastoralist villagers and the Witu Nyangoro ranch official told us that some herders do take their livestock to Galana during the short rains (Figure 3). But pastoralist spokesmen rejected the idea as completely unsuitable for dry-season grazing (Kweyu 2008, Nyassy 2008). One group of Orma women we interviewed pointed out ironically that they could not benefit from the much-promised jobs from the sugar project if they relocated to Galana.

Some agropastoralists have another argument against the TISP. Instead of plantation agriculture, they say, people should be assisted in acquiring title and expanding their own *shambas*. This alternative argument could indicate a class and/or generational divide within Orma society, as it was put forward by agropastoralists who rely on farming and are seen as poorer than pastoralists with large herds, and by some young men who expressed disaffection with the pastoralist way of life, their views (or 'life-worlds') having been shaped by personal experience of cattle being washed away during El Niño floods. However, the two positions are complementary, and both result in rejection of the TISP proposal.

The only discursive position in favour of the TISP during interviews was found in Pokomo villages 3 and 4. Respondents here hoped agricultural investment would stimulate development and employment. 'We welcome new ideas if they'll help us move on,' one Pokomo elder said. These residents planned to keep their *shambas* even if they worked on the TISP scheme, thus trying to avoid a depeasantisation process (Van der Ploeg 2010). The villagers do not currently use TISP land and have little sympathy for pastoralists losing access to pasture. This has to do with experiences of the conflict between Pokomos and Orma nearby in 2001–2002. One woman said she had been too intimidated to return to her *shamba* since that time. It was also maintained that Orma had redirected the flow of the river near Matomba Brook, causing a water crisis in the villages (see IRIN 2008). The reduced flow and a lack of rain are blamed for poor harvests in recent years, meaning that farmers can typically grow only enough produce for subsistence. This has slowed the circulation of money in the local economy and increased reliance on food aid. The prospect of employment on the TISP plantation and outgrower scheme offered a solution, and there is frustration that pastoralists have stalled progress. Indeed, it does appear that the practical and symbolic importance of the delta has so far outweighed these villagers' preference for development.

As a more arid environment, the ranches (aside from Kon-Dertu) do not seem to be as valuable to pastoralists and fisherfolk in the sense of a haven or home. Livestock is regularly moved through ranches to and from the hinterland (the Bedford Biofuels ESIA estimates 30,000 cattle graze across the entire project area in peak season). But there are few farming villages on ranches owing to the distance from the floodplain and river. This compounds the message in the ESIA that the land is under-utilised yet overgrazed, and helps justify its conversion to *Jatropha*. There are no comparable villages to the cultural lodestones of Ngao (Pokomo) or

Dida Warede (Orma), which are located in the delta. Although many people – technically squatters – do call the ranch land home, their presence has not slowed the progress of the *Jatropha* proposal. In some cases their needs have been accommodated in the project design via Bedford Biofuels' consultations. Others are marginalised at the bottom of the ranch stakeholder hierarchy. We found little evidence for organised opposition of the Bedford Biofuels deal among squatters, suggesting that grievances are piecemeal and lack the coherent storyline of dispossession through enclosure that unites people who resist the TISP. The more dominant voice comes from longer established villages with connections to the ranches, where Bedford Biofuels and its promised jobs are eagerly awaited.

In considering the varied local response to the two deals, then, it is useful to take a historical perspective. The threat of eviction or losing access to valuable land is a fairly immediate concern. Yet discursive positions for or against the deals are also shaped by the sense of solidarity with neighbours, the perceived threat from outsiders, past grievances and exposure to earlier development schemes.

4.7. Politics and activism

One of the factors that have affected the progress of the TISP has been the nature of its business model as a public–private joint venture. In 2006, the project was halted by a lawsuit brought against TARDA and Mumias by TARDA's previous partner MAT International (Kenya Law Reports 2006). On winning the case, Mumias faced the challenge of raising USD 400 million (Mugambi 2009) and it is not clear if it will have the funds to take the TISP forward. However, a greater cause of delay has been the emergence of resistance coalitions that have counteracted voices of support.

The discursive position of the villagers who support the TISP closely matches that of the government. Seeing ASALs as an overflow for farming, the belief in poverty reduction through commercial agriculture and a determination to bring irrigation to Tana Delta are all visible in Kenya's Vision 2030 development blueprint (GoK 2008). In May 2011, the Prime Minister reiterated commitment to the TISP, stating that 'areas which are suitable for large-scale agriculture production need to be used for those purposes' (Plenary Hansard 2011, 37). TISP proponents emphasise the conservatism of pastoralists, and Tana Delta residents in general, in order to reframe local resistance as backwardness (see Nyassy 2008, RedOrbit 2008). Despite the force of this rhetoric, local resistance has combined with wider activism to obstruct the TARDA–Mumias vision.

The typical strategies of residents for opposing development in Tana Delta are to refuse to move and threaten violence, write letters to local or national politicians, and protest through the courts. After refusing to evacuate the settlement of Gamba on TDIP land, for example, Wardei residents wrote to the Prime Minister when they were forcibly evicted in 2010 (RSPB 2011a). Residents who oppose the TISP have also used (occasionally violent) protest strategies such as demonstrating with placards and threatening TARDA vehicles, according to interview testimonies and news reports (Ndunda 2004, Ringa 2004, The Standard 2006, Ringa 2008). What is crucial in the TISP case is that the grievances of Orma and Wardei have been aired in public meetings and taken up by a local pastoralist elite with links to the political scene (Muchai 2005, Nyassy 2008). In 2004, a nearby MP and the former MP for Garsen wrote against the proposal in a national newspaper, questioning TARDA's connections with the government (Wario and Shambaro 2004). During public meetings in 2007 and 2008, pastoralist spokespeople disputed TARDA's land ownership and claimed the right to grow sugar themselves. The TISP was also debated in parliament. The proposal became politicised, and it was in this context that TARDA's original partner, a Spanish firm, exited and NEMA declined to grant TARDA–Mumias a full licence (NEMA 2008; interview, local activist, 26 July 2011).

The TISP was also opposed by Kenyan and international environmentalists, who had been advocating for protection of the delta under the Ramsar Convention on Wetlands. When details of TARDA's sugar proposal emerged in 2004, environmentalists, notably the East African Wild Life Society, immediately objected (KWF 2005). The ESIA process provided an opportunity for opposition via public hearings and position statements, and it was at this time that the international NGOs BirdLife and RSPB became involved. In 2008 Nature Kenya began a campaign of awareness-raising and Ramsar advocacy among villages. The environmentalist narrative is of an ecosystem in peril. It stresses the delta's importance as a habitat for birds, primates and other wildlife. BirdLife International (2010) maintains that 'Tana delta is important not only to Kenya but also to the global community', thereby staking a claim to what happens there (Roe 1995). The environmentalists' way-of-seeing is different from the way local respondents view the delta. Their primary objective is biodiversity conservation (interview, Nature Kenya, 24 August 2011) whereas most villagers perceive wildlife as dangerous pests, constantly threatening crops (baboons, buffalo), livestock (hyenas, lions) and people (hippos, crocodiles). It was only in an interview with the LTDCT, a community organisation that earns money from tourism, that wildlife was described as an asset.

Despite these differences, environmentalists have been able to form a 'discourse coalition' (Hajer 1995) with pastoralist activists and residents who oppose the TISP by combining ecological concerns with social issues. In a 2005 report, for example, Kenya Wetlands Forum argued that local people had an indigenous claim to TARDA land (KWF 2005). The most significant outcomes of this coalition have been two lawsuits brought against TARDA–Mumias.[9] In the first, a court injunction halted progress on the project from July 2008 to July 2009. The second lawsuit makes more of a human rights argument, based on Kenya's new Constitution (interview, Kituo cha Sheria, 20 September 2011), which aligns it more closely with local opposition discourse. It mirrors the growing concern with land grabbing as an abuse of rural people's rights, which has characterised recent coverage of the Tana Delta deals in the Kenyan media and abroad (FIAN 2010, Aluanga 2011, McVeigh 2011).

The environmentalists' position was able to accommodate the Bedford Biofuels proposal when it emerged in 2009. Indeed, their narrative works best when the delta is perceived to be facing multiple threats. NGOs argue against the viability of *Jatropha* and advocate strategic planning for the whole of Tana Delta (RSPB 2011b). As with their anti-TISP campaigning, they also incorporate social issues; specifically, the threat of eviction mentioned in the Bedford Biofuels ESIA. However, there is no obvious local resistance movement for this to attach to. Since Bedford Biofuels is not associated with the government as TARDA is, the project has not

[9]An earlier example is a group statement made by environmentalists, pastoral groups and a minority rights lawyer in 2005 (Mayoyo 2005).

become politicised to the same degree. Pastoralist rhetoric is weaker because some of the ranch officials that have signed the deal are Orma and grazing will be allowed to continue. Whereas the environmentalists' plea to save the delta for biodiversity conservation complements the local desire to save the delta for farming, grazing and fishing, several respondents believe that the campaigning by NGOs such as Nature Kenya against the *Jatropha* project is spoiling Tana Delta's chance for development. Several respondents, including ranch officials and two interviewees from the otherwise supportive Pokomo villages 3 and 4, were wary of *Jatropha* (in two cases because they had seen crops fail). However, as we have argued above, apparently this has not led to wide-scale questioning of the project, owing to a lack of information, the decision-making power of the ranches and the overriding desire for jobs. Furthermore, there is no unified opposition to Bedford Biofuels from squatters and other threatened resource users across the six ranches. Tana Delta lacks effective civil society to unify their varied interests. Environmentalists are not best suited to campaigning for land rights, and without civil society support, it is difficult to mobilise resistance. Three male respondents said they lacked power or know-how to oppose development, while the interviewee from Nature Kenya suggested that some residents came together to protest against the *Jatropha* project using legal means, but could not afford the legal fees. The forces of conservationist concern, pastoralist political rhetoric and local unity in the face of state alienation of commons have therefore not combined in the same way when it comes to Bedford Biofuels.

5. Discussion and concluding remarks

The previous section has allowed us to identify a series of access mechanisms that TARDA–Mumias and Bedford Biofuels have used to acquire land (Table 4). They have access to capital, authority and knowledge, which eased their acquisition of legal title through bureaucratic channels. Their discursive strategies are similar, promising jobs, fuel security and modernisation, but Bedford Biofuels' unblemished status as an outside investor and its negotiations with local owners gave it an advantage over the TISP. Illegal mechanisms, such as bribery, have allegedly been employed by some project proponents to support and facilitate claims. We can also consider the mechanisms used by Tana Delta residents to re-gain or defend access to the project land. Under the 'legal' category are their customary claims, as well as lawsuits and court injunctions, the key means of local opposition. They also use illegal mechanisms such as trespass and intimidation. Despite the constraints of illiteracy and economic marginalisation, some residents are able to register opposition with figures in authority. However, others lack the necessary know-how or funds. In fact, limited access to capital, technology and markets has stymied agricultural productivity, and this indirectly encourages the land deals by strengthening the perception that the area needs economic development. In the absence of power and capital, the ability to maintain access via social networks become more important, benefiting local land users with status and connections. Even actors who are relatively powerless may find support from electioneering politicians or conservationists who happen to choose their village for advocacy. On all sides, access is further moderated by aspects of the wider institutional and socio-economic setting, such as the national procedure for ESIAs and the weak rural economy in which local actors are embedded.

Table 4. Key mechanisms and institutions for gaining, maintaining and re-gaining access to land in the deals' early stages (adopted from Ribot and Peluso 2003).

	Mechanisms and *wider socio-economic factors* (in italics) shaping access to land	
Interest group	Weak or constraining	Strong or enabling
Local land users (aggregated)	**Lack of legal ownership** **Poor access to capital, technology and markets** **Limited access to information** hinders people's ability to evaluate deals ***Weak rural economy*** ***Policy environment***	**Customary claims** **Illegal mechanisms** and threat of violence **Discourse** to justify access and weaken competing claims **Access to justice system,** meditated by access to capital, knowhow and civil society **Access to authority** (e.g. chiefs, MPs) **Social relations** with pastoralist elite, conservationists ***ESIA regulations*** *provided opportunity to negotiate access through hearings and consultations* ***New Constitution*** *strengthens court case against TARDA–Mumias* Ability to benefit from the above greatly depends on **social identity** (e.g. gender, ethnicity, indigeneity, elder status, ranch membership)
TARDA–Mumias	**Social identity** (poor reputation) **Poor social relations** with political opponents Need to raise **capital** ***ESIA regulations*** *have slowed progress, although EIA standards are low* ***New Constitution*** *has created a more sympathetic environment for debating land rights* ***Justice system*** *(court cases)*	**Legal ownership** **Alleged illicit mechanisms** **Access to information** and expertise **Discourse** and policy narratives **Social identity** as 'insider' with state support **Social relations** in a position of local power ***Policy environment*** ***Land system*** *that historically favours top-down acquisition*
Bedford Biofuels	**Social identity** as 'outsider' in discussions with authorities and negotiations with ranches ***ESIA regulations*** ***New Constitution***	**Legal ownership** **Alleged illicit mechanisms** **Access to capital** **Access to information** and expertise **Discourse** **Social relations** (e.g. negotiations with ranches, concessions for local pastoralists) **Social identity** as 'outsider' distances it from state
Ranch owners	Ranches suffered from **poor access to capital, technology and markets**	**Legal ownership** **Social identity** (ethnicity, elite status)

(*continued*)

Table 4. (*Continued*).

Interest group	Mechanisms and *wider socio-economic factors* (in italics) shaping access to land	
	Weak or constraining	Strong or enabling
	Imperfect **access to information** weakens bargaining position **Weak social relations** with some locals Political *pressure to re-develop or sub-lease 'idle' land*	**Social relations** with villagers, Bedford Biofuels, authorities **Discourse**
Conservationists	**Lack legal claims** **Social identity** as 'interfering' outsiders	**Discourse** **Access to information** and expertise **Access to capital** through donors **Social relations** with local people ***ESIA regulations*** ***Justice system***

A useful way to conceptualise the difference between the two deals is Hall's (2011) distinction of 'secure' and 'insecure' tenure conditions. The TISP case is insecure, meaning that the proposal does not fully acknowledge or compensate existing land use. The ranches are secure, meaning that the investor – Bedford Biofuels – must respect the existing ownership. Bedford Biofuels' more collaborative approach, and much of its relative popularity, stems from having to negotiate with existing private owners. This appears to support the argument that providing local users with clearly defined property rights makes it difficult for their presence to be ignored. Certainly, interview respondents expressed strong demand for title deed. However, most large-scale deals target unfarmed commons (Alden Wily 2011), so we are not talking about mass registration of small plots. The kind of property rights that could protect users in this setting is probably formally recognised community access, incorporating multiple land uses – incidentally, this is the aim of the 'community land' category in Kenya's new Constitution. Deininger (2011), however, acknowledges that it is difficult to design community-wide rights systems that represent all stakeholders and prevent elites from benefiting disproportionately. In the case of the ranches analysed here, it appears that their establishment has deepened or formalised access inequalities and given ranch officials a decisive say in who is included in their 'community', even though their efforts to safeguard land may be well intentioned.

Furthermore, the existence of formally recognised property rights can actually speed up land acquisition. As Deininger writes, 'recording rights provides outside investors with somebody to talk to' (236). Bedford Biofuels and other prospective investors have actually helped ranches to acquire paperwork so that land can be traded. Because Bedford needed the leases and licence, ranch officials and others who were consulted were able to extract concessions from the company. Although this may have made the bargaining positions more even, the ranch owners were, in turn, disadvantaged by their need to clear debts, as well as a lack of information about

Jatropha and land prices (interview, ranch member, 31 July 2011). There is, in any case, a leap from acquisition to implementation, and being able to acquire *de jure* rights does not guarantee investors progress, as TARDA and Mumias Sugar have discovered. This is the experience of other large-scale acquisitions in Africa (Poulton *et al.* 2008, Cotula and Vermeulen 2011). Land deals where tenure conditions are insecure can be forced through, but this creates the conditions for widespread resistance because the large number of people that may be affected provides the foundations for collective organisation and response (McCarthy and Cramb 2009, Vermeulen and Cotula 2010b, Hall *et al.* 2011).

Research has suggested that policy discourse facilitates large-scale land acquisition (McCarthy and Cramb 2009, Ariza-Montobbio *et al.* 2010, Mançano Fernandes *et al.* 2010). So it is striking that the TISP deal has stalled despite the policy vision of irrigated sugarcane and the undertone of doomed pastoralism. Perhaps the influence of discourse has been overestimated. Another explanation is that this policy vision is not hegemonic (Adger *et al.* 2001). In 1993, for instance, President Moi intervened to prevent Kon-Dertu being developed for aquaculture and directed that it be protected as a wetland reserve, leaving confusion today over whether the land is legally owned by the thwarted developer, the ranch company or the community, who understood the land to have been returned to them (Eriksen *et al.* 1996; interview, LTDCT, 3 September 2011). Similarly, the 2008–2012 Tana Delta District development plan includes an objective of conserving the lower delta alongside the commitment to sugar irrigation on the floodplain (RoK 2009). Secondly, the policy has met robust counter-movements, as Polanyi termed them, which is not surprising given the threat to access relations that TARDA's proposal involves (Hall *et al.* 2011). On the subject of doomed pastoralism, we found that Tana Delta's pastoralists are closer to the government position than expected. For some Orma and Wardei respondents, livestock has become a liability. Young men can find life as a herder boring and lonely. These pastoralists are open to agriculture – however, they object to the passive role that they have been given in the storyline of modernising development, especially because they already resent TARDA.

People in Tana Delta evaluate narratives for or against the deals by seeing if they fit with their worldviews and interests. To use rhetorical terminology (Hajer 1995), many people have rejected the TISP proposal for lacking *logos* (i.e. a reasonable answer to 'where will the animals go?'), *ethos* (trustworthiness) and *pathos* (empathy with existing land users). TARDA's attempt to override the insecure tenure conditions has failed – so far. Personal history influences people's attitudes towards the deals, and some of these individual opinions escape the simplified discursive position that seems to be shaped by elders and held collectively by villages. To date, critical land grab research has rarely studied this formation of subject-positions and the way they are distilled into counter-narratives (Davies and Harré 1990). Everyone has their own discursive position – what has slowed the TISP is that people have formed coalitions, linking local residents with activists farther afield. This capacity for activism depends on access to information and civil society, and a certain social cohesion between stakeholder groups, as Pye (2010) discovered in his study of resistance against biofuel developments. Bedford Biofuels' project, apparently more just than the TISP, has not been tested by a comparable counter-movement.

To conclude, it is possible to make two more general observations relevant to the land grab debate. The first is that neither deal in this case study is archetypal. One is a domestic affair, originating earlier than the current global land rush. The other is an acquisition of private ranch land, which, though similar to conversions of ranches in Latin America (CIFOR 2010, Mançano Fernandes *et al.* 2010), is unusual for large-scale land deals in sub-Saharan Africa (Cotula *et al.* 2011). Although the Bedford Biofuels proposal is a better example of a stereotypical neo-colonialist land grab, it has received a warmer reception than the domestic TISP project. As Borras *et al.* (2010) contend, proposals for large-scale agriculture are not necessarily opposed at the grassroots. Bedford Biofuels conducted more democratic consultations, and made more concessions to local resource needs, than TARDA–Mumias. Rather than being a benevolent alternative to neo-colonialism, domestic investors may be tied more closely to a corrupt political or bureaucratic system, and subject to lower expectations of transparency, than international developers.

Tana Delta itself is a strikingly contested site for large-scale land acquisition. Our study found land at varying stages of privatisation, multiple *de facto* arrangements and remote claims to the land made by conservationists and policy planners. Further work is needed to investigate the relationship between social heterogeneity, tenure complexity and the progress of large-scale land deals. Several authors forecast an increase in inter-community tensions and claims based on ethnicity or indigeneity at times of pressure on land (Peters and Kambewa 2007, Berry 2009, Ndahinda 2011). Divisions along lines of class, ethnicity and indigeneity in Tana Delta have affected people's strategic responses to the deals; however, the responses are complex, and in the latest court case against TARDA–Mumias, residents from diverse livelihoods and ethnicities have united in opposition.

The second observation is that, in Kenya at least, land grabbing is not a recent phenomenon. Indeed, the two current deals in Tana Delta are built upon a political openness to commercial agriculture, a longstanding commitment to monoculture plantations and the easy acquisition of land by elites. Local responses are partly shaped by a resentment of earlier appropriation and by a willingness in certain villages to engage in large-scale farm labour. Whereas foreign researchers and campaigners frame recent events discursively as land grabbing, residents of Tana Delta are able to put the deals in context of what has gone before, culturally, politically and ecologically. Far from being the *tabula rasa* described by project proponents, the landscape can be read, and each person reads it differently. Here is where animals are taken to the hinterland. Here is a dormant ranch. Here is where my father lived before the *Shiftas* came. Here is where the river used to flow. Given the collapse of previous top-down projects that are also imprinted into the landscape, we cannot assume that the two current deals will prosper even if they do go ahead. Tana Delta, and the people that live and work there, have proved to be resilient survivors of projects that don't suit them.

In this paper, we have unveiled the complex and evolving socio-ecological conditions in which the Tana Delta deals have been negotiated, developed or brought to a standstill. Land tenure, social identity and landscape values play a key role in shaping actor responses, both individual and collective, to such deals, and contribute to the articulation of discourses of support and dissent across scales. Our insights may help to explain why so many large-scale agricultural projects are not yet operational. The World Bank surmises that delays are due to 'unrealistic objectives, price changes, and inadequate infrastructure, technology, and institutions' (World Bank 2011, xiv). To this must be added the obstructive role of local opposition that arises if projects are not designed in harmony with land users' priorities and values. Conversely, though Tana Delta residents remain vulnerable to displacement and deprived access, the challenges of resistance coalitions are chipping away at sedimented policy discourse. NEMA recently withdrew its support for Bedford Biofuels, citing scientific concerns about *Jatropha* (Muchangi 2011). In September 2011, ministry officials, TARDA and other stakeholders agreed on a new 'Tana Delta Planning and SEA Initiative', committing to produce a long-term, sustainable land-use plan for the area by March 2013 (GoK 2011). Overall, these are signs that the 1950s-era received wisdom of top-down development and profligate pastoralists is changing and that, like the Tana River, policy narrative could be redirected to take a new course.

References

Abercrombie, F.D. 1974. *Range development and management in Africa.* Agency for International Development (USAID). University of Minnesota: Agency for International Development.

ABF (African Business Foundation). 2010. Environmental and social impact assessment study report: the proposed establishment of jatropha plantations in Garsen Division, Tana Delta District, Kenya. Prepared by African Business Foundation (consultant) and Bedford Biofuels (project proponent). Available from: http://www.nema.go.ke/index2.php?option=com_docman&task=doc_view&gid=360&Itemid=35 [Accessed 6 April 2011].

Adams, W.M. 1990. How beautiful is small? Scale, control and success in Kenyan irrigation. *World Development*, 18(10), 1309–23.

Adger, W.N., T.A. Benjaminsen, K. Brown and H. Svarstad. 2001. Advancing a political ecology of global environmental discourses. *Development and Change*, 32, 681–715.

Akram-Lodhi, A.H. 2008. (Re)imagining agrarian relations? The World Development Report 2008: Agriculture for Development. *Development and Change*, 39(6), 1145–61.

Alden Wily, L. 2011. 'The law is to blame': the vulnerable status of common property rights in Sub-Saharan Africa. *Development and Change*, 42(3), 733–57.

Aluanga, L. 2011. Land crisis in the Tana River Delta. *The Standard*, 7 October.

Amechi, E.P. 2009. Poverty, socio-political factors and degradation of the environment in sub-Saharan Africa: the need for a holistic approach to the protection of the environment and realisation of the right toe. *Law, Environment and Development Journal*, 5(2), 107–28.

Anyanzwa, J. 2004. Spanish firm to spend 2b in sugar project. *The Standard*, 30 October.

Ariza-Montobbio, P., L. Sharachchandra, G. Kallis and J. Martinez-Alier. 2010. The political economy of Jatropha plantations for biodiesel in Tamil Nadu. *Journal of Peasant Studies*, 37(4), 875–97.

A Rocha. 2011. Public hearing for jatropha biofuels in Tana River Delta. Available from: http://arochakenya.wildlifedirect.org/2011/02/07/public-hearing-for-jatropha-biofuels-in-tana-river-delta/ [Accessed 22 December 2011].

Bates, R.H. 2005. *Beyond the miracle of the market, new edition. The political economy of agrarian development in Kenya.* New York: Cambridge University Press.

Bedford Biofuels. 2011. Bedford in Kenya. *Generations* newsletter, 2, Spring 2011, 3.

Berry, S. 2009. Property, authority and citizenship: land claims, politics and the dynamics of social division in West Africa. *Development and Change*, 40(1), 23–45.

BirdLife International. 2010. BirdLife position statement on Tana. Available from: http://www.birdlife.org/action/campaigns/tana_delta/ [Accessed 14 October 2011].

Bisset, R. 2000. Methods of consultation and public participation. In: N. Lee and C. George, eds. *Environmental assessment in developing and transitional countries: principles, methods and practice.* London: Wiley, pp. 149–60.

Borras Jr, S. and J. Franco. 2010. Towards a broader view of the politics of global land grab: rethinking land issues, reframing resistance. ICAS Working Paper Series No. 001. Available from: http://www.tni.org/paper/towards-broader-view-politics-global-land-grabbing [Accessed 18 October 2011].

Borras, S.M., P. McMichael and I. Scoones. 2010. The politics of biofuels, land and agrarian change: editors' introduction. *Journal of Peasant Studies*, 37(4), 575–92.

Bratton, M. 1977. Structural transformation in Zimbabwe: comparative notes from the neo-colonisation of Kenya. *The Journal of Modern African Studies*, 15(4), 591–611.

CIFOR. 2010. The local social and environmental impacts of biofuel feedstock expansion: a synthesis of case studies from Asia, Africa and Latin America. *Infobrief*, 34. Available from: http://www.cifor.org/online-library/browse.html [Accessed 12 March 2012].

Cochet, H. and M. Merlet. 2011. Land grabbing and share of the value added in agricultural processes: a new look at the distribution of land revenues. Paper presented at the International Conference on Global Land Grabbing, 6–8 April, IDS, University of Sussex, UK.

Collett, D. 1987. Pastoralists and wildlife: image and reality in Kenya Maasailand. In: D. Anderson and R. Grove, eds. *Conservation in Africa: people, politics and practice*. Cambridge: Cambridge University Press, pp. 129–48.

Cotula, L. 2011. *Land deals in Africa: what is in the contracts?* London: IIED.

Cotula, L. and S. Vermeulen. 2011. Contexts and procedures for farmland acquisitions in Africa: what outcomes for local people? *Development*, 54(1), 40–8.

Cotula, L., S. Vermeulen, R. Leonard and J. Keeley. 2009. *Land grab or development opportunity? Agricultural investment and international land deals in Africa*. London/Rome: IIED/FAO/IFAD.

Cotula, L., S. Vermeulen, P. Mathieu and C. Toulmin. 2011. Agricultural investment and international land deals: evidence from a multi-country study in Africa. *Food Security*, 3(Suppl. 1), S99–113.

Cuffaro, N. and D. Hallam. 2011. 'Land grabbing' in developing countries: foreign investors, regulation and codes of conduct. Working paper 3, Dipartimento di Scienze Economiche, Università di Cassino. Available from: http://dipse.unicas.it/files/wp201103.pdf [Accessed 26 October 2011].

Daily Nation. 2004. All parties to have a say in Sh24bn sugar project. *Daily Nation*, 11 December.

Da Via, E. 2011. The politics of 'win-win' narratives: land grab as development opportunity? Paper presented at the International Conference on Global Land Grabbing, 6–8 April, IDS, University of Sussex.

Davies, B. and R. Harré. 1990. Positioning: the discursive production of selves. *Journal of the Theory of Social Behaviour*, 20, 43–65.

Deininger, K. 2011. Challenges posed by the new wave of farmland investment. *Journal of Peasant Studies*, 38(2), 217–47.

Deininger, K. and H.P. Binswanger. (World Bank). 1995. Rent seeking and the development of large-scale agriculture in Kenya, South Africa, and Zimbabwe. *Economic Development and Cultural Change*, 43, 493–522.

De Schutter, O. 2011. How not to think of land-grabbing: three critiques of large-scale investments in farmland. *Journal of Peasant Studies*, 38(2), 249–79.

Dever, J.D. 2009. Small-farm access to high-value horticultural markets in Kenya. In: P. Pinstrup-Andersen and F. Cheng, eds. *Case studies in food policy for developing countries: domestic policies for markets, production and environment*. Ithaca, NY: Cornell University Press, pp. 61–71.

Diao, X., P. Hazell and J. Thurlow. 2010. The role of agriculture in African development. *World Development*, 38(10), 1375–83.

Dorondel, S. 2009. Forest restitution, ethnic groups and patronage in post-socialist Romania. In: D. Fay and D. James, eds. *The rights and wrongs of land restitution*. Abingdon: Taylor & Francis, pp. 43–66.

EAWLS (East Africa Wildlife Society). 2008. Still no respite for Tana Delta. Available from: http://eawls.wildlifedirect.org/2008/06/06/still-no-respite-for-tana-delta/ [Accessed 22 December 2011].

EcoSystems Ltd. 1985. *Tana Delta ecological impact study. Final report*. Nairobi: EcoSystems Ltd.

Ellis, F. and S. Biggs. 2001. Evolving themes in rural development, 1950s–2000s. *Development Policy Review*, 19(4), 437–48.

Ensminger, J. and A. Rutten. 1991. The political economy of changing property rights: dismantling a pastoral commons. *American Ethnologist*, 18(4), 683–99.

Eriksen, S. and J. Lind. 2009. Adaptation as a political process: adjusting to drought and conflict in Kenya's drylands. *Environmental Management*, 48, 817–35.

Eriksen, S., E. Ouko and N. Marekia and 1996. Land tenure and wildlife management. In: C. Juma and J.B. Ojwang, eds. *In land we trust: environmental private property and constitutional change*. Nairobi: Initiatives Publishers and London: Zed Books, pp. 199–230.

FIAN. 2010. Land grabbing in Kenya and Mozambique. A report on two research missions – and a human rights analysis of land grabbing. Heidelberg: FIAN International Secretariat.

Franco, J., L. Levidow, D. Fig, L. Goldfarb, M. Honicke and M.L. Mendonca. 2010. Assumptions in the European Union biofuels policy: frictions with experiences in Germany, Brazil and Mozambique. *Journal of Peasant Studies*, 37(4), 661–98.

Friends of the Earth. 2010. Africa: up for grabs. The scale and impact of land-grabbing for agrofuels. Available from: www.foeeurope.org [Accessed 1 July 2011].

Gasper, D. and R. Apthorpe. 1996. Introduction: discourse analysis and policy discourse. *European Journal of Development Research*, 8(1), 1–15.

Glenday, J. 2005. Preliminary assessment of carbon storage and the potential for forestry based carbon offset projects in the lower Tana River forests. Available from: http://coastalforests.tfcg.org/pubs/Tana_Carbon_Study.pdf [Accessed 9 October 2011].

Global IDP. 2004. Profile of internal displacement: Kenya. Available from: http://www.internal-displacement.org/8025708F004BE3B1/(httpInfoFiles)/A7C2D6322766F7B1802570BA00529A80/$file/Kenya%20-November%202004.pdf [Accessed 3 March 2012].

Glover, D. and K. Kusterer. 1990. *Small farmers, big business: contract farming and rural development*. Basingstoke and London: Macmillan.

GoK (Government of Kenya). 2008. Kenya Vision 2030. First medium term plan, 2008–2012. Available from: http://www.planning.go.ke/index.php?option=com_docman&task=cat_view&gid=64&Itemid=69&limitstart=5 [Accessed 25 October 2011].

GoK. 2010. Agricultural sector development strategy 2010–2010. Available from: http://www.ascu.go.ke/DOCS/ASDS%20Final.pdf [Accessed 29 December 2011].

GoK. 2011. Draft communiqué of the Inter-Ministerial Consultative Meeting on sustainable development of deltas in Kenya. Malindi, 15 September.

Gow, J. and K. Parton, 1995. Evolution of Kenyan agricultural policy. *Development Southern Africa*, 12(4), 467–79.

GRAIN. 2008. Seized! The 2008 land grab for food and financial security. GRAIN Briefing October 2008. Available from: www.grain.org/go/landgrab [Accessed 1 July 2011].

GRAIN. 2012. Global land grab data set. Available from: http://www.grain.org/attachments/2453/download] [Accessed 1 March 2012].

Grajales, J. 2011. The rifle and the title: paramilitary violence and land control in Colombia. *Journal of Peasant Studies*, 38(4), 771–92.

Hajer, M.A. 1995. *The politics of environmental discourse: ecological modernization and the policy process*. Oxford: Clarendon Press.

Hall, D. 2011. Land grabs, land control, and Southeast Asian crop booms. *Journal of Peasant Studies*, 38(4), 837–57.

Hall, D., P. Hirsch and T.M. Li. 2011. *Powers of exclusion: land dilemmas in Southeast Asia*. Singapore: National University of Singapore Press.

Hamerlynck, O., J. Nyunja, Q. Luke, D. Nyingi, D. Lebrun and D. Duvail. 2010. The communal forest, wetland, rangeland and agricultural landscape mosaics of the Lower Tana, Kenya: A socio-ecological entity in peril. In: C. Bélair, K. Ichikawa, B.Y.L. Wong and K.J. Mulongoy, eds. *Sustainable use of biological diversity in socio-ecological production landscapes. Background to the 'Satoyama Initiative for the Benefit of Biodiversity and Human Well-being'*. Montreal: Secretariat of the Convention on Biological Diversity.

Haralambous, S., H. Liversage and M. Romano. 2009. The growing demand for land risks and opportunities for smallholder farmers. Discussion paper prepared for the round table organized during the thirty-second session of IFAD's governing council, 18 February 2009. Available from: www.ifad.org/events/gc/32/roundtables/2.pdf [Accessed 6 April 2011].

Henry, W. 2008. No end in sight for TARDA project. *The People Daily*, 10 July.

Hinderink, J. and J.J. Sterkenburg. 1985. Agricultural policy and the organization of production in sub-Saharan Africa. *Journal of Rural Studies*, 1(1), 73–85.

Hodgson, D.L. and R.A. Schroeder. 2002. Dilemmas of counter-mapping community resources in Tanzania. *Development and Change*, 33(1), 79–100.

Huggins, C. 2011. A historical perspective on the 'global land rush'. Prepared for the International Land Coalition. Available from: http://www.landcoalition.org/publications/historical-perspective-global-land-rush [Accessed 8 September 2011].

Hughes, F. 1987. Conflicting uses for forest resources in the Lower Tana River basin of Kenya. In: D. Anderson and R. Grove, eds. *Conservation in Africa: people, politics and practice*. Cambridge: Cambridge University Press, pp. 211–28.
Hughes, F. 1990. The influence of flooding regimes on forest distribution and composition in the Tana River floodplain, Kenya. *Journal of Applied Ecology*, 27(2), 475–91.
Hunsberger, C. 2010. The politics of Jatropha-based biofuels in Kenya: convergence and divergence among NGOs, donors, government officials and farmers. *Journal of Peasant Studies*, 37(4), 939–62.
HVA International and MA Consulting. 2007. Environmental impact assessment study report for the proposed Tana Integrated Sugar Project in Tana River and Lamu Districts, Coast Province, Kenya. Available from: http://www.tanariverdelta.org/tana/g1/impacts.html [Accessed 2 April 2011].
IFPRI (Von Braun, J. and R. Meinzen-Dick.). 2009. 'Land grabbing' by foreign investors in developing countries: risks and opportunities. IFPRI Policy Brief 13. Available from: www.ifpri.org/sites/default/files/publications/bp013all.pdf [Accessed 19 October 2011].
ILCA (International Livestock Centre for Africa). 1979. *District ranch development briefs*. Working document. Nairobi: ILCA.
ILO (International Labour Office). 2000. *Traditional occupations of indigenous and tribal peoples: emerging trends*. Geneva: ILO Publications.
IRIN. 2008. Thousands affected as a river runs through it no more. 22 September. Available from: http://www.unicnairobi.org/Tana%20river.asp [Accessed 27 October 2011].
Irungu, P. 2000. Cattle keeping practices of the Orma people: a household survey in Tana River District, Kenya. KETRI-ILRI collaborative study. Available from: http://www.ilri.org/biometrics/Publication/Full%20Text/Irungu.pdf [Accessed 31 December 2011].
Jaffee, S.M. 1994. Contract farming in the shadow of competitive markets: the experience of Kenyan horticulture. In: P.D. Little and M.J. Watts, eds. *Living under contract: contract farming and agrarian transformation in Sub-Saharan Africa*. Madison: University of Wisconsin Press, pp. 97–139.
Kanyinga K. 1998. Politics and struggles for access to land: 'grants from above' and 'squatters' in coastal Kenya. *The European Journal of Development Research*, 10(2), 50–69.
Kanyinga, K., O. Lumumba and K.S. Amanor. 2008. The struggle for sustainable land management and democratic development in Kenya: a history of greed and grievances. In: K.S. Amanor and S. Moyo, eds. *Land and sustainable development in Africa*. London: Zed Books, pp. 100–23.
Kenya Land Alliance. 2002. Land, environment and natural resources. Submission to the Constitution of Kenya Review Commission. Available from: http://www.oxfam.org.uk/resources/learning/landrights/downloads/klackrc.rtf [Accessed 4 March 2012].
Kenya Law Reports. 2006. Available from: http://www.kenyalaw.org [Accessed 1 October 2011].
Kimenyi, M.S. and N.S. Ndungu. 2005. Sporadic ethnic violence: why has Kenya not experienced a full-blown civil war? In: P. Collier and N. Sambanis, eds. *Understanding civil war. Volume 1: Africa*. Washington, DC: IBRD/World Bank, pp. 123–56.
Kithi, N. 2004. Tarda launches campaign to 'educate' stakeholders. The East African, 6 December.
Klopp, J.M. 2000. Pilfering the public. *Africa Today*, 47(1), 7–26.
KNBS (Kenya Nationa Bureau of Statistics). 2009. Available from: http://marsgroupkenya.org.
Kolstad, I. and T. Soreide. 2009. Corruption in natural resource management: Implications for policy makers. *Resources Policy*, 34(4), 214–26.
Kweyu, D. 2008. Hue and cry over Tana Sugar project. *The Nation*, 20 August.
KWF (Kenya Wetlands Forum). 2005. Development, conservation and people's livelihood at crossroads: the proposed sugarcane project in the Tana Delta, Tana River District – Coast Province, Kenya. A technical paper. Available from: http://www.tanariverdelta.org/tana/10218-DSY/version/default/part/AttachmentData/data/The%20Proposed%20Sugarcane%20Project%20In%20The%20Tana%20Delta%20-%20A%20Technical%20Paper.pdf [Accessed 18 October 2011].

Land Research Action Network. 2011. Introduction: global land grabs: investments, risks and dangerous legacies. *Development*, 54(1), 5–11.
Lebrun, D. 2009. The importance of flexibility: an analysis of the large-scale Tana Delta irrigation project in Kenya, implemented under estate system. Master's Thesis, University of Gent, Unpublished.
Lebrun, D, O. Hamerlynck, S. Duvail and J. Nyunja. 2010. The importance of flexibility: an analysis of the large-scale Tana Delta Irrigation Project in Kenya, implemented under an estate system. In: B. Calas and C. Mumma Martinon, eds. *Shared water, shared opportunities. Hydropolitics in East Africa*. Dar es Salaam: Mkuki na Nyota Publishers Ltd., pp. 261–82.
Lenaola, I., H.H. Jenner and T. Wickert. 1996. Land tenure in pastoral lands. In: C. Juma and J.B. Ojwang, eds. *In land we trust: environmental private property and constitutional change*. Nairobi: Initiatives Publishers and London: Zed Books, pp. 234–57.
Li, T.M. 2007. *The will to improve*. Durham, NC: Duke University Press.
Li, T.M. 2011. Centering labor in the land grab debate. *Journal of Peasant Studies*, 38(2), 281–98.
Little, P.D. 1987. Comparative analysis of institutional experiences with river basin development in Africa: the case of the Tana Basin, Kenya. Prepared for the Africa Bureau, US Agency for International Development. Washington, DC: USAID.
Loewenson, R. 1992. *Modern plantation agriculture: corporate wealth and labour squalor*. London: Zed Books.
Lukudu, C., U. Weber, J.K. Muthai, A.O. Owiro, K. Kanyinga and J.K. Muteti, eds. 2000. *Alienation of public land in Kenya*. Nairobi: The Catholic University of East Africa.
Mahmoud, H.A. 2010. Livestock trade in the Kenyan, Somali and Ethiopian borderlands. Chatham House briefing paper. Available from: http://www.chathamhouse.org/sites/default/files/public/Research/Africa/0910mahmoud.pdf [Accessed 26 October 2011].
Mançano Fernandes, B., C.A. Welch and E. Constantino Gonçalves. 2010. Agrofuel policies in Brazil: paradigmatic and territorial disputes. *Journal of Peasant Studies*, 37(4), 793–819.
Mandere, N.M., S. Anderberg, F. Ato Armah and S.W. Abaya. 2011. Assessing the contribution of alternative agriculture to poverty reduction and employment creation: a case study of sugar beet cultivation in Kenya. *African Journal of Agricultural Research*, 6(2), 440–50.
Mango, C. 2005. MP warned over sugar project. *The Standard*, 25 January.
Mayoyo, P. 2005. Lobby opposes Tarda projects. *Daily Nation*, 14 February.
McCarthy, J.F. and R.A. Cramb. 2009. Policy narratives, landholder engagement, and oil palm expansion on the Malaysian and Indonesian frontiers. *The Geographical Journal*, 175(2), 112–3.
McMichael, P. 2011. The food regime in the land grab: articulating 'global ecology' and political economy. Paper presented at the International Conference on Global Land Grabbing, 6–8 April, IDS, University of Sussex.
McVeigh, T. 2011. Biofuels land grab in Kenya fuels talk of war. *The Observer*, 2 July.
Miller, D.L. 1981. Social formations in transition: social and economic change in the Lower Tana Valley, Kenya, 1850–1939. PhD Dissertation, Syracuse University.
Ministry of Lands. 2009. Land adjudication and settlement. Available from: http://www.lands.go.ke/index.php?option=com_content&task=view&id=14&Itemid=29 [Accessed 22 December 2011].
Muchai, E. 2005. Watchdog approves Tana River project. *Daily Nation*, 8 September.
Muchangi, J. 2011. Nema directors suspended over illegal jatropha licences. *Nairobi Star*, 26 August.
Mugambi, K. 2009. Mumias seeks to unlock Sh30bn Tana Delta funds. *The Daily Nation*, 18 July.
Mulaa, J. 1981. The politics of a changing society: Mumias. *Review of African Political Economy*, 20, 89–107.
Muthee, A.M. 2006. Kenya livestock sector study: an analysis of pastoralist livestock products, market value chains and potential external markets for live animals and meat. A report by Deloitte Consulting Ltd for USAID Kenya. Available from: www.value-chains.org/dyn/bds/docs/552/KenyaLivestockValuechainReport.pdf [Accessed 1 June 2011].
Mutiso, S.K. 1995. Adaptive strategies for sustainable livelihoods: a review and analysis of development policy for the arid and semi-arid areas of Kenya. IISD/KENGO. Available from: http://www.iisd.org/casl/ASALProjectDetails/Kenyapol.txt [Accessed 1 June 2011].

Mwangi, W.M. 1981. Alternatives for improving production, employment and income distribution in Kenyan agriculture. Discussion paper No. 271. Nairobi: Institute for Development Studies.

Nature Kenya. 2008. The economic valuation of the proposed Tana Integrated Sugar Project (TISP), Kenya. Available from: http://www.rspb.org.uk/Images/tana_tcm9-188706.pdf [Accessed 10 March 2012].

Nature Kenya. 2011. Tana River Delta: important bird area. Leaflet. Nairobi: Nature Kenya.

Ndahinda, F.M., 2011. Marginality, disempowerment and contested discourses on indigenousness in Africa. *International Journal on Minority and Group Rights*, 18, 479–514.

Ndunda, E. 2004. Tarda suspends River Tana cane project. *The Standard*, 10 December.

Ndurya, M. 2009. Lobby groups call for legislation over Tana Delta land. *Daily Nation*, 2 April.

NEMA (National Environment Management Authority). 2008. Due process in Tana Delta sugar project. Available from: http://www.nema.go.ke/index.php?option=com_content &task=view&id=203 [Accessed 28 October 2011].

NES (National Environment Secretariat, Ministry of Environment and Natural Resources). 1985. *Lower Tana River environmental assessment report*. Nairobi: Government of Kenya.

Neumann, R.P. 1995. Local challenges to global agendas: conservation, economic liberalisation and the pastoralists' rights movement in Tanzania. *Antipode*, 27(4), 363–82.

Nyassy, D. 2008. Minister drums up support for project. *Daily Nation*, 9 July.

OECD. 2008. *Business for development: promoting commercial agriculture in Africa*. Paris: OECD Publishing.

Ogolla, B.D. with Mugabe, J, 1996. Land tenure systems and natural resource management. In: C. Juma and J.B. Ojwang, eds. *In land we trust: environmental private property and constitutional change*. Nairobi: Initiatives Publishers and London: Zed Books, pp. 85–116.

Ominde, S.H. 1971. The semi-arid and arid lands of Kenya. In: S.H. Ominde, ed. *Studies in East African geography and development*. Berkeley and Los Angeles: University of California Press, pp. 146–61.

Omondi. 2008. NEMA clarifies on Mumias project in Tana. *Business Daily*, 26 June.

Opala, K. 2006. Bitter fight over mystery sugar project. *The East African*, 20 June.

Osoro, S., S. Murakaru, P.J. Slabbers and P. Slot. 1983. Report on the Joint Kenyan–Dutch review and evaluation of the Lower Tana Village Irrigation Programme. Kenya. Available from: http://library.wur.nl/isric/fulltext/isricu_i17089_001.pdf [Accessed 2 October 2011].

Owiro, A.O. 2000. The state and public land: an analysis. In: C. Lukudu, U. Weber, J.K. Muthai, A.O. Owiro, K. Kanyinga and J.K. Muteti, eds. *Alienation of public land in Kenya*. Nairobi: The Catholic University of East Africa, pp. 18–31.

Oya, C. 2012. Contract farming in sub-Saharan Africa: a survey of approaches, debates and issues. *Journal of Agrarian Change*, 12(1), 1–33.

Peters, P.E. 2004. Inequality and social conflict over land in Africa. *Journal of Agrarian Change*, 4(3), 269–314.

Peters, P.E. 2009. Challenges in land tenure and land reform in Africa: anthropological contributions. *World Development*, 37(8), 1317–25.

Peters, P.E. and D. Kambewa. 2007. Whose security? Deepening social conflict over 'customary' land in the shadow of land tenure reform in Malawi. *Journal of Modern African Studies*, 45, 447–72.

Plenary Hansard. 2011. National Assembly Official Report, Wednesday, 18th May. Motion: Adoption of Report on Challenges Facing Sugar Sector, p. 37. Available from: www.parliament.go.ke [Accessed 8 May 2012].

Poulton, C., G. Tyler, P. Hazell, A. Dorward, J. Kydd and M. Stockbridge. 2008. *All-Africa review of experiences with commercial agriculture: lessons from success and failure*. World Bank/FAO report. Available from: http://siteresources.worldbank.org/INTAFRICA/Resources/257994-1215457178567/CCAA_Success_failure.pdf [Accessed 10 June 2011].

Pye, O. 2010. The biofuel connection – transnational activism and the palm oil boom. *Journal of Peasant Studies*, 37(4), 851–74.

RedOrbit. 2008. Kenya biofuel project causes dissension. Available from: http://www.redorbit.com/news/science/1487744/kenya_biofuel_project_causes_dissension/index.html [Accessed 24 March 2011].

Ribots, J. and N. Peluso. 2003. A theory of access. *Rural Sociology*, 68(2), 153–81.
Ringa, M. 2004. Pastoralists claim cane project will destroy rich Tana habitat. *The Standard*, 23 December.
Ringa, M. 2008. Tana Delta sugar project hearing starts. *Daily Nation*, 7 May.
Robbins, P., K. McSweeney, T. Waite and J. Rice. 2006. Even conservation rules are made to be broken: implications for biodiversity. *Environmental Management*, 37(2), 162–9.
Robertson, B. and P. Pinstrup-Andersen. 2010. Global land acquisition: neo-colonialism or development opportunity? *Food Security*, 2, 271–83.
Roe, E.M. 1995. Except-Africa: postscript to a special section on development narratives. *World Development*, 23(6), 1065–9.
RoK (Republic of Kenya). 1989. *Tana River District development plan 1989–1993*. Nairobi: Government printer.
RoK (Republic of Kenya). 2006. Arid Lands Resource Management Project (phase II): community driven development component. Tana River District: progress report, 2003–2006. Available from: http://www.aridland.go.ke/CDD/tana_river_cdd_progress_report.pdf [Accessed 31 December 2011].
RoK (Republic of Kenya). 2009. *Tana Delta District development plan 2008–2012*. Nairobi: Government printer.
Rowntree, K. 1990. Political and administrative constraints on integrated river basin development: an evaluation of the Tana and Athi Rivers Development Authority, Kenya. *Applied Geography*, 10, 21–41.
RSPB. 2011a. Gamba Manyatta residents letter to PM. In 'Casework: Tana River Delta'. Available from: http://www.rspb.org.uk/ourwork/casework/details.aspx?id=tcm:9-228564 [Accessed 22 December 2011].
RSPB. 2011b. Letter to the Acting Director General of the National Environmental Management Authority on NEMA's decision on the Bedford Biofuels (Tana Delta) Limited Jatropha plantation project, Garsen District, Tana River Delta. 6 June. Available from: https://www.rspb.org.uk/Images/nemadecision_tcm9-291829.pdf [Accessed 31 December 2011].
RSPB. 2011c. Timeline: January 2011. In 'Casework: Tana River Delta'. Available from: http://www.rspb.org.uk/ourwork/casework/details.aspx?id=tcm:9-228564 [Accessed 22 December 2011].
Rutten, M. 1992. Selling wealth to buy poverty: the process of individualization of landownership among the Maasai pastoralists of Kajiado District, Kenya, 1890–1990. PhD Dissertation, Catholic University Nijmegen.
Schutz, A. and T. Luckmann. 1973. *The structures of the life-world*. R.M. Zaner and H.T. Engelhardt, trans. London: Heinemann.
Schwartz, K.Z.S. 2006. Masters in our native place: the politics of Latvian national parks on the road from Communism to Europe. *Political Geography*, 25, 42–71.
Sikor, T. and C. Lund. 2009. Access and property: a question of power and authority. *Development and Change*, 40(1), 1–22.
Simpson, M.C. 1973. Alternative strategies for range land development in Kenya. Rural development study 2. Leeds: University of Leeds, Department of Agricultural Economics.
Southall, R. 2005. The Ndungu report. *Review of African Political Economy*, 32(103), Briefings, 142–51.
Syagga, P.M. and A.K. Mwenda. 2010. *Political economy and governance issues surrounding policy interventions in the land sector in Kenya*. Final report. Prepared for the World Bank, March 2010.
The Standard. 2006. Minister defends Mumias over Tana. *The Standard*, 24 November.
The Standard. 2007. Public endorses Tarda, Mumias sugar plant. *The Standard*, 27 August.
Thurston, A. 1987. *Smallholder agriculture in colonial Kenya: the official mind and the Swynnerton Plan*. Cambridge African Monographs 8. Cambridge: African Studies Centre.
Umar, A. (Kenya Pastoral Forum). 1997. Resource utilisation, conflict and insecurity in pastoral areas of Kenya. A paper for the USAID organised seminar on conflict resolution in the Horn of Africa, Nairobi, 27–29 March 1997.
UNCTAD. 2011. *World investment report 2011: non-equity modes of international production and development*. New York and Geneva: United Nations.

UNDP (United Nations Development Programme). 2010. *Kenya national human development report 2009*. Nairobi: UNDP Kenya.

Van der Ploeg, J.P. 2010. The peasantries of the twenty-first century: the commoditisation debate revisited. *Journal of Peasant Studies*, 37(1), 1–30.

Van Eijk, T. 1998. Farming systems research and spirituality: an analysis of the foundations of professionalism in developing sustainable farming systems. PhD Thesis, Wageningen Agricultural University.

Vermeulen, S. and L. Cotula. 2010a. Over the heads of local people: consultation, consent, and recompense in large-scale land deals for biofuels projects in Africa. *Journal of Peasant Studies*, 37(4), 899–916.

Vermeulen, S. and L. Cotula. 2010b. *Making the most of agricultural investment: a survey of business models that provide opportunities for smallholders*. London/Rome/Bern: IIED/FAO/IFAD/SDC.

Von Blanckenburg, P. 1994. *Large commercial farmers and land reform in Africa: the case of Zimbabwe*. Aldershot: Avebury.

Von Kaufmann, R. 1976. The development of the range land areas. In: J. Heyer, J.K. Maitha and W. M. Senga, eds. *Agricultural development in Kenya: an economic assessment*. Nairobi: Oxford University Press, pp. 225–87.

Wario, A. and M. Shambaro. 2004. There's something fishy about Garsen deal. *Daily Nation*, 10 November.

Wegulo, F.N. and H.W. Obulinji. 2001. The interface between farm and non-farm activities among the Mumias sugar cane growers. In: P.O. Akila and P.O. Pedersen, eds. *Negotiating social space: East African micro enterprises*. Asmara: African World Press, pp. 223–52.

Wetherell, M. 2007. A step too far: discursive psychology, linguistic ethnography and questions of identity. *Journal of Sociolinguistics*, 11(5), 661–81.

Winter-Nelson, A. 1995. A history of agricultural policy in Kenya. In: Pearson *et al.*, eds. *Agricultural policy in Kenya*. Ithaca and London: Cornell University Press, pp. 31–48.

World Bank. 2009. Awakening Africa's sleeping giant: prospects for commercial agriculture in the Guinea Savannah zone and beyond. Agricultural & Rural Development Notes 48, June 2009. Available from: http://siteresources.worldbank.org/EXTARD/Resources/336681-1231508336979/SleepingGiantFinal.pdf [Accessed 1 July 2011].

World Bank. 2011. *Rising global interest in farmland: can it yield sustainable and equitable benefits?* Washington, DC: IBRD/World Bank.

Wylie, J. 2007. *Landscape*. Abingdon: Routledge.

Rebecca Smalley has completed an MSc in Environment and International Development at the University of East Anglia's School of International Development. She has worked for the secretariat of the United Nations Framework Convention on Climate Change and has spent time on grassroots projects in Kenya.

Esteve Corbera is a Senior Research Fellow at the Institute of Environmental Sciences and Technology (ICTA) and the Department of Economics and Economic History, Universitat Autònoma de Barcelona. His research focuses on the governance of land-use management options for climate mitigation across scales, including the analysis of large-scale agriculture for biofuels production and of climate-policy and biodiversity conservation related instruments, such as carbon offset projects and Reducing Emissions from Deforestation and Forest Degradation (REDD).

Tree plantations, politics of possession and the absence of land grabs in Vietnam

Thomas Sikor

Viewed from the lens of 'land grabs', Vietnam's fast-growing tree plantations look like an anomaly; many of them are tiny, owned and operated by rural households. In contrast, private companies and transnational corporations have not been able to get much of a foot into Vietnam's plantation sector. This paper identifies the practices and processes underlying the apparent anomaly. On the basis of fieldwork in four villages, it points to the concrete mechanisms by which households have gained access to land, finance, and wood markets. Government policy emerges as a critical factor enabling household access, not in the sense of a coherent policy package but understood as sedimented outcomes of everyday processes of state formation over the past three decades. A central element in contestations over the state is what I call 'politics of possession': possession, referring to entitlement and control, has been closely tied to ideas about the state. The paper uses these empirical observations to contribute towards theoretical understandings of land grabs and exclusion. Land grabs may have to say as much about dynamics of state formation as processes of commodification and market expansion.

Introduction

Looking through the lens of 'land grabs', Vietnam's tree plantations appear to be an anomaly. They are a far cry from the images conjured by the terms 'land grabs', 'large-scale land acquisitions' or 'industrial tree plantations'[1] which are commonly used to depict the purchase or lease of land by transnational corporations for growing food, fiber or fuel crops in the Global South (Carrere and Lohmannn 1996, Borras *et al.* 2011). Many of Vietnam's tree plantations are owned and operated by rural households on plots measuring anything from less than a single hectare (ha) to a few ha. As tiny as they are individually, together they account for nearly two thirds

I thank the villagers of Tan Quang, Thuan Phong, Village 5 and Xom Lam for their support, and Hoang Lien Son, Le Van Cuong and Nguyen Hoang Nghia from the Forest Science Institute of Vietnam for collaboration, particularly in the implementation of the household survey. Two anonymous reviewers provided very helpful comments. This research is financed by the Gesellschaft für Internationale Zusammenarbeit (GIZ) on behalf of the government of the Federal Republic of Germany through a grant awarded to the Center for International Forestry Research (CIFOR).

[1]Although these are problematic terms, I will avoid using quotation marks in the following.

of all productive tree plantations in Vietnam (Forest Protection Department 2010). In contrast, private plantation companies and transnational corporations have not been able to get much of a foot into Vietnam's tree plantation sector.

Smallholders plant fast-growing trees not only in Vietnam but also in many other countries.[2] In fact, smallholders own an increasing share of global tree plantations, surpassing the amount of land under corporate ownership. The FAO estimates that smallholders have raised their share in productive plantations (including fast-wood exotic and natural trees) from 12 percent in 1990 to 32 percent in 2005 (Del Lungo *et al.* 2006, 24). Smallholders have driven the dramatic expansion of tree plantations worldwide, which the FAO estimates to amount to an annual rate of five million ha in recent years (FAO 2010). In Vietnam, the eucalyptus and acacia trees harvested from smallholder plantations increasingly feed chip mills and paper plants and are becoming a significant source of raw materials for construction sectors and furniture industries.

I do not offer these observations to contradict the significance of land grabs or industry-scale tree plantations worldwide. Instead, I want to suggest that they point to a special opportunity for examining the processes and practices driving land grabs, leading to the occurrence of grabs in some places and their absence in others. Empirical insights from Vietnam's tree plantations may illuminate under what circumstances smallholders can engage in the production of export commodities, and how transnational corporations and other actors gain the upper hand in others (cf. Borras and Franco forthcoming, Fairhead *et al.* forthcoming). In addition, they may shed light on key processes of exclusion in Southeast Asia (cf. Hall *et al.* 2011). Considering the marginal role played by private companies and transnational corporations, one is prompted to ask about the factors that have enabled presumably disadvantaged rural households to exclude presumably powerful players from land.

In this paper, I therefore examine the practices and processes underlying smallholders' engagement with tree plantations in Vietnam. I analyze households' access to productive resources and wood markets at the local level, i.e. their ability to get the necessary land and finance for growing trees and to find suitable market outlets. Moving beyond the local level, I seek to identify the larger forces enabling and conditioning households' access to productive resources and product markets in the sense of agrarian political economy (White and Dasgupta 2010).[3] As a central element in this agrarian political economy, I am particularly interested in the significance of 'everyday processes of state formation' (Sikor and Lund 2009, Sikor *et al.* 2009). My interest is not with any of the few high-profile cases of conflict over tree plantations, but with the everyday practices that are available to the large

[2]Fast-growing trees are often problematic, especially if planted in monocultures, for causing detrimental effects on local environments and biodiversity (Carrere and Lohmann 1996, Cossalter and Pye-Smith 2003). A closer look at the environmental effects of fast-growing tree plantations in Vietnam or conflicts surrounding plantations worldwide is beyond the scope of this paper, however.

[3]The paper is admittedly partial, as it focuses squarely on the dynamics that enable households to grow trees. The analysis ignores differences among households, which I analyze in another paper (Sikor and Baggio, under review) as well as differences between villages and regions. Also, it neglects the benefits people may derive from working as hired laborers on plantations. The focus is on the tiny plantations owned and operated by rural households in an attempt to explore politics of possession operating at local and national levels.

majority of rural people. When they grow trees, they not only help to generate biomass but also contribute to dynamics constituting the state, particularly ideas about the state (Abrams 1988).[4]

My primary interest in this essay is to explain smallholders' involvement in tree plantations, to unearth the political economic dynamics enabling involvement, and to relate the latter to wider 'politics of possession'. Such politics of possession, I suggest, is a key element in everyday processes of state formation in Vietnam. Politics of possession operates at the local level, e.g. when households grow trees or government officials promote tree plantations via targeted loans, and at the national level, e.g. when villagers nationwide set out to occupy land or apply for land titles, or when the central government institutes a nationwide program of land allocation. Villagers act politically when they acquire the land required for tree plantations, raise the necessary finance to plant trees and sell wood on markets. Local state officials are equally political when they issue land titles, approve loan requests or issue licenses for wood traders. Villagers' and local officials' practices are political in the sense that these actors negotiate the opportunities afforded by central government policy as well as the constraints imposed by central regulation on the basis of their conceptions of what the state ought to do, or should not do. Villagers' access to productive resources and product markets, consequently, is as much about their ability to benefit from tree plantations as about dynamics of state formation (cf. Sikor and Lund 2009). Simply stated, as villagers and local state officials deal with land, loans and wood sales, they simultaneously negotiate the authority attributed to the state, the latter understood as an image or idea. More broadly, I argue that politics of possession is a central element in Vietnam's agrarian political economy. Changing ideas of possession underlie the practices employed by villagers, government officials and other actors when they negotiate the role of the state in the economy. Ideas of possession motivate state officials' and villagers' actions directed at productive resources and product markets, shaping specific policies and villagers' reaction to them. The central role of possession thus produces dynamics of state formation tied to smallholder production despite the presence of powerful actors associated with land grabs in other circumstances.

The paper draws on fieldwork in four villages of Vietnam, two located in northern Phu Tho province and two in southern Binh Dinh province (see Figure 1). The two provinces stand for different plantation trends. Phu Tho has long been known as the site of Bai Bang Paper Mill, a mill constructed by Vietnam's government with Swedish support in the late 1970s and connected with efforts of promoting tree plantations since the early 1980s. In Binh Dinh, tree plantations have expanded rapidly only over the past decade in response to rising international wood prices and the construction of a few chip mills. In each province, fieldwork took place in two villages selected with the purpose of capturing a wide range of conditions under which households grow trees.[5] None of the villages stands out as exceptional within the Vietnamese context, as all four villages can be considered

[4]See Gerber (2011) for a review of high-profile conflicts over tree plantations worldwide, not including any cases from Vietnam. See Kröger (2011) for an analysis of organized politics surrounding large-scale fast-wood plantations in Brazil.

[5]In each village, fieldwork included a survey of randomly selected households in late 2008, follow-up interviews on access to land and finance with purposively chosen sub-sets of survey households in 2009, and key informant interviews with local government officials, traders and bank staff in 2008 and 2009.

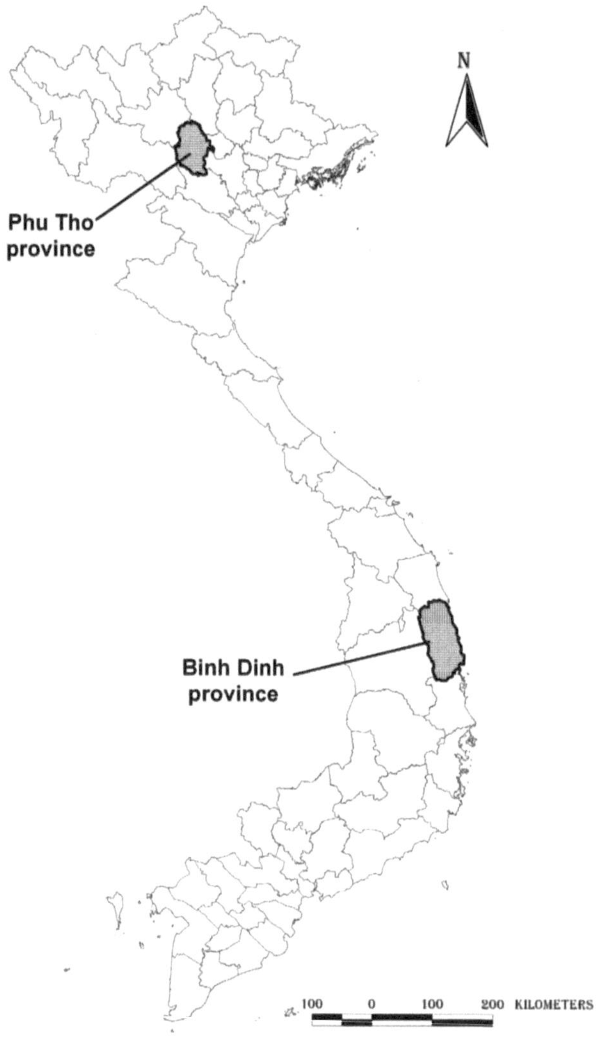

Figure 1. Location of Binh Dinh and Phu Tho.
Source: Map prepared by Dao Minh Truong.

agrarian in the sense that villagers combine crop cultivation, livestock husbandry, and limited off-farm work within their locality with migration of individual household members to urban areas. In international comparison, however, they may stand out for representing fairly successful transitions away from subsistence-oriented agricultural production towards a diversified rural economy.

In the following, I begin with a brief review of the literature on crop booms in Southeast Asia. I then turn to the empirical analysis, first analyzing household access to land, finance and wood markets in the four study sites with a look at the question on what villagers' practices imply for everyday dynamics of state formation. I proceed to examine the wider agrarian political economy enabling household access to land, finance and wood markets in an effort to link the four villages to key rural policies and agrarian events of national significance. I subsequently draw on these empirical insights for an account of politics of possession in Vietnam, before I conclude with broader implications for our understanding of land grabs and exclusion.

Crop booms in Southeast Asia

Over the past decade, research on rural transformations in Southeast Asia has increasingly looked at commodity markets and their effects on local livelihoods and landscapes (Hall *et al.* 2011). Much of the interest has centered on crop booms, i.e. rapid increases in the production of certain crops for export. Some of these crop booms, it has been noted, have been driven by dramatic expansions in smallholder production, such as coffee, shrimps, rubber and cocoa. Other boom crops, such as oil palm and fast-growing trees, have been found to mostly involve large-scale plantations (Hall 2011). Oil palm plantations have grown rapidly in Malaysia and Indonesia, and fast-growing tree plantations have spread all over the region (Hall 2011).

The general dynamics characterizing the tree and oil palm booms in Southeast Asia conform to the notion of land grabbing in significant ways. First, underlying the booms has been rising global demand for food and fiber (Hall 2011). Second, new regulations in consumer and producer countries have aided the global expansion of demand and its penetration into the Global South via commodity markets. The European Union, for example, set concrete targets for the use of renewable resources as transportation fuels, thereby boosting demand for palm oil (Pye 2010). On the side of producer countries, governments have enabled market expansion through regulatory changes (Hall *et al.* 2011, 112). For example, Malaysian and Indonesian approaches to agricultural development have recently reduced many barriers of entry for international investment and trade (McCarthy and Cramb 2009). Third, these regulatory changes and market expansion have received the support of the international financial institutions (Barney 2008).

Most importantly, these crop booms have been associated with large-scale land acquisitions by domestic companies and transnational corporations (Hall 2011). They have involved forms of exclusion that are different from those predominant in the production of other crops (Hall *et al.* 2011). Fast-growing tree plantations, in particular, are often considered to require large tracts of land of at least a few hundred ha (Cossalter and Pye-Smith 2003, Hall 2003). Investors have been able to acquire such parcels in areas with insecure 'land control relations', leading to radical transformations of previously existing land tenure relations (Hall 2011). These areas are different from areas with secure land control relations, where 'secure' booms would leave 'the basic tenure relations that existed before the boom' intact (839). An example for the latter comes from the booms in coffee and cocoa, in which smallholder production prevailed (Li 2002).

Even where smallholders have held on to their land initially, McCarthy (2010) suggests, they may lose it over time in a process of 'adverse incorporation'. Smallholders may face problems to stay on their land due to difficult access to state-of-the-art technology and high levels of debts (Hall 2011, 839). High debts forced Indonesian cocoa producers and Vietnamese coffee farmers into distress sales of land (Li 2002, Ha and Shively 2008). Expensive inputs have made oil palm a 'rich farmer's crop' (McCarthy 2010, 824). Smallholders may not be able to finance the required investments, as it takes too many years for these to bear fruit, as in the case of eucalyptus and acacia notwithstanding their fast-growing nature (Hall 2003, 255). Or they may have to enroll into contract farming on unfavorable terms (Carrere and Lohmann 1996, 239–240).

Despite these general tendencies, the literature attests to significant variation in the concrete dynamics and outcomes of the crop booms. The nature of booms and their effects on livelihoods and landscapes do not simply depend on the specific biophysical features of particular crops. Fast-growing tree plantations, for example, are not invariably large, as illustrated by the numerous smallholders who have established small-scale tree plantations in Thailand (Hall 2003). Instead, the literature indicates, micro-processes of agrarian change combine with larger political economic processes to produce diverse boom dynamics and outcomes, which may or may not fit the general tendencies. The causes underlying variation are connected with the particular mechanisms creating and sustaining access to land and finance, among others (White and Dasgupta 2010, Hall *et al.* 2011). They also originate from variable access to commodity markets over time and space despite the powerful influence of international demand (Sikor and Pham 2005, McCarthy 2010). The resulting paths of agrarian change are uneven over time and space, as they depend on the specific terms under which various actors engage with green crops.[6]

Thus, there may be room to maneuver for smallholders in crop booms, room that is conditioned by particular agrarian political economies at certain times and places. These political economies are likely to vary between countries, as suggested by McCarthy and Cramb (2009). Their comparison between Jambi (Indonesia) and Sarawak (Malaysia) reveals significant differences in the mechanisms of access to land, finance and product markets due to different state policies and practices, even though both regions have experienced massive expansions of private sector investment in large-scale oil palm plantations. Moreover, political economies may be highly variable within particular countries or regions (McCarthy 2010). The state, it appears, assumes a critical role in these political economies, as state practices influence agrarian dynamics and outcomes. In Indonesia, the shift from a strategy of upgrading smallholder productivity to a regime inviting participation by companies played a key role in the expansion of oil palm plantations (McCarthy 2010, 844). In Laos, tree plantations spread once Laotian policy moved away from promoting production for domestic purposes to world market integration (Barney 2008).

In this paper, I thus seek to examine how larger political-economic dynamics open up and constrain smallholders' room of maneuver. Crop booms may display an overarching tendency towards large-scale production. Yet there is also sufficient evidence that attests to the influence of agrarian political economies and the state on the paths of agrarian change. I am particularly keen to explore how everyday processes of state formation shape agrarian political economies and, more specifically, smallholders' access to productive resources and product markets. The dynamics of state formation may bring about large-scale plantations or mass dispossessions in some settings, yet they may also be supportive of smallholders in other circumstances. Of particular interest to me are the ideas various kinds of actors, including state officials and villagers, attributed to the state (cf. Abrams 1988). In Vietnam, I suggest, many people's notions of the state have been closely wedded to ideas of possession. Many state officials and villagers alike have

[6]Looking beyond Southeast Asia and the primary interest of this paper in everyday dynamics of state formation, interesting insights emerge from research on agrarian politics in Brazil, as one of the anonymous reviewers has pointed out (Kröger 2011).

understood the provision of possession as a key state task, even though the concrete meanings of possession and notions of the state have changed over time. This suggestion does not attempt to deny that people are in different positions to engage in politics of possession or benefit from material opportunities afforded by the politics (e.g. Sikor and Nguyen 2007, Sikor 2011b); nevertheless, it seeks to foreground a feature of state formation that may set Vietnam apart from other countries.[7]

Household access to land, finance and wood markets

I begin my empirical investigation by examining smallholders' access to productive resources and wood markets in the four study sites. Looking at access to land, finance and product markets, I seek to establish how smallholders were able to grow trees, with a particular interest in how various agencies, organizations and people associated with the state figured in the practices creating and controlling access.

Smallholders were a major player in tree plantations in all four villages at the time of the fieldwork in 2008, even though they grew acacia and eucalyptus trees in small stands only. Households operated tree plantations on 202 ha in Village 5 and 189 ha in Xom Lam, the two northern villages, the average household size being 2.1 and 1.3 ha, respectively (see Figure 2). The southern villages included plantations of 696 ha (Tan Quang) and 804 ha (Thuan Phong), the average household working 2.8 and 2.0 ha, respectively. Moreover, nine out of 10 households grew trees in three of the four villages.[8]

Access to land

Villagers were able to grow trees because they had the required land in their possession (see Figure 2). In all four sites, households' average plantation size was roughly equivalent to their forestry land.[9] Possession of such land was the single most critical precondition for growing trees, as people had planted trees on all the land.

Villagers had held the land used for tree plantations in their possession for some twenty years already. They told us that they had gone out to occupy hilly and sloping land in the late 1980s and early 1990s when the control over land exercised by agricultural cooperatives and State Forest Enterprises weakened. They originally grew agricultural crops and grazed their livestock on the land. The need to cultivate the land for meeting subsistence needs provided their land claims with a strong justification. It was no surprise consequently that all villages ended up with roughly two ha of hilly land: this was about as much as the average household could work using their own labor only.

In the southern village of Thuan Phong, a plantation company had also seized on the opportunity to acquire land in the early 1990s. The Binh Dinh People's

[7]I thank one of the anonymous reviewers for the encouragement to highlight this point.

[8]Tan Quang stood out among the four villages for rather pronounced differences between households in their ability to grow trees, something I have discussed in another paper (Sikor and Baggio under review). Only slightly less than half of all households planted trees.

[9]I use the term 'forestry land' to reflect the Vietnamese term 'dat lam nghiep', which highlights the legal classification as land to be used for forestry. It is different from the term used to denote forested land ('dat co rung').

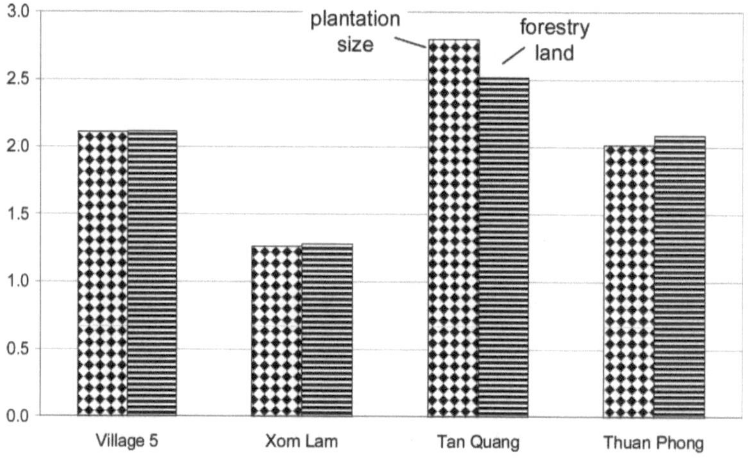

Figure 2. Average plantation size and forestry land (ha/household).

Committee had issued a 35-year lease for a total of 9,900 ha to Quy Nhon Plantation Company, a subsidiary fully owned by the transnational corporation Nissho Iwai.[10] The provincial allocation included many parcels spread out over the whole province, among them some 107 ha of forestry land directly adjacent to Thuan Phong village. It took the Company seven years to actually acquire the allocated land area, as the province had problems in identifying suitable parcels on the land held by State Forest Enterprises and transferred the land only gradually. Once the Company had acquired land, it hired the local company PISICO owned by the provincial People's Committee to manage the tree plantation in Thuan Phong.

The land frontier closed by the mid-1990s, making it virtually impossible for villagers or companies to acquire new land in the four sites. One of the key actors limiting access to new land was the omnipresent State Forest Enterprises and their successors, the Forest Companies. The Enterprises and Companies remained powerful actors in all study sites, holding on to virtually all of the land left to them after the land rush. For example, pressure from the provincial People's Committee enticed Phu Cat Enterprise to transfer forestry land to Thuan Phong village twice, yet each transfer involved very small parcels of a few dozen ha only. The transfers thus accounted for minor additions to the roughly 650 ha held by villagers already. In Village 5, Doan Hung Enterprise/Company successfully retained the 270 ha in and around Village 5 that the district People's Committee had allocated to the Enterprise in 1998. Villagers never saw a chance to claim the land, even though the Company holdings exceeded the aggregate size of tree plantations in the village.

The land allocations conducted by local government officials in all four villages did not introduce any significant modifications to the overall distribution of land. Land allocations generally took the form of mere registrations of plots already held by households. For example, in Village 5, villagers eventually received temporary land certificates, so-called Green Books, for the land that they had occupied or acquired in other ways earlier on from the district-level Forest Protection Unit in the

[10]Nissho Iwai was later taken over by Sojitz Corporation, another transnational corporation with headquarters in Japan.

mid-1990s. In subsequent years, some of them requested the issuance of formal land certificates, the so-called Red Books, but by 2008, more than half of all plots had yet to be certified in the officially required manner. The situation was similar in Tan Quang, as only two out of five plots were formally registered in Red Books by 2008. It was different in Thuan Phong, where four out of five plots had received the required documentation. The high rate of formal documentation was due to the operation of a World Bank-funded project, which required households to obtain Red Books as a precondition for access to plantation loans and had supported the local government to implement land allocation.

Government agencies, state-owned enterprises and state officials, therefore, played a role in villagers' access to land, even though it was long an accommodating one. When villagers set out to occupy land, they did so in plain disregard of government regulations prohibiting their practices. Local officials eventually came around to recognize villagers' land holdings and formalize them in land certificates. At the same time, local implementation of central government policy also prevented any further expansion of villagers' land, as local officials granted land certificates to State Forest Enterprises and a plantation company.

Access to finance

Villagers not only had suitable land in their possession, but they also commanded the required capital to invest into trees. More importantly, many households employed tree-growing practices that allowed them to keep the investment requirements to a minimum. They planted fast-growing acacia and eucalyptus on short rotations, harvesting trees after only three or four years. They also reduced the cash expenditures required for plantations by buying cheap tree seedlings from small local nurseries, using little fertilizer, hiring little labor, and harvesting manually. Only a few households invested higher amounts in plantations, purchasing high-quality seedlings and fertilizer, hiring workers, and managing their acacia and eucalyptus stands under medium rotations of six to seven years. Overall investment levels were consequently low (see Figure 3).[11]

Tree plantations had become a desirable target for the use of household surplus over the previous five years. They were not as popular as livestock, particularly pigs and cattle, which remained the preferred investment objects. Yet households considered tree plantations to be very complementary, as trees required little labor in contrast to the labor-intensive livestock husbandry. As a result, households used an increasing share of the surplus generated from other activities for investments in trees. Internal surplus became the main source of finance for tree plantations, particularly so in Village 5 and Xom Lam (see Figure 3). In these villages, households covered virtually all of the necessary start-up expenditures from cash savings or the sale of livestock.

Access to external finance was not a necessary precondition for households to grow trees, but it allowed them to raise investment levels (see Figure 3). In Tan Quang and particularly Thuan Phong, households used bank loans to purchase high-

[11]Households invested as little as VND VND 2.6 (USD USD 130) and 3.0 (USD USD 150) million per ha in the northern villages. Investment levels in the South were higher (VND VND 3.8–4.3 million/ha, or USD USD 190–215/ha), but still very low in comparison with the rate of VND VND 20 million (USD USD 1,000) per ha recommended by the World Bank.

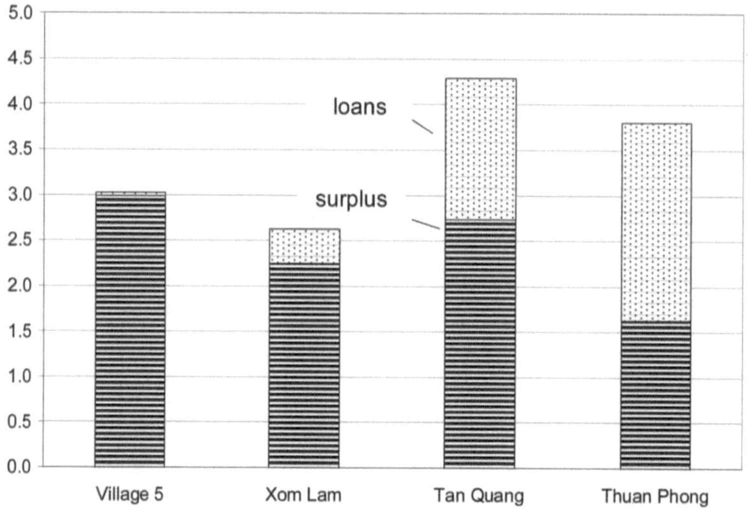

Figure 3. Levels and sources of investment (VND million/ha).

quality seedlings and fertilizer, in contrast to their peers in the northern villages. The villagers of Tan Quang enjoyed good access to loans through four Savings and Loans Groups operated by the district branch of the Bank for Social Policies. The Groups facilitated four out of five households in the village to take out loans of modest size (around VND 15 million, or USD 750), which they used in parts to purchase livestock, and in parts to invest in tree plantations. Households preferred taking out modest-sized loans as the small amount allowed them to repay the loans from internally-produced surplus in the following years. Consequently, it did not matter to them that the loans had duration of three years only, as they expected to repay the loans from the surplus made in all kinds of activities on the farm. In addition, they appreciated the low interest rates charged by the Bank, which were as low as 0.65 percent/month and thereby below the inflation rate in November 2008 (Sikor 2011a).

Access to external finance was easiest for the villagers of Thuan Phong due to the operation of the World Bank-financed project. Many villagers were able to borrow significant amounts of capital for investments in acacia plantations. Starting in 2006, the program offered sizable loans of up to VND 15 million per ha with duration of seven years. As a result, many households financed a much larger share of their plantation investments through external sources than in the other villages (see Figure 3). They were willing to incur relatively high debts, and ready to accept the risks associated with plantations to a much larger degree than their peers in the other three villages. At the same time, these households also made sure to invest a significant proportion of the plantation loans into other productive activities, as a simple comparison between the average investment level of VND 4 million/ha (USD 200/ha) and loan size of VND 15 million/ha (USD 750/ha) demonstrates.[12]

[12]Households' uptake of bank loans and their readiness to incur high levels of debt raises valid questions about their ability to repay the loans (cf. Barney 2008). In the case of the World Bank-financed loans, our fieldwork came too early as no household had reached the end of the loan period yet. Yet experience from Tan Quang, where two or three households had to sell their houses to repay overdue bank loans, suggests that high debts may become an issue for some households in the near future.

Thus, a bank set up by Vietnam's government figured in villagers' access to finance. Similar to my observations on land, it was not a coercive agency that implemented central policy in a top-down manner. The state-owned Bank for Social Policies offered villagers loans, some of them financed from a World Bank project. Nevertheless, the loans made the state omnipresent in the villages. Villagers commonly referred to the Bank as 'the state' when we asked them about the source of their loans in interviews.

Access to wood markets

Finally, villagers had no problem selling their trees, and they usually had a choice to whom they wanted to sell. In 2008, every villager ready to sell trees in any of the four villages easily found a buyer, such as another villager purchasing trees locally, a wood trader from the surrounding area, an agent licensed by a chip mill, or a company procuring wood directly from growers. Between 2004 and 2008, the average village household earned between VND 1.5 million (USD 75, Thuan Phong) and VND 2.7 million (USD 135, Village 5) (see Figure 4).[13] Earnings from wood sales thus contributed between five percent (Tram Than) and 12 percent (Tan Quang) of annual household earnings. In the southern villages, earnings from tree plantations were on a rapid rise due to the recent expansion of plantations. In Tan Quang, for example, annual earnings were as high as VND 5.3 million (USD 265) in 2007 and 2008, accounting for nearly a quarter of total household earnings.

Even the two villages in Phu Tho enjoyed ready access to wood markets despite their location far away from the coast. They had no problem selling their eucalyptus and acacia trees, as the Bai Bang Paper Mill built by the Vietnam government with Swedish assistance in the late 1970s was always looking for raw materials. Since the

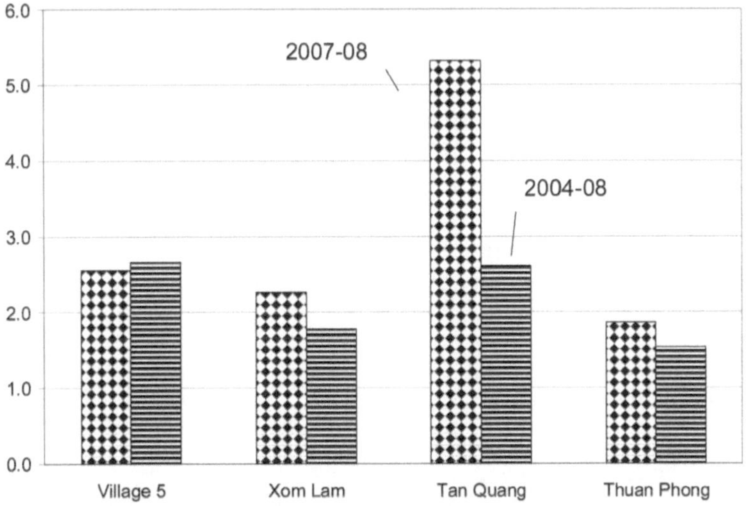

Figure 4. Earnings from tree plantations (VND million/year).

[13]These figures refer to gross earnings, from which one could derive net earnings (income) by subtracting the costs of purchased inputs and household labor.

mid-1990s, the Mill had provided a ready outlet for wood produced on household plantations. In recent years, another small paper mill and numerous small wood processing and furniture factories located in Phu Tho provinces had started to compete with Bai Bang for the wood coming from Village 5 and Xom Lam. In addition, some traders had started to ship medium-sized logs on the river connecting the province to the coast. Yet despite the competition, wood prices remained much lower in Phu Tho than in Binh Dinh. The price of a ton of wood was VND 720,000 (USD 36) at the factory gate in Binh Dinh, double the price of VND 358,000 (USD 18) in Phu Tho.

Intermediaries facilitated households' easy access to wood markets, yet they also ate away at the overall revenues. Most villagers sold their wood to traders from their own village or the surrounding area. In Thuan Phong, for example, a trader licensed by Quy Nhon Pulpwood Company managed to buy up around four out of five tons of wood harvested in the village. As wood passed through several hands before it reached the factory gate, many more people derived a share of the ultimate revenue, particularly in the two northern villages. In Village 5, for example, traders make about VND 134,000 (USD 6.80) per ton of wood, equivalent to a share of 37 percent in total product value. In addition, most villagers sold their standing trees to traders for the production of either pulp/paper or wood chips. This common practice often allowed traders to underestimate the volume of wood, thereby lowering the amount of money they had to pay tree growers (Nguyen 2011). Half of all sales in Village 5 and Thuan Phong and 90 percent of all sales in Xom Lam involved transactions of standing trees.

By way of a preliminary summary, I note that households in all four villages were able to grow trees. They were in a position to grow trees because their endowments and capacities had enabled them to acquire the required land in the past, keep finance requirements low through corresponding silvicultural techniques, source the necessary finance from other productive activities on the farm and externally available credit, and find ready outlets on local wood markets. Tree plantations had become a stable source of earnings in all four villages and accounted for a rapidly increasing share of total earnings in the two southern villages. Households had been able to involve themselves in tree plantations despite the local presence of powerful actors, i.e. State Forest Enterprises, Vietnamese plantation companies and transnational corporations. Moreover, tree plantations had developed into a foundation for everyday dynamics of state formation in the villages. Households' ability to benefit from tree plantations was closely tied to villagers' and state officials' ideas about and experiences of the state, as I now set out to show in the remainder of this paper.

The political economy of agrarian change

The empirical observations presented above highlight the critical role of particular government policies in defining household access to land, finance and wood markets in the four study sites: the paper mill in Phu Tho was a central government project, land allocation was a key government program in the 1990s, the external finance available to households originated from state-owned banks, and the chip mills located along Binh Dinh's coast resulted from efforts by Vietnam's government to attract Foreign Direct Investment (FDI). This section takes a closer look at these four policy fields to examine the larger political economic dynamics that facilitated household access to land, finance and wood markets not only in the four villages but across Vietnam. To demonstrate the linkages between local and national dynamics, I will continue to make reference to events in the four villages.

Socialist economy and international solidarity: Bai Bang

There is not much left from Vietnam's planned economy in the four villages. The village economies have moved miles away from the annual production plans and procurement quotas of the 1970s and 1980s. Cooperative buildings have all but disappeared from the village landscapes and the plantations date back to more recent years. Nevertheless, there is one legacy from the socialist past that continues to exert strong influence on the landscapes of the two northern villages: Bai Bang Paper Mill. Without the mill, Phu Tho's plantations would be located at an unfavorable distance from coastal wood markets. With the mill, tree growers in Phu Tho had a ready outlet for their wood since the 1990s.

Bai Bang was one of the government investments in heavy industry at the core of socialist Vietnam's import substitution policy. The mill was also a flagship project supported by the Swedish government to demonstrate international solidarity with Vietnam, a project that long attracted the attention of high-level leaders up to the General Secretary of Vietnam's Community Party and Prime Minister of Sweden (Jerve *et al.* 1999). Having a design capacity of 55,000 tons of paper and 48,000 tons of pulp per year, the mill long produced about a quarter of Vietnam's domestic paper. It also became 'Sweden's largest, most expensive and longest running development corporation project (Blower *et al.* 1999, xvii). At times, up to 700 Swedes, Finns and other foreigners would work on the construction site to make the transfer of state-of-the-art technology happen (xvi). The plant eventually began operations in 1981, reached its design capacity by the mid-1990s and became the most profitable paper mill in Vietnam (7). It was profitable because a tariff of 40 percent and occasional import prohibitions protected it against competition from cheap paper produced in neighboring countries (xvii).

As soon as the plant started operations, the need for new plantations became clear. Bai Bang required some 250,000 tons of wood per year, because Vietnam's government insisted that the mill should produce its own pulp instead of importing it from other countries (Jerve *et al.* 1999, 144–145). Vietnam's government initially assigned State Forest Enterprises in several provinces to produce raw materials for Bai Bang (Blower *et al.* 1999, 46). Yet Vietnam's government and its Swedish partners realized soon that the Enterprises alone would not be able to establish the required plantation area and began to involve agricultural cooperatives through tree-growing contracts. In the late 1980s, they further expanded the tree plantation program to make direct contracts with households through Bai Bang's extension system, the cooperatives and local governments. By 1996, some 15 percent of the logs already came from household eucalyptus plantations, and households made a profit from selling the wood to Bai Bang (8, 46).

In 2008, Bai Bang was no longer the sole purchaser of wood coming from Phu Tho, but it remained the single largest outlet for wood and continued to be profitable. The mill now required some 350,000 tons of wood per year after the 2003 expansion, about half of which came from household plantations. Just as in previous years, it enjoyed various forms of government support, as is common around the world (Kröger 2011). Domestic paper production continued to be protected by a tariff on imported paper, even though the government had to reduce the tariff rate to 20 percent in 2003 under Asian Free Trade Association (AFTA) rules (Karsenty 2005). Also, three quarters of the plantation funds allocated by the central government to Phu Tho province under the 661 Program in 2008 (i.e. USD 2.5 million) went to Forest Companies producing wood for Bai Bang. Thus, overall, the mill's operations remained profitable (Ernst & Young/A&C Sandwell 2004 cited in Karadjis 2011).

Nevertheless, Vietnam's government faced increasing difficulties to justify the support given to Bai Bang, leaving the management of Bai Bang squeezed between international competition and the mandate to procure trees from households. As Vietnam's trade regime became increasingly liberalized after access to AFTA and the World Trade Organizations, there was further pressure on the government to reduce the tariff on imported paper to five percent (Karsenty 2005). Mill managers saw themselves confronted with the task of achieving significant cost reductions and efficiency increases (Blower *et al.* 1999, xxv, Karsenty 2005). Pressure mounted on the mill to stop buying wood from households as that was considered inefficient. An audit conducted by the international companies Ernst & Young and A&C Sandwell concluded that Bai Bang could only become profitable if it purchased logs from large plantations (Ernst and Young/A&C Sandwell 2004 cited in Karadjis 2011).

The discussion about Bai Bang's future reflected a larger debate about Vietnam's political economic model and the role of the state in the economy (Karadjis 2011). At the core of the debate were two radically different ideas about the state. Some argued in favor of state-owned industry as a provider of secure employment and a pole of regional development. Others preferred privatizing state-owned industry (equitization in Vietnam's terminology) for the sake of profitability and international competitiveness. The two camps clashed head on about Bai Bang, as the mill retained the symbolic significance it had originally gained as one of Vietnam's flagship investments.[14]

Land rush, land allocation and the persistence of state enterprises

Just as observed in the four villages, people set out all over rural Vietnam in the late 1980s and early 1990s to abandon collective work and cultivate the land surrounding their houses on their own account (Fforde and de Vylder 1996, Kerkvliet 2005). Outside the rice-growing deltas, villagers expanded their fields up neighboring hills and mountain slopes or used those for livestock husbandry (Sikor and Dao 2002). Problems with cooperative management had quickly swept through the countryside, prompting many cooperative leaders to devolve land management to individual households in practice (Ban Bien Tap 1987). Moreover, villagers claimed significant tracts of land formally designated as forestry land and under the formal control of State Forest Enterprises or local governments for cultivation and livestock husbandry (Sikor 2001). This was a period in which the control exerted by the once-powerful agricultural cooperatives and State Forest Enterprises eroded drastically. It was also the time in which villagers' actions laid the foundations for the forestry land they came to hold later on, i.e. the 25 percent of total forestry land

[14]The political significance of the discussion finds reflection in the fact that the Ministry of Finance and World Bank never made the Ernst and Young/A&C Sandewall report public.

and 45 percent of tree plantations considered to be held by households at the end of 2009 (Forest Protection Department 2010).

Vietnam's government responded to the land rush through a reformed land policy and land allocation (Sikor 2001). The Land Law, passed by Vietnam's National Assembly in 1993, mandated the government to formally allocate land to households under long-term use arrangements. In the case of forestry land, allocation entitled land users to use the land for 50 years, enjoy its benefits, transfer use rights to other entities, and use allocated rights as collateral, as specified in Decree 02 of 2004. Yet allocation not only legalized the *de facto* use rights households had acquired over land, but also sought to institute government control over land matters. Land allocation came along with the creation of a new specialized government agency for land administration with subordinate units at the provincial and district levels. The General Department of Land Administration was tasked to issue the new 'land use right certificates' to households, demarcate and register all plots, administer cadastral records, assist land classification, register and approve changes in land holdings, and arbitrate in land disputes.

Similar efforts by Vietnam's central government to reform the State Forest Enterprises met very limited success due to resistance by the Enterprises themselves and parts of the forestry bureaucracy (Artemiev 2003, McElwee 2004, World Bank 2005). The government instituted a number of policy initiatives to close down loss-making Enterprises and entice all other Enterprises to reduce the land in their possession, such as Decree 200 in 2004 and Decree 178 in 2006. Whereas early initiatives had caused some Enterprises to shut down or pass some of their land to local governments for allocation to people, most Enterprises defied central government attempts at a more radical reform and could count on the support of provincial leaders in their defiance. The underlying structures and land holdings remained very much in place even though Decree 178 may have done away with the Enterprises on paper by forcing them to re-register as either commercially-oriented Forest Companies or government-funded Protection Forest Management Boards. Thus, by the end of 2009, nearly half of Vietnam's forestry land and a third of the tree plantations remained under the formal control of Forest Companies and Management Boards.

This resilience of State Forest Enterprises and their successors to withstand central reform initiatives finds vivid testimony in Phu Tho and Binh Dinh provinces. In Phu Tho, the Vietnam Paper Corporation managed to retain all 21,500 ha of forestland zoned for productive purposes in its possession, even though it had to transfer protection and special use forests to the provincial People's Committee in reaction to Decision 178. Doan Hung State Forest Enterprise became Doan Hung Forest Company, but avoided passing any land on to the local People's Committees. The Company continues to hold the 2068 ha allocated to the Enterprise in 1998. In Binh Dinh, the four State Forest Enterprises managed to retain all production forest when they were converted into Forest Companies in 2008. They had to separate themselves from protection forests in the process, which were put under the formal control of various Management Boards. Yet just as in Phu Tho, not a single hectare was passed on to villagers in the process.

Other domestic companies, private or state-owned, and transnational corporations faced often insurmountable hurdles when they sought to acquire land. Nationwide, they had received long-term land use rights to less than one percent of forestry land by the end of 2009 (Forest Protection Department 2010). The only way for them to obtain land for tree plantations was through land leases granted by the national or provincial governments. Sizable land leases, such as the 9,900 ha given to Nissho Iwai by the Binh Dinh People's Committee, remained highly exceptional, as illustrated by land statistics from the province: the total area of land leased in the province was just 1,900 ha larger than the Nissho Iwai lease. In Phu Tho, there was not a single lease to a plantation company until 2008. Other international investors searching for land had to return home with empty hands. The President of IKEA, it was reported, had even tried his luck with Vietnam's Prime Minister asking for a lease of 200,000 ha for plantation development, but failed just as badly as other transnational corporations (IUCN et al. 2009).

The land rush of the years around 1990 and subsequent land allocation stand to indicate key dynamics of state formation in Vietnam. Villagers transformed the country's agrarian political economy including the role of the state through their everyday practices (Kerkvliet 2005). As the state faced a massive crisis of authority over rural areas in the late 1980s, the central government was ready to concede significant land rights to villagers. The recognition of villagers' de facto land rights via the land allocation program not only allowed villagers to strengthen their access to land but also aimed to resurrect the authority attributed to the state over land matters and, more broadly, rural affairs. At the same time, the state-owned forest companies remained critical players in this political economy, similar to state-owned wood processing industries like Bai Bang. Private companies, domestic and transnational alike, were relegated to a relatively marginal position with regards to land, particularly in international comparison.

The politics of rural finance: the Bank for Social Policies

The access to credit enjoyed by households in the four villages was exceptional by international standards. Outstanding loans ranged VND 6–14 million (USD 300–700) per households. The high loan rates reflected the remarkable access villagers had to state-owned rural banks, in particular the Bank for Social Policies.

Vietnam's government set up the Bank for Social Policies in 2002 to take over the tasks previously handled by the Vietnam Bank for the Poor, i.e. to support poverty alleviation. Since then, the Bank operated a number of central government programs targeted at poor people, including loan schemes for the poor, people living in remote areas, members of ethnic minority groups, students, etc. It quickly expanded the amount of finance available for lending, total outstanding loans reaching VND 34,300 billion (USD 1.8 billion at current rates) at the end of 2007 (Sikor 2011a).[15] Because of its primarily social goals, the Bank did not operate under commercial principles and was not expected to generate financial profits. The difference between the interest rates applied to loans and savings sufficed to cover the Bank's operational costs. The Bank did not generate sufficient profits to increase its capital stock at a standard market rate, however (K.A. Nguyen 2008).

The Bank for Social Policies built up the most extensive network of outlets and savings and loans groups among all banks in the country. Every district People's Committee housed a Bank branch and virtually every commune People's Committee reserved a room for the Bank to meet customers on a regular basis through its 9,800

[15]To put this number in perspective, the Agri-Bank, another state-owned bank lending to households as well as state companies and private businesses, had total outstanding loans of VND VND 200,000 billion (USD USD 10.7 billion) at the end of 2007 (Sikor 2011a).

mobile transaction units. In addition, the Bank set up around 180,000 savings and loans groups around the country, thus exceeding the number of villages in Vietnam (estimated at between 60,000 and 100,000). The groups helped loan applicants to fill in the required forms, verified their financial capacity, organized counter-signatures by fellow villagers on loan applications, and collected members' monthly interest payments.

The Bank for Social Policies ran four successful savings and loans groups in Tan Quang in cooperation with the Women's Union and Farmer Association. The groups managed to include 200 out of the 249 households in the village, allowing each to borrow between VND 5 and VND 30 million (USD 250–1,500) for up to three years. Households thus held outstanding loans for a total of VND 3.2 billion (USD 160,000) through the four groups in 2007. Coverage was much lower in the other sites, particularly in Village 5, which had only a single savings and loans group that included 31 out of the 96 village households. Most households borrowed VND 7–8 million (USD 350–400) only, yielding a total amount of outstanding loans of VND 274 million (USD 13,700) in November 2008.

The Bank was popular with villagers because it offered easy access to loans, and because it offered favorable conditions on loans. Most importantly, borrowers did not require collateral for most loans, in stark contrast with commercial banks that usually demanded households to deposit their Red Books with them. In addition, the interest rates charged by the Bank were significantly lower than the rates offered by other banks. Up to April 2009, most programs, such as the Loans for the Poor Program, applied a monthly interest rate of 0.65 percent. In comparison, the Agri-Bank charged 1.25 percent in Nov 2008, i.e. almost double. In May 2009, the government even lowered the rate applicable to the Loans for the Poor program to 0.32 percent/month.

Interestingly enough, the Bank was not only the Bank of choice for the government's poverty alleviation programs but also chosen for the implementation of the World Bank-financed plantation project including Thuan Phong. The project offered plantation loans to households on a total of 66,000 ha under a loan agreement signed with the World Bank in April 2005. The agreement covered a total investment of USD 59 million, including USD 33 million allocated to the Bank for disbursement to households until 2010.

The World Bank-funded plantation project was remarkable in many ways. First, it targeted numerous households growing trees on tiny plantations, plantations that were even smaller than the smallholder plantation projects funded by donors and governments in surrounding countries (cf. Barney 2008). Second, the interest rate charged to households was the same that the Bank for Social Policies charged on many of their other programs: 0.65 percent per month. This put the interest rate below inflation rates, despite obligations by Vietnam's government to repay the loan to the World Bank and stop subsidies to state-owned banks under the rules of access to the World Trade Organization (Vien Chien Luoc Ngan Hang 2008). Last, the plantation project reversed the conventional rationale promoted by the World Bank for land titling. Households did not become eligible for bank loans by virtue of their possession of Red Books, as the World Bank's stated rationale would have it (e.g. Deininger 2003). Instead, they were expected to take up loans, and the World Bank and government made sure that they got a Red Book in the process.

The apparent twists in the World Bank's logic, I suggest, reveal underlying negotiations over the state, rural livelihoods, land rights and credit. The provision of credit was as important to rural households as to Vietnam's central government. They considered the supply of credit into every remote part of the country and villagers' uptake of credit an essential element in the state's rural operations. The World Bank did not have much choice but to comply with this rationale.

FDI promotion: the chip mills of Binh Dinh

Binh Dinh's wood markets had been very vibrant since about 2002, as noted above. Households in the two villages got a price for their wood that was about double the price in the two northern villages. The villagers of Tan Quang and Thuan Phong were in an advantaged position because they were located close to the coast, and because several chip mills had opened recently along the coast to buy up plantation wood.

As international wood prices rose around 2000, Vietnam's FDI promotion policy had lured the chip mills to Binh Dinh. From the early 1990s, Vietnam's government had instituted a number of measures to liberalize foreign trade and induce FDI. As integration into the international economy accelerated after 2000, the central government decentralized FDI approval and regulation to provinces and provided various tax breaks and exemptions for enterprises located in Industrial Zones and Export Production Zones (Karsenty 2006). As a result, FDI in the forest sector rose significantly from a mere USD 7 million in 2000 to USD 40 million in each 2004 and 2005 (MARD 2007). By 2009, 16 percent of Vietnam's wood processing enterprises were those with FDI (GSO 2011).

Vietnam's policy for promoting FDI in wood processing as well as actual investment levels mirrored the general trend, as chip mills rapidly expanded their productive capacity after 2000. In 2001, the government removed the export tax on wood chips of five percent (Karsenty 2006, 2). Provincial governments had established Industrial and Export Production Zones along the coast in favorable locations for mills, as virtually the entire wood chip production was destined for export (Karsenty 2006). The wood chips were exported mainly to China, Japan, Taiwan and South Korea, where they were processed into pulp for paper production (Wheatley and Peters 2008a). Chip mills, consequently, mushroomed along Vietnam's Central Coast, most of them owned by foreign owners. Production capacity expanded from 400,000 Bone Dried Tons (BDTs) in 2000 to 635,000 BDTs in 2003 (Barney 2005, 6) and 1,400,000 BDTs in 2006 (Wheatley and Peters 2008b).

Binh Dinh became a center of wood chip production in Vietnam. In 2008, four wood chip facilities operated in the province, buying up the wood coming out of the plantations in the two study villages. Most of the wood from Tan Quang ended up at Binh Dinh Chip Corporation, a joint venture between Quy Nhon Wood Company and Quy Nhon Plantation Company, the latter set up by Nissho Iwai to manage their local plantations (Nguyen 2011). Its production capacity was 120,000 BDTs per year, the output being earmarked for Japanese Oji Paper (Karsenty 2006). In contrast, most of the logs coming out of Thuan Phong reached Duc Hai Company Ltd., a domestic company that had been in operation since 2006 and had exported around 45,000 BDTs of wood chips to China, Taiwan and Japan in 2007.

As a form of synthesis, I want to highlight the critical role attributed to the state for enabling household access to land, finance and wood markets in Vietnam. Moreover, the particular policies facilitating household access were critical elements of changing policy regimes in Vietnam. Thus, government involvement was not only a critical precondition for household access, but the particular policies enabling government involvement got to the core of government policy over the past three decades. Nevertheless, government involvement did not come about in the form of a coherent policy package. The effects of government policy on household tree plantations appeared almost coincidental at the surface, especially if one compares the apparently contradictory government investment in Bai Bang versus the promotion of FDI in wood processing, or Bai Bang's beneficial contributions to rural incomes versus the detrimental effects that state-owned forest companies had on villagers' access to land. Nevertheless, driving government policy was a broader underlying dynamic, which I explore in the following.

Dynamics of state formation

Government policy and villagers' reaction to policy, I suggest, reflected the influence of everyday practices and processes of state formation. Both policy and reactions operated on the grounds of ongoing, albeit shifting, dynamics of state formation, particularly at the level of ideas regarding what the state is and what it ought to be about. In the following, therefore, I provide a brief, somewhat speculative sketch of everyday processes of state formation in Vietnam, how these influenced the agrarian political economy discussed above, and how agrarian change in turn fed into larger politics of the state. Drawing on the conceptual framework set out in Sikor and Lund (2009), I employ the four study sites as a microcosm to trace shifting negotiations over the idea of the state, negotiations that centered on an image of the state as a provider of possession.

Government investment in Bai Bang illustrates how ideas about the state were closely tied to possession in the socialist economy (Fforde 1989, Fforde and de Vylder 1996).[16] The state was directly involved in matters of production and exchange through agricultural cooperatives and state enterprises. State Forest Enterprises, for example, were considered rural development poles, providing work and social services to villagers, and therefore given control over vast areas of land outside the agricultural core zones. Heavy Industry, such as Bai Bang, assumed a central role in the production of goods for the country. Not only government officials but also average people expected the state to provide economic entitlements through administrative allocation. Villagers and officials alike saw the state emerge from this bureaucratic provision of entitlements. Bureaucratic provision made the link between possession and state very direct.

The land rush observed at all four sites reflected the rapid decline in the capacity to provide possession on the side of the organizations and people associated with the state in the 1980s. As agricultural cooperatives, state enterprises and administrative distribution were no longer able to meet people's expectations, farmers and workers increasingly set out to produce and trade things on their own account. Decollectivization began when many villagers across the country decided to take things into their own hands (Kerkvliet 2005). Similarly, 'fence-breaking' activities precipitated the abolishment of bureaucratic controls on industrial production and exchange (Fforde and de Vylder 1996). Thus, state authority unraveled rapidly in the

[16] Another key arena of contestations over the state was, of course, national security. It would go beyond the scope of this paper, however, to discuss the dynamics of state formation in a comprehensive manner.

1980s when simultaneously occurring changes in many localities undermined the idea of the state as enabling economic entitlements. State authority waned as people pursued venues of production and exchange no longer fitting the image of the providing state. The close link between state and possession was severed.

The 1993 Land Law and subsequent land allocation were key elements in the central leadership's efforts to react to the erosion of state authority, efforts that are commonly referred to as the much heralded *doi moi* (renovation) program. From 1986 onwards, the Communist Party's Congress, National Assembly and other organs of the central government enacted a series of resolutions, laws, and decrees to legalize and regularize the changes that had already taken place in many locations (Fforde and de Vylder 1996). The flurry of new legislation increased in the first half of the 1990s, as the push for legislative reforms began to receive support from the World Bank and International Monetary Fund, among others. International financial institutions and donors considered law-making an imperative in Vietnam's move toward a market economy. Vietnam needed a solid legal framework for its emergent market economy, and it required a state to first enact and then enforce laws. Public discourse increasingly moved towards ideas of a 'law-governed' state (*nha nuoc phap quyen*) (Balme and Sidel 2007). A focus on citizen's rights served to recreate the link between possession and state, as rights presumably required the state to secure those. The reforms thus projected the image of a state that emerged through the rights it granted to people, something that Wells-Dang (2010) tries to capture through the term 'rice-roots democracy'.

Land rights were a key arena of not only policy-making but also of the everyday dynamics of state formation, as I have argued elsewhere (Sikor 2011b, forthcoming). Land rights were not merely about who was entitled to use land but were also intimately connected with the exercise of state control. In a similar vein, land allocation was not just about the formalization of people's use rights but implied a massive re-assertion of state control over land affairs in the wake of the land rush, as perhaps best illustrated in the requirement imposed on people in Thuan Phong to obtain land certificates if they wanted to apply for loans. Negotiations over land and, more generally, property were intimately tied to ideas about the state, as many invoked the state as the ultimate arbiter in disputes and definer of rights. This implied that many villagers contributed to the social constitution of the state when they requested the allocation of land, as observed in the four study sites. Villagers also called upon state officials to solve land disputes or requested them to enforce their land claims against competing ones. Land allocation and everyday practices of land use thus were a key influence on the authority people attributed to the Vietnamese state. Villagers and officials saw the state in close connection with possession, possession involving rights of use as well as control over use rights.

Moreover, the empirical insights from the four study sites indicate that possession was not only about property in the sense of legal rights, but also property understood as economic rights. People's expectations that the state would bring about legal and economic rights intimately tied possession to the state. In other words, even after the demise of the planned economy, people recognized the state through the provision of economic entitlements. People no longer expected the state to bring about economic entitlements through bureaucratic allocation but by other means, primarily economic growth and poverty reduction. The promotion of FDI, as witnessed in Binh Dinh's wood processing sector, was a critical element in state efforts to facilitate economic growth.

The Bank for Social Policies perhaps provides the best illustration of the perceived close tie between the state and the provision of economic entitlements. The state-owned Bank was an important vehicle by which people's expectations for possession became closely tied to ideas about the state. The Bank provided rural credit at an exceptional scale in comparison with other countries, thereby helping rural people to expand their economic entitlements (Sikor 2011a). While giving out loans, the Bank also made sure that people saw them in close connection with the state. In fact, the Bank was the state for many villagers, as indicated by their tendency to refer to the Bank as 'the state'. At the same time, villagers also understood that they could not take access to the loans for granted. Not every poor household received a loan, but only those deemed worthy by state officials (V.C. Nguyen 2008). Moreover, the central government made sure that all poverty alleviation loans would be provided by state-owned banks, primarily the Bank for Social Policies. It did not allow the up-scaling of autonomous micro-finance schemes, as successful those might have been in economic terms (Rankin 2008).

Thus, I surmise that politics of possession was a central moment in everyday processes of state formation in Vietnam. Like Gainsborough (2010), I place emphasis on continuity in Vietnamese politics and highlight the intimate linkages between state and possession. Yet I suggest that people saw the state not only as a vehicle for public office holders to enrich themselves, as Gainsborough concludes, but also as a critical source of economic entitlements for everyone. For many people, state officials and average people alike, the idea of the state was wedded closely to possession. State authority arose, expanded, deepened, weakened and waned with possession, or the lack thereof. The concrete meanings attached to possession changed over time. Neither were all people equally able to participate in politics of possession, or benefit from the material opportunities afforded by the politics (Sikor and Nguyen 2007, Sikor and Baggio under review).[17] Nonetheless, possession was central to the image of the state.

Conclusions: politics of possession

Vietnam's rural households have enjoyed remarkable access to land, finance and wood markets, allowing them to grow trees and benefit from tree plantations. Households' access has resulted from favorable government policy over the past three decades, even though the positive influence of policy has not taken the form of a coherent policy package. Instead, the positive influence of policy reflects the sedimented outcomes of everyday dynamics of state formation over the past three decades. In particular, households' ability to grow trees reflects the operation of ongoing, albeit changing politics of possession, which has been a central element in wider contestations over the state in Vietnam. These shifting politics of possession underlie efforts by Vietnam's central government to provide land rights to rural households, grant them access to rural finance, and open up product markets for them. They also explain why villagers expected the state to get involved in not only land matters but also issues of finance and industrial investment.

Politics of possession have been central to processes of state formation in Vietnam, and they may well be in other countries and settings as well. Possession

[17]In addition, people may seek to tie possession to other politico-legal institutions than the state, such as customary regulations, in some places (Sikor 2011b).

tends to connect with ideas about the state because possession is about benefits as well as control (Sikor and Lund 2009). Possession is not merely about the legal rights people have to land. Nor it is simply about their ability to translate land rights into tangible benefits through access to complementary productive resources and markets. Possession also relates to control as rights and entitlements always require politico-legal institutions to sanction them. Possession thus always involves an element of control, as rights and entitlements situate a social actor in relation to the authorizing powers of politico-legal institutions. Amongst the most critical politico-legal institutions is typically the state, because the authority to enable, enhance and defend some forms of possession and to weaken, undermine and erase other forms appears to be a key feature of 'state-like' institutions in many settings. Politics of possession thus may be tied intimately to the dynamics of state formation, as highlighted for Vietnam above.

My account highlights politics of possession as a central dynamic of state formation in Vietnam over other possible explanations. For example, I could have set out to examine everyday practices and processes of state formation through the lens of accountability, accountability understood as people's rights to make claims, demand a response to their claims and rely on mechanisms enforcing response and sanctioning the lack thereof (cf. Wells-Dang 2010). A focus on the politics of accountability would most likely resonate with the new language of rights and a law-governed state in Vietnam (cf. Newell and Wheeler 2006). Nevertheless, I do not consider struggles over accountability to be as central to everyday dynamics of state formation in Vietnam's agrarian political economy as contestations over possession. The difference between these two types of politics is that possession involves a responsive but not necessarily accountable state. Notions of legal redress, answerability and enforceability have not been as present as demands for entitlements in my four cases or Vietnam's agrarian political economy. Central elements in the dynamics of state formation have been negotiations over access to productive resources and product markets, mechanisms which I take to be associated with politics of possession, possession in the sense of entitlement and control.

Even though dynamics of state formation may take various forms, I want to argue that they may have profound implications for the occurrence of land grabs beyond Vietnam and tree plantations. Processes of state formation may explain why land grabs take place in some instances of commodity booms, such as where ideas about the state are not tied to possession, or where certain social groups are excluded from effective participation in politics of possession (e.g. McCarthy and Cramb 2009). In other instances, they may include politics of possession (or accountability) that provide room to maneuver for smallholders, just as reported here. Also, politics of possession may attribute a central role to land rights in some settings, as observed for contemporary Vietnam above. Under other, de-agrarianizing conditions, possession may have less to do with land rights than access to jobs and social services. Similarly, politics of possession or accountability may lead to organized mobilizations and other forms of contentious agency in some settings (e.g. Kröger 2011), but employ more everyday and hidden forms in others. Attention to the dynamics of state formation may even suggest new interpretations for the contradictory roles of international institutions in financing large corporate plantations in some countries, but assisting tiny smallholder plantations in others. Politics of possession may act as a counterforce to 'green neoliberalism' (Goldman 2005) under certain conditions but be made quite compatible through populist approaches in others (Barney 2008).

More broadly, I surmise that the significance of processes of state formation has implications for our theoretical understanding of land grabs. Narrow readings of commodificaton, market expansion or accumulation by dispossession may not fully explain the occurrence of land grabs in some circumstances, and their absence in others. Neither may constructs such as adverse incorporation offer sufficiently robust explanations for the inclusion of smallholders in crop booms. Understanding practices and processes driving or limiting land grabs requires attention to dynamics of state formation. Land grabs may have to say as much about dynamics of state formation and transnational governance as about processes of market penetration and commodification.

Finally, possession may offer an important conceptual lens for understanding exclusion. As Hall *et al.* (2011) have observed, exclusion is unavoidable. Possession by one person almost always implies that someone else is excluded. Nevertheless, in order to understand practices and processes of exclusion, one may need to look at politics of possession, whatever concrete forms they may take in a particular instance. It is often the urge for possession – understood as entitlement *and* control – that drives exclusion, and not the other way round.

References

Abrams, P., 1988. Notes on the difficulty of studying the state. *Journal of Historical Sociology*, 1, 58–89.

Artemiev, I. 2003. *State forestry enterprise reform in Vietnam: unlocking the potential for commercial wood growing*. Washington, DC: World Bank.

Balme, S. and M. Sidel. 2007. Vietnam in comparative communist and postcommunist perspectives. *In*: S. Balme and M. Sidel, eds. *Vietnam's new order*. New York: Palgrave Macmillan, pp. 1–10.

Ban Bien Tap. 1987. Ve Khoan San Pham Trong Nong Nghiep. *Tap Chi Cong San*, 12, 53–58.

Barney, K. 2005. *Central plans and global exports: tracking Vietnam's forestry commodity chains and export links to China*. Washington, DC: Forest Trends.

Barney, K. 2008. Local vulnerability, project risk, and intractable debt: the politics of smallholder eucalyptus promotion in Salavane province, southern Laos. *In*: D.J. Snelder and R.D. Lasco, eds. *Smallholder tree growing for rural development and environmental services*. New York: Springer, pp. 263–286.

Blower, N., M.H. Ngo, J. Allan, M. Adam, P. David, Q.H. Pham, Q. Derek, Q. Nguyen, T. Mandy, T.B. Do, V.H. Hoang, W. Bob and V. David. 1999. *Paper, prices, and politics: an evaluation of the Swedish support to the Bai Bang project in Vietnam*. Stockholm: Swedish International Development Cooperation Agency.

Borras, J. and D. Franco. Forthcoming. Global land grabbing and trajectories of agrarian change: a preliminary analysis. *Journal of Agrarian Change*, 12(1).

Borras, J., R., Hall, I., Scoones, B. White, and W. Wolford. 2011. Towards a better understanding of global land grabbing: an editorial introduction. *Journal of Peasant Studies*, 38(2), 209–216.

Carrere, R. and L. Lohmann. 1996. *Pulping the South: industrial tree plantations & the world paper economy*. London: Zed Books.

Cossalter, C. and C. Pye-Smith. 2003. *Fast-wood forestry: myths and realities*. Bogor: Center for International Forestry Research.

Deininger, K. 2003. *Land policies for growth and poverty reduction*. Washington, DC: World Bank and Oxford University Press.

Del Lungo, A., J. Ball and J. Carle. 2006. *Global planted forests thematic study: results and analysis*. Planted Forests and Trees Working Paper 38, Rome: FAO.

Fairhead, J., M. Leach and I. Scoones (eds.). Forthcoming. Green grabbing: A new appropriation of nature? *Journal of Peasant Studies*, 39(3).

FAO (Food and Agriculture Organisation of the United Nations). 2010. *Global forest resources assessment 2010*. Rome: FAO.

Fforde, A. 1989. *The agrarian question in North Vietnam 1974–1979: a study of cooperator resistance to state policy*. Armonk, NY: M.E. Sharpe.
Fforde, A. and S. de Vylder. 1996. *From plan to market: economic transition in Vietnam 1979–1994*. Boulder, CO: Westview Press, Boulder.
Forest Protection Department. 2010. *Yearly statistics on forest development 2009*. Hanoi: Forest Protection Department.
Gainsborough, M. 2010. *Vietnam: rethinking the State*. London: Zed Books.
Gerber, J.-F. 2011. Conflicts over industrial tree plantations in the South: who, how and why? *Global Environmental Change*, 21(1), 165–176.
Goldman, M. 2005. *Imperial nature: the World Bank and struggles for social justice in the age of globalization*. New Haven: Yale University Press.
GSO (General Statistic Office). 2011. Survey Data Survey on Industry. Available at: http://www.gso.gov.vn. [Accessed 12/10/2011].
Ha, D.T. and G. Shivley. 2008. Coffee boom, coffee bust and smallholder response in Vietnam's Central Highlands. *Review of Development Economics*, 12(2), 312–326.
Hall, D. 2003. The international political ecology of industrial shrimp aquaculture and industrial plantation forestry in Southeast Asia. *Journal of Southeast Asian Studies*, 34(2), 251–264.
Hall, D. 2011. Land grabs, land control, and Southeast Asian crop booms. *Journal of Peasant Studies*, 38(4), 837–857.
Hall, D., P. Hirsch, and T.M. Li. 2011. *Powers of exclusion: land dilemmas in Southeast Asia*. Singapore: National University of Singapore Press.
IUCN. 2009. Ministry of Agriculture and Rural Development and the Vietnam Timber and Forest Product Association workshop. Nam Dinh, October 8–9. Hanoi.
Jerve, A.M. 1999. *A leap of faith: a story of Swedish aid and paper production in Vietnam – the Bai Bang project, 1969–1996*. Stockholm: Swedish International Development Cooperation Agency.
Karadjis, M. 2011. State enterprise workers: 'masters' or 'commodities'?. *In*: A. Chan, ed. *Labour in Vietnam*. Singapore: Institute of Southeast Asian Studies, pp. 46–90.
Karsenty, A. 2005. *Study on the impacts of market and investment liberalization on Vietnam's pulp and plantations sector, with particular reference to China*. Bogor: Center for International Forestry Research.
Kerkvliet, B.J.T. 2005. *The power of everyday politics: how Vietnamese peasants transformed national policy*. Ithaca, NY: Cornell University Press.
Kröger, M. 2011. Promotion of contentious agency as a rewarding movement strategy: evidence from the MST-paper industry conflicts in Brazil. *The Journal of Peasant Studies*, 38(2), 435–458.
Li, T.M., 2002. Local histories, global markets: cocoa and class in upland Sulawesi. *Development and Change*, 33, 415–437.
MARD (Ministry of Agriculture and Rural Development). 2007. Vietnam forest development strategy 2006–2020. Hanoi: Agriculture Publisher.
McCarthy, J.F. 2010. Processes of inclusion and adverse incorporation: oil palm and agrarian change in Sumatra, Indonesia. *Journal of Peasant Studies*, 37(4), 821–850.
McCarthy, J.F. and R. A. Cramb. 2009. Policy narratives, landholder engagement, and oil palm expansion on the Malaysian and Indonesian frontiers. *The Geographical Journal*, 175(2), 112–123.
McElwee, P. 2004. You say illegal, I say legal: the relationship between 'illegal' logging and land tenure, poverty, and forest use rights in Vietnam. *Journal of Sustainable Forestry*, 19(1/2/3), 97–135.
Newell, P. and J. Wheeler (eds.). 2006. *Rights, resources and the politics of accountability*. London: Zed Books.
Nguyen K.A. 2008. Vai suy nghi ve hoat dong cua cac dinh che tai chinh cung cap tin dung doi voi khu vuc nong nghiep, nong thon Viet Nam. State Bank of Vietnam and the Ford Foundation.
Nguyen, Q.T. 2011. *Plantation wood market chain in Binh Dinh and Phu Tho provinces: an analysis of benefit and cost distribution along the commodity line*. Discussion Paper no. 65, Bogor: Center for International Forestry Research.

Nguyen V.C. 2008. Is a governmental micro-credit program for the poor really pro-poor? Evidence from Vietnam. *The Developing Economies*, 46(2), 151–187.
Pye, O. 2010. The biofuel connection: transnational activism and the palm oil boom. *Journal of Peasant Studies*, 37(4), 851–874.
Rankin, K.N. 2008. Manufacturing rural finance in Asia: Institutional assemblages, market societies, entrepreneurial subjects. *Geoforum*, 39(8), 1965–1977.
Sikor, T. 2001. The allocation of forestry land in Vietnam: Did it cause the expansion of forests in the Northwest? *Forest Policy and Economics*, 2(1), 1–11.
Sikor, T. 2011a. *Financing household tree plantations in Vietnam: current programmes and future options*. Discussion Paper No. 69, Bogor: Center for International Forestry Research.
Sikor, T. 2011b. Land allocations in Vietnam's uplands: negotiating property and authority. *In*: T. Sikor, P.T. Nghiem, J. Sowerwine and J. Romm, eds. *Upland transformations in Vietnam*. Singapore: National University of Singapore Press, pp. 146–162.
Sikor, T. Forthcoming. Property and state in Vietnam and beyond. Forthcoming in H.-T.H. Tai and M. Sidel (eds.), *Property, state and society in Vietnam*. London: Routledge.
Sikor, T. and J. Baggio. Under review. Do smallholder tree plantations contribute to poverty alleviation? An Entitlements Analysis from Vietnam.
Sikor, T. and M.T. Dao. 2002. Agricultural policy and land use changes in black Thai villages of Northern Vietnam, 1952–1997. *Mountain Research and Development*, 22(3), 248–255.
Sikor, T. and C. Lund. 2009. Access and property: a question of power and authority, *Development and Change*, 40(1), 1–22.
Sikor, T. and Q.T. Nguyen. 2007. Why may forest devolution not benefit the rural poor? Forest entitlements in Vietnam's central highlands, *World Development*, 35(11), 2010–2025.
Sikor, T. and T.T.V. Pham. 2005. The dynamics of commoditization in a Vietnamese uplands village, 1980–2000, *Journal of Agrarian Change*, 5(3), 405–428.
Sikor, T., J. Stahl, and S. Dorondel. 2009. Negotiating post-Socialist property and state: struggles over forests in Albania and Romania. *Development and Change*, 40(1), 171–193.
Vien Chien Luoc Ngan hang (Strategy Institute for Banking). 2008. Tin dung nong nghiep nong thon – Thuc trang va dinh huong phat trien sau khi gia nhap WTO (Agricultural and rural credit – Situation and development trends after WTO entry). Unpublished report. Hanoi.
Wells-Dang, A. 2010. Political space in Vietnam: a view from the 'rice-roots'. *The Pacific Review*, 23(1), 93–112.
Wheatley, C. and D. Peters. 2008a. Acacia value chain in the North Central Region of Vietnam. Hanoi: Netherlands Development Organisation.
Wheatley, C. and D. Peters. 2008b. Acacia species in North Central Vietnam: desk study. Hanoi: Netherlands Development Organisation.
White, B. and A. Dasgupta. 2010. Agrofuels capitalism: a view from political economy. *Journal of Peasant Studies*, 37(4), 593–607.
World Bank. 2005. State forest enterprise reform in Vietnam: review of policy and implementation framework for Decree 200. Washington, DC: World Bank.

Thomas Sikor is Reader in International Development at the University of East Anglia, UK. He has published numerous articles on property, resource governance and agrarian change. Sikor is co-editor of *The politics of possession: property, access and authority* (2009), *Upland transformations in Vietnam* (2011) and *Forests and people: property, governance and human rights* (2011), and guest-editor of special journal issues in *World Development* (2009) and *Development and Change* (2009).

Index

References in **Bold** represent illustrations.